UJF NATO ASI

LES HOUCHES

Session LXXV

2001

PHYSICS OF BIO-MOLECULES AND CELLS

PHYSIQUE DES BIOMOLÉCULES ET DES CELLULES

CONFÉRENCIERS

R.H. AUSTIN
W. BIALEK
R. BRUINSMA
M. DOGTEROM
T. DUKE
E. EVANS
J. HOWARD
A. LIBCHABER
M. MAGNASCO
J. PROST
E. SACKMANN
E. SIGGIA
P. WILLIAMS

ÉCOLE DE PHYSIQUE DES HOUCHES – UJF & INPG – GRENOBLE

a NATO Advanced Study Institute

LES HOUCHES

SESSION LXXV

2-27 July 2001

Physics of bio-molecules and cells

Physique des biomolécules et des cellules

Edited by

H. FLYVBJERG, F. JÜLICHER, P. ORMOS and F. DAVID

EDP SCIENCES

Les Ulis, Paris, Cambridge

Springer

Berlin, Heidelberg, New York,
Barcelona, Hong Kong, London
Milan, Paris, Tokyo

Published in cooperation with the NATO Scientific Affair Division

ISSN 0924-8099 print edition
ISSN 1610-3459 online edition

ISBN 3-540-44132-8 Springer-Verlag Berlin Heidelberg New York
ISBN 2-86883-618-6 EDP Sciences Les Ulis

Printed in France

LES HOUCHES - ÉCOLE DE PHYSIQUE
ÉCOLE D'ÉTÉ DE PHYSIQUE THÉORIQUE

SERVICE INTER-UNIVERSITAIRE COMMUN
À L'UNIVERSITÉ JOSEPH FOURIER DE GRENOBLE
ET À L'INSTITUT NATIONAL POLYTECHNIQUE
DE GRENOBLE, SUBVENTIONNÉ PAR LE MINISTÈRE
DE L'ÉDUCATION NATIONALE, DE LA RECHERCHE
ET DE LA TECHNOLOGIE, LE CENTRE NATIONAL
DE LA RECHERCHE SCIENTIFIQUE ET LE COMMISSARIAT
À L'ÉNERGIE ATOMIQUE

ÉCOLE D'ÉTÉ DE PHYSIQUE THÉORIQUE
SESSION LXXV

INSTITUT D'ÉTUDES AVANCÉES DE L'OTAN
NATO ADVANCED STUDY INSTITUTE

EURO SUMMER SCHOOL
ÉCOLE THÉMATIQUE DU CNRS
DANISH SCHOOL OF BIOPHYSICS

2-27 juillet 2001

Directeurs Scientifiques de la session : H. FLYVBJERG, *Graduate School of Biophysics, Copenhagen, and Risø National Laboratory, Denmark ;* F. JÜLICHER, *Institut Curie, Paris, France ;* P. ORMOS, *Institute of Biophysics, Szeged, Hungary ;* François DAVID, *SPhT, CEA Saclay, France*

SESSIONS PRÉCÉDENTES

1951 - 2001

I	1951	Quantum mechanics. Quantum field theory
II	1952	Quantum mechanics. Statistical mechanics. Nuclear physics
III	1953	Quantum mechanics. Solid state physics. Statistical mechanics. Elementary particle physics
IV	1954	Quantum mechanics. Collision theory. Nucleon-nucleon interaction. Quantum electrodynamics
V	1955	Quantum mechanics. Non-equilibrium phenomena. Nuclear reactions. Interaction of a nucleus with atomic and molecular fields
VI	1956	Quantum perturbation theory. Low temperature physics. Quantum theory of solids; dislocations and plastic properties. Magnesium; ferromagnetism
VII	1957	Scattering theory. Recent developments in field theory. Nuclear interaction; strong interactions. High energy electrons. Experiments in high energy nuclear physics
VIII	1958	The many body problem
IX	1959	The theory of neutral and ionized gases*
X	1960	Elementary particles and dispersion relations*
XI	1961	Low temperature physics*
XII	1962	Geophysics; the earth's environment*
XIII	1963	Relativity groups and topology*
XIV	1964	Quantum optics and electronics*
XV	1965	High energy physics*
XVI	1966	High energy astrophysics*
XVII	1967	Many body physics*
XVIII	1968	Nuclear physics*

* Sessions ayant reçu l'appui du Comité Scientifique de l'OTAN.

XIX	1969	Physical problems in biological systems*
XX	1970	Statistical mechanics and quantum field theory*
XXI	1971	Particle physics*
XXII	1972	Plasma physics*
XXIII	1972	Black holes*
XXIV	1973	Fluids dynamics*
XXV	1973	Molecular fluids*
XXVI	1974	Atomic and molecular physics and the interstellar matter*
June Inst.	1975	Structural analysis of collision amplitudes
XXVII	1975	Frontiers in laser spectroscopy*
XXVIII	1975	Methods in field theory*
XXIX	1976	Weak and electromagnetic interactions at high energy*
XXX	1977	Nuclear physics with heavy ions and mesons*
XXXI	1978	Ill-condensed matter*
XXXII	1979	Membranes and intercellular communication*
XXXIII	1979	Physical cosmology*
XXXIV	1980	Laser plasma interaction
XXXV	1980	Physics of defects*
XXXVI	1981	Chaotic behaviour of deterministic systems*
XXXVII	1981	Gauge theories in high energy physics*
XXXVIII	1982	New trends in atomic physics*
XXXIX	1982	Recent advances in field theory and statistical mechanics*
XL	1983	Relativity, groups and topology*
XLI	1983	Birth and infancy of stars*
XLII	1984	Cellular and molecular aspects of developmental biology*
XLIII	1984	Critical phenomena, random systems, gauge theories*
XLIV	1985	Architecture of fundamental interactions at short distances*
XLV	1985	Signal processing*
XLVI	1986	Chance and matter*
XLVII	1986	Astrophysical fluid dynamics*
XLVIII	1988	Liquids at interfaces*
XLIX	1988	Fields, strings and critical phenomena*
L	1988	Oceanographic and geophysical tomography*
LI	1989	Liquids, freezing and glass transition*
LII	1989	Chaos and quantum physics*
LIII	1990	Fundamental systems in quantum optics*
LIV	1990	Supernovae*

* Sessions ayant reçu l'appui du Comité Scientifique de l'OTAN.

LV	1991	Particles in the nineties*
LVI	1991	Strongly interacting fermions and high T_c superconductivity*
LVII	1992	Gravitation and quantizations*
LVIII	1992	Progress in picture processing*
LIX	1993	Computational fluid dynamics*
LX	1993	Cosmology and large scale structure*
LXI	1994	Mesoscopic quantum physics*
LXII	1994	Fluctuating geometries in statistical mechanics and quantum field theory*
LXIII	1995	Quantum fluctuations*
LXIV	1995	Quantum symmetries*
LXV	1996	From cell to brain*
LXVI	1996	Trends in nuclear physics, 100 years later
LXVII	1997	Modélisation du climat de la terre et de sa variabilité*
LXVIII	1997	Particules et interactions : le modèle standard mis à l'épreuve*
LXIX	1998	Topological aspects of low dimensional systems*
LXX	1998	Infrared space astronomy, today and tomorrow*
LXXI	1999	The primordial universe
LXXII	1999	Coherent atomic matter waves*
LXXIII	2000	Atomic clusters and nanoparticles*
LXXIV	2000	New trends in turbulence*

* Sessions ayant reçu l'appui du Comité Scientifique de l'OTAN.
Publishers: Session VIII: Dunod, Wiley, Methuen; Sessions IX & X: Herman, Wiley – Session XI: Gordon and Breach, Presses Universitaires – Sessions XII-XXV: Gordon and Breach – Sessions XXVI-LXVIII: North-Holland – Session LXIX-LXXIV: EDP Sciences, Springer.

ORGANIZERS

H. FLYVBJERG, Graduate School of Biophysics, Niels Bohr Institute, Blegdamsvej 17, 2100 Copenhagen, Denmark and Materials Research Department, Risø National Laboratory, DK-4000 Roskilde, Denmark

F. JÜLICHER, Institut Curie,11 rue Pierre et Marie Curie,75231 Paris Cedex 05, France

P. ORMOS, Institute of Biophysics, Biological Research Centre Szeged, Temesvari krt 62, P.O. Box 521, Hungary

F. DAVID, SPhT, CEA Saclay, 91191 Gif-sur-Yvette Cedex, France

LECTURERS

R.H. AUSTIN, Department of Physics, Princeton University, Princeton, NJ 08544, U.S.A.

W. BIALEK, NEC Research Institute Inc., 4 Independance Way, and Department of Physics, Princeton University, Princeton, NJ 08540, U.S.A.

R. BRUINSMA, Institut Lorentz for Theoretical Physics, Universiteit Leiden, P.O. Box 9506, 2300 RA Leiden, The Netherlands

T. DUKE, Cavendish Laboratory, Madingley Road, Cambridge CB OHE, U.K.

E. EVANS, Biomedical Engineering, Boston University, Boston, MA 02215, U.S.A., and Physics, University of British Columbia, Vancouver V6T 2A6, BC Canada

J. HOWARD, MPI Molecular Cell Biology & Genetics, Pfotenhauerstrasse 108, 01307 Dresden, Germany

M. MAGNASCO, Mathematical Physic Laboratory, Rockefeller University, Box 212, 1230 York Avenue, NY 100021, U.S.A.

J. PROST, Institut Curie, Section de Recherche, 26 rue d'Ulm, 75005 Paris, France

E. SACKMANN, Physik Department E2, Technische Universitaet Muenchen, 85747 Garching, Germany

E. SIGGIA, Center for Studies in Physics & Biology, Box 25, Rockefeller University, 1230 York Avenue, NY 10021, U.S.A.

SEMINAR SPEAKERS

M. DOGTEROM, FOM/AMOLF, Kruislaan 4078, 1098 SJ Amsterdam, The Netherlands

A. LIBCHABER, Physics and Mathematics, Rockefeller University, 1230 York Avenue, New York 10021, U.S.A.

P. WILLIAMS, Laboratory of Biophysics & Surface Analysis, School of Pharmaceutical Sciences, The University of Nottingham, Nottingham NG7 2RD, U.K.

PARTICIPANTS (STUDENTS & AUDITORS)

D. ANDOR, Theory of Condensed Matter, Cavendish Laboratory, Madingley Road, Cambridge CB3 OHE, U.K.

A. BANCAUD, Institut Curie, Laboratoire PCC/UMR 168, 11 rue Pierre et Marie Curie, 75231 Paris Cedex 05, France

V. BAULIN, SI3M (SPrAM), DRFMC, CEA Grenoble, 17 rue des Martyrs, 38054 Grenoble, France

K. BERG-SØRENSEN, Niels Bohr Institute & Nordita, Blegdamsvej 17, 2100 Copenhagen, Denmark

M. BETTERTON, Courant Institute, 251 Mercer St., New York, NY 10012, U.S.A.

D. BIRON, Weizmann Institute, Rehovot 76100, Israel

K. CHRISTENSEN, The University of Southern Denmark, Physics Department, Campusvej 55, Odense M-5230, Denmark

I. COLUZZA, FOM-AMOLF, Kruislaan 407, 1098 SJ Amsterdam, The Netherlands

M. DAHAN, Laboratoire Kastler Brossel, ENS, 24 rue Lhomond, 75005 Paris, France

D. DAN, Institute of Physics, Sachivalaya Marg, Bhubaneswar 751005, India

N. DEKKER, Laboratoire de Physique Statistique, ENS, 24 rue Lhomond, 75005 Paris, France

H. DELANOE, Laboratoire de Spectrométrie Physique, bâtiment E45, Domaine Universitaire, BP. 87, 38402 Saint-Martin-d'Hères, France

I. DERENYI, Institut Curie, UMR 168, 11 rue Pierre et Marie Curie, 75231 Paris Cedex 05, France

Z. FRENKEL, Department of Metal Physics & Computer Technologies, St. Petersburg State Technical University, Polytechnicheskaya 29, 195251 St. Petersburg, Russia

C. GABEL, Harvard University, 16 Divinity Ave., Cambridge, MA 02138, U.S.A.

F. GABEL, Institut de Biologie Structurale, 41 rue Jules Horowitz, 38027 Grenoble, France

N. GADEGAARD, Danish Polymer Centre, Risø National Laboratory, Frederiksborgvej 399, 4000 Roskilde, Denmark

P. GALAJDA, Institute of Biophysics, Szeged, Temesvari Krt. 62, 6726, Hungary

N. GODDARD, Rockefeller University, 1230 York Ave., Box. 67, New York 10021, U.S.A.

J. GOIDIN, Institut Curie, UMR 168 du CNRS/IC, 11 rue Pierre et Marie Curie, 75231 Paris Cedex, France

C. GORBA, Max Planck Institute, Kennedyallee 70, 60596 Frankfurt, Germany

Z. GUEROUI, Laboratoire de Physique, ENS, 46 allée d'Italie, 69364 Lyon Cedex 07, France

P.H. HENNINGSEN, Niels Bohr Institute, Blegdamsvej 17, 2100 Copenhagen, Denmark

J. HENRIKSEN, Technical University of Denmark, Department of Chemistry, Building 206, 2800 Lyngby, Denmark

R. HORVATH, Eötvös University, Department of Biological Physics, Budapest, Pazmany P. Stny 1A, 1117, Hungary

V. HVASS, Niels Bohr Institute, Blegdamsvej 17, 2100 Copenhagen, Denmark

T. KAASGAARD, Department of Chemistry, Building 206, Technical University of Denmark, 2800 Lyngby, Denmark

R. KAPON, Department of Condensed Matter Physics, Weizmann Institute, Rehovot 76100, Israel

I. KOSINSKA, Jagellonian University, Marian Smoluchowski, Institute of Physics, 30-059 Krakow 16, Reymonta 4, Poland

G. KOSTER, AMOLF, Kruislaan 407, 1098 SJ Amsterdam, The Netherlands

Z. KOZA, Institute of Theoretical Physics, University of Wroclaw, 50-204 Wroclaw, Poland

K. KRUSE, L.D. Landau Institute, 2 Kosygina, Moscow 117940, Russia

T. LIVERPOOL, Condensed Matter Theory, Blackett Laboratory, Imperial College, Prince Consort Road, London SW7 2BZ, U.K.

T. McLEISH, Department of Physics & Astronomy, University of Leeds, Leeds LS2 9JT, U.K.

S. MANLEY, Harvard University, 40 Oxford St., Cambridge, MA 02138, U.S.A.

I. MESROPYAN, Tbilisi, Ijavakhetski Pereulok Korp 7, KV1, 380061, Georgia

F. MOHAMMAD RAFIEE, Institute for Advanced Studies in Basic Sciences (IASBS), Gava Zang 45195-159, Zanjan, Iran

M. MULLER, LPTMS, Université de Paris XI, Bât. 100, 91405 Orsay Cedex, France

V. MYKHAYLOVSKYY, Institute for Nuclear Research, 47 Prospect Nauki, Kiev 03680, Ukraine

I. NEMENMAN, NEC Research Institute, 4 Independence Way, Princeton, NJ 08540, U.S.A.

A. NICOLAS, CEA, 17 rue des Martyrs, 38054 Grenoble Cedex, France

H. NØRGAARD, Politiken's House, Raadhusplabsen, 1785 Copenhagen, Denmark

Y. ORLOV, Institute of Cytology & Genetics, Lavrentiev Ave. 10, Novosibrik 630090, Russia

K. PENC, Research Institute for Solid State Physics and Optics, P.O. Box 49, 1525 Budapest, Hungary

T. PILIZOTA, University of Zagreb, Bijenicka 32, 10000 Zagreb, Croatia

F. POUTHAS, LPMC, ENS, 24 rue Lhomond, 75005 Paris, France

C. PRINZ, Princeton University, Department of Physics, Jardwin Hall, P.O. Box 708, Princeton, NJ 08544, U.S.A.

N. PSHECHENKO, Entuziastov Str.1-46, 61108 Kharkov, Ukraine

I. RIEDEL, Max Planck Institute, Pfotenhauer St. 8, 01307 Dresden, Germany

T. RISLER, Unité Physico-Chimie Curie, UMR 168, 11 rue Pierre et Marie Curie, 75248 Paris Cedex 05, France

K. SAS, Eötvös University, Budapest, Pazmany P. Stny 1/A, Budapest 1117, Hungary

M. SIU, 8411-42 Ave., Edmonton, AB T6K 1C9, Canada

S. STILL, NEC Research Institute, 4 Independence Way, Princeton, NJ 08540, U.S.A.

C. TANASE, Fom Institute for Atomic and Molecular Physics, Kruislaan 407, 1098 SJ Amsterdam, The Netherlands

J. TEGENFELDT, Princeton University, Department of Physics, Jadwin Hall, P.O. Box 708, Princeton, NJ 08544-0708, U.S.A.

N. TER-OGANESSIAN, Biophysics Laboratory E22, Technical University of Munich, James-Frank-Strasse, 85748 Garching, Germany

I. TESSMER, Laboratory of Biophysics & Surface Analysis, University of Nottingham, Nottingham NG7 2RD, U.K.

I.M. TOLIC-NØRRELYKKE, Institute "Ruder Boskovic", Bijenicka 54, 10000 Zagreb, Croatia

S. TOLIC-NØRRELYKKE, Niels Bohr Institute, Blegdamsvej 17, 2100 Copenhagen, Denmark

K. VERMEULEN, Vrije Universiteit Amsterdam, Fac. FEW/Div. N&S, De Boelelaan 1081, 1081 HV Amsterdam, The Netherlands

A. VILFAN, Cavendish Laboratory, University of Cambridge, Madingley Road, Cambridge CB3 0HE, U.K.

M. WEISS, EMBL, Cell Biology & Cell Biophysics Programme, Meyerhofstr.1, 69117 Heidelberg, Germany

C. WIGGINS, Courant Institute, 251 Mercer St., New York, NY 10012, U.S.A.

A. ZILMAN, Weizmann Institute, Department of Materials & Interfaces, 71600 Rehovot, Israel

Préface

La matière existe dans de nombreux états, dont ceux de « matière molle », qu'elle soit inerte ou vivante. Cette dernière est très éloignée de l'équilibre thermodynamique, et semble suivre ses propres règles. Cette différence trouve son origine dans la complexité de la matière vivante, complexité apparue lors de l'évolution vers des états qui acquièrent et manipulent l'information, s'autorépliquent, etc.

Notre connaissance de cette complexité s'est accrue de façon spectaculaire ces dernières décennies grâce à la biologie moléculaire et cellulaire. Cette connaissance est le fondement de la physique des systèmes biologiques, qui se développe rapidement et contribue désormais elle même à cette compréhension. Le rôle de cette dernière en biologie représente un merveilleux défi : faire la distinction entre nécessité et possibilité, entre les conséquences des lois immuables de la physique et les résultats de l'évolution qui peuvent être spécifiques à la seule histoire naturelle qui nous soit accessible. Après tout, l'étude de la Vie est similaire à l'ingénierie inverse[1] (reverse engineering). Mais à quel système fascinant cette étude s'intéresse ! Plus on accède à ses détails, plus elle devient captivante : les systèmes vivants ont « été conçus » à l'envers, si bien que certaines des questions les plus importantes sur la vie sont cachées dans ses plus petits composants.

Cette 75e école d'été des Houches était consacrée à la physique des molécules biologiques et des cellules. Elle s'intéressait aux questions qui, dans les systèmes biologiques allant des biomolécules simples jusqu'aux cellules et aux systèmes plus larges, ont besoin pour leur analyse et leur compréhension des concepts et les méthodes de la physique. L'école s'est ouverte par deux séries de cours en parallèle par Robijn Bruinsma et Jonathon Howard. Le cours sur *la physique des interactions protéines-ADN* par R. Bruinsma est parti de la structure de l'ADN et des molécules qui lui sont associées, pour arriver à la discussion des interactions électrostatiques entre ces protéines et l'ADN, et à la recherche diffusive des sites spécifiques de liaison. Le cours de J. Howard sur *la mécanique des protéines motrices* a traité des propriétés mécaniques des molécules individuelles, des moteurs moléculaires et des structures complexes du cytosquelette. En même temps, une série plus courte de leçons par Evan Evans sur *l'utilisation des forces pour sonder la chimie des liaisons*

---

[1] Ingénierie inverse : processus consistant à analyser un système afin d'identifier ses composantes et leur interrelation, et de créer une représentation de ce système sous une autre forme ou à un plus haut niveau d'abstraction. (E.J. Chikofsky and J.H. Cross, II. *IEEE Software* 7 (1990) 13-17.)

biomoléculaires et les transitions structurales a permis d'explorer la grande richesse des comportements dynamiques présents lors de la rupture des liaisons biomoléculaires individuelles. Ces cours furent suivis de la discussion par Erich Sackmann de la *microrhéométrie des réseaux d'actine et des réseaux cellulaires*. Il a présenté une introduction aux membranes cellulaires et au cytosquelette, et une discussion des propriétés mécaniques des cellules et de la physique de l'adhésion cellulaire. R. Bruinsma a complété les cours de E. Sackmann par des cours théoriques sur la *mécanique statistique et la bioadhésion*. La deuxième moitié de l'école comprenait deux longues séries de cours en parallèle par Thomas Duke et par Bill Bialek. T. Duke a complété le cours de J. Howard en traitant des approches théoriques de la *modélisation des systèmes de protéines motrices*. Partant des modèles physiques des protéines motrices, il a discuté des aspects physiques des cils et des flagelles, et a montré l'importance des phénomènes physiques actifs à l'échelle cellulaire dans l'audition. Un autre exemple de création de mouvement a été discuté par Jacques Prost dans son cours sur la *physique de la propulsion de la Listéria*, où il a décrit de façon générale comment la polymérisation contrôlée d'un gel d'actine peut être utilisée pour la propulsion. La longue série de cours de Bill Bialek sur *réfléchir sur le cerveau* a présenté une introduction aux principes qui gouvernent les systèmes sensitifs et nerveux. Partant d'exemples simples de traitement de l'information dans le système visuel de la mouche, il est passé à des questions fondamentales sur ce traitement dans le système nerveux. Dans un cours sur *bioinformatique et mécanique statistique* Eric Siggia a présenté une revue sur le décodage de l'information génétique obtenu dans les projets de génomique. Suivaient les cours de Bob Austin sur *micro- et nanotechnologie et physique en biotechnologie*, ces nouvelles technologies qui rendent possible l'étude et la manipulation de biomolécules dans des réseaux et des structures artificielles. Marcello Magnasco a excellemment mené cette école à sa conclusion grâce à ses *trois leçons sur les réseaux biologiques*, qui traitaient du dénouage de l'ADN, de l'analyse des données des puces à ADN, et de l'expression des gènes lors de l'apprentissage chez les canaris, le tout dans un style qui su captiver le public jusqu'à la dernière minute de ces quatre semaines.

Ces cours principaux étaient complétés par des séminaires donnés par Albert Libchaber (*Polymérisation de RecA sur des molécules d'ADN simple brin* et *Évolution dirigée : une étude moléculaire*) et Marileen Dogterom (*Polymerization Forces*). Deux conférences à destination du grand public ont été données au village des Houches, l'une par A. Libchaber (*Qu'est-ce que la Vie* ?), l'autre par T. Duke (*Les moteurs de la Vie*). Phil Williams a donné un exposé invité lors des cours de E. Evans, et Tom McLeish et Chris Wiggins ont contribué par deux séminaires : *Le cas mystérieux des feuillets beta supplémentaires* et *modèles physiques de biopolymères*. Pendant les temps libres, T. McLeish a donné sur la pelouse de l'école une introduction pédagogique – très suivie – sur *le passage de barrière activé thermiquement*, avec en arrière plan le

Mont Blanc comme barrière massive et des plus illustrative. Nous avons aussi organisé seize mini séminaires d'étudiants, répartis sur quatre soirées, et deux sessions de présentation de posters, pour un total de seize ; les titres et les intervenants sont donnés à la fin de l'ouvrage. Les étudiants ont fait preuve de beaucoup d'énergie et d'enthousiasme, et – de façon remarquable au vu de la densité du programme – se sont accrochés jusqu'à la fin de la session.

Nous avons eu pour cette session trois fois plus de demandes que de places dans l'amphi, et nous avons dû malheureusement refuser beaucoup d'excellents candidats. Nous espérons pour eux que ce livre palliera, dans une certaine mesure, à cette malencontreuse limitation. La solitude relative de l'école de physique sur le versant en face du Mont Blanc est parfaite pour l'étude et les contacts. Tout autant le sont les longues randonnées des week-ends. Des amitiés durables s'y nouent, comme nous pouvons en témoigner : c'est aux Houches lors une école d'été que deux des présents organisateurs se sont rencontrés pour la première fois. Si cette session a formé et inspiré ses participants comme celle d'alors, nous aurons bien travaillé !

Remerciements

Ces quatre semaines passées aux Houches se sont très bien déroulées, grâce au secrétariat de l'école : Ghislaine d'Henry, Isabel Lelièvre et Brigitte Rousset. Cette école était financée par l'OTAN comme *Institut d'Études Avancées*, par l'Union Européenne comme *École d'Été Européenne*, par le CNRS comme *École Thématique*, et par la Danish Research Agency au travers de la *Danish School of Biophysics*. La NSF a financé les voyages de certains participants des USA. Nous remercions toutes ces institutions grâce auxquelles cette école a été rendue possible. Nous sommes persuadés que ce livre permet de présenter des exemples remarquables de physique des systèmes biologiques, et nous remercions encore une fois tous ceux qui y ont contribué par leurs cours et leur mise en forme.

Henrik Flyvbjerg
Frank Jülicher
Pál Ormos
François David

Preface

Matter has many states, including soft condensed, inert or alive. The latter is far from thermodynamic equilibrium, and apparently has an agenda of its own. Yet the same physical laws apply to all matter. The difference is in the complexity to which living systems have evolved, to states that gather and process information, replicate themselves, etc.

Molecular and cell biology have dramatically expanded our knowledge about this complexity in the last decades. This knowledge is the foundation of biological physics, which is currently expanding rapidly and is itself adding to this knowledge. Its role in biology is a wonderful challenge: to draw the line between necessity and possibility, between results of immutable physical laws and results of evolution that may be specific to the one natural history we have access to. The study of life is, after all, similar to reverse engineering[1]. What fascinating engineering it describes, however! The deeper one gets into the details, the more captivating the study becomes: these systems were "designed" bottom-up, so answers to some of the biggest questions about Life are hidden in their smallest parts.

The 75th Les Houches summer school addressed the physics of biomolecules and cells. In biological systems ranging from single biomolecules to entire cells and larger biological systems, it focused on aspects that require concepts and methods from physics for their analysis and understanding. The school opened with two parallel lecture series by Robijn Bruinsma and Jonathon Howard. *Physics of Protein-DNA Interaction* by Robijn Bruinsma started from the structure of DNA and associated proteins, and lead to discussions of electrostatic interactions between proteins and DNA, and the diffusive search for specific binding sites. Joe Howard's lectures on *Mechanics of Motor Proteins* discussed mechanical properties of individual proteins and motors, and of complex cytoskeletal structures. Simultaneously, Evan Evans' shorter series *Using Force to Probe Chemistry of Biomolecular Bonds and Structural Transitions* explored the rich dynamic behaviors of rupturing individual biomolecular bonds. These lectures were followed by Erich Sackmann's discussion of *Micro-rheometry of Actin Networks and Cellular Scaffolds*. He gave an introduction to membranes and the cytoskeleton and discussed the mechanical properties of cells and the physics of cell adhesion. Robijn Bruinsma

[1] Reverse engineering: "the process of analysing a subject system to identify the system's components and their interrelationships and create representations of the system in another form or at a higher level of abstraction". (E.J. Chikofsky and J.H. Cross, II. *IEEE Software* **7** (1990) 13-17.)

complemented Erich Sackmann's lectures with theoretical lectures on *Statistical Mechanics and Bioadhesion*. The second half of the school contained two long, parallel lecture series by Thomas Duke and Bill Bialek. Tom Duke complemented Joe Howard's course with *Modelling Motor Protein Systems*, which focused more on theoretical approaches. Starting with physical models for motor proteins, he discussed physical aspects of cilia and flagella and showed that active physical phenomena on the cellular scale are important in hearing. Another example of motion generation was discussed in Jacques Prost's lectures *Physics of Listeria Propulsion*, which provided a general description of how the controlled polymerization of an actin gel can be used for propulsion. Bill Bialek's long series of lectures *Thinking About the Brain* gave an introduction to the principles governing sensory and nervous systems. Starting from simple examples of information processing in the visual system of the fly, he moved to fundamental questions on how nervous systems process information. In *Bioinformatics and Statistical Mechanics* Eric Siggia reviewed decoding of genetic information obtained from genome projects. It was followed by Bob Austin's lectures on *Micro- and Nanotechnology-Physics in Biotechnology*, new technologies which make it possible to study and manipulate biomolecules in artificial arrays and structures. Marcelo Magnasco wrapped up the school excellently with his *Three Lectures on Biological Networks*, which covered the unknotting of DNA, the analysis of gene chip data, and studies of gene expression in learning canaries, all in a style that kept the attention of the audience to the last minute of four weeks.

The lectures were complemented by invited seminars given by Albert Libchaber (*RecA Polymerization on Single-Stranded DNA* and *Directed Evolution: A Molecular Study*) and Marileen Dogterom (*Polymerization Forces*). Two public lectures were given in the town of Les Houches, by Albert Libchaber (*Qu'est-ce que la vie* ?) and Thomas Duke (*Les moteurs de la vie*). Furthermore, Phil Williams gave an invited lecture within Evan Evan's series, and Tom McLeish and Chris Wiggins contributed with seminars: *The Mysterious Case of Too Many β-Sheets* and *Into Physical Models of Biopolymers*, respectively. During a study period, Tom McLeish gave a well-attended tutorial on thermally activated barrier crossing, on the school's lawn, with Mt. Blanc as a backdrop and most illustrative barrier. We also organized sixteen short student presentations over four evenings, and two poster sessions with a total of seventeen posters; see titles and presenters at the end of this volume. The students had great energy and enthusiasm, and, amazingly in view of their schedule, kept it up till the very end.

This school had three times as many applicants as there are seats in the lecture hall, and we had to turn down many strong applicants. We hope this book to some extent makes up for this unfortunate restriction on admission.

The relative isolation of the Les Houches Physics School on the mountain side vis-à-vis Mt. Blanc is perfect for learning and interacting. As are long hikes in the mountains on weekends. Life-long friendships are formed, we know: Two of this school's organizers first met as students in a Les Houches summer school. If the present school has taught and inspired its participants as much as that school did years ago, we have done well.

Acknowledgements

The four weeks spent in Les Houches went smoothly, thanks to the staff of the school: Ghislaine D'Henry, Isabel Lelièvre and Brigitte Rousset. The school was sponsored by NATO as an *Advanced Study Institute*, by the EU as a *Eurosummerschool*, by CNRS as an *École Théematique*, and by the Danish Research Agency through its *Graduate School of Biophysics*. NSF covered travel costs for some US residents. We thank them all for making the school possible. We are convinced that this book presents outstanding examples of biological physics, and thank the contributors again for their great efforts of lecturing and writing.

Henrik Flyvbjerg
Frank Jülicher
Pál Ormos
François David

CONTENTS

Course 4. Dynamic Force Spectroscopy

by E. Evans and P. Williams *145*

Course 6. Physics of Composite Cell Membrane and Actin Based Cytoskeleton

by E. Sackmann, A.R. Bausch and L. Vonna *237*

Course 7. Cell Adhesion as Wetting Transition?

by E. Sackmann and R. Bruinsma *285*

COURSE 1

PHYSICS OF PROTEIN-DNA INTERACTION

R.F. BRUINSMA

Department of Physics and Astronomy, University of California, Los Angeles, CA 90024, USA, and Instituut-Lorentz for Theoretical Physics, Universiteit Leiden, Postbus 9506, 2300 Leiden, The Netherlands

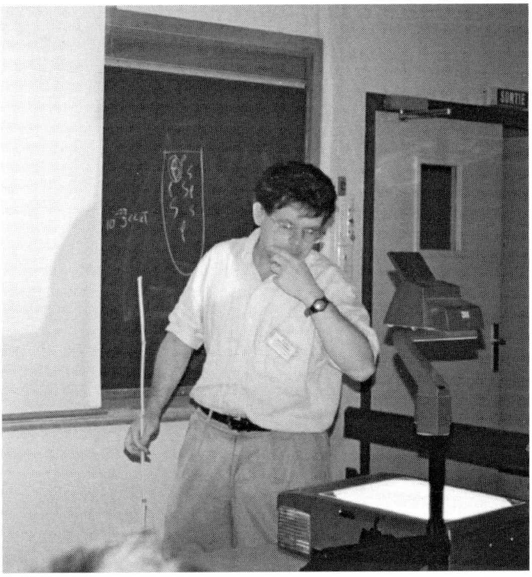

Contents

PHYSICS OF PROTEIN-DNA INTERACTION

R.F. Bruinsma

1 Introduction

1.1 The central dogma and bacterial gene expression

1.1.1 Two families

Life is based on a symbiotic relationship between two families of biopolymers: DNA and RNA, constituted of nucleic-acids, and proteins, constituted of amino-acids [1]. Proteins are the active agents of the cell. As *enzymes*, they control the rates of biochemical reactions taking place inside the cell. They are responsible for the *transcription* of the genetic code, *i.e.*, the production of copies of short segments of the genetic code that are used as blue-prints for the production of new proteins, and for the *duplication* of the genetic code, *i.e.*, the production of a full copy of the genetic code during cell division. Synthesis of other macromolecules, such as lipids and sugars, is carried out by proteins, the mechanical force of our muscles is generated by specialized proteins adept at "mechano-chemistry", they detect light, sound, and smell, and maintain the structural integrity of cells.

If we view the cell as a miniature chemical factory that simultaneously runs many chemical processes, then the proteins form the control system of the factory, turning reactions on and off. The control system obeys orders from the central office: the cell nucleus. The DNA inside the nucleus can be considered as the *memory* of the computer system of the central office: it is the *information storage system* of the cell. Blueprints for the synthesis of proteins are stored in the form of DNA base-pair sequences, much like strings of zero's and one's store information in digital computers. A *gene* is the data string required for the production of one protein (actually, multiple variants of a protein can be produced from the same gene). The beginning and end points of a gene are marked by special "start" and a "stop" signals. When a protein has to be synthesized, a specialized copying protein, RNA polymerase, transcribes a copy of a gene beginning at the start signal and ending at the stop signal (see Fig. 1).

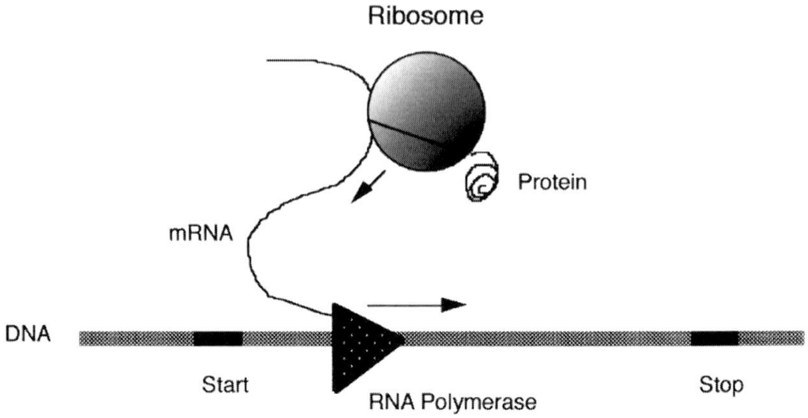

Fig. 1. Gene transcription.

This copy is in the form of an RNA strand known as *m*RNA (or "messenger" RNA). A huge molecular machine, the *Ribosome*, synthesizes the protein from the *m*RNA blueprint. Interestingly, these Ribosomes are compound constructs of RNA strands (known as *r*RNA) and proteins, with the active biochemistry carried out not by the protein part, as you might have expected, but by the RNA part. Indeed, unlike DNA, RNA strands are in fact capable to act as enzymes.

The information stream is strictly *one way*: DNA contains the information required for the synthesis of proteins. The genetic code is not altered by the transcription, and RNA strands do not insert their code into DNA. We call this basic principle of biochemical information flow the "**central dogma**". We know next to nothing about how this elaborate relationship between the nucleic and amino acids developed. The basic chemical structure of the two families is quite different. The molecular biology of living organisms is all highly similar and based on the central dogma and we do not know of the existence of more primitive molecular information and control systems from which we could somehow infer a developmental history (though we suspect that once upon a time both information storage and enzymatic activity was based purely on RNA since RNA is able to carry out enzymatic activity as we saw). The central dogma applies to *living* organisms. Retroviruses are able to insert their RNA code into host DNA, using a special enzyme called "reverse transcriptase". This looks like an exception to the central dogma but viruses are not considered living organisms since they are not able to reproduce themselves independently nor do they carry out metabolic activity, the two defining requirements of a living organism.

It is reasonable to ask why *bio*polymers should be of special interest to physicists. The physics of polymers – particular synthetic polymers – has been studied for decades and an elegant, general theoretical framework is available. The motivation behind a study of the interaction between DNA and proteins is quite different from that of a study of synthetic polymers. In polymer physics, we want to compute the free energy and correlation functions of a *typical* polymer in a solution or melt, with results that are as much as possible independent of the detailed molecular structure of the polymers. That philosophy does not apply to biopolymers where we are dealing with highly *a-typical* molecules that carry out certain *functions*. Their structure presumably evolved under the adaptive pressures exerted on micro-organisms that relied for their survival on efficient performance of the functions these molecules are involved with. A molecular biophysicist tries to shed light on how functional molecular devices work and how their design constraints are met. These are of course very complex systems, so it is a good strategy to focus as much as possible on basic principles of physics of general validity and relying as little as possible on assumptions concerning the detailed molecular structure. The hope is that this will provide us with constraints on the design and operation of functional biopolymers in the way that the Second Law of Thermodynamics constrains the maximum efficiency of steam engines.

In order to illustrate this approach, we will focus on two special cases that have been particularly important in the development of our understanding of protein-DNA interaction, the *lac repressor* and the *Nucleosome Complex*. These two systems have been studied in such detail that we may hope to understand how they "work" as molecular devices. In these lectures, we will see what insights thermodynamics, statistical mechanics, elasticity theory, and electrostatics can provide us in this respect.

1.1.2 Prokaryote gene expression

How does an organism "know" when to turn gene transcription on and when to turn it off? We divide cells in two groups: eukaryotes and prokaryotes. The cells of animals and plants – the eukaryotes – have their DNA sequestered inside a nucleus and the cell has a complex set of internal "organs" called organelles. Gene expression of eukaryotic cells, the focus of much current research, is a complex affair, which we will discuss in a later section. Bacteria, prokaryotes, lack a nucleus and organelles and their gene expression is much better understood [3]. We will discuss a simple example: the expression of the "lac" gene of the bacterium Escirichia Coli (E.Coli for short) [4].

Large numbers of the E.Coli parasitic bacteria live inside your intestines ("colon"). When you drink a glass of milk, part of it will be metabolized not by you but by your E.Coli bacteria. The first step is the breakdown of *lactose*, sugar molecules consisting of two linked molecular rings. Lactose is broken down into two single-ring glucose molecules. This chemical reaction requires an enzyme, called "β Galactosidase", to proceed because lactose does not dissociate spontaneously (an enzyme speeds up a reaction by lowering the activation energy barrier). First though, the lactose molecules must be transferred from the exterior of the bacterium to the cell interior (or "cytoplasm") across the membrane that surrounds E.Coli. This is done by another protein, called "Permease". Finally, a third protein, called "Transacetylase", is required for chemical modification of the sugar molecules.

The DNA of E.Coli carries three separate genes for the production of these three enzymes: lacY, lacZ, and lacA. Expression of the three genes starts when the environmental lactose concentration rises, and it stops when the lactose concentration drops (to avoid wasteful use of precious macromolecular material). The three genes are located right behind each other on the DNA, and – sensibly – they are transcribed collectively. Such a cluster of functionally connected genes is called an "operon". The lac operon also contains three *regulatory* sequences:

a) Promoter Sequence

   This sequence is "recognized" by RNA Polymerase. By that we mean that RNA Polymerase molecules in solution bind to Promoter Sequences on the DNA but not to other sequences. From this start site, RNA polymerase can transcribe RNA *in either direction*. In one direction, "downstream", it produces the RNA code of our three enzymes. In the other direction, "upstream", it transcribes the neighboring "Regulator" sequence.

b) Regulator Sequence

   The Regulator sequence is the code of a fourth protein: *lac repressor*. The lac repressor, which is *not* involved in the metabolic of lactose, plays a key regulatory role in turning the gene "on" or "off".

c) Operator Sequence

   The operator sequence is a DNA sequence that is recognized by lac repressor. If lac repressor is bound to the operator sequence, then downstream gene expression is blocked. The Figure 2 shows how this "genetic switch" works.

First, assume that the concentration of lactose in the environment is high. Lactose molecules bind reversibly to the repressor protein. For high lactose concentrations, the lactose-bound form is favored under conditions of chemical equilibrium. In the lactose-bound (or "induced") form, the repressor

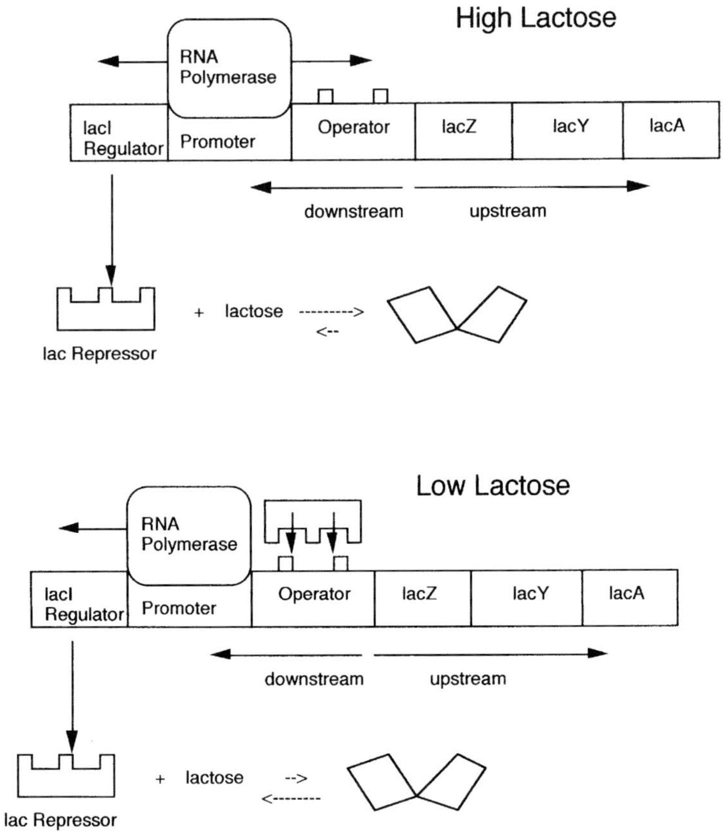

Fig. 2. The lac operon and gene regulation.

has a different structure in which it does *not* bind to the operator sequence. RNA polymerase proteins binding to the promoter sequence are free to transcribe in the down-stream (and up-stream) direction. Along the downstream direction, it will produce an RNA copy of the genes of the three enzymes required for lactose breakdown. Transcription along the up-stream direction produces an RNA copy of the lac repressor gene. Production of repressor proteins at a low level is necessary to maintain their concentration since proteins have a finite lifetime (after a certain period, a protein receives a molecular "tag" targeting it for future breakdown as part of the scheduled maintenance program of the cell).

Next, assume that the lactose concentration has dropped. The chemical equilibrium now favors the lactose-free conformation of the repressor. Lac repressor binds to the operator sequence and downstream gene transcription is blocked. Genetic switches of this type are used by E.Coli (and other bacteria) to respond to changes in temperature, salinity, acidity, and the oxygen level. Efficiency of these switches clearly is a matter of life and death for the bacterium so we should expect that the structure of proteins like the lac repressor has been "sharpened" by natural selection for optimal performance. If you would put yourself the task of designing a lac repressor protein some obvious minimum engineering requirements would be:

Specificity: the lac repressor must be able to recognize the operator sequence. Repressor proteins must be able to efficiently "read" the DNA code.

Reversibility: the lac Repressor must bind reversibly to lactose or else gene expression could not be turned off. Similarly, it must bind reversibly to DNA or else gene expression could not be turned on.

Reactivity: The lac Repressor must locate the operator sequence within minutes after the lactose concentration drops. If it takes too long to turn a genetic switch then the bacterium could be dead before it had the change to respond to the changing environment.

In the next sections, we will see what thermodynamics, statistical mechanics, and elasticity theory have to say about these requirements. First, we have to learn more about the molecular structure of the two biopolymer families [2].

1.2 Molecular structure

1.2.1 Chemical structure of DNA

The basic monomer unit – the polymer repeat unit – of double-stranded DNA is shown in Figure 3.

The parts marked B and B* are large, planar organic groups consisting of one or two 5-atom *aromatic rings*. They resemble benzene and, like benzene, these groups do not dissolve very easily in water. The symbols B and B* stand either for the smaller single- ring Cytosine and Thymine (the "pyrimidines"), or the larger two- ring Guanine and Adenine (the "purines"). We will use the notation G, T, C, and A for short. The four groups all have the chemical character of a *base* (*i.e.*, they are proton acceptors).

Not every combination of bases is permitted: in particular only B-B* pairs of purines and pyrimidines are possible. The Watson–Crick base-pairing consists of combining A with T and G with C. An A-T pair is connected by two *hydrogen bonds* and a G-C pair by three hydrogen bonds, so they have a higher binding energy. Other purine-pyrimidine pairings

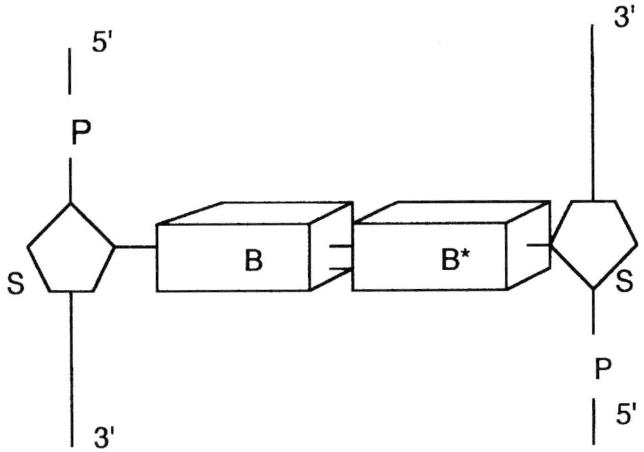

Fig. 3. Double-stranded DNA repeat unit.

(like G with T) are possible, but they do have a lower binding energy. The genetic code of an organism, or "genome", is simply a listing of the different base-pairings along the DNA sequence of that organism. Note that if you know the sequence of bases of one strand, you always can reconstruct the other "complementary" strand, assuming that Watson–Crick base pairing is valid.

The bases are connected to sugar groups (indicated by S in the figure). Sugars have the general formula $(CH_2O)_n$ and usually are water soluble. The particular sugar of DNA belong to the group of *pentoses*, 5-atom sugar rings, and is known as *deoxyribose*. The deoxyriboses of two adjacent bases are connected together by tetrahedral *phosphate groups* (PO_4^-) to form together the sugar-phosphate "backbone". Adjacent sugar groups are separated by 6 Å. The backbone strands have a *directionality*: they start with a deoxyribose at the $3'$ end and end with a phosphate at the $5'$ end. The backbone has two important physical characteristics for our purposes: it is *highly flexible* and, in water at room temperature, it is *highly charged*. The negative charge of the backbone is due to the fact that the phosphate groups in water at physiological acidity levels are fully dissociated. Charged molecular groups are usually soluble in water and the sugar-phosphate backbone is indeed highly soluble in water. The flexibility is due to the fact that the covalent P-O bonds can freely rotate around so adjacent PO_4^- tetrahedra and ribose rings along the backbone can rotate around their joining axis. We can describe the backbone as a *charged, freely jointed chain*.

RNA molecules are similar to single stranded DNA molecules with two differences. First, the base Thymine is replaced by another base, Uracil, and second, the sugar group has an extra OH group and is called a ribose.

Intermezzo: *Hydrogen bonding and the hydrophobic force*

Hydrogen bonding provides the binding mechanism between complementary bases. Hydrogen bonding plays in general a central role whenever macromolecules are dissolved in water. The hydrogen bond is an electrostatic bond with a positively charged proton from one molecular group associating with a negatively charged atom of another molecular group, usually an oxygen (O^-), Carbon (C^-) or nitrogen (N^-) atom. The cohesion of water is due to hydrogen bonding between water molecules, with the proton of one water molecule binding to the oxygen of another water molecule. The characteristic energy scale of the hydrogen bond is of the order of the thermal energy $k_B T$, so it is a relatively weak bond. At room temperature, a thermally fluctuating network of hydrogen bonds connects the water molecules.

Molecules such as alcohol that are easy to dissolve in water are called "hydrophilic" while molecules, such as hydrocarbons, that are not soluble in water are called "hydrophobic" [5]. Hydrophobic molecules cannot be incorporated in the thermally fluctuating network of the hydrogen bonds. They are surrounded by a shell of water molecules that have a reduced entropy, since they have fewer potential partners for the formation of a hydrogen bonding network. As far as the water molecules are concerned, the surface of a large hydrophobic molecule resembles the air-water surface, which has a surface energy γ of about 70 dynes/cm. We thus can estimate the solvation free energy – the free energy cost of inserting a molecule in a solvent – as the surface area of the hydrophobic molecule times γ. If we wanted to dissolve a certain number of hydrophobic molecules we could reduce the total exposed surface area in order to minimize the free energy cost by collecting the hydrophobic molecules in dense clusters. This effect is known as the "*hydrophobic interaction*", though it obviously is not a pair-wise interaction between molecules. Ultimately, the clustering leads to phase-separation, which you can observe when you try to mix oil with water. An important thermodynamic characteristic of the hydrophobic interaction is that it is predominantly entropic in nature.

1.2.2 Physical structure of DNA

The physical structure of double-stranded DNA is determined by the fact that it is neither hydrophobic nor hydrophilic. It belongs to a special intermediate group the "*amphiphiles*" that share properties from both classes: one part of DNA – the backbone – is hydrophilic and another part of DNA – the bases – is hydrophobic. This frustrated, amphiphlic character of DNA, plus the flexibility of the backbone, produces the famous double-helical structure of DNA. To see how, imagine DNA stretched out like a (straight) ladder (Fig. 4).

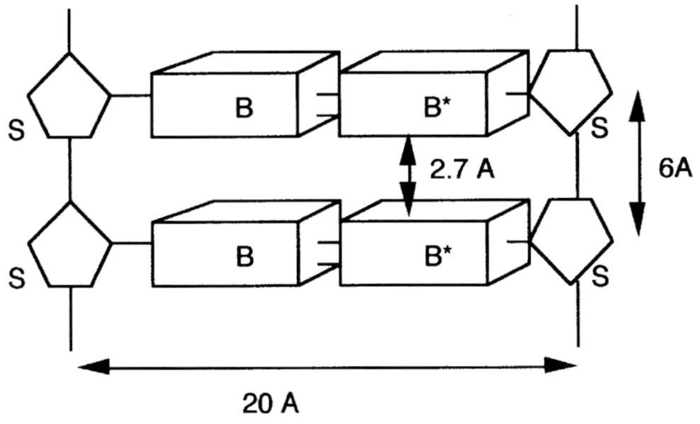

Fig. 4. Geometry of stretched DNA.

It turns out that the gaps between the rungs of the ladder, 2.7 Å, are wide enough to allow water molecules to slip in between the bases. Under the action of the hydrophobic force, the bases attract each other. The fixed 6 Å spacing between the sugar groups prevents a contraction of the ladder, but there is another way to bring the bases in contact. Imagine that we gradually *twist* the ladder, thereby forming a double spiral. This *is* possible because of the flexibility of the backbone. As we increase the twisting, the bases are brought into closer contact and the water molecules are squeezed out. For a twist angle T of about 32 degrees between adjacent bases, the gap is completely closed. This produces the classical double helix shown below. The repeat length is $360/T$ bases, or about 11 bases. The repeat distance along the helix, or pitch, is about 35 Å (using Fig. 4, compute T yourself).

The DNA double-helix is thus held together by the hydrophobic attraction between bases, sometimes called the stacking interaction, and the hydrogen bonding between complementary bases. The double-helix is not very stable. If you heat DNA, the two strands start to fall apart for temperatures in the range of 70–80 °C. In addition, a number of different variants of the double helix can be realized by modest changes in the environmental conditions. Under conditions relevant for the life of cells, the dominant structure is the "B form", a right-handed helix with a 24 Å diameter. Increasing the salt concentration somewhat weakens the electrostatic repulsion between the two backbones. A new structure, known as "A DNA", appears,

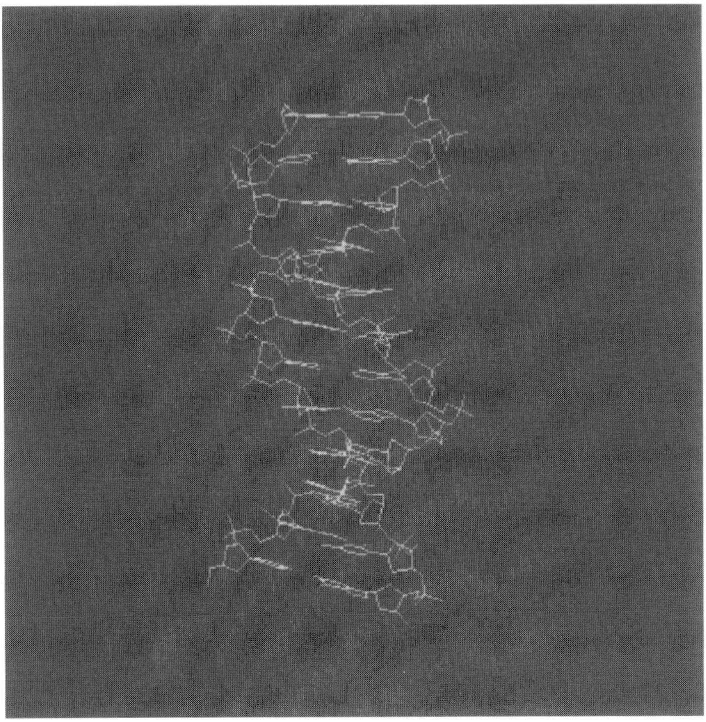

Fig. 5. B DNA.

with a smaller 18 Å diameter for the double helix and larger pitch of about 45 Å. This structural flexibility of DNA is actually essential for its function: in order for the genetic code to be "read" by RNA Polymerase and other proteins, you must be able to "open-up" the double-helix. Storing the genetic code in an overly rigid and stable storage device would be like having a library with no doors.

1.2.3 Chemical structure of proteins

There are 20 different monomer units, or "residues", that can be used to construct protein biopolymers. These are the naturally occurring *amino-acids*. Amino-acids have the form of a tetrahedron with a Carbon atom at the center, denoted by C_α. Recall that carbon has 4 electrons available to make chemical bonds. For the central C_α atom, these four electrons occupy four electronic orbitals (the "sp^4" orbitals) directed towards the vertices of a tetrahedron (as in diamond). At the four vertices are placed, respectively, a hydrogen atom, an NH_2 "amino" group, an acidic COOH "carboxyl" group, and finally one of 20 different side groups denoted by "R". There are two distinct ways of distributing the four molecular groups over the tetrahedron. The one shown in the next figure is known as the *left-handed* or L form. If the amino and hydroxyl groups are exchanged, we obtain the right-handed or R form. In proteins, only the L form is encountered. The side group determines the chemical character of an amino-acid. It can be neutral or charged, hydrophobic or hydrophilic, acidic or basic.

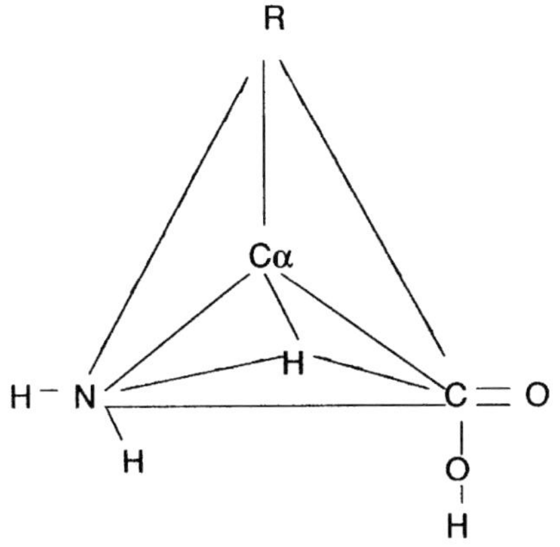

Fig. 6. Amino-Acid.

The simplest case is Glycine, with R a hydrogen atom. There are two abbreviations for Glycine: Gly and G. Each amino acid has two of these abbreviations (which biochemists know by heart). Lysine for instance is a positively charged amino-acid with R equal to $(CH_2)_4NH_2^+$ having the abbreviations Lys and K.

Nature uses its own abbreviation for the 20 amino-acids when it stores the information required to produce a protein. Three adjacent DNA bases code for one amino-acid. For instance, the triplet AAA is the code for the amino-acid "Phenylalanine" while TTT is the code for Lysine. We call such a triplet a "codon". You can construct $4^3 = 64$ different codons from such a triplet, more that enough for the 20 natural amino-acids. Finding the complete set of codons of all the amino-acids was one of the great landmark achievements of molecular biology.

To construct a protein, we must hook together these amino-acids by a polymerisation reaction. This takes place between the amino group of one amino acid and the hydroxyl group of another amino acid creating a covalent bond – known as a *peptide bond* – between a Carbon and a nitrogen atom under release of a water molecule. The link between amino-acid complexes is not rigid: the peptide bond allows considerable freedom of motion. When repeated over and over, this reaction produces a flexible string of amino acids – a polypeptide – that starts with an amino group, the so-called "N-terminal" and that ends with a hydroxyl group, the "C-terminal".

We saw that the base-pair sequence of the DNA of an organism is a code for the production of amino-acid strings with three adjacent base-pairs coding for one amino-acid. The following DNA sequence for instance will produce a simple polypeptide of four amino-acids, starting with an N terminal and ending with a C terminal:

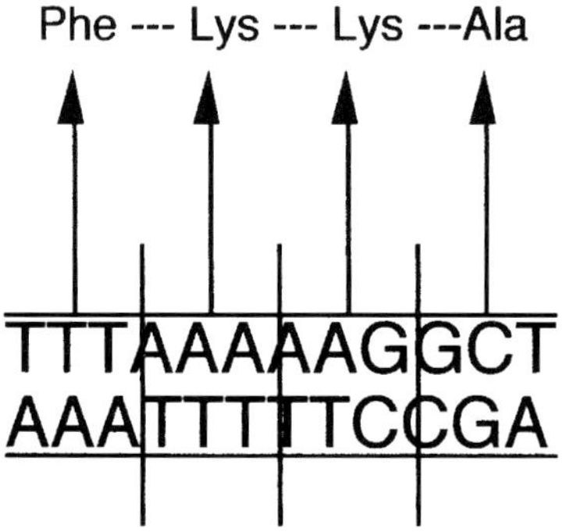

Fig. 7. Codons and amino-acids.

("Ala" stands for "Alanine", a hydrophobic amino-acid.) Note that only one of the two DNA strands is actually used for the production of proteins, the "coding" strand.

1.2.4 Physical structure of proteins

The physical structure of protein is determined by two physical mechanisms. On the one hand, proteins are again amphiphiles. Among the string of residues making up a protein, certain will have side chains that are hydrophobic, like Ala, and certain that are hydrophilic, like Lys. When dissolved in water, the string will try to fold up into a ball, with the hydrophobic residues hidden in the interior and the hydrophilic residues on the exterior surface. Such a ball is called a "globule", with a radius of the order of 2–3 nm.

The second important effect is the ability of amino-acids to establish hydrogen bonds. The oxygen atom of the $C = O$ group at one of the corners of the amino-acid tetrahedron of one residue can act as a proton receptor for the C–H or N–H group of another residue. Linus Pauling first proposed that for a *helical* polypeptide string having the right pitch and diameter, known as the *α-helix*, every residue can establish a hydrogen bond with a

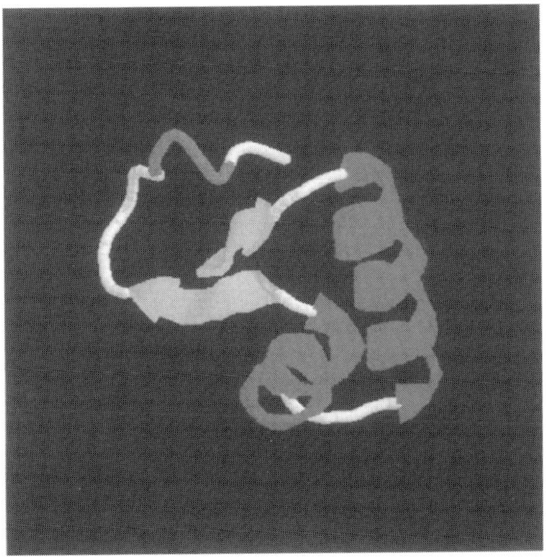

Fig. 8. A small protein with two α-helices and a β sheet.

residue further up (or down) the helix. Not every residue of a protein "likes" to be part of an *α-helix* because certain side groups may interfere with each other by steric hindering. Hydrogen-bonding also can be used to link two straight polypeptide strands that run either parallel or anti-parallel, which is known as a *β-sheet.*

The actual structure of a protein is mainly determined by the combined effects of α-helix/β-sheet formation and the requirement to keep hydrophobic residues inside the protein interior. Drawings of protein structures show the α-helical regions as spirals and the β sheets as arrows. Figure 8 shows an example of a very simple protein with two α helices and a β sheet. Note that not all amino acids are part of α helices or β sheets.

What is remarkable about natural proteins compared with a random polypeptide chain, is that over a certain range of temperatures, the minimum free energy state is a unique, folded structure with most of the atoms of the protein occupying well-defined positions. Only in this folded state can proteins act as molecular machines. This functional, folded state of a protein is not very stable. Formation of the folded structure involves a significant loss of entropy. Heating indeed unfolds, or "denatures" proteins. The "folding energy" – the difference between the properly folded state and the denatured state – is only of order 10 $k_{\mathrm{B}}T$ or so. Moderately raising (or lowering!) temperature, changing salt concentration or acidity level can produces unfolding. However, it is precisely the fragility of folded proteins that allows them to adopt multiple configurations, which permits their use as switching devices, catalysts, and detection devices. For a molecule to act as a molecular machine, it must have "moving parts".

The following website has a nice tutorial on the chemical structure of DNA and proteins

`http://www.clunet.edu/BioDev/omm/exhibits.htm#displays`

2 Thermodynamics and kinetics of repressor-DNA interaction

2.1 Thermodynamics and the lac repressor

The first branch of physics that we will bring to bear on the design of a repressor protein is thermodynamics/statistical mechanics. We will apply the principles of thermodynamics to understand how the *specificity* and *reversibility* requirements are met for the interaction between the lac repressor and DNA.

2.1.1 The law of mass action

Prepare an aqueous solution containing a certain low concentration of short, identical DNA strands and the repressor proteins. The base-pair sequence of the DNA strands may or may not contain the operator sequence. We can describe the reversible binding of the repressor to the DNA as an associative chemical reaction:

$$R + \mathrm{DNA} \leftrightarrow R|\mathrm{DNA}$$

where $R|\mathrm{DNA}$ stands for a repressor-DNA complex. The concentration of DNA strands with no repressor is indicated by [DNA], that of free repressor by $[R]$, and that of the complexes by $[R|\mathrm{DNA}]$. Concentrations can be measured by filtration methods and the results are expressed in "Molar", or moles per liter (symbol M). Salt water has, for instance, a salt concentration of about 0.1 M while one molecule per micron^3 (volume of a bacterium) equals 10^{-9} M.

Thermodynamic processes inside cells normally take place under conditions of (nearly) fixed temperature and pressure. Under these conditions, the Gibbs Free Energy G must be minimized according to the Second Law of Thermodynamics. The Gibbs Free energy can be expressed as:

$$G = N_{\mathrm{DNA}}\,\mu_{\mathrm{DNA}} + N_R \mu_R + N_{R|\mathrm{DNA}}\,\mu_{R|\mathrm{DNA}}. \tag{2.1}$$

Here, N and μ are the number of molecules and the chemical potential of each of the three species (actually, we also should add a term for the water molecules). At low concentrations, the chemical potential $\mu([C])$ of "solute" molecules in a solvent like water has the following general form:

$$\mu([C]) = k_{\mathrm{B}}T \ln([C]\nu_{\mathrm{C}}) + \mu_{\mathrm{C}}. \tag{2.2}$$

The first term, with ν_{C} the volume of the molecule, is very similar to the free energy per particle of an ideal gas and it is indeed due to the translational degrees of freedom of the solute particles. The second term, the "standard chemical potential", can be viewed as the intrinsic free energy per solute particle meaning that it depends on the type of solute molecule, and on temperature and pressure, but not on concentration.

Assume that there is a very small variation δN in the number of $R|$DNA complexes. According to the reaction scheme $R + \mathrm{DNA} \leftrightarrow R|\mathrm{DNA}$, there must be corresponding variation of $-\delta N$ in the number of uncomplexed DNA and repressor molecules. The change in G equals:

$$\delta G = [\mu([R|\mathrm{DNA}]) - \mu([\mathrm{DNA}]) - \mu([R])]\,\delta N. \tag{2.3}$$

The Second Law of Thermodynamics demands that $\delta G = 0$ so $\mu([R|\mathrm{DNA}])$ $= \mu([\mathrm{DNA}) + \mu([R)]$. Using equation (2.2) and this condition gives:

$$\frac{[R][\mathrm{DNA}]}{[R|\mathrm{DNA}]} = \frac{1}{\nu}\exp(-\Delta G_0/k_{\mathrm{B}}T) \tag{2.4}$$

where we introduced the following two quantities:

$$\begin{aligned}\Delta G_0 &= \mu_R + \mu_{\mathrm{DNA}} - \mu_{R|\mathrm{DNA}}\\ \nu &= \frac{\nu_R\,\nu_{\mathrm{DNA}}}{\nu_{R|\mathrm{DNA}}}\,.\end{aligned} \tag{2.5}$$

The energy scale ΔG_0 is called the "Standard Free Energy Change" of the reaction. At the intuitive level, you can think of it as the free energy gain when a repressor combines with a DNA strand, the *binding energy* in other words. The quantity ν has dimensions of a volume. You can think of it as the "*reaction volume*": if the repressor is located inside this volume, then it can bind to DNA.

Equation (2.4) is a special case of a fundamental principle of chemical thermodynamics: the *Law of Mass Action.* The Law of Mass Action is such an important principle that the right hand side of equation (2.4) has it's own name and symbol: the *equilibrium constant* K_{eq}

$$K_{\mathrm{eq}} = \frac{1}{\nu}\exp(-\Delta G_0/k_{\mathrm{B}}T). \tag{2.6}$$

Equilibrium constants of associative reactions have dimensions of concentration, so they are expressed in Molar. Using the Law of Mass Action, the equilibrium constant can be obtained by measuring the concentrations, and hence the standard free energy change. The beauty is that we obtain this way an important microscopic quantity, the standard free energy change, by measuring purely macroscopic quantities.

When such an experiment is performed in a test-tube ("*in vitro*") on a DNA/repressor solution [6], one finds that the result is very sensitive to the absence or presence of the operator sequence on the DNA:

$$K_{\mathrm{eq}} \approx \begin{cases} 10^{-10} & \text{operator} \quad \text{DNA} \\ 10^{-4} & \text{non} - \text{operator} \quad \text{DNA.} \end{cases} \tag{2.7}$$

This large difference between the specific and non-specific equilibrium constants is the thermodynamic signature of the ability of repressor proteins to read DNA sequences.

We call the interaction between lac repressor and operator DNA the "specific" protein-DNA interaction and that with non-operator DNA the "non-specific" interaction. You might expect the equilibrium constant for the non-specific interaction to be independent of the DNA sequence but it actually can vary over two orders of magnitude when the non-operator sequence is varied. Later, this will turn out to be a quite important effect. From equations (2.6) and (2.7), one finds that the standard free energy change for the operator case ΔG_0 (specific) is of the order of 20–25 $k_{\mathrm{B}}T$ while for the non-operator case ΔG_0 (non-specific) is of the order of 5–10 $k_{\mathrm{B}}T$.

What happens if we apply the Law of Mass Action to conditions relevant to the crowded interior of E.Coli (rather than test-tubes)? The genome of E.Coli contains about 10^7 base-pairs (or "bp") restricted to a volume of the order of one μ^3. Let's approximate the *non-operator* part of the bacterial genome as a fairly concentrated solution of short (10 bp) DNA sequences having a concentration of the order of $10^6/\mu^3$ (10 milliMolar). First suppose that the lac repressors all are bound to lactose molecules, so they will not recognize the operator sequence. Let F be the fraction of unbound lac repressors. This "free fraction" can be related to the equilibrium constant through the Law of Mass Action:

$$\begin{aligned} F &= \frac{[R]}{[R] + [R|\mathrm{DNA}]} \\ &= \frac{[R][\mathrm{DNA}]/[R|\mathrm{DNA}]}{[R][\mathrm{DNA}]/[R|\mathrm{DNA}] + [\mathrm{DNA}]} \\ &= \frac{K_{\mathrm{eq}}}{K_{\mathrm{eq}} + [\mathrm{DNA}]} \,. \end{aligned} \tag{2.8}$$

The Law of Mass Action was used in the last step. If we insert the measured value of the equilibrium constant (for the non-specific interaction) and our estimated value for [DNA], we find that F is of the order of 10^{-2} (the non-specific equilibrium constant K_{eq} is measured in the absence of lactose so we are assuming here that lactose binding does not affect the non-specific interaction). That is an interesting result! Induced lac repressors in E.Coli still "live" most of the time on DNA even though they do not recognize the operator sequence.

Intermezzo: *Energy scales in molecular biochemistry*

This 25 $k_{\mathrm{B}}T$ value for ΔG_0 (specific) is a typical energy scale for the complexation of biological macromolecules. On the one hand, this energy scale must be sufficiently *high* compared with the thermal energy scale $k_{\mathrm{B}}T$ so thermal fluctuations do not break up the complex. On the other hand, the energy scale must be sufficiently low so the binding is reversible and can be easily disrupted when required for the signaling process. In molecular biology, the universal "energy currency" for driving thermodynamically unfavorable processes is the hydrolysis of an ATP molecule: ATP $+\mathrm{H_2O} \rightarrow$ ADP + Pi + H, which delivers about 10 $k_{\mathrm{B}}T$ in free energy. A 25 $k_{\mathrm{B}}T$ value for the binding energy is thus quite reasonable. Protein complexes are in general maintained by multiple "weak bonds", such as the van der Waals attraction, hydrogen bonds, and the "polar" interaction (*i.e.* screened electrostatic interaction), all of the order of $k_{\mathrm{B}}T$. Spatial patterns of these weak links provide a basis for highly specific "lock-and-key" type recognition between proteins. This must be contrasted with the covalent "strong bonds" (of the order of a hundred $k_{\mathrm{B}}T$) that maintain the structural integrity of the macromolecules.

2.1.2 Statistical mechanics and operator occupancy

Now assume that the lactose concentration has dropped so the lac repressor proteins can bind to the operator sequence. Efficient design requires a high probability for the operator site to be occupied (to avoid unwanted gene transcription). We will compute the operator occupancy probability P using elementary statistical mechanics. Let there be M copies of the lac repressor distributed over N possible sites of the bacterial genome (with N, of the order of 10^7, large compared to M). We will neglect the small fraction of free repressors. There are then $A(N,M) = N.(N-1)..(N-M)$ ways to distribute the M proteins over the N non-operator sites and there are $C(N,M) = M[N.(N-1)....(N-(M-1))]$ ways of choosing one of the M proteins to occupy the operator site and distribute the remaining $M-1$ proteins over the non-operator sites, treating the proteins as classical, distinguishable objects. Let the Boltzmann factor of a protein occupying an operator site, respectively, a non-operator site, be $B_{\mathrm{s,ns}} \equiv \exp(+\Delta G_0(\mathrm{specific, non-specific})/k_{\mathrm{B}}T)$. The occupation probability is then

$$P = \frac{C(N,M)B_{\mathrm{s}}(B_{\mathrm{ns}})^{M-1}}{C(N,M)B_{\mathrm{s}}(B_{\mathrm{ns}})^{M-1} + A(N,M)(B_{\mathrm{ns}})^{M}} \cdot \tag{2.9}$$

This simplifies to

$$P = \frac{1}{1 + \left(\frac{N}{M}\right)\exp(-\Delta\Delta G_0/k_{\mathrm{B}}T)} \tag{2.10}$$

where $\Delta\Delta G_0 = \Delta G_0$ (specific)$-\Delta G_0$ (non-specific) is the difference between the specific and non-specific binding energies [7].

When we put in the "numbers" for the binding energy obtained earlier something interesting shows up: the very large number N and the very small number $\exp(-\Delta\Delta G_0/k_{\mathrm{B}}T)$ nearly cancel each other ($N\exp(-\Delta\Delta G_0/k_{\mathrm{B}}T)$ is about 10). Suppose we wanted to make sure that the operator is at least 99% of the time occupied. According to equation (2.10), that requires the number of copies M of the lac repressor to exceed 10^3. The actual number of lac repressors of an E.coli bacterium is maintained at a comparable value (about 10^2). There is thus a "design connection" between the values of the specific and non-specific binding energies on the one hand and the number of repressor copies maintained by the cell on the other hand. Simple statistical mechanics arguments provide us with insight how the "working parameters" are set for bacterial gene expression. The most important lesson is that the value of quantities such as ΔG_0 (specific), ΔG_0 (non-specific), N and M must be understood in the light of the functioning of the bacterium as an *integrated system.*

What is puzzling at this stage is why we need the non-specific interaction in the first place. According to equation (2.10), if we turned off the non-specific interaction, we would only need about 10 repressor copies. We will return to that question in the discussion of the kinetics.

2.1.3 Entropy, enthalpy, and direct read-out

The Gibbs Free Energy is defined as

$$G = H - TS. \tag{2.11}$$

It is the sum of an "energetic" term: the enthalpy $H = E + PV$ (E is the internal energy) and an "entropic" term. The change in Gibbs Free Energy ΔG_0 that takes place when a repressor molecules binds to DNA can be obtained from the equilibrium constant. Can we obtain the separate enthalpic and entropic contributions as well? Under conditions of fixed pressure and temperature, the change in enthalpy equals:

$$\Delta H = \Delta E + P\Delta V = \Delta Q \qquad (2.12)$$

with ΔQ the heat released (which is why we call the enthalpy also the "heat function"). We call chemical reactions *exothermic* if $\Delta Q > 0$ and *endothermic* if $\Delta Q < 0$. Endothermic reactions are interesting because the driving mechanism is entropy increase rather than reduction of the potential energy of interaction between molecules. The heat released by a reaction can be measured by calorimetry so the change in enthalpy can be found. Since the total change of the Gibbs Free Energy is known, we can also deduce the change in entropy.

When the enthalpic and entropic contributions ΔH and $-T\Delta S$ are determined in this manner for the interaction between the lac repressor and DNA, one finds the following results [8].

Specific interaction

The dominant contribution to ΔG_0 is entropic. As a function of temperature, $-T\Delta S$ decreases significantly with T. ΔH is negative so the reaction is endothermic.

Non-specific interaction

The dominant contribution to ΔG_0 is again entropic, but $-T\Delta S$ now does not depend significantly on temperature. The enthalpic contribution is again negative.

Both are surprising results. To see why, we turn to the results of structure determinations of protein-DNA complexes. It is possible to grow crystals of repressor proteins complexed with short bit of DNA, known as "co-crystals". X-ray diffraction experiments on these crystals allow us to determine atomic positions with a resolution of 2A, and sometimes even better than that [9]. Below we show the result of such an experiment for case of "cro" a very simple bacterial repressor (unlike the lac repressor).

The first panel shows the pattern of chemical bonds. There is a C_2 rotation symmetry. This symmetry is a characteristic of many prokaryote repressor proteins. The DNA operator sequence has a corresponding (approximate) rotation symmetry. Simple repressor proteins like cro address the DNA with "reading heads". A reading head is an α-helix that can be inserted into the major or minor groove of the DNA double helix (usually the major groove). The second panel is a cartoon of the cro repressor/DNA complex showing the α-helices of the protein. There are two reading heads

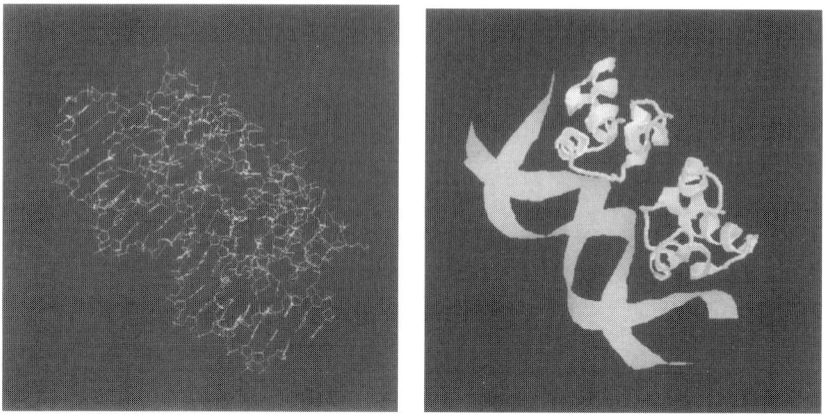

Fig. 9. Cro-repressor/DNA Complex. First panel: chemical bonds. Second panel: cartoon showing reading heads.

visible, one near the top and one near the bottom. The ends of certain side chains of the reading head can establish specific links with certain DNA bases. An example is the interaction between the amino-acid Arginine and the base Guanine shown in Figure 10 below.

The Arg side chain terminates with two N-H pairs. The two hydrogen atoms are positively charged and they "fit" exactly with negatively charged nitrogen and oxygen atoms of the Guanine base. The nitrogen and oxygen atoms act as proton acceptors so hydrogen bonds can be established, indicated in the figure by the two ovals. Base-pairs are surrounded by a unique combination of proton donors and proton acceptors that can be read by specific amino-acids. For instance, the amino-acid Glutamine "recognizes" an A-T pair in the major groove of DNA, just as Arginine recognizes a G-C pair in the major groove, while Asparagine recognizes a G-C pair in the minor groove.

We call this the "Direct Read-Out" mechanism [10] and it is based on hydrogen bonding between amino-acids and nucleic acids.

Intermezzo: *The second code*

Molecular Biologists have established long lists detailing contacts between the amino-acids of DNA-binding proteins and DNA base-pairs [11]. They originally hoped they could determine a "second code". By this they mean a one-to-one relation between amino-acids and base-pairs so they could predict to which base-pair sequence a given repressor protein would bind. That would enable design of highly specific drugs turning on or off particular

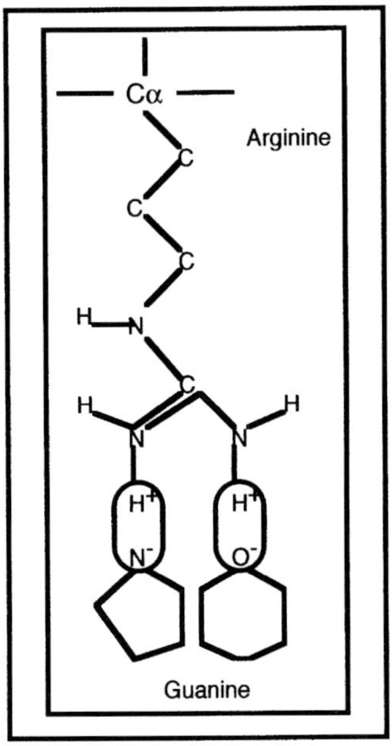

Fig. 10. Direct read-out.

genes. Unfortunately, there appears to be no universal second code. DNA-associating proteins come in different design forms. The same amino-acids interact differently in different types of proteins.

There is an obvious discrepancy between the Direct Read-Out model and the results obtained from thermodynamics. If hydrogen bonding between the reading heads and DNA really was the dominant binding mechanism, then DNA/repressor binding should have been enthalpic in nature and formation of the complex would be associated with a loss of entropy. The puzzle is that there can be little doubt that Direct-Read Out is an important mechanism for the reading of DNA sequence by proteins.

2.1.4 The lac repressor complex: A molecular machine

The resolution of this paradox comes from X-ray structural studies of lac repressor/DNA co-crystals [12] shown below.

The actual structure responsible for the repression of gene transcription is a complex consisting of two lac repressor protein dimers, so four copies in all. They bind pair-wise to two separate operator sequences; note the four reading heads. The four reading heads are pair-wise attached to the body of the complex by a linker unit that undergoes an *order-disorder*

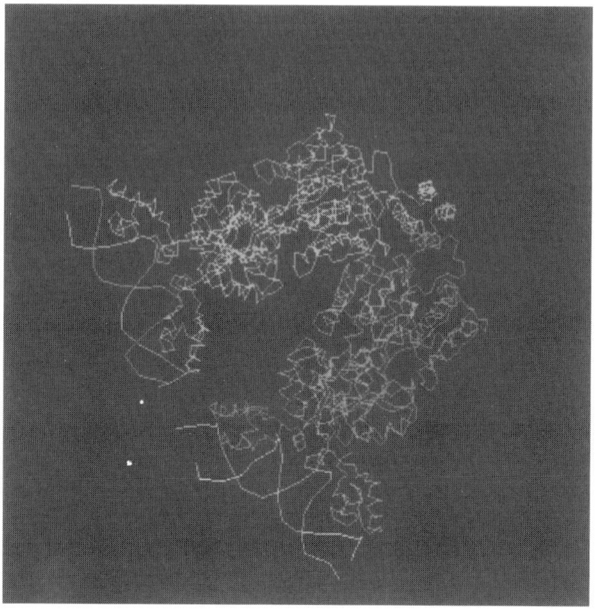

Fig. 11. The lac repressor complex.

transition upon lactose binding. In the presence of lactose, the complex adopts a structure in which the linker unit is disordered and the reading heads can not be inserted into the DNA major groove. Release of the lactose produces ordering of the linkers and allows insertion of the reading heads into DNA. In addition, the transition brings two *hydrophobic surfaces*, belonging to the two dimers, into close contact. It seems reasonable to assume that if lactose-free repressor monomers or dimers move along non-operator DNA, locate the operator sequence, and form the full four-protein repressor complex, then the hydrophobic attraction plays a central role as well, so we can understand at least qualitatively why the specific binding of the lac repressor has an entropic character. The intervening DNA sequence between the two operator sequences loops around as shown in Figure 12. Interestingly, another protein, known as CAP, binds to the DNA sequence *inside* the loop. This stabilizes the loop but once the loop opens, it also stimulates gene expression!

The reading heads thus are only a small part of the lac repressor complex. We could view the complex as a molecular *detector* and *amplifier*. The binding of lactose to the repressor complex triggers a large structural transition that breaks up the complex and opens the loop. Release of the lactose closes the loop and restores the complex. Note that there is an analogy between the operation of the lac repressor complex and the molecular motors discussed in J. Howard's lectures where the binding/release of ATP and ADP drives a cyclical structural transition that performs work.

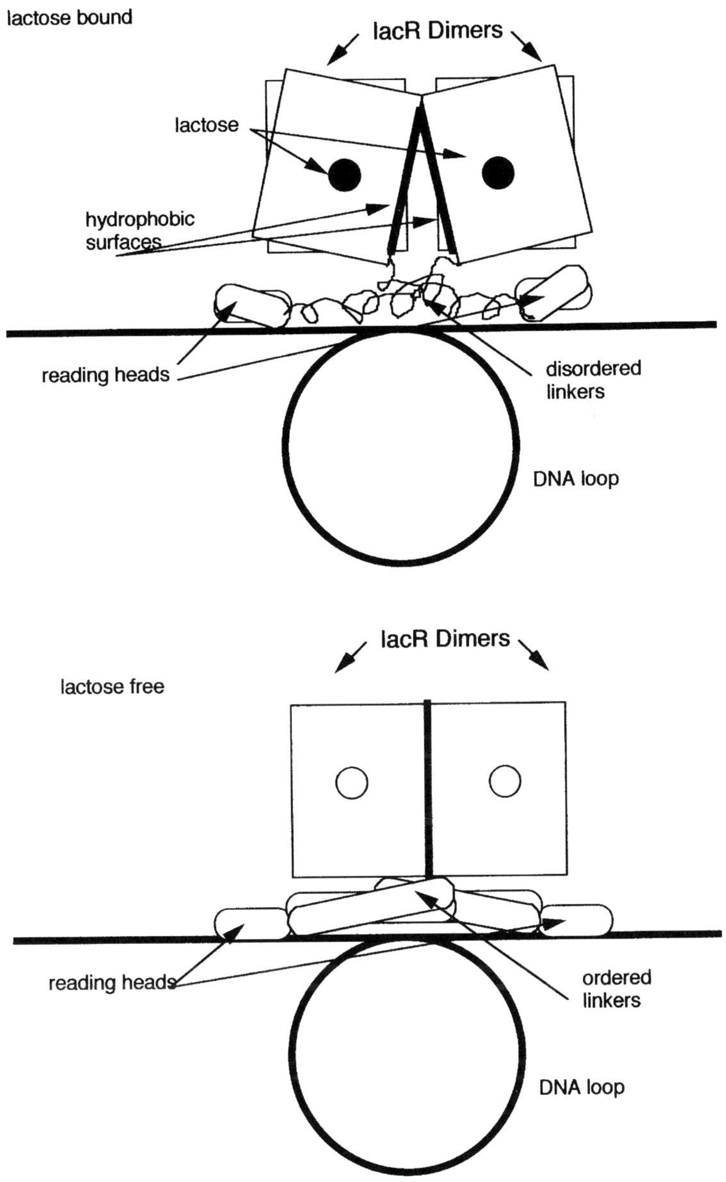

Fig. 12. Order-disorder Transition of the lac repressor complex.

The thermodynamics of the non-specific interaction also is puzzling. The binding free energy had a different temperature dependence, indicating that the hydrophobic interaction is not the interaction that dominates the thermodynamics, which is indeed the case. The structural studies tell us that the linker units connecting the reading heads of the lac repressor to the body of the protein are positively *charged* and interact with the adjacent minor groove: the non-specific interaction is electrostatic in origin. We must understand how electrostatic attraction can have an entropic character. We will postpone addressing this question to Section 4.

At this point, you should go the following Web site where you will find an elegant tutorial on the structural changes of the lac repressor tetramer and its interaction with DNA.

`http://www.worthpublishers.com/lehninger3D/index_title.html`

2.2 *Kinetics of repressor-DNA interaction*

We now turn to the third engineering requirement: *reactivity*. How quickly does a lac repressor respond to environmental changes, such as a reduction in lactose concentration? We start again with a discussion of *in vitro* experiments.

2.2.1 Reaction kinetics

The rate of change with time of the concentration of a repressor-DNA complex is the sum of two terms. A positive contribution due to complex formation between a previously unbound DNA molecule and a previously free repressor, and a negative contribution due to complex break-up. At sufficiently low concentrations, the first term must be proportional to the probability of finding a free DNA molecules and a free repressor molecule at the same site, and the second term must be proportional to the concentration of the complex:

$$\frac{\mathrm{d}}{\mathrm{d}t}[R|\mathrm{DNA}] = k_{\mathrm{a}}[R][\mathrm{DNA}] - k_{\mathrm{d}}[R|\mathrm{DNA}]. \tag{2.13}$$

The proportionality constants k_{a} and k_{d} are called, respectively, the "*on-rate*" and the "*off-rate*". These constants are supposed not to depend on concentration though they can be quite strongly temperature dependent. The off-rate really does have dimensions of a rate but the (so-called) on-rate has dimensions of Volume/Time (chemists and biologists have a free-and-easy attitude to units). The on-rate and the off-rate have a surprising connection. Under conditions of thermodynamic equilibrium, the concentrations of the reactants obviously must be constant, so the left hand side must be zero. That means that in equilibrium the following relation must hold:

$$\frac{[R][\mathrm{DNA}]}{[R|\mathrm{DNA}]} = \frac{k_\mathrm{d}}{k_\mathrm{a}} \,. \tag{2.14}$$

This is just the Law of Mass Action so the right hand side must equal the equilibrium constant:

$$\frac{k_\mathrm{d}}{k_\mathrm{a}} = K_\mathrm{eq}. \tag{2.15}$$

Because the on and off rates do not depend on concentration, *this relation must hold also away from thermal equilibrium*! That means that we only need to determine one of the two rates, the other rate follows from equation (2.15). *In vitro* experiments on repressor-DNA solutions (containing the operator sequence) report that for the lac repressor k_a is of order 10^{10} M^{-1} s^{-1} under standard conditions.

Using this information, let's apply equation (2.13) to a colony of E.Coli bacteria. Suppose that at times $t < 0$ there are no complexes because the environmental concentration of lactose is high. At time $t = 0$, the lactose concentration drops to zero. How long will it take the activated lac repressors to locate the operator sequence and switch-off gene expression? There are only a few operator sequences per E.Coli. Assuming a volume of $1\mu^3$, the (initial) concentration of unoccupied operator sequences is of order $1/\mu^3$ or about 10^{-9} M. According to equation (2.13), for early times t, the concentration of occupied operator sequences in the colony will grow linearly in time as:

$$\frac{\mathrm{d}}{\mathrm{d}t} = [R|\mathrm{DNA}] \approx (k_\mathrm{a}[\mathrm{DNA}])[R] \tag{2.16}$$

keeping in mind that at $t = 0$, $[R|\mathrm{DNA}] = 0$. It follows that we can identify $\tau = 1/(k_\mathrm{a}[\mathrm{DNA}])$ as the characteristic time scale for a free repressor to locate the operator sequence, the *switching time* in other words. For the measured value of k_a, this switching time is of order 0.1 s. This is a sensible result from the viewpoint of design: the actual switching time should be less than a minute or so for genetic switching to be a relevant response to a changing environment. Our estimate of the switching time must be viewed as a *lower bound*, because the cell environment is quite crowded. The actual on-rate inside a cell must be significantly less than this *in vitro* value. This means that the *in vitro* on rate *must* be of order 10^{10} M^{-1} s^{-1} (or higher) for reasonable *in vivo* repressor reactivity.

2.2.2 Debye–Smoluchowski theory

Let's try to compute this on-rate. The classical theory of the on-rate of diffusion-limited chemical reactions is due to Debye and Smoluchowski (DS). Assume a spherical container (the cell) of radius R. Place the operator sequence at the center of the container. Let $C(\vec{r}, t)$ be the concentration of free repressors. The concentration field obeys the diffusion equation:

$$\frac{\partial C}{\partial t} = D_3 \nabla^2 C \tag{2.17}$$

with D_3 the diffusion constant of the lac repressor in water. It is about 3×10^{-7} cm^2/s *in vitro* though under the crowded conditions of the bacterial interior, the effective diffusion constant is likely to be smaller.

We now want to know when the operator is occupied *for the first time* by a repressor. Assume that this will happen when a diffusing repressor enters for the first time a small sphere, of radius $b \ll R$, at the origin (b is a molecular length scale). You can view this sphere as the reaction volume of the Law of Mass Action.

Aside: you can estimate the diffusion constant for proteins using the formula $D = \frac{k_B T}{6\pi\eta r}$ for the diffusion constant of a sphere of radius r of order a few nanometer in a fluid with viscosity η (for water $\eta = 10^{-2}$ poise).

We actually will solve an easier problem by assuming that the small sphere at the origin acts as an *absorber*, so whenever a diffusing particle hits the small sphere, it disappears. The concentration at the outer radius R is kept at a fixed value $c(\infty)$. This is an easier problem because under these conditions, a time-independent steady-state current I is established of repressor molecules diffusing from the outer to the inner sphere. To obtain this current, we must solve Laplace's Law:

$$\nabla^2 c = 0 \tag{2.18}$$

with the boundary conditions $c(R) = c(\infty)$ and $c(b) = 0$ (because diffusing particles disappear at $r = b$). The only solution of Laplace's Law with spherical symmetry is the monopole field. Assuming $b \ll R$, and imposing the boundary conditions:

$$c(r) = c(\infty)\left[1 - \frac{b}{r}\right]. \tag{2.19}$$

The diffusion current density $\vec{J} = -D_3 \vec{\nabla} c$ is along the radial inward direction, so the diffusion current I equals $J(r)$ times the surface area $4\pi r^2$:

$$I = -4\pi D_3 b c(\infty). \tag{2.20}$$

Now compare this result with equation (2.13) for the case $k_d = 0$. The left-hand side of equation (2.13) is the number of complexes forming per second. That must equal (minus) the incoming current I. On the right-hand side we can identify $c(\infty)$ with the repressor concentration $[R]$ far from the operator. This leads to:

$$k_a = 4\pi D_3 b \tag{2.21}$$

known as the Debye–Smoluchowski (DS) rate. If we use for the "target radius" b the typical size of a protein, say 4 nm, we find that on-rate is of order 10^9 M^{-1} s^{-1}.

It turns out that this is a "hard" upper bound. Actual on-rates are nearly always smaller than the DS rate because it takes a certain time for the protein to line up with the target. Associative reactions involving proteins able to achieve on-rates approaching a limiting value of 10^9 M^{-1} s^{-1} are said to have reached "*kinetic perfection*". Now recall that for repressor/DNA association, an on-rate of 10^{10} M^{-1} s^{-1} was obtained, *an order of magnitude larger than kinetic perfection.* We also saw that this high on-rate was quite essential to support a reasonable response time of the bacterial gene-transcription system. If the bacterium had to make do with a typical protein-protein association on-rate it would be living a live on the Razor's Edge.

How does the lac repressor manage this phenomenally high rate? It was suggested by M. Eigen in the 1970's that the non-specific protein-DNA interaction may provide the answer. If inactive repressors are mostly located on the DNA, then diffusion is a predominantly *one-dimensional* process, not three-dimensional as assumed in the DS theory. This ought to speed up the on-rate since less time is wasted searching empty space. Eigen's idea can be tested. If we could somehow reduce the non-specific repressor-DNA interaction, we should find that the on-rate decreases and approaches the DS value, since one-dimensional diffusion is replaced by three-dimensional diffusion. This is actually possible: increasing the *salt concentration* reduces the non-specific binding energy ΔG_0 (non-specific) since this interaction is predominantly electrostatic. Experimentally, one finds the following dependence of the non-specific equilibrium constant on salt concentration:

$$-\log_{10} K_{\mathrm{eq}}(\mathrm{non-specific}) \cong -10 \log_{10}[K\mathrm{Cl}] - 2.5. \tag{2.22}$$

It follows from the definition of the equilibrium constant that $-\log_{10}(\nu K_{\mathrm{eq}}) = 0.43\frac{\Delta G_0}{k_B T}$, so the non-specific binding energy decreases monotonically with the salt concentration $[K\mathrm{Cl}]$. If Eigen's idea is correct, we would expect that the on-rate decreases monotonically as well. Actually, the dependence of the on-rate on $[K\mathrm{Cl}]$ in laboratory experiments is highly non-monotonic.

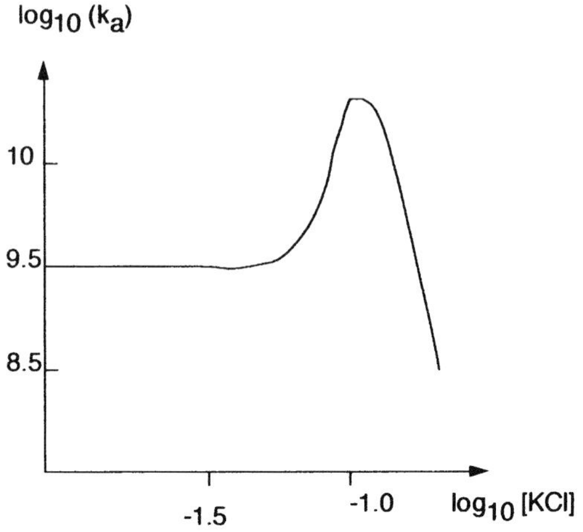

Fig. 13. Dependence of the on-rate on salt concentration.

As shown in Figure 13, there is sharp *maximum* for salt concentrations around 0.1 M (which happens to be the physiological salt concentration).

There are also functional objections against Eigen's idea. DNA does not really provide a nice one-dimensional "track" for a rondom walh. Under realistic conditions, the repressor will encounter many obstacles such as other repressor proteins bound to their respective operator sites or structural proteins that keep the DNA properly folded. These obstacles would quickly terminate a one-dimensional search. We could not hope to have free "runs" for one-dimensional diffusion of more than a few hundred bp.

2.2.3 BWH theory

Our present understanding of the on-rate for protein-DNA interaction is based on the work of Berg *et al.* [13] (BWH). Assume that at time $t = 0$ a repressor protein, located somewhere on the highly convoluted genome of an E.Coli bacterium, is activated due to the release of its bound lactose molecule. How long will it take for the repressor to locate the operator site, assuming that there are no other repressors? Let $L(t)$ be the length of DNA searched by the repressor at time t. The characteristic time T^* for the protein to locate the operator sequence is obtained by equating $L(T^*)$ with the total length $L_{\rm tot}$ of the genome.

To find $L(t)$, recall that we learned earlier that a non-specifically bound repressor spends most of its time on the DNA, say 99%. *Most* of the time the repressor motion is thus restricted to the DNA. When however the repressor is released from the DNA, it starts a three-dimensional random walk – as

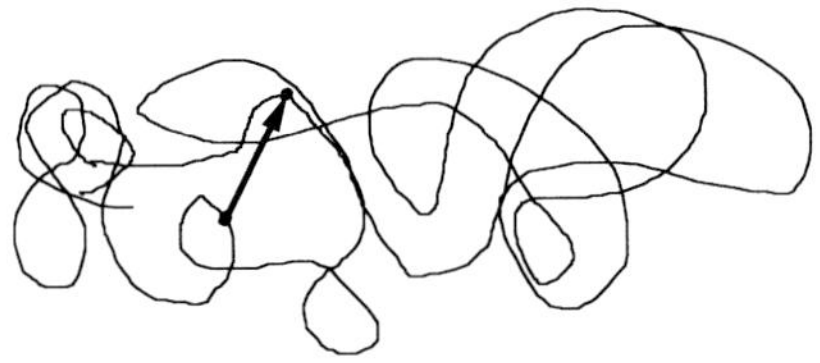

Fig. 14. Two points close in space but distant along the DNA.

in the classical DS theory – that terminates when the protein comes once again in contact with bacterial DNA. The key idea of the BWH theory is that even though the *Euclidean Distance* between the start and end of the three-dimensional random walk is likely to be short – since the interior of a bacterium is densely crowded with DNA – the *Base-Pair Distance* is likely to be very large since the bacterial genome is highly convoluted.

This means that, as long as $L(t) \ll L_{\text{tot}}$, it is very likely that the new section of DNA that will be explored by the repressor following a three-dimensional jump has not yet been explored after the repressor was activated. Assume that at time t_0, the repressor is just starting a new one-dimensional random walk. At time t, it has explored a length of DNA equal to:

$$\Delta L(t) \approx \sqrt{D_1(t - t_0)} \tag{2.23}$$

with D_1 the one-dimensional diffusion constant. Let k_{d}^* be the repressor dissociation rate for *non-operator* DNA, which can be measured experimentally, just as for the case of operator DNA. The typical duration of the one-dimensional random walk is thus about $1/k_{\text{d}}^*$ seconds so the length of DNA section searched equals:

$$\Delta L \approx \sqrt{D_1/k_{\text{d}}^*}. \tag{2.24}$$

Since repressors spend only a small fraction of their time away from DNA, the duration of the three-dimensional random walk must be short compared to that of the one-dimensional random walk. This means that after $T \gg 1/k_{\text{d}}^*$ seconds, there have been of order $k_{\text{d}}^* T$ one-dimensional random walks. The total DNA length visited equals ΔL times $k_{\text{d}}^* T$ or:

$$L(T) \approx T\sqrt{D_1/k_{\text{d}}^*} \tag{2.25}$$

valid as long as $L(T)$ is less than L_{tot}.

That is a surprising prediction. Even though the molecule is performing a random walk, the length of searched DNA grows *linearly* in time. The total search time T^* required to locate the operator is found by equating $L(T^*)$ with the total length L_{tot}:

$$T^*(L_{\mathrm{tot}}) \approx \frac{L_{\mathrm{tot}}}{\sqrt{D_1/k_{\mathrm{d}}^*}T} \tag{2.26}$$

so the search time is proportional to the length of the genome.

Let's put in the numbers. The one-dimensional diffusion constant for a sphere in water confined to a cylindrical surface with the appropriate dimensions is about 10^{-9} cm^2/s (considerably less than the three-dimensional diffusion constant). *In vitro* measurements of the non-specific dissociation rate show that k_{d}^* is quite sensitive to the salt concentration but in the physiological range, it is of the order of 10 s^{-1} for lac repressor. The distance ΔL of DNA searched per jump (Eq. (2.24)) is then of the order of 1000 Å, and the total search time for a genome of 10 microns is about 10 s. For a purely one-dimensional search, the corresponding search time $T^*(L_{\mathrm{tot}}) \approx L_{\mathrm{tot}}^2/D_1$ would have been about 1000 s. Note that the search-time would be proportional to the *square* of the total DNA length for purely one-dimensional diffusion.

These results are quite encouraging. The one-dimensional part of the search process extends only over stretches of the order of 1000 Å, a few hundred bp, and the total *single*-protein search time of 10 s is reasonable. Keep in mind that there could be of order 100 copies of the repressor searching at the same time. "Mixed diffusion" works much better as a search strategy than either purely one-dimensional diffusion or three-dimensional diffusion. The "on-rate" can be calculated as a function of k_{d}^* and, using the measured dependence of k_{d}^* on salt-concentration, one indeed finds that the on-rate of the lac repressor has a maximum around the physiological value of 0.1 M representing the cross-over from one-dimensional diffusion to three-dimensional diffusion.

We now begin to appreciate the biological role of the non-specific protein-DNA interaction: it significantly speeds up the search kinetics. Recall that when we discussed the equilibrium properties, the non-specific interaction only had a "nuisance value" since it required the bacterium to maintain an extra number of lac repressor copies to assure high operator occupancy. We can speculate that for bacteria the adaptive value of rapid genetic switching outweighs the metabolic cost of maintaining extra copies of the repressor.

2.2.4 Indirect read-out and induced fit

Apart from the Direct Read-Out mechanism, there actually is a *second* mechanism enabling repressors to read the DNA sequence [14]. Recall that the non-specific equilibrium constant K_{eq} (non-specific) depends on the bp sequence: it can vary by two orders of magnitude if the bp sequence is varied. It turns out that the non-specific binding energy is maximized when the DNA sequence approaches the operator sequence.

How is the lac repressor able to identify the operator sequence, at least in a crude way, without "addressing" directly the bp's? The geometrical parameters characterizing the DNA double helix and the local *deformability* depend on the base-pair sequence. When a protein binds to DNA, the DNA structure is deformed. If you look carefully at the structure of the cro-DNA complex shown in Figure 9, you will see that the DNA is *bent.* Transcription regulation proteins indeed usually induce a local bend or kink in the DNA structure [15]. As a result, certain sequences allow a better structural fit between the repressor and DNA than others (even if they do not contain the precise operator sequence). The idea that, apart from Direct Read-Out, the local structural and elastic properties of the DNA operator sequence must present a good fit for the repressor is known as the "Induced-Fit" model [16].

What is the point of a second read-out mechanism? The indirect read-out mechanism is much less sensitive than Direct Read-Out (for specific recognition, the equilibrium constant of the operator sequence is a factor 10^6 smaller than that of a random sequence while for the non-specific part the variation is only a factor 10^2). Consider how much time the lac repressor has available to make sure that it is or is not at the operator site. The lac repressor should be within a bp of the target site for the reading heads to be able to swing in place. The time τ spent by lac repressor on one bp is of the order $\tau \approx \frac{a^2}{D_1}$ with a the distance between bp's (say 3 Å) and D_1 the one-dimensional diffusion constant. This is of the order of a micro-second, taking our earlier value for D_1. Now recall that we know from the structural and thermodynamics studies that full recognition of the operator by lac repressor involves a significant structural change. The characteristic time-scale for large structural changes of a protein is in the micro-to-milli second range, so there is not enough time to "test" each and every DNA site by continually swinging the reading heads in and out of position all the time.

We thus can speculate that the lac repressor is *slowed down*, by induced fitting, on DNA sequences that structurally resemble the operator sequence. The extra time available provides the opportunity for the full Direct Read-Out mechanism to test whether the sequence actually is the operator sequence. If correct, this would mean make the engineering design of the lac repressor even more impressive.

3 DNA deformability and protein-DNA interaction

3.1 Introduction

We have seen that when a protein binds to DNA, the DNA structure deforms in return and that sequence dependent structural flexibility provides the second read-out mechanism for protein-DNA interaction. A second example of the importance of structural flexibility of DNA for protein-DNA interaction is connected to "DNA Condensation": the folding of DNA by histone proteins in eukaryotes. There are two sorts of descriptions of DNA deformability. The simpler "Worm-Like Chain" (WLC) model, popular among physicists, is inspired by continuum elasticity theory and focuses on the response of DNA to bending and twisting deformation at large length scales. It is particularly useful to help us understand how DNA condensation works. The more sophisticated "RST" model focuses on the connection between DNA deformability and translation and rotation of the bases and it provides insight how the indirect read-out mechanism works. We first will consider the WLC, starting with a brief discussion of gene expression in eukaryotes.

3.1.1 Eukaryotic gene expression and Chromatin condensation

The DNA of eukaryotic cells is sequestered inside the nucleus where the gene transcription takes place. The mRNA strands are exported through gateways in the nuclear membrane called "nuclear pores". Unlike bacteria, only a certain part of the genome is accessible for transcription, depending on cell type. Differentiation of eukaryotic cells takes place by the progressive "silencing" of certain genes and enhanced expression of other genes.

After bacterial gene expression had been clarified, it was expected that gene expression of animal and plant cells quickly would be understood as well, so it was a great disappointment when it was found that eukaryotic gene transcription was far more complex. In particular, when eukaryotic DNA sequences and RNA Polymerase molecules are placed in a solution of nucleotides, test-tube gene transcription does *not* take place. Gene transcription in the eukaryotes requires the presence of a large regulatory cluster of proteins known as the "Pre-Initiation Complex" (or PIC) [17], of which RNA Polymerase is a part. Wrapped around the PIC is a large loop of upstream DNA, containing enhancer and repressor sequences that affect the structure of the PIC. The resulting PIC structure determines the *rate* of gene expression by RNA Polymerase, possibly by differential adjustments of the RNA Polymerase binding energy to the PIC. The PIC can in fact be considered as a universal *molecular computer* that calculates for a given gene the appropriate level of gene expression based on input coming from the DNA in the form of the pattern of upstream enhancer and repressor sequences. Figuring out exactly how the PIC computer works would be a major step forward, but so far there have not been any serious studies of the physical properties of PIC's.

Silencing of genes can take place by (irreversible) binding of represser proteins at the transcription start sites or by chemical alterations (methylation). A third – and for our purpose most interesting – mechanism [18] is related to the fact that eukaryotic DNA material is highly condensed. Human DNA consists of pairs of 23 separate DNA strands, the *chromosomes*, each a few centimeters long. The total DNA length is thus of the order of one meter. The volume of a meter of DNA with a "hard-core" diameter of 20 Å is about 3 μ^3, only a little less than the actual volume of the nucleus. DNA inside the nucleus is thus nearly close-packed. In the highly condensed regions of the nucleus ("euchromatin"), there is no room for the assembly of the PIC. Active genes are indeed mostly located near the nuclear pores, where the DNA is partially *de*condensed ("heterochromatin"). Cell differentiation is thus in part a question of smart DNA folding so as to locate the right genes close to the nuclear membrane.

Proteins are responsible for the folding of DNA. The nucleus contains a considerable amount of protein and the combined nuclear DNA-Protein material is known as "chromatin" [19]. The main component of the nuclear proteins is the *histone* family denoted by H1, H2A and H2B, H3, and H4. These proteins – which are nearly the same for different species – are characterized by an usually large positive charge. It is possible to chemically extract the chromatin material from the nucleus and de-condense it in solutions of low salinity. Electron micrographs of decondensed chromatin reveal a linear, necklace-like array of beads separated by linkers as shown in the second panel of Figure 15 [20].

The beads, known as nucleosomes [21], have a diameter of about 10 nm, so the necklace is often called the "10 nm fiber". As the salinity is reduced [22] the 10 nm fiber condenses into a thicker fiber with a diameter of about 30 nm called the "30 nm fiber" (first panel). Recent simulation [23], SFM [24] and cryo-TEM studies [25] have addressed its internal structure, but there is as yet no consensus: a superhelical solenoid [26] has been proposed, as well as an "in-out" zig-zag structure [27]. The overall length of the DNA strand is reduced by a factor of about 40 in the 30 nm fiber, still much less than the total required amount of condensation. The organization of chromatin at larger length scales is even more controversial [28]. We should stress that *local* de-condensation of chromatin in the nucleus could not take place by changes in the *global* salinity level of the nucleus. Instead chemical

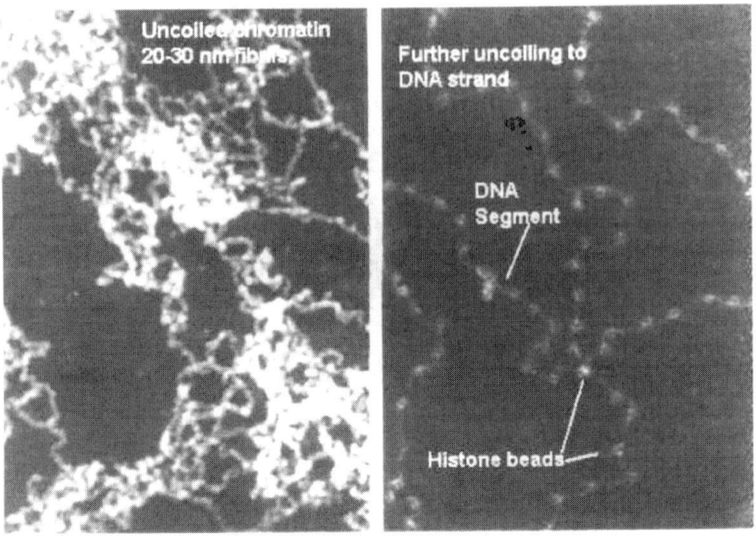

Fig. 15. Electron micrographs of the 30 nm fiber (first panel) and the 10 nm fiber (second panel). From reference [20].

modification of the histones, in particular *acetylation*, changes the binding affinity of DNA.

Certain proteins (such as DNaseI) cut DNA strands that are not complexed with proteins. Such "digestion" experiments on chromatin produced DNA strands with lengths *quantized* in units of the order of about 200 bp, depending on the species. Combined with the electron microscopy results, the natural explanation is that about 200 bp's are associated with each nucleosome. Additional digestion by another enzyme, micrococcal nuclease, showed that the DNA directly associated with the nucleosome itself has a *fixed* length of 140 bp, while the linker length connecting nucleosomes varied from species to species.

Remarkably, it proved possible to produce *crystals* of the nucleosome complexes. In a landmark achievement of X-ray crystallography, it proved possible to perform diffraction experiments on nucleosome crystals and hence to determine their atomic structure [29], shown in Figure 16.

It was found this way that the nucleosome consists of a core of 8 histones, two copies each of H2A, H2B, H3, and H4, in the form a of cylinder with a height of 55 Å, a diameter of 110 Å, and a 2-fold symmetry axis perpendicular to the cylinder. The core is also called the *histone octamer*. DNA is wrapped around the core in the form of a spiral with 1.75 turns, while the H1 histone straddles the incoming and outgoing DNA strands as

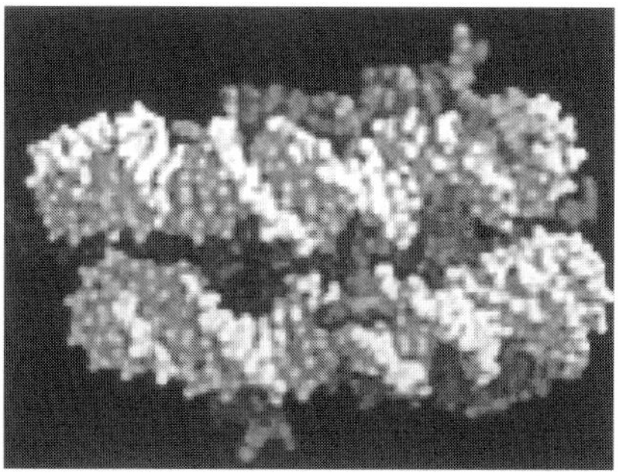

Fig. 16. Nucleosome structure (Ref. [29]).

a safety pin. As a consequence of the nucleosome structure, the 10 nm fiber has a spiral structure:

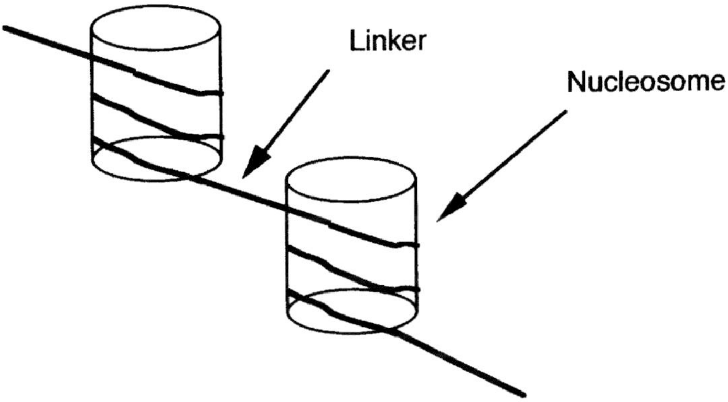

Fig. 17. The 10 nm fiber.

3.1.2 A mathematical experiment and White's theorem

Figure 18 shows a *mathematical experiment* inspired by DNA condensation. Wrap a belt N spiral turns around the surface of a cylinder so that the belt is not twisted. If you slide a ruler along the belt, the ruler should not rotate. Now remove the cylinder and stretch the belt, *without allowing the endpoints to rotate*, for instance by first closing the belt into a loop using the buckle.

When the belt is straightened out, slide the ruler again along the belt. The ruler now makes N turns: we have transformed N turns in space (the

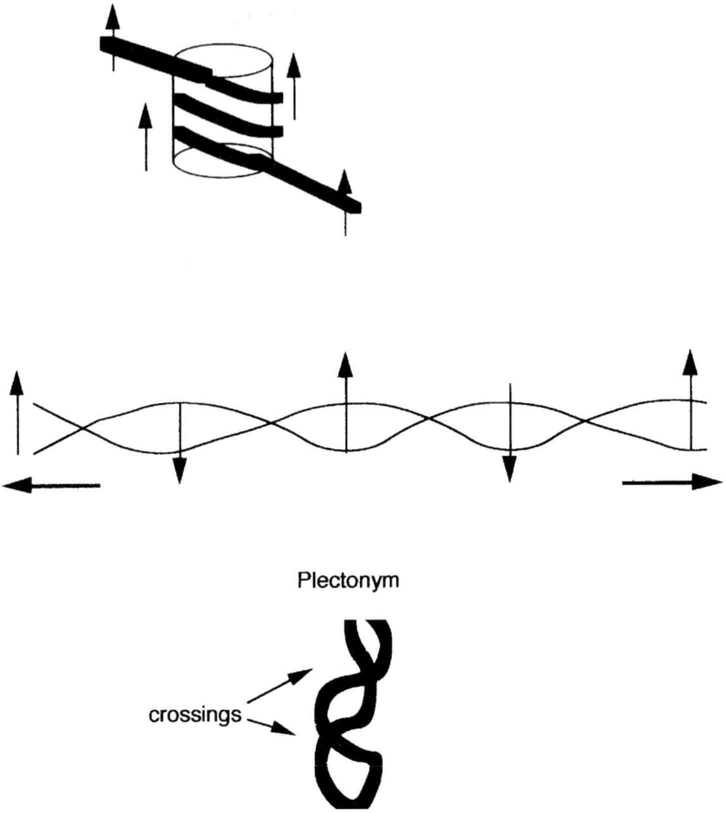

Fig. 18. A mathematical experiment. Spiral writhe is converted to twist, which in turn is converted to a plectonym.

spiral) into the N twist turns of the straight belt. We call the spatial turns "Writhe" and the turns of the straight belt "Twist". Finally, bring the endpoints together, still not allowing end-point rotation. After the tension is removed, the belt is twisted up like a telephone wire into a shape known either as "interwound" or "plectonymic". Using the ruler, verify that for large N the number of twists of the belt in the interwound state is much less than that for the straight belt. In this step, we reconverted twist into writhe, but the writhe is in a form different from the spiral we started out with. Note that you had to perform *work* to convert writhe into twist when you stretched the belt. This indicates that writhe is energetically less costly than twist.

The spiral (or "toroidal") structure we started with is characteristic of DNA condensation in eukaryotes while the interwound state is characteristic of *bacterial* DNA folding: DNA extracted from bacteria has a branched plectonymic structure. Our mathematical experiment shows that there is no fundamental topological difference between the two ways of folding DNA; they represent different forms of writhe and can be inter-converted.

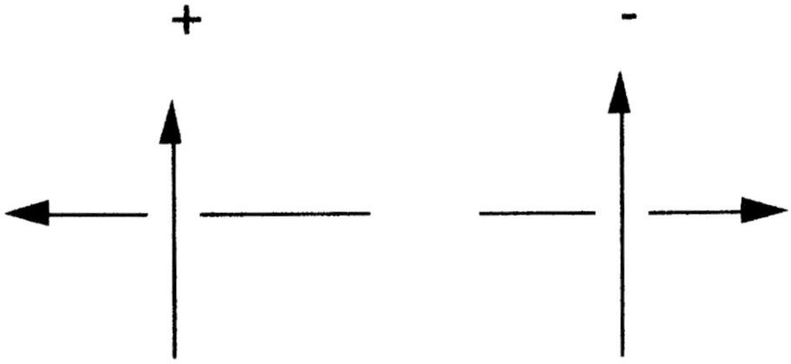

Fig. 19. Sign convention for computing Wr.

The concept of writhe of an interwound curve can be quantified as follows. Move along the curve in a given direction. Assign each crossing of the curve with itself $+1$ or -1, according to the convention given in the picture. The writhe Wr of the curve is the sum of these numbers over all the crossings. The twist Tw is defined as the number of turns of the ruler in radians divided by 2π. If the mathematical experiment is repeated, you wil discover that – provided you do not open the buckle – the *sum* of Tw and Wr is a constant, known as the *Linking Number Lk*:

$$Lk = Tw + Wr. \tag{3.1}$$

This is known as *White's Theorem*, and it can be demonstrated mathematically [30]. You can freely convert twist into writhe and *vice versa* but the sum remains fixed. If the belt is closed, then the linking number can be identified as the number of turns of the belt before you close it. You must be aware that the $Lk = 0$ state sometimes is defined with respect to the relaxed B DNA state and sometimes with respect to fully untwisted DNA with no double helix.

How do you compute the writhe if the belt is not a plectonym but, say, a spiral that twists in space without self-crossings? There exists a more precise definition of writhe. For a closed loop, Wr is a double path integral along the contour C:

$$Wr = \frac{1}{4\pi} \oint_C \oint_C \frac{[\mathrm{d}\vec{r}_1 \times \mathrm{d}\vec{r}_2] \cdot \vec{r}_{12}}{|\vec{r}_{12}|^3} \,. \tag{3.2}$$

If you compute this integral, you find that the writhe of a spiral with N spiral turns and a pitch angle α equals:

$$Wr = \pm N(1 + \sin\alpha) \tag{3.3}$$

with the $+$ sign for a right handed spiral and the $-$ sign for a left handed spiral (try to prove this yourself). What is the Lk? We already know that *after* stretching the belt you obtain N twisting turns. In this stretched state, there are no spatial turns so $Wr = 0$. It then follows from White's Theorem that the linking number must be $+N$ or $-N$. (the convention is that right handed twist is positive). Applying again White's Theorem, we find the twist:

$$Tw = \pm N \sin\alpha. \tag{3.4}$$

Let's now apply White's Theorem to the 10 nm fiber. Assume that the DNA of the 10 nm fiber is wrapped around the nucleosomes like a belt around a cylinder. In that case, we can simply adopt the preceding formulae. The Linking Number of the 10 nm fiber should then be approximately 1.75 times the number of nucleosomes N. When you measure Lk, you find that Lk is only about 1.0 times N. This is known as the "*Linking Number Paradox*". It is not really a paradox: it only shows that that DNA is *not* wrapped around the nucleosome the way a belt is wrapped around a cylinder. White's Theorem tells us that the nucleosome surface twists up the DNA to produce a negative contribution of about -0.75 per nucleosome. A mathematical theorem is providing us with detailed, microscopic information on the association of DNA with histones!

3.2 The worm-like chain

We now introduce the WLC model for DNA deformation [31]. It is based on the idea that the energy cost of small deformations that do not seriously alter the local DNA structures can be computed by elasticity theory. According to elasticity theory the energy cost of an infinitesimal deformation of a tube with uniform cross-section made of an isotropic elastic material is given by:

$$H = \frac{1}{2}\int \mathrm{d}s \left\{ \kappa \frac{1}{R^2(s)} + C\left(\frac{\mathrm{d}\theta}{\mathrm{d}s}\right)^2 \right\}. \tag{3.5}$$

The line integral runs along the center of the tube, with s the arc-length. In the first term, $R(s)$ is the curvature radius of the center line with s the arclength along the center line (see Fig. 20, $1/R$ is called the curvature). If $\hat{t}(s)$ is the tangent unit vector of the center line then $(1/R(s))^2 = (\mathrm{d}\hat{t}/\mathrm{d}s)^2$.

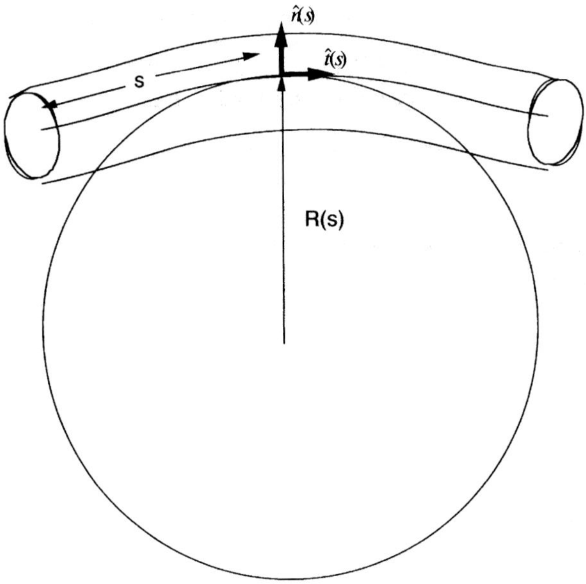

Fig. 20. The Worm-Like chain.

If this relation is used in the first term of equation (3.3) we obtain the well-known *one-dimensional non-linear sigma model.* The material parameter κ – with dimensions of Energy Length – is called the *bending modulus* (also denoted by A). The torque τ required to bend the tube along a circle of radius R is equal to κ/R.

The second term of equation (3.3) describes the twist energy. Let $\hat{n}$ be a unit vector attached everywhere along the axis of the *undeformed* tube in a direction perpendicular to the tube axis. In the deformed state, $\mathrm{d}\theta/\mathrm{d}s$ is then the rate of rotation of $\hat{n}(s)$ along s. ($\hat{n}(s)$ is just the ruler of our mathematical experiment). Note that the twist modulus C has the same dimensions as the bend modulus.

DNA is of course *not* well described as being composed of an isotropic elastic material so why does equation (3.5) make any sense? That is because the mathematical form for H can be valid even if elasticity theory does not apply. Assume that if DNA is only very weakly deformed, the energy cost must be an integral along s of some function of the bend, twist, and torsion. If you assume that this expression is an *analytic* function of bend, twist, and torsion and if you expand this function in a Taylor series to lowest order, you again obtain equation (3.5) if you demand that DNA has no *intrinsic* bend, twist, or torsion.

3.2.1 Circular DNA and the persistence length

As a first application of the WLC, consider a closed strand of DNA that lies along a circle of radius R (no twisting). The elastic energy cost (Eq. (3.5)) equals

$$H = 2\pi\kappa/R. \tag{3.6}$$

For increasing R, H becomes smaller than the thermal energy $k_{\mathrm{B}}T$ for $R \gg \xi$ with $\xi = k_{\mathrm{B}}T/\kappa$ known as the *"persistence length"*. If $R \gg \xi$, then thermal fluctuations will deform the large circle by introducing smaller loops and turns with a characteristic size equal to ξ. On the other hand, if $R \ll \xi$, then it is not possible for thermal fluctuations to deform the circle. We conclude that thermal fluctuations are not able to deform DNA on length-scales less than ξ. It is possible to microscopically observe the thermal fluctuations – or Brownian Motion – of DNA strands by attaching fluorescent markers and measure the persistence length ξ (about 500 Å). As an application, we can calculate the bending energy cost of the lac repressor loop. The distance between the two lac repressor operator sequences is about 300 Å. Assuming that the DNA spans a 300 Å perimeter circle between the operator sequences and using equation (3.6) gives a bending energy H for the loop of more than 30 $k_{\mathrm{B}}T$. This is an overestimate, but the elastic energy cost of the loop is indeed substantial compared with the typical protein-DNA binding energy.

By the way, the precise definition of the persistence length is as a *correlation length* for orientational fluctuations:

$$\lim_{s\to\infty} \langle \hat{t}(0)\cdot\hat{t}(s)\rangle \propto \exp(-s/\xi) \tag{3.7}$$

where $\langle\ldots\rangle$ denotes a thermal average and $\hat{t}(s)$ is again the tangent unit vector.

3.2.2 Nucleosomes and the Marky–Manning transition

We now apply the WLC to the nucleosome. Let L be the length of DNA wrapped around the core, with L less than or equal to 140 bp. Treat the core as a cylinder of radius R equal to 5 nm. The bending energy cost is then $H = \frac{1}{2}\kappa L/R^2$ according to equation (3.5) (there is some twist energy too but we will neglect it). Assuming $\kappa/k_{\mathrm{B}}T = 500$ Å, one finds that wrapping the full 140 bp around the core requires about 36 $k_{\mathrm{B}}T$ in energy per nucleosome. The total human genome is about 10^9 bp, requiring about $10^9/200$ nucleosomes. The total amount of DNA elastic energy stored in the nucleus is about 10^8 $k_{\mathrm{B}}T$. That is serious energy. In the figure below [32], you see what some of this energy does when you decondense the chromatin by chemical treatment: it causes a DNA explosion!

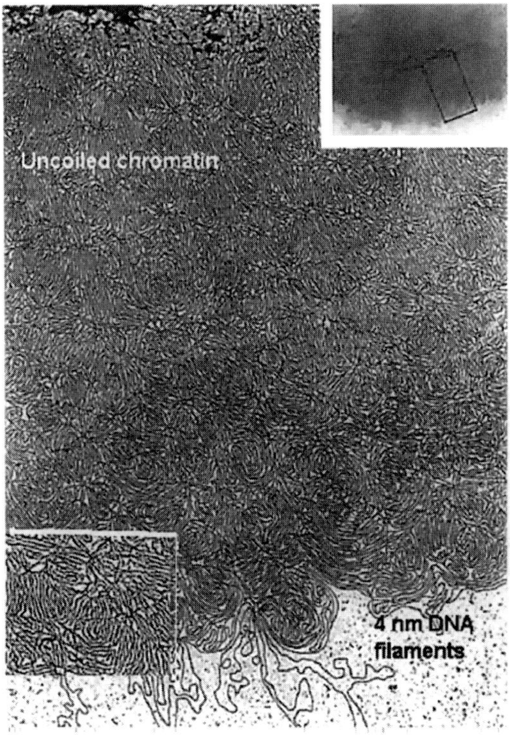

Fig. 21. Decondensed Chromatin (from Ref. [32]).

This bending energy cost must be compensated by some form of attractive interaction between the core and the DNA. This attraction is due to electrostsatics and will be discussed further in Section 4. Here we only need the fact that the binding between DNA and histones is nearly sequence-insensitive so the binding energy should mainly depend on *how much* DNA is associated with the nucleosome, *i.e.*, on L. We thus introduce an (unknown) adhesion energy λ *per unit length*, with λ to be determined experimentally. Adding bending and adhesion energies gives:

$$E(L) = \left(\frac{1}{2}\kappa/R^2 - \lambda\right) L. \tag{3.8}$$

There are now two possibilities: if $1/2\kappa/R^2$ does not exceed λ, then DNA will wrap around the core to the maximum amount since $E(L)$ decreases with increasing L. If $1/2\kappa/R^2$ exceeds λ, then $E(L)$ is minimized by $L = 0$ so there is no complexation. This "unwrapping" transition of Marky and Manning [33] was studied experimentally by Yaeger and von Holde [34] (see

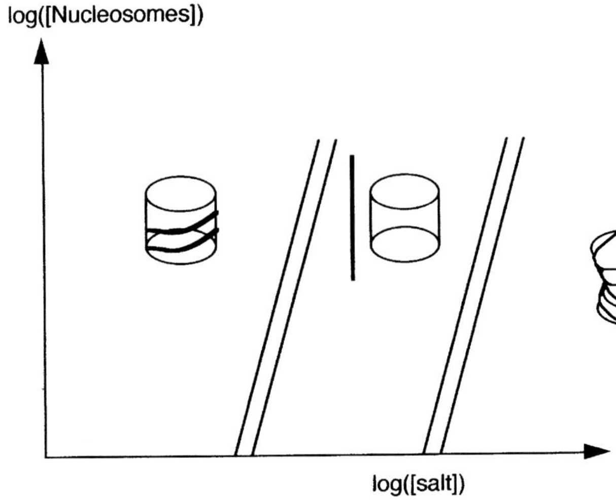

Fig. 22. Phase Diagram of a Nucleosome Solution.

Fig. 22). They changed the salt concentration in a dilute solution of nucleosome particles – in order to change λ – and found a first-order unwrapping transition for salt concentrations somewhat above the physiological level of 0.1 M. At even higher salt concentrations, the core fell apart. That indicates that nucleosome particles are rather close to the unbinding transition under physiological conditions. Since we know the bending modulus of DNA, we can estimate this way that λ is of order κ/R^2, which is about 0.2 $k_{\mathrm{B}}T/A$, or about 0.6 $k_{\mathrm{B}}T/A$ per bp.

It is actually a very reasonable design strategy for nucleosomes to be close to the unwrapping transition. Chromatin decondensation can then take place by minor chemical changes of the nucleosome core (*e.g.*, by reducing the positive charge of the histones through acetylation).

Polach and Widom developed a clever way of measuring λ in an independent way [35]. Recall that digestion enzymes can cut DNA strands that are not protected by proteins. If the nucleosome "operating point" indeed is close to the Marky–Manning transition, we might expect significant *thermal fluctuations* in the wrapping length L. This can be checked by allowing the nucleosome solution to be incubated for a while with digestion enzymes. According to the Boltzmann Distribution, the probability distribution $P(L)$ of the wrapping lengths should be an exponential function of

L (using Eq. (3.6)):

$$P(L) \propto \exp\left\{\left(\frac{1}{2}\kappa/R^2 - \lambda\right)(L_{\text{max}} - L)/\kappa_{\text{B}}T\right\} \tag{3.9}$$

with L_{max} the maximum wrapping length. The distribution of wrapping lengths was indeed found to be exponential, which allowed a separate measurement of λ (obtaining a value of the same order as our crude estimate).

The fact that there are significant fluctuations in the DNA length is interesting by itself for a number of reasons. The possibility of formation by thermal fluctuations of small, open DNA loops along the wrapped length provides a mechanism for the mobility of nucleosomes along DNA without the nucleosome becoming detached from the DNA [36], which indeed has been observed experimentally.

3.2.3 Protein-DNA interaction under tension

Our next example of the WLC concerns what happens with DNA associated proteins when the DNA is placed under *tension*. We saw earlier that the binding of many proteins to DNA produces a bend or a kink in the DNA. Tension will increase the energy cost of producing the kink. Let's compute this energy cost? Assume that the protein binds at the midpoint of a strand of length L. Let the kink angle be $\pi - 2\alpha$ (see figure). Now place the DNA strand under a tension F.

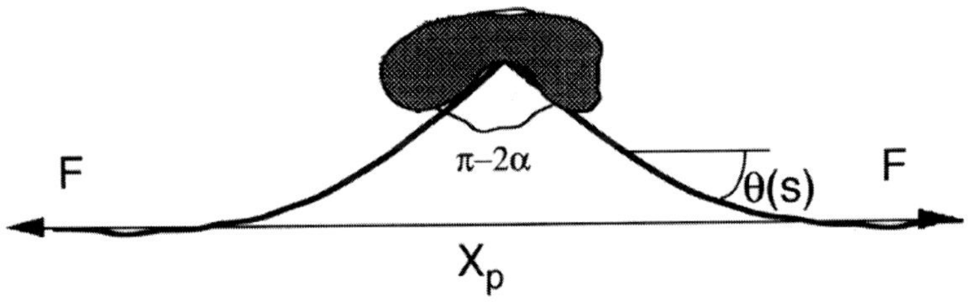

Fig. 23. A protein/DNA complex under tension.

Let s be the arc-length along the DNA, with $s = 0$ the kink location. The curvature $1/R(s)$ at s is then $\mathrm{d}\theta/\mathrm{d}s$, with θ the angle between the tangent and the direction along which the tension is applied. The boundary conditions on $\theta(s)$ for large L are $\theta(0) = \alpha$ and $\theta(\infty) = 0$. The total energy is the WLC bending energy plus the work $-F\ X_{\text{p}}$ performed *by* the DNA *against* the tension, with X_{p} the length of the projection of the DNA along the coordinate axis:

$$E = 2\int_0^{L/2} \mathrm{d}s \left\{ \frac{1}{2}\kappa \left(\frac{\mathrm{d}\theta}{\mathrm{d}s}\right)^2 - F\cos\theta \right\}. \tag{3.10}$$

Equation (3.10) is a famous 19'th century problem of elasticity known as the "Elastica". The Euler–Lagrange equation $\delta E/\delta\theta = 0$ minimizing E reads:

$$\kappa \frac{\mathrm{d}^2\theta}{\mathrm{d}s^2} - F\sin\theta = 0. \tag{3.11}$$

This happens to be the equation of motion for an *ideal pendulum* of length l, mass M if we make the following identifications:

$$\kappa \to \mathrm{Ml}$$
$$F \to \mathrm{Mg}$$
$$s \to \mathrm{time}$$

and $\theta(s)$ the angle between the pendulum and the upwards vertical direction (so $\theta = \pi$ is the equilibrium position). The particular pendulum solution we are interested in starts in an upside-down position and then rotates clockwise until the pendulum makes an angle α with the vertical:

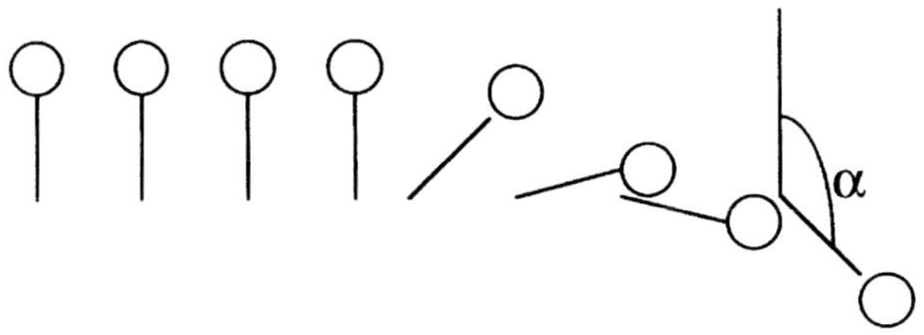

Fig. 24. The Pendulum Analog.

This must be matched with a similar solution that makes a counter-clockwise turn. Although the general solution can be expressed in terms of ellipitic integrals, we will restrict ourselves to the case that $\alpha \ll 1$. In that case, we can linearize the Euler–Lagrange equation:

$$\kappa \frac{\mathrm{d}^2\theta}{\mathrm{d}s^2} - F\theta \cong 0. \tag{3.12}$$

Note that this is *not* the harmonic oscillator equation, and the solution is an exponential, not a trigonometric function:

$$\theta(s) = \pm\alpha \exp(-|s|/\lambda)$$
$$\lambda = \sqrt{\frac{\kappa}{F}}. \tag{3.13}$$

The length scale λ tells us how far away from the protein we still find a noticeable deformation of the DNA. Note that λ diverges when we allow the tension F to vanish. For a tension of one pN (one pN equals 10^{-12} Newton), λ is about 7 nm. If we insert equation (3.10) into the energy expression and perform the integral, we obtain:

$$\Delta E = 2\alpha^2 \sqrt{\kappa F} \tag{3.14}$$

for the bending energy contribution to E [37].

If ΔE exceeds the binding energy of the protein, then it would be energetically favorable to remove the protein from the DNA. In other words, when the binding energy is less than the work done by the tension T upon removing the protein, the protein will "pop" from the DNA. Writing ΔE as $2\alpha^2 k_{\mathrm{B}}T(\xi/\lambda)$, it follows that for a modest one pN tension and $\alpha = 0.5$, ΔE is equal to 3.6 $k_{\mathrm{B}}T$. This already is not negligible compared to the typical DNA/protein binding energy of a protein (20–30 $k_{\mathrm{B}}T$). Since ΔE scales as $F^{1/2}$, a tension of 100 pN would effectively prevent protein/DNA complexation.

That is interesting since DNA *is* placed under tension during cell division. When a cell divides, two duplicates of the chromosome are pulled apart towards opposite "poles" of the cell by two bundles of fibers (the mitotic spindles) that generate force through motor proteins. If this would produce tensions of the order of 100 pN or larger on individual DNA strands making up the two chromosomes, then this would lead to massive loss of DNA- associating proteins, including the nucleosome cores required to maintain the structural integrity of chromosomes. Fundamental physical considerations place limits on the operating parameters of the cell machinery.

3.2.4 Force-Extension Curves

Let's compute the projected length X_{p} as a function of the applied tension:

$$\begin{aligned} X_{\mathrm{p}} &= \int \mathrm{d}s \cos\theta(s) \\ &\cong 2\int_0^{L/2} \mathrm{d}s \left(1 - \frac{1}{2}\theta^2(s)\right) \\ &= L - \frac{1}{2}\alpha^2\sqrt{\kappa/F} \end{aligned} \tag{3.15}$$

where we used equation (3.13) in the last step. This is known as a "*force-extension curve*". Force extension curves play an important role in modern single-molecule biophysics because they can be measured experimentally. It is possible to chemically attach silicon beads to the end of a DNA strand and capture these beads in the focus of a lens. By measuring the position of the bead in the trap, one can measure the force F on the bead and hence, by Newton's Third Law, the force exerted by the bead on the strands. Pulling the two beads apart, one obtains a relation between the distance X_p between the beads and F.

It is usual to plot the dependence of F on the extension X_p, as shown in Figure 25. In the present case, F diverges as $1/(L - X_p)^2$. This divergence

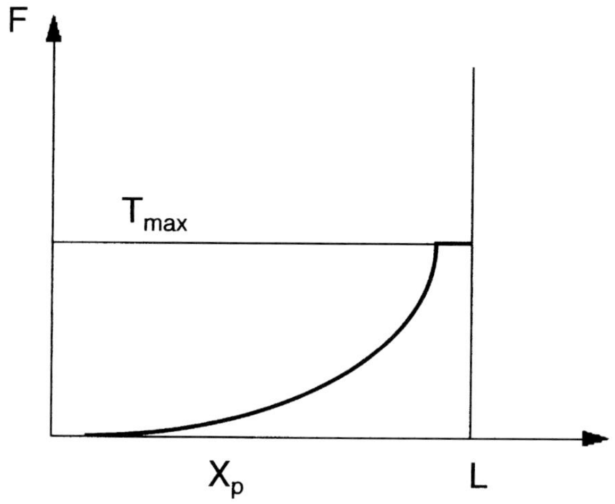

Fig. 25. Force-Extension Curve of a DNA/protein complex.

is truncated when the tension reaches the maximum value T_{max} where the protein pops off. In principle, both the kink angle and the binding energy can be obtained from the force extension curve.

Measuring force extension curves is extremely popular: they have been measured for DNA itself [38] and the 30 nm fiber [39] (as well as for RNA strands and large proteins). In practice, the interpretation of force extension curves in terms of molecular-level events is complex beacuse the measurements are characterized by irreversibility and "waiting-time" dependence as you go up and down the force- extension curve. The lectures of Evans address the fact that applying direct mechanical forces to macromolecule complexes indeed must produce, for fundamental reasons, highly "rate-dependent" results.

Why would the force extension curve of a DNA strand be of such interest? In the calculation leading up to equation (3.15) we assumed that DNA is inextensible. This is not a good approximation. In the harmonic approximation, the bending energy of a strand under tension (without associated proteins) is:

$$
\begin{aligned}
E &= \int_{-L/2}^{L/2} \mathrm{d}s \left\{ \frac{1}{2}\kappa \left(\frac{\mathrm{d}\theta}{\mathrm{d}s} \right)^2 - F\cos\theta \right\} \\
&\cong TL + \frac{1}{2} \int_{-L/2}^{L/2} \mathrm{d}s \left\{ \kappa \left(\frac{\mathrm{d}\theta}{\mathrm{d}s} \right)^2 + F\theta^2 \right\} \\
&\cong TL + \frac{1}{2} L \sum_q \left[\kappa q^2 + F \right] |\theta_q|^2 .
\end{aligned} \tag{3.16}
$$

In the last step, we performed a Fourier Transform $\theta(s) = \sum_q \theta_q \mathrm{e}^{iqs}$. Equation (3.16) is not quite right because it actually assumes that the DNA is restricted to a plane, but this only affects numerical factors.

The energy is the sum of *decoupled* harmonic degrees of freedom for each Fourier mode. That means that we can apply the Equipartition Theorem, taking the Fourier amplitudes as the displacement variables. The effective "spring constant" for θ_q equals $\kappa q^2 + F$ so it follows that:

$$
\left\langle |\theta_q|^2 \right\rangle = \frac{k_\mathrm{B} T}{\kappa q^2 + F} \tag{3.17}
$$

from the Equipartition Theorem. Using Parseval's Theorem:

$$
\begin{aligned}
\langle \theta^2 \rangle &= \sum_q \left\langle |\theta_q|^2 \right\rangle \\
&= \sum_q \frac{k_\mathrm{B} T}{\kappa q^2 + F} \\
&= \frac{L}{2\pi} \int_0^\infty \mathrm{d}q \frac{k_\mathrm{B} T}{\kappa q^2 + F} .
\end{aligned} \tag{3.18}
$$

In the last step we changed from summation to integration. If you introduce the variable $x = (\kappa/F)^{1/2} q$, then this integral becomes a numerical factor:

$$
\langle \theta^2 \rangle = C \frac{k_\mathrm{B} T}{\sqrt{\kappa F}} \tag{3.19}
$$

with C some number. The amplitude of the angle fluctuations is thus proportional to $1/F^{1/2}$. Using again equation (3.15), we can obtain a force-extension curve:

$$\begin{aligned} X_{\mathrm{p}} &= \int \mathrm{d}s \cos\theta s \\ &\cong 2\int_0^{L/2} \mathrm{d}s \left(1 - \frac{1}{2}\theta^2(s)\right) \\ &= L\left(1 - \langle\theta^2\rangle\right) \\ &= L\left(1 - C\frac{k_{\mathrm{B}}T}{\sqrt{\kappa F}}\right). \end{aligned} \tag{3.20}$$

Solving for F, we find that the tension is again proportional to $1/(L - X_{\mathrm{p}})^2$:

$$F \approx \frac{k_{\mathrm{B}}T}{\xi}\left(\frac{1}{1 - X_{\mathrm{p}}/L}\right)^2 \quad \text{(high tension).} \tag{3.21}$$

This equation tells us that when you pull on DNA against the restoring force due to thermal fluctuations, DNA will be stretched out for tensions large compared to kT/ξ, the ratio of the thermal energy and the persistence length. This force is about 0.01 pN. That means that if you do force extension measurements using DNA as your "leads", as for instance for the protein pop-off case, then for tensions of less than, say, a few pN, you must allow for this "entropic" elasticity of the DNA strands to properly interpret your results. Actually, equation (3.21) curve is not right for tensions of order 0.01 pN or less, since the harmonic approximation breaks down. In that regime (not particularly relevant for our purposes) the force extension curve can be shown to be linear:

$$F \approx \frac{k_{\mathrm{B}}T}{\xi}\left(\frac{X_{\mathrm{p}}}{L}\right)^2 \quad \text{(low tension).} \tag{3.22}$$

3.3 The RST model

The worm-like chain is a great favorite among physicists working on DNA but it has a serious drawback. It treats DNA as a generic flexible tube so it is of no help to understand the induced-fit mechanism of protein-DNA interaction and the remarkable dual information storage capability of DNA. There exists a second model, the RST model, that is more suitable to address these issues.

3.3.1 Structural sequence sensitivity

Recall that a DNA base-pair consist of a larger purine and a smaller pyrimidine base. Treat these two groups as a smaller and a larger rectangular plate. Importantly, the two plates do not lie in the same plane: they are rotated by a certain angle known as the *propeller twist*. In order to satisfy the hydrophobic interaction, we must stack these twisted pairs on top of each other, each one rotated with respect to the previous one by an angle of about 30 degrees, the twist angle T. There is no particular problem to stack a pyrimidine-purine pair on top of a pyrimidine-pair as shown in the next figure. However, when you place a pyrimidine-purine pair on top of a purine-pyrimidine (such as in a TATA sequence) you run into a "frustration" problem because of the propeller twist: there is no way for the two lower plates *both* to be in contact with the upper plates. By sliding the lower pair in one direction (see figure), you can improve the overlap between the two purine plates. Alternatively, by sliding it in the opposite direction and rotating along an axis perpendicular to the double helix ("roll"), you can increase somewhat the overlap between a purine and a pyrimidine (show this yourself). Clearly, there is considerable structural flexibility in the second case, so it is indeed reasonable that both the local structure and the local deformability of DNA depend strongly on sequence.

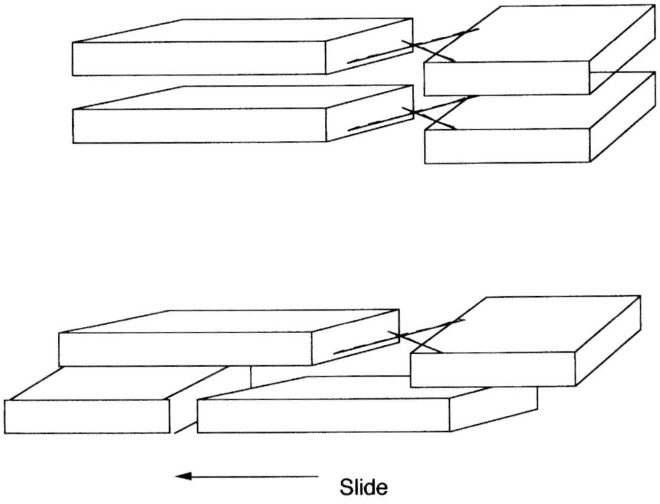

Fig. 26. A purine/pyrimidine pair does not stack easily on top of a pyrimidine/purine pair.

Suppose for simplicity that for every pair of bases (like AT followed by GC), we can identify preferred values for the relative Roll (R), slide (S), and Twist (T), which are known as the Calladine Parameters (see Fig. 27). Imagine the base-pair sequence as a deck of cards stacked on top of each other. Let R, S, and T specify the relative orientation of subsequent pairs (see figure). For R = S = 0, and fixed T, you obtain a spiral. If R has a fixed non-zero value, then you obtain a *super-spiral*: the axis of the original spiral is itself a spiral. In general, by specifying R, S, and T for each pair of cards, we obtain a unique three-dimensional structure that reflected the base-pair code [40].

Actually, the relative twist, slide, and roll of two base-pairs on top of each other are *not* uniquely defined by the base-pair identity of the two pairs, though there is a statistical distribution of preferred values [41], presumably related to longer-range correlations.

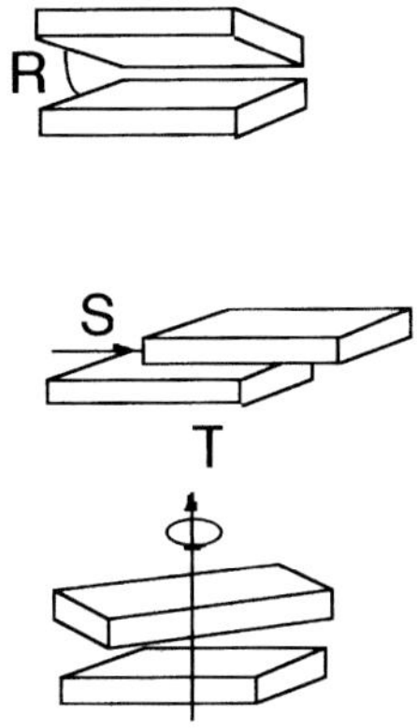

Fig. 27. Calladine parameters.

3.3.2 Thermal fluctuations

The interior structure of DNA is far from rigid and the R, S, and T values actually undergo quite dramatic thermal fluctuations. The next figure shows an example of a DNA configuration obtained by a 140 picosecond molecular dynamics (MD) simulation [42] compared with the ideal B DNA structure. Long-time MD simulations of DNA molecules in a "box" of water molecules lead to RMS fluctuation angles for R and T of order 5 and 9 degrees [43] in the ns time-window. The mean base-pair spacing also shows large-amplitude fluctuations. When the R, S, and T variables are treated as collective *harmonic degrees of freedom*, then the respective stiffness constants can be obtained either from these MD simulations (or from analysis of thermal diffuse X-ray scattering of DNA crystals).

Structural fluctuations in the ps to nanosecond (ns) time-window have been observed [44] experimentally as dynamic Stokes shifts in the fluorescence spectrum of DNA. These local fluctuations are extraordinary strong compared with those due to thermally excited phonon modes in crystalline materials. The RST model clearly gives a more interesting and realistic

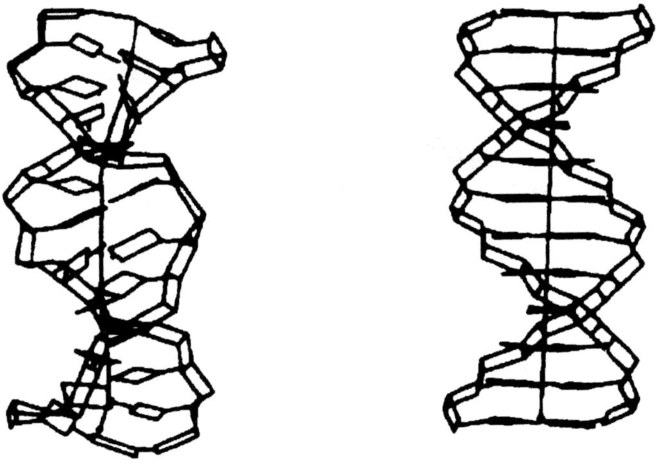

Fig. 28. DNA structure after a 140 ps simulation (first panel) compared with the DNA crystal structure (second panel). From reference [42].

account of the structural properties of DNA than the Worm-Like Chain. However, it is mathematically complex and the structural, elastic, and statistical mechanics properties of the RST model so far have not been investigated by physicists.

4 Electrostatics in water and protein-DNA interaction

We noted that electrostatics provides the basis both for the non-specific repressor-DNA interaction and for the stabilization of the nucleosome complex. Electrostatics is known to be of general importance the organization of chromatin [45]. DNA is about the most highly charged polymer you are likely to come across: it has a negative charge per unit length λ of -0.6 elementary charges per Angström (the negatively charged phosphate groups discussed in Sect. 1). There is a fundamental functional reason for all this charge: DNA is highly concentrated inside cells and bacteria and the Coulomb self-repulsion prevents unwanted *aggregation* by the van der Waals attraction. Other important biopolymers such as actin, microtubules, and hyaluronic acid also are negatively charged and also often come in high concentrations. The fact that charged biopolymers all have a *negative* charge prevents unwanted *complexation* between biopolymers of opposite charge.

The high charge per unit length produces counter-intuitive effects that are of considerable importance for DNA-protein interaction and they form the focus of this section. We will emphasize fundamental aspects that do not depend on making highly specific assumptions concerning the atomic structure that would obscure the elegant, underlying physics of aqueous electrostatics. First, we will introduce the physics of aqueous electrostatics as it applies to macro-ions and their interactions [46].

4.1 Macro-ions and aqueous electrostatics

Assume a collection of highly charged macromolecules ("Macro-Ions") placed in water at fixed positions. Treat the water molecules as a continuous medium of polar molecules, characterized by a large dielectric constant ε (about 78 at 25 °C, this is a reasonable description only for length-scales large compared to the size of a water molecule). In order to determine the electrostatic potential ϕ surrounding the macro-ions, we must solve Poisson's Law:

$$\nabla^2 \phi = -\frac{4\pi}{\varepsilon}\rho \tag{4.1}$$

with ρ the charge density. This charge density includes both the charge density of the small mobile ions and of the fixed macro-ions. Macro-ions have a low dielectric constant interior, while their charges normally are located on the outer surface in contact with water. Under these conditions, the electrical field just outside the macro-ions is determined by Gauss' Law:

$$\partial\phi/\partial n = -\frac{4\pi\sigma}{\varepsilon} \tag{4.2}$$

with σ the macro-ion surface charge density. We now can treat equation (4.2) as a boundary condition to the solution of equation (4.1) restricted to the space in between the macro-ions. To obtain the charge density of the "small" salt ions we assume the Boltzmann–Poisson (BP) approximation in which the concentration profile of the small ions is determined by the Boltzmann Distribution:

$$c_i(\vec{r}) = c_i \exp(-e z_i \phi(\vec{r})/k_{\mathrm{B}}T). \tag{4.3}$$

Here, z_i is the small ion valence, and c_i the ion concentration far from the macro-ions. The total charge density is $\rho_{\mathrm{ions}}(\vec{r}) = e \sum_{\mathrm{species}} z_i c_i(\vec{r})$. The condition of charge neutrality requires that the charge density of the salt solution far from the macro-ions vanishes so: $e \sum_{\mathrm{species}} z_i c_i = 0$.

If we insert $\rho_{\text{ions}}(\vec{r}) = e \sum_{\text{species}} z_i c_i(\vec{r})$ in equation (4.1) assuming equation (4.2), we are left with a well-posed mathematical problem: we must solve a second-order non-linear differential equation for the electrical potential with boundary condition equation (4.2), plus the condition that the potential vanishes at infinity.

From the solution of the BP equation, we can calculate the forces on the macro-ions by computing first the *free energy*:

$$F = \int \mathrm{d}^3 r \left\{ k_{\mathrm{B}} T \sum_{\text{species}} c_i(\vec{r}) \ln c_i(\vec{r}) + \frac{1}{2}\rho(\vec{r})\phi(\vec{r}) \right\}. \tag{4.4}$$

The first part, with the sum over the different ion species, is entropic in origin. It assigns each species the free energy density of an ideal, non-interacting solution. The second term is purely enthalpic: it is the electrostatic energy density of a collection of charges. The *electrostatic* or *charging* free energy is defined as the difference in free energies of F and the free energy of the system with all charges set to zero. The free energy $F(\vec{R}_1, \vec{R}_2, ...)$, computed as a function of the positions of the macro-ions, can be treated as an effective potential energy of the macro-ions.

Why is this the correct free energy density for the BP equation? The derivative of the free energy density with respect to $c_i(\vec{r})$ should equal the chemical potential of the i-th species, μ_i, according to the principles of thermodynamics. Using $\rho_{\text{ions}}(\vec{r}) = e \sum_{\text{species}} z_i c_i(\vec{r})$, you find:

$$k_{\mathrm{B}} T(\ln c_i(\vec{r}) + 1) + 1/2 e z_i \phi(\vec{r}) \stackrel{?}{=} \mu_i.$$

This is not quite right since the potential in equation (4.4) depends on the ion concentration through Poisson's Law, equation (4.1). If you allow for this dependence, by expressing first the potential in terms of the charge density, you find that the factor 1/2 turns into one. If you then solve for the concentration, you obtain Boltzmann's Law equation (4.3). Equation (4.4) is thus the "right" variational free energy.

The BP equation has been solved only for a few special case. The mathematical difficulties of the BP theory simplify however in the *high-temperature limit*. If the electrostatic energy per ion always is small compared to the thermal energy, we can linearize the exponential in equation (4.3) to obtain:

$$c_i(\vec{r}) = c_i \exp(-e z_i \phi(\vec{r})/k_{\mathrm{B}} T) \cong c_i(1 - e z_i \phi(\vec{r})/k_{\mathrm{B}} T). \tag{4.5}$$

Inserting equation (4.5) in equation (4.1):

$$\nabla^2\phi \cong -\frac{4\pi}{\varepsilon}\left(e\sum_{\text{species}} z_i c_i (1 - e z_i \phi(\vec{r})/k_{\text{B}}T)\right)$$

$$= \left\{\frac{4\pi e^2}{\varepsilon k_{\text{B}}T}\sum_{\text{species}} z_i^2 c_i\right\}\phi(\vec{r}) \tag{4.6a}$$

using again $e\sum_{\text{species}} z_i c_i = 0$. This is the famous Debye–Hückel (DH) equation. In terms of the Debye Parameter $\kappa \cong \sqrt{\frac{4\pi e^2}{\varepsilon k_{\text{B}}T}\sum_{\text{species}} z_i^2 c_i}$ (dimensions of $1/L$), equation (4.6) has the form of the Helmholtz equation:

$$\nabla^2\phi = \kappa^2\phi \tag{4.6b}$$

which is much easier to solve than the BP equation. For instance, for a point charge q the solution of equation (4.6) is $(q/\varepsilon r)\exp(-\kappa r)$ while there is no analytical form for the potential of a spherical charge in the BP equation.

The DH equation tells us that the characteristic length-scale over which the potential decays to zero is $1/\kappa$, the famous Debye Screening length. This length-scale depends on the salt concentration. For concentrations in the physiological range around 0.1 M, it is of the order of one nanometer, while at the lowest salt concentrations that can be achieved ("millipore water"), it is of the order of microns. The crucial question concerning the applicability of DH theory to macro-ions will be discussed shortly.

From Poisson's Law equation (4.1) and the DH equation equation (4.6), it follows that the charge density ρ is proportional to the potential: $\rho \cong -\frac{\varepsilon\kappa^2}{4\pi}\phi$. Using this in equation (4.4) with equation (4.5), you can find the electrostatic free energy:

$$\Delta F_{\text{DH}} = \frac{\varepsilon\kappa^2}{8\pi}\int \mathrm{d}^3 r\,\phi^2(\vec{r}). \tag{4.7}$$

The DH electrostatic *free* energy is thus positive and equal to *minus* the electrostatic energy: the positive entropic contribution is twice as large as the negative enthalpic contribution.

4.2 *The primitive model*

The "Primitive Model" is a simple toy model that is helpful to study the electrostatics of cylindrical macro-ions like DNA. The macro-ion is represented as an infinitely long, negatively charged rod. The radius of the rod is denoted by a and the charge per unit length by λ (see Fig. 29). If we write λ

as $-e/b$, with b the spacing per elementary charge, then the surface charge density σ equals $-e/2\pi ab$. The charged rod is immersed in a cylindrical container with radius $R \gg a$ filled with water (you also could consider R as the typical distance to other rod-like macro-ions).

4.2.1 The primitive model: Ion-free

We start with the case of ion-free water. Poisson's Law reduced to Laplace's equation $\Delta\phi = 0$ in a cylindrical geometry with solution $\phi(r) = \alpha \ln r + \beta$. Imposing the boundary condition equation (4.2) gives α:

$$\phi(r) = -\frac{2\lambda}{\varepsilon} \ln r/R \tag{4.8}$$

(demanding $\phi(R) = 0$). The first thing you learn from equation (4.8) is that there is a characteristic *voltage scale* $\Delta V = -\lambda/\varepsilon$ in the problem. How does this voltage scale compare with the thermal energy? First, introduce the *Bjerrum Length* l_B:

$$\frac{e^2}{\varepsilon l_\mathrm{B}} = k_\mathrm{B}T. \tag{4.9}$$

Two monovalent ions that are one Bjerrum length apart have an electrostatic energy equal to the thermal energy (the Bjerrum length is about 7 Å under standard conditions). In terms of the Bjerrum Length, the ratio $\xi = e\Delta V/kT$ of the electrostatic and thermal energies equals:

$$\xi = \frac{e\Delta V}{k_\mathrm{B}T} = \frac{l_\mathrm{B}}{b}\,. \tag{4.10}$$

For DNA, this ratio – which is known as the *Manning Parameter* – is about 5, so the electrostatic energy dominates over the thermal energy near the surface of DNA.

4.2.2 The primitive model: DH regime

Now let's add ions and assume the high temperature limit where we can apply DH theory. We must solve the Helmholtz Equation $\Delta\phi = \kappa^2\phi$ in cylindrical geometry. The appropriate solution vanishing at infinity is the K_0 Bessel Function: $\phi(r) = \alpha K_0(\kappa r)$. Using the boundary condition equation (4.2) to fix the intgeration constant α gives $\alpha = \frac{2\lambda}{\varepsilon\kappa a K_0'(\kappa a)}$ so

$$\phi(r) = \frac{2\lambda}{\varepsilon\kappa a K_0'(\kappa a)} K_0(\kappa r). \tag{4.11}$$

For small arguments, $K_0(x)$ can be approximated by $-\ln x$, and we recover the ion-free case (up to a constant). For large arguments, $K_0(x)$ can be approximated by $x^{-1/2}\exp -x$, so the potential is screened exponentially distances large compared to the Debye screening length. From the relation $\rho \cong -\frac{\varepsilon\kappa^2}{4\pi}\phi$ it follows that there is a cloud of positive ions surrounding the negatively charged rod (also called the "bilayer"). The total the charge per unit length of this cloud equals:

$$\lambda^+ = -\frac{\varepsilon\kappa^2}{4\pi}\int_a^R 2\pi r \mathrm{d}\,r\left\{\frac{2\lambda}{\varepsilon\kappa a K_0'(\kappa a)}K_0(\kappa r)\right\} = -\lambda. \tag{4.12}$$

The positive charge per unit length of the bilayer is just equal to the negative charge per unit length so the net enclosed charge is zero. You thus can consider the charged rod as a *cylindrical capacitor.*

The number of positive and negative small ions is thus not equal. The extra (positive) ions required to neutralize the rod are called the "counter-ions", the other positive and negative ions the "co-ions".

4.3 Manning condensation

4.3.1 Charge renormalization

Is DH theory valid for DNA under physiological conditions? It depends. For DNA, κa is of the order of one, so it follows from equation (4.11) that the characteristic voltage scale at the surface is still $\Delta V = -\lambda/\varepsilon$. We know that in that case the ratio ξ of the electrostatic energy and thermal energy is large compared to one so DH theory fails seriously near the surface of DNA. Yet, far from the surface the potential decreases to zero. In the asymptotic regime far from the macro-ion where the electrostatic energy is small compared to the thermal energy, the DH equation must be valid. What this really amounts to is that when we solve the DH equation, *we are not allowed to use the boundary condition equation (4.2) to determine α!* The integration constant α must be determined by a full solution of the non-linear Boltzmann–Poisson (BP) equation (or even some more accurate theory). This leads to a change in the value of α which we will incorporate by the introduction of a "renormalized" charge per unit length λ^*:

$$\lim_{r\to\infty}\phi(r) = \frac{2\lambda^*}{\varepsilon\kappa a K_0'(\kappa a)}K_0(\kappa r). \tag{4.13}$$

We thus can use the nice, linear DH theory to compute, for example, the *asymptotic* force of interaction between two highly charged rods, for separations large compared to the Debye screening length, provided we replace everywhere the bare charge by the renormalized charge. All the "hard" part of the problem is hidden in obtaining the renormalized charge, but that can be determined experimentally if necessary! We can use in general DH theory to study the phase behavior of *dilute solutions* of macro-ions of the same sign of charge and it has been used with success to study dilute polyelectrolyte and colloidal solutions. It is however not appropriate for a study of the *short-range* interaction between macro-ions. The interaction between DNA strands in chromatin and the interaction between macro-ions of *opposite* charge, such as in DNA/protein association, are beyond DH theory as we shall see shortly. Our next step is to look for a physical description that can apply to these cases without excessive mathematical complications.

4.3.2 Primitive model: Oosawa theory

First, let's try to estimate the renormalized line charge. Assume that the Manning Parameter is large compared to one so the electrostatic energy of an ion near the surface of the negatively charged rod exceeds $k_\mathrm{B}T$. Some of the positive ions will be attracted by the negative charges of the rod and establish a close association with the rod. For the first of these "condensed" ions, the thermal energy is small compared to the electrostatic energy so thermal fluctuations will not break up the association. The condensed ions are known as the "Manning Cloud" [47] by the way. Condensed ions are restricted to a thin sheet surrounding the rod and should be considered as a separate "population" from the free ions further out in the Debye Cloud (see Fig. 29).

As more and more positive ions become associated with the rod, the effective charge per unit length decreases and the ratio of electrostatic and thermal energy decreases as well. We will estimate the equilibrium value λ^* of the effective charge by demanding that the ions of the two populations are in *chemical equilibrium* with each other. Note that ions can freely exchange between the two populations. This point of equilibrium is set by the condition that the electrostatic energy per ion in the Manning Cloud must be comparable to thermal energy: the energy gain per ion is then balanced by the entropy loss. Using equation (4.10), this condition gives $\frac{e\Delta V}{k_\mathrm{B}T} \approx \frac{-\lambda^* e/\varepsilon}{k_\mathrm{B}T} \approx 1$, or $\lambda^* \approx -e/l_\mathrm{B}$. The prediction is thus that the renormalized charge per unit length should be of order one charge per Bjerrum Length. That is curious: the renormalized charge should not depend on the bare charge! Of course, if the bare charge per unit length is so small that the Manning parameter is less than one, then there would be no charge renormalization in the first place and $\lambda^* = \lambda$.

We can make this argument more precise in the limit $\kappa a \ll 1$. Let C_C be the ion concentration of the condensed fraction and C_DH the free ion concentration far from the rod but inside the DH cloud. If you placed a microscopic voltmeter with one electrode at the edge of the Debye layer and the other just inside the Manning cloud, you would measure an electrical voltage difference $\Delta\Phi$ between the two populations:

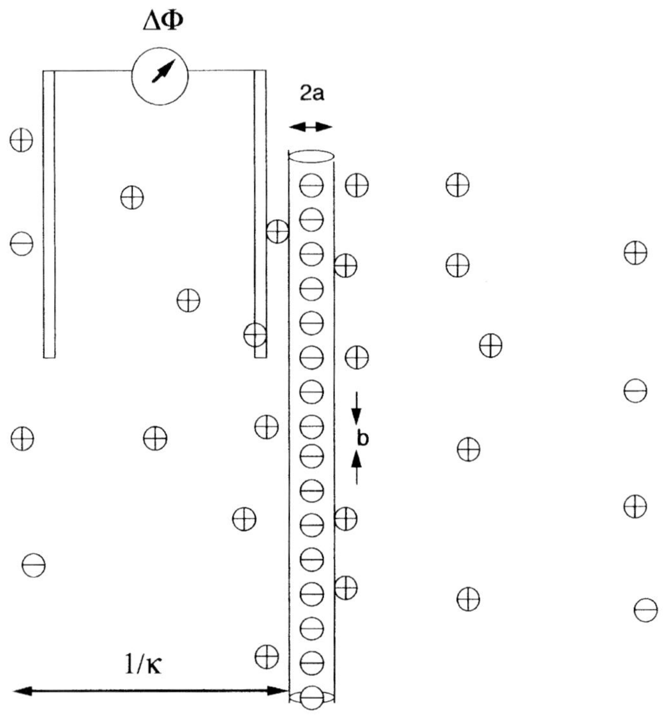

Fig. 29. Measuring the potential of the primitive model.

We can use the DH expression equation (4.12) for the asymptotic part of the electrostatic potential to compute $\Delta\Phi$:

$$\Delta\Phi \cong -(2\lambda^*/\varepsilon)\ln(1/\kappa a). \tag{4.14}$$

Since the two populations are in chemical equilibrium with each other, their concentration ratio must obey the Boltzmann Distribution:

$$\frac{C_{\mathrm{C}}}{C_{\mathrm{DH}}} = \exp(-e\Delta\Phi/k_{\mathrm{B}}T). \tag{4.15}$$

Next, we know from DH theory (Eq. (4.12)) that the free ion concentration of the Debye cloud must cancel the *effective* charge of the rod:

$$e\pi\kappa^{-2}C_{\mathrm{DH}} \approx -\lambda^*. \tag{4.16}$$

Overall charge neutrality requires that the total number of free ions per unit length in the Debye cloud plus that of the condensed ions must cancel the bare charge per unit length:

$$eC_{\mathrm{C}}\pi a^2 + eC_{\mathrm{C}}\pi\kappa^{-2} = -\lambda. \tag{4.17}$$

Now eliminate the free and condensed ion concentrations in the Boltzmann condition using the last two equations. Inserting equation (4.13), you then obtain a *self-consistency condition* for the renormalized charge λ^*:

$$\left(\frac{|\lambda| - |\lambda^*|}{|\lambda^*|}\right)(\kappa a)^2 \cong \exp\left(-\frac{2e|\lambda^*|}{\varepsilon k_{\mathrm{B}}T}\ln(1/\kappa a)\right). \tag{4.18}$$

That does not seem very interesting since it is only an approximate expression and we already have an estimate for λ^*. Here is the subtle part though: the approximations involved the "fuzzyness" of length scales κ^{-1} and a. The only thing we are going to assume concerning these two length scales is that $\kappa a \ll 1$. If you look at the self-consistency condition, you see that the *functional dependence* on κa imposes a condition on the exponent of κa, no matter what these two length scales precisely are. The exponent equals two on the left-hand side so it must equal two on the right-hand side as well! This means that (at least in the limit of low salt) $\lambda^* = -e/l_{\mathrm{B}}$ exactly.

Actually, this is the solution only if the renormalized charge *differs* from the bare charge per unit length, which is the case as long as $|\lambda| > e/l_{\mathrm{B}}$. There is another solution to the self-consistency equation, namely $\lambda = \lambda^*$! Check yourself that this is the solution if $|\lambda| < e/l_{\mathrm{B}}$. There is thus a *mathematical singularity* in the dependence of the renormalized charge on the bare charge at the point $|\lambda| = e/l_{\mathrm{B}}$.

It is possible to compute the renormalized charge per unit length analytically from the non-linear BP equation [48] in the limit of zero salt and it confirms this result. From numerical solutions of the BP equation at finite salt concentration, we know that this mathematical singularity is "smeared-out" as the salt concentration increases.

You can apply this method – due to Oosawa [49] – also to a charged sphere of radius a, charge Z. Show yourself that in the large Z limit, the renormalized charge Z^* of a sphere is of order (a/l_{B}) [50], the ratio of the radius and the Bjerrum length.

4.3.3 Primitive model: Free energy

If the analytical BP solution for an infinite rod is inserted in the free energy expression equation (4.4), you obtain in the strong- coupling limit of large

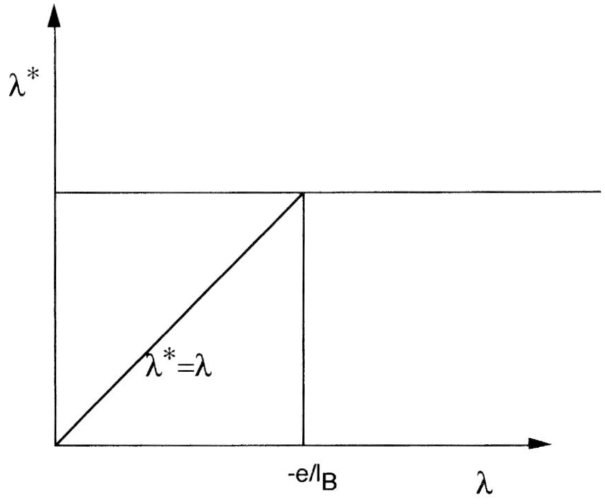

Fig. 30. Dependence of the renormalized charge per unit length on the bare charge per unit length.

Manning Parameter ξ for the electrostatic free energy:

$$\lim_{\xi \to \infty} f(\xi) = \frac{k_{\mathrm{B}}T}{b} \ln\left\{2\xi^2/(\kappa a)^2\right\} \tag{4.19}$$

in the limit of low salt concentration (this result, due to Lifson and Katchalsky [51], dates from 1954!). Like the weak-coupling DH expression equation (4.7), the charging free energy is positive. That is surprising: you would expect that in the strong coupling limit the electrostatic free energy should be *negative* due to the free energy gain obtained when you bring positive and negative charges together after the charge is turned on. Such an enthalpic term is indeed present, but it is a factor of order $1/\xi$ smaller. The electrostatic free energy for large ξ in BP theory is dominated by the entropic contribution. Its origin lies in the *entropy* loss that is suffered by the ions drawn into the Manning cloud: when you "turn on" the charges from zero, counter-ions are forced to condense on the rod once ξ exceeds one. The condensed ions lose $k_{\mathrm{B}}T$ per ion in the charging process, so the entropic contribution to the free energy per unit length is of order $k_{\mathrm{B}}T$ per charge or $k_{\mathrm{B}}T/b$, which is indeed the order of magnitude of f. We sometimes call this effect "counterion confinement".

It is interesting to compare the entropic free energy of the counterions and the conformational free energy of a polymer. For DNA, the latter is of order $k_{\mathrm{B}}T$ per persistence length. Since the persistence length is of order 500 Å, we see that the conformational free energy of DNA is completely negligible compared to the counter-ion free energy.

4.4 Counter-ion release and non-specific protein-DNA interaction

4.4.1 Counter-ion release

Assume a positively charged macro-ion with a charge Z of order 10 or so approaching a highly negatively charged rod. For distances large compared to $1/\kappa$, the electrostatic potential can be approximated by the asymptotic DH form. However, when the positive charge enters the Debye cloud unusual things start to happen. Recall that the counter-ions of the Manning cloud carry a free energy cost of about $k_{\mathrm{B}}T$ per counter-ion. When the positive macro-ion charges are brought into close contact with the negative macro-ion charges, a certain number of condensed counter-ions *can be released*. If the number of released counterions equals $Z' < Z$, then the free energy gain ΔF_{CR} upon binding is:

$$\begin{aligned}\Delta F_{\mathrm{CR}} &= Z' b f(\xi) \\ &= Z'\, k_{\mathrm{B}}T \ln\left\{2\xi^2/(\kappa a)^2\right\} \cdot\end{aligned} \tag{4.20}$$

This is a predominantly entropic binding mechanism. You might think of the highly charged rod with it condensed cloud of counter-ions as an entropic electro-chemical battery and the binding of the positive charge as a local discharge of this battery.

Let's apply the idea of counter-ion release to protein-DNA interaction [52]. Most proteins associating with DNA have a small total charge, but a quite significant *dipole moment*. The protein surface that fits on the DNA is lined with positive charge and the surface facing away with negative charge. Z' is then the number of positive charges on the DNA-binding surface of the protein. Note that this clever arrangement assists the "docking", the proper orienting of the protein, as it approaches DNA, while it avoids generic association of the protein with other negatively charged biopolymers.

In Section 2, we discussed the fact that the non-specific interaction between the lac repressor and DNA is predominantly entropic. We now understand this at a fundamental level [53]: the entropy gain of the non-specific interaction is due to release of the counter-ions, just like the entropic hydrophobic interaction can be viewed as being due to release of water molecules. Since the Debye Parameter is proportional to $[\mathrm{Salt}]^{1/2}$, ΔF_{CR} should be proportional to $-\ln([\mathrm{Salt}])$ according to equation (4.19). This indeed agrees with the dependence on salt concentration of the non-specific interaction between the lac repressor and DNA given by equation (2.21). Comparing equations (2.21) and (4.19), we even can estimate the number Z' of released counter-ions of the lac repressor to be about ten.

The predominantly entropic nature of the non-specific protein-DNA interaction informs us that the DH approximation is inappropriate for treating protein-DNA complexation because in the DH approximation, entropic and enthalpic contributions are comparable in magnitude, in disagreement with the experimental observations on the non-specific binding energy discussed in Section 2. In general, the use of DH potentials with renormalized charges is thus *not* appropriate for solutions containing macro-ions with opposite sign of charge, since that will lead to formation of complexes of macro-ions of opposite charge and counter-ion release of condensed counterions.

4.4.2 Nucleosome formation and the isoelectric instability

The most dramatic example of the non-specific protein-DNA interaction is the complexation of the positively charged nucleosome core with negatively charged DNA discussed in the previous section. The total positive charge of the core, Z_{core}, is about 245 elementary charges and the total negative charge of the wrapped DNA equals $Z_{\mathrm{DNA}} = -292$. That is curious: the net charge Z^* of the nucleosome is positive, about -47. It is not obvious why *more* DNA should be wrapped around the nucleosome than required to neutralize the core charge, assuming that electrostatics is the dominant association mechanism. If anything, it would be expected that Z^* is *negative*, because of the free energy penalty incurred when we bend the DNA around the nucleosome.

Let's apply the concept of counter-ion release to nucleosome formation. We discussed, in Section 3, the Marky–Manning model for the formation of the nucleosome. In this model, the electrostatic attraction between DNA and the nucleosome core is characterized by an adhesion energy per unit length. If the binding of DNA to the nucleosome core is due to counter-ion release, then we can simply identify the adhesion energy per unit length with the $f(\xi)$ given by equation (4.18). Within the counter-ion release model we would expect to release no more counter-ions than required to compensate the positive charge of the core. That again leads to the (intuitive) result that Z^* should be zero (or negative). Such neutralized, charge compensated complexes are known as "isoelectric" by the way.

Let's return to the Oosawa description of counter-ion condensation and measure the electrostatic potential around an isoelectric nucleosome complex with $Z^* = 0$, using again a microscopic voltmeter. One of the electrodes of the voltmeter is placed in contact with the Manning cloud of the DNA far from the nucleosome and the other electrode is placed near the surface of the nucleosome (see figure). Let $\Delta\Phi = V(\mathrm{DNA}) - V(\mathrm{Nucleosome})$ be the read-out of the voltmeter. The Boltzmann distribution informs us that the ratio of (small) positive ion concentration at the two electrode locations

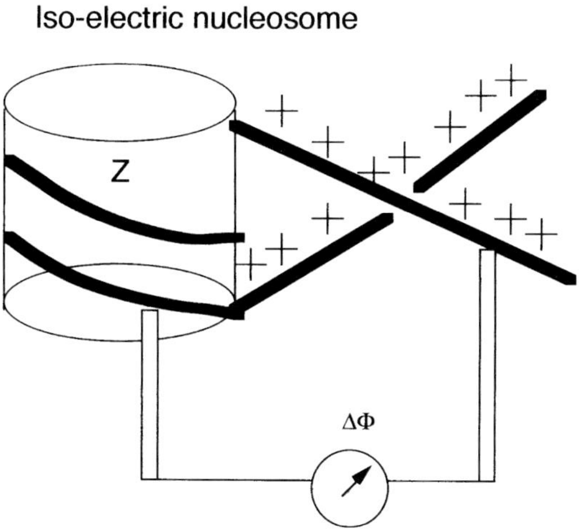

Fig. 31. Measuring the electrical potential of a charge-neutral nucleosome.

under conditions of chemical equilibrium must equal:

$$\frac{C_{\mathrm{DNA}}}{C_{\mathrm{Nucleosome}}} = \exp(-e\Delta\Phi/k_{\mathrm{B}}T). \tag{4.21}$$

Since the complex is charge neutral, the concentration of positive ions at the nucleosome surface must be low (not quite zero since the electrostatic potential surrounding the complex will contain higher order multipole contributions), while the ion concentration inside the Manning cloud is high as we saw (of order 1 M in fact). That means that the left-hand side of equation (4.20) must be large, so the voltage difference $\Delta\Phi$ must be *large and negative* to insure chemical equilibrium of the positive ions. The electrical potential at the nucleosome surface thus must exceed the potential at the DNA surface by a significant amount. This electrical potential difference will "pull-in" extra DNA material leading to a deviation from charge neutrality, in fact to *charge-reversal*: by that we mean that the effective charge Z^* of the complex has a different sign from the bare charge. Counter-ions are *not* released to infinity when this happens but instead remain associated with the charged nucleosome complex, but this still increases their entropy. As you wrap more and more DNA over the complex, the capacitive charging energy cost of the complex grows and eventually the wrapping process stops.

Let's try to make this argument quantitative [54]. Assume a spherical nucleosome of radius a with bare charge Z and assume that a length L of DNA has been wrapped around it. The net charge $Z^*(L)$ of the nucleosome equals:

$$Z^*(L) = Z - (L/b) \tag{4.22}$$

with b the spacing between charges along DNA. At the isoelectric point $Z^* = 0$, the wrapping length is $L_{\text{iso}} = Zb$. The free energy is the sum of three terms. The bending energy cost, the adhesion energy per unit length, and the capacitive charging contribution just mentioned. The first two terms were already present in the Marky–Manning model of Section 3, while the last term will be handled by treating the nucleosome as a sperical capacitor with capacitance $1/\varepsilon a$:

$$\Delta F = \left\{ \frac{\kappa}{a^2} - f(\xi) \right\} L + \frac{e^2 Z^*(L)^2}{\varepsilon a} \tag{4.23}$$

(we are not concerned here about numerical factors). The second term has a quadratic minimum at $L = L_{\text{iso}}$, while the first linear term favors large L (at least in the relevant regime where the adhesion energy exceeds the bending energy). The optimal value L^* clearly must exceed L_{iso}. After minimization with respect to L, we obtain:

$$Z^* \approx \frac{a}{l_{\text{B}}} \left[\ln(1/(\kappa a)^2) - \kappa b / k_{\text{B}} T a^2 \right]. \tag{4.24}$$

The effective net charge is predicted to be of the order of the ratio of the radius of the nucleosome and the Bjerrum length. Note that at the Marky–Manning unwrapping point, the effective nucleosome charge vanishes. Interestingly, this effective charge is of order the typical renormalized charge of a highly charged sphere in BP theory. We actually might have guessed our result right away!

This charge-reversal effect is our final example of proper design. A classical biophysics experiment is the precipitation of proteins at the iso-electric point (by varying pH). A solution of isoelectric complexes aggregates under the action of the van der Waals attraction. If the nucleosomes would have been charge-neutral, they would have stuck together, *even though DNA is highly charged*. The effective charge of the nucleosome prevents this unwanted aggregation. A subtle and counter-intuitive physical mechanism, counter-ion release, is recruited in the efficient management of the macroion household of the cell.

RB would like to thank the Les Houches center for its hospitality. I have enormously benefited in my work in this area from interaction with numerous colleagues, in particular William Gelbart, Itamar Borukhov, Andrea Liu, Jay Mashl, Stella Park, Fyl Pincus, Joseph Rudnick, Cyrus Safinya, and Helmut Schiessel.

References

[1] For a general introduction to gene expression: Alberts, *et al.*, *Molecular Biology of the Cell* (Garland, NY, 1994).

[2] For an excellent introduction to DNA from the viewpoint of physics: C. Calladine and H. Drew, *Understanding DNA*, Second Edition (Academic Press, 1997).

[3] For a very readable introduction: M. Ptashne, *A Genetic Switch* (Blackwell Scientific Publishing, Cambridge Mass, 1992).

[4] F. Jacob and J. Monod, *J. Mol. Biol.* **3** (1967) 318.

[5] J.N. Israelachvili, *Intermolecular and Surface Force* (Academic Press Inc., San Diego, 1985).

[6] P. von Hippel, *et al.*, *Proc. Nat. Acad. Sci.* **71** (1974) 4808.

[7] For an adsorbing surface exposed to a gas atmosphere, a similar relation between surface coverage and gas pressure is known as the "Langmuir Isotherm".

[8] R. Spolar and T. Record, *Science* **263** (1994) 777.

[9] High resolution studies are done at synchrotron facilities such as at Argonne National Laboratory.

[10] A. Travers, *DNA-Protein Interactions*, Chap. 3 (Chapman & Hall, 1994).

[11] D.M.J. Lilley, *DNA-Protein Structural Interactions* (IRL Press, Oxford, 1995).

[12] Lewis, *et al.*, *Science* **271** (1996) 1247.

[13] Berg, Winter and von Hippel, *Biochemistry* **20** (1981) 6929.

[14] Z. Otwinowski, *et al.*, *Nature* **355** (1988) 321.

[15] M. Werner, A. Gronenborn and M. Clore, *Science* **271** (1996) 778.

[16] R. Spolar and T. Record, *Science* **263** (1994) 777.

[17] For a review: M. Hampsey, *Micr. Mol. Biol. Rev.* **62** (1998) 1.

[18] M. Grunstein, *Ann. Rev. Cell. Biol.* **6** (1990) 643.

[19] For a general discussion of nucleosome and chromatin structure and dynamics, see J. Widom, *Ann. Rev. Biophys. Biomol. Struct.* **27** (1998) 285, and references therein.

[20] `http://cellbio.utmb.edu/cellbio/nucleus2.htm`

[21] R.D. Kornberg, *Ann. Rev. Biochem.* **46** (1977) 931; R.D. Kornberg and A. Klug, *Sci. Amer.* **244** (1981) 52.

[22] J. Widom, *J. Mol. Biol.* **190** (1986) 411.

[23] C.L. Woodcock, S. Grigoryev, R. Horowitz and N. Whitaker, *Proc. Natl. Acad. Sci.* **90** (1993) 9021.

[24] S. Leuba, *et al.*, *Proc. Natl. Acad. Sci.* **91** (1994) 11621.

[25] J. Bednar, *et al.*, *J. Cell. Biol.* **131** (1995) 1365.

[26] J. Finch and A. Klug, *Proc. Natl. Acad. Sci.* **73** (1976) 1897. J. Widom and A. Klug, *Cell* **43** (1985) 207.

[27] J. Bednar, *et al.*, *Proc. Natl. Acad. Sci.* **95** (1998) 14171.

[28] K. van Holde and J. Zlatanova, *J. Biol. Chem.* (1995) 8373.

[29] K. Luger, *et al.*, *Nature* **389** (1997) 251.

[30] J. White, *Am. J. Math.* **91** (1969) 603; F. Fuller, *PNAS* **75** (1987) 3557.

[31] P. Hagerman, *Biopolymers* **20** (1981) 1503; M. Doi and S. Edwards, *The Theory of Polymer Dynamics* (Clarendon Press, Oxford, 1986).

[32] Bloom and Fawcett, *A Textbook of Histology* (Chapman and Hall, 12'th Ed.).

[33] N. Marky and G. Manning, *Biopolymers* **31** (1991) 1543.

[34] T.D. Yager, C.T. McMurray and K.E. van Holde, *Biochemistry* **28** (1989) 2271.

[35] K.J. Polach and J. Widom, *J. Mol. Biol.* **254** (1995) 130; **258** (1996) 800.

[36] H. Schiessel, *et al.*, *Phys. Rev. Lett.* **86** (2001) 4414.

[37] J. Marko and E. Siggia, *Biophysical J.* **73** (1997) 2173; J. Rudnick and R. Bruinsma, *Biophysical J.* **76** (1999) 1725.

[38] C. Bustamante, J.F. Marko, E.D. Siggia and S. Smith, *Science* **265** (1994) 1599.

[39] Y. Cui and C. Bustamante, *Proc. Natl. Acad. Sci. USA* **97** (2000) 127.

[40] C. Calladine and H. Drew, *Understanding DNA*, Second Edition (Academic Press, 1997).

[41] C. Calladine and H. Drew, *J. Mol. Biol.* **178** (1984) 773.

[42] S. Swaminathan, G. Ravishankar and D. Beveridge, *J. Am. Chem. Soc.* **113** (5027) 1991.

[43] Y. Duan, P. Wilkosz, M. Crowley and J. Rosenberg, *J. Mol. Biol.* **272** (1997) 553.

[44] E. Brauns, *et al.*, *J. Am. Chem. Soc.* **121** (1999) 11644.

[45] D.J. Clark and T. Kimura, *J. Mol. Biol.* **211** (1990) 883.

[46] The best introduction to macro-ion electrostatics is: F. Oosawa, *Polyelectrolytes* (Marcel Dekker, New York, 1971).

[47] G.S. Manning, *J. Chem. Phys.* **51** (1969) 924.

[48] M. Le Bret and B. Zimm, *Biopolymers* **23** (1984) 287.

[49] F. Oosawa, *Polyelectrolytes* (Marcel Dekker, New York, 1971).

[50] S. Alexander, P. Chaikin, G. Morales, P. Pincus and D. Hone, *J. Chem. Phys.* **80** (1984) 5776.

[51] S. Lifson and A. Katchalsky, *J. Polym. Sci.* **13** (1954) 43.

[52] T. Record, *et al.*, *Q. Rev. Biophys.* **11** (1978); C.F. Anderson and M.T. Record, *J. Phys. Chem.* **97** (1993) 7116, and references therein.

[53] P. de Haseth, T. Lohman and T. Record, *Biochemistry* **16** (1977) 4783.

[54] S.Y. Park, R.F. Bruinsma and W.M. Gelbart, *Europhys. Lett.* **46** (1999) 454.

COURSE 2

MECHANICS OF MOTOR PROTEINS

J. HOWARD

Max Planck Institute of Molecular Cell Biology & Genetics, Pfotenhauerstrasse 108, 01307 Dresden, Germany

Contents

MECHANICS OF MOTOR PROTEINS

J. Howard

1 Introduction

Motor proteins are molecular machines that convert the chemical energy derived from the hydrolysis of ATP into mechanical work used to power cellular motility. In addition to specialized motile cells like muscle fibers and cellular processes like cilia, all eukaryotic cells contain motor proteins (Fig. 1). The reason is that eukaryotic cells are large and their cytosols are crowded with filaments and organelles; as a result, diffusion is too slow to efficiently move material from one part of a cell to another (Luby-Phelps *et al.* 1987). Instead, the intracellular transport of organelles such as vesicles, mitochondria, and chromosomes is mediated by motor proteins. These proteins include myosins and dyneins that are relatives of the proteins found in the specialized muscle and ciliated cells, as well as members of a third family of motor proteins, the kinesins, which are distantly related to the myosin family.

The focus of this chapter is on how motor proteins work. How do they move? How much fuel do they consume, and with what efficiency? How do chemical reactions generate force? What is the role of thermal fluctuations? These questions are especially fascinating because motor proteins are unusual machines that do what no manmade machines do–they convert chemical energy to mechanical energy directly, rather than *via* an intermediate such as heat or electrical energy. Tremendous insight into this chemomechanical energy transduction process has come from technical developments over the last ten years that allow single protein molecules to be detected and manipulated. The goal of this review is to provide a framework within which to understand these new observations: how do mechanical, thermal, and chemical forces converge as a molecular motor moves along its filamentous track. For background, the reader is directed to *Molecular Biology of the Cell* (Alberts *et al.* 2002) for an introduction to the biology of cells and molecules, to *Cell Movements* (Bray 2000) for a broad review of cell motility, and to *Mechanics of Motor Proteins and the Cytoplasm* (Howard 2001) for more detailed discussion of the mechanics of molecular motors and the cytoskeleton.

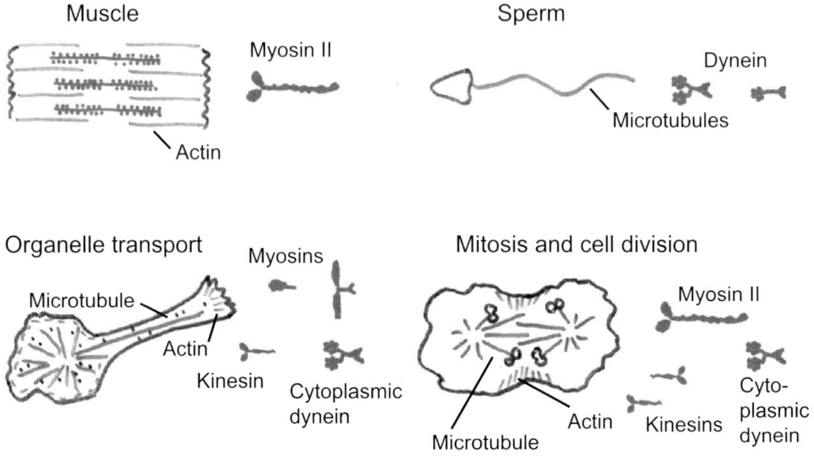

Fig. 1. Motor proteins and cellular motility.

2 Cell motility and motor proteins

The study of motor proteins begins with myosin, which drives the contraction of muscle. Myosin was first isolated as a complex with actin filaments by Kühne (Kühne *et al.* 1864), though it was not until the 1940s that the complex was dissociated into the separate proteins, myosin and actin (Straub 1941-2; Szent-Gyorgyi 1941-2). The discovery of the myosin crossbridges by H.E. Huxley in 1957 (Huxley 1957b, Fig. 2A) provided a molecular basis for the contraction of muscle: the bending or rotation of these crossbridges causes the actin-containing thin filaments to slide relative to the myosin-containing thick filaments, and the sliding of these filaments, in turn, leads to the shortening of the muscle, as had been demonstrated a few years earlier (Huxley & Hanson 1954; Huxley & Niedergerke 1954).

Since its initial discovery, the crossbridge (also called a head) has proven to be central to the mechanism of cell motility. Dynein, which drives the beating of cilia, was identified in the 1963 (Gibbons 1963). The dynein crossbridges cause the adjacent doublet microtubules to slide with respect to each other (Fig. 2B). Because shear between the microtubules at the base of the cilium (*e.g.* near the head of the sperm) is prevented by strong linkages, the sliding is converted into bending of the microtubules along the length of the cilium. In this way the sperm undergoes its snake-like propulsion through solution. Kinesin, which moves organelles along microtubules, was purified in 1985 (Brady 1985; Vale *et al.* 1985). Attached to the cargo at one end, the crossbridges at the other end of the kinesin molecule walk along the surface of the microtubule (Fig. 2C).

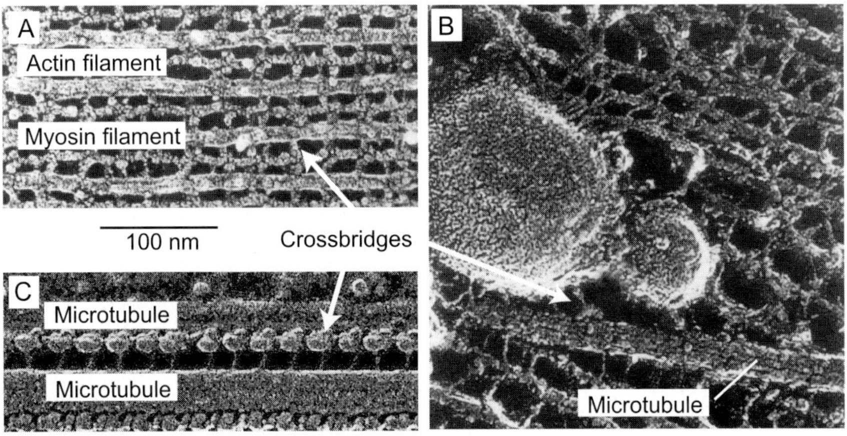

Fig. 2. Crossbridges formed by the motor domains of motor proteins drive motion along cytoskeletal filaments. A. Muscle. Myosin crossbridges protruding from the thick, myosin-containing filament drive the sliding of the thin, actin-containing filaments. B. Kinesin crossbridges walk along microtubules carrying organelles. C. Sperm, dynein crossbridges cause the sliding of adjacent microtubules.

In all cases studied in detail, the motion of a motor protein is directed. Actin filaments and microtubules are polar structures made of asymmetric protein subunits, and a given motor always moves towards a particular end of the filament. The myosin, dynein and kinesin families have a large number of members. For example, humans have 33 genes which code for proteins of similar amino acid sequence to the heavy chain of muscle myosin (`http://www.gene.ucl.ac.uk/nomenclature/`, `http://www.mrc-lmb.cam.ac.uk/myosin/myosin.html`), and they have 21 dynein heavy-chain genes and 45 kinesin genes (Miki *et al.* 2001; `http://www.blocks.fhcrc.org/ kinesin/index.html`). Interestingly, different myosins go in different directions along actin filaments and different kinesins go in different directions along microtubules. This is important because the orientation of actin filaments and microtubules in cells is tightly controlled: thus, by using differently directed motors cells are able to move cargoes from one part of the cell to another (and back) in order to organize the cell's internal structure.

3 Motility assays

The study of motor proteins was revolutionized by the development of *in vitro* motility assays in which the motility of purified motor proteins along purified cytoskeletal filaments is reconstituted in cell-free conditions.

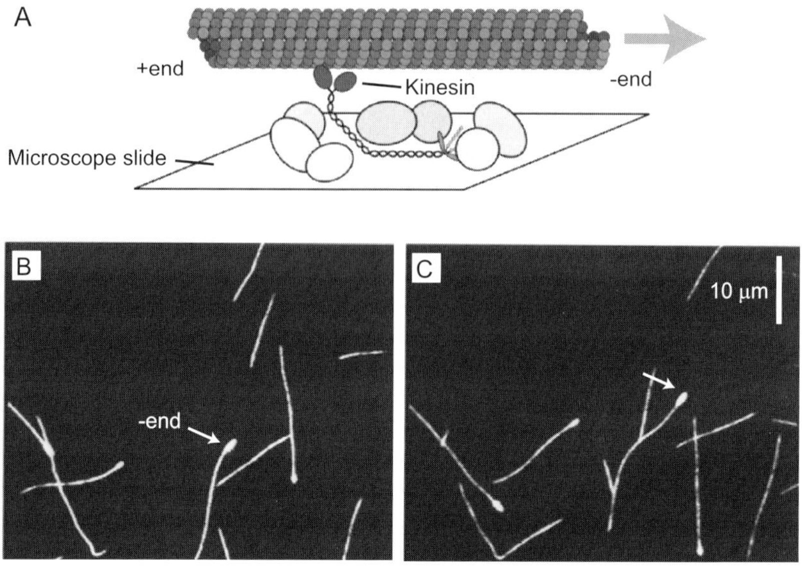

Fig. 3. A motility assay in which the motor is bound to a surface and the filament is observed to glide over the surface in the presence of ATP. B, C. Fluorescently labeled microtubules gliding across a kinesin-coated surface (15 s between frames). The arrowed microtubule has a bright mark on its slowly growing end (called the minus end): this end leads, showing that kinesin is a plus-end-directed motor.

An important milestone in this development was the visualization of fluorescent beads coated with purified myosin moving along actin cables in the cytoplasm of the alga *Nitella* (Sheetz & Spudich 1983). This was quickly followed by the first completely reconstituted assay in which motor-coated beads were shown to move along oriented filaments made from purified actin that had been bound to the surface of a microscope slide (Spudich *et al.* 1985). Though "threads" of actin and myosin had been known to contract in the presence of ATP (Szent-Gyorgyi 1941), this contraction was very slow. The significance of the new findings was that they proved that myosin (together with actin) was sufficient to produce movement at rates consistent with the speeds of muscle contraction and cell motility.

There are two geometries used in *in vitro* motility assays: the gliding assay and the bead assay. In the gliding assay, the motors themselves are fixed to the substrate, and the filaments are observed under a light microscope as they diffuse down from solution, attach to, and glide along the motor-coated surface (Fig. 3) in the presence of ATP (Fig. 4). In the bead assay, filaments are fixed to a substrate, such as a microscope slide, and

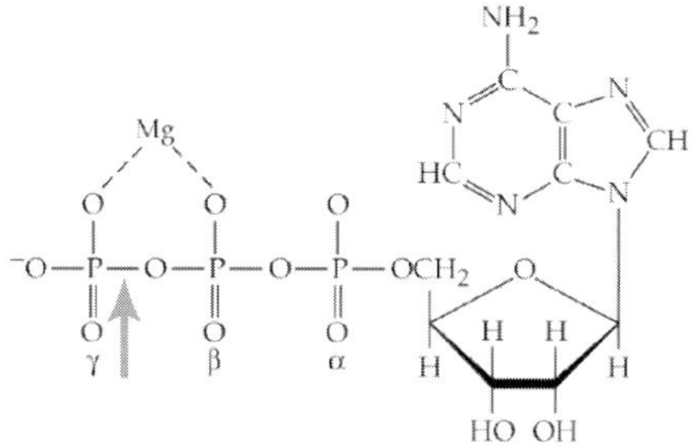

Fig. 4. The hydrolysis of the gamma-phosphate bond (arrowed) of ATP can be summarized by the following reaction: ATP $\leftrightarrow$ ADP + Pi

$$\Delta G = \Delta G_0 - kT \ln \frac{[\mathrm{ATP}]_c}{[\mathrm{ADP}]_c[\mathrm{P_i}]_c} \qquad \Delta G_0 = kT \ln \left[\frac{[\mathrm{ATP}]_{eq}}{[\mathrm{ADP}]_{eq}[\mathrm{P_i}]_{eq}}\right] = -54 \times 10^{-21}\ \mathrm{J}$$

where the subscript c refers to cellular, the subscript eq refers to equilibrium and the concentrations have units moles per liter. In cells, the reaction is very far from equilibrium with typical concentrations [ATP] = 1 mM, [ADP] = 0.01 mM and [Pi] = 1 mM; this makes the free energy very large and negative, $\Delta G \cong -100 \times 10^{-21}$ J.

motors are attached to small plastic or glass beads with typical diameters of 1 μm. The motions can be recorded and the speed measured by tracking the centroid of the bead or the leading edge of the filament (see Scholey 1993 for detailed methods). There is good overall agreement between the speed of a motor protein *in vitro* and the speed of the cellular motion that is attributed to the motor (Table 1).

4 Single-molecules assays

The progress of research on motor proteins has gone hand-in-hand with increases in the sensitivity of light microscope techniques. Single protein molecules can now be observed and manipulated. A crucial development was the visualization of individual actin filaments by darkfield microscopy (Nagashima & Asakura 1980). This was followed by visualization of microtubules by differential interference contrast microscopy (Allen *et al.* 1981) and actin filaments by fluorescence microscopy (Yanagida *et al.* 1984). Further refinement of the motility assays led to detection of movement by single motor molecules (Howard *et al.* 1989). With improved fluorescence sensitivity, it was even possible to image individual fluorescently labeled motors (Funatsu *et al.* 1995) (rather than the much larger filaments), and to watch the motors individually while they move along filaments (Vale *et al.* 1996; Yajima *et al.* 2002). The combination of these assays with increasingly sophisticated optical and mechanical techniques such as optical tweezers

Table 1. Motor speeds *in vivo* and *in vitro*.

Motor	Speed[a] *in vivo* (nm/s)	Speed[b] *in vitro* (nm/s)	ATPase[c] (s^{-1})	Function
Myosins				
1. Myosin IB	500	200	6	Amoeboid motility, hair cell adaptation
2. Myosin II	6000	8000	20	Fast skeletal muscle contraction
3. Myosin II	200	250	1.2	Smooth muscle contraction
4. Myosin V	200	350	5	Vesicle transport
5. Myosin VI	ND	−58	0.8	Vesicle transport?
6. Myosin XI	60 000	60 000	ND	Cytoplasmic streaming
Dyneins				
7. Axonemal	−7000	−4500	10	Sperm and cilial motility
8. Cytoplasmic	−1100	−1250	2	Retrograde axonal transport, mitosis, transport in flagella
Kinesins				
9. Conventional	1800	840	44	Anterograde axonal transport
10. Nkin	800	1800	78	Transport of secretory vesicles
11. Unc104/KIF	690	1200	110	Transport of synaptic vesicle precursors and mitochondria
12. Fla10/KinII	2000	400	ND	Transport in flagella, axons, melanocytes
13. BimC/Eg5	18	60	2	Mitosis and meiosis
14. Ncd	ND	−90	1	Meiosis and mitosis

ND not determined or known.

(Svoboda *et al.* 1993) has allowed measurement of the stepwise movement of motors along their filaments (Fig. 5) and the measurement of the force generated by a single motor protein (Fig. 11).

Single-molecule mechanical and optical techniques are now being applied to many biochemical processes mediated by other molecular machines; these include ATP synthesis (Noji *et al.* 1997), DNA transcription (Wang *et al.* 1998), and DNA replication (Wuite *et al.* 2000). The folding of individual proteins (Deniz *et al.* 2000) and RNA (Liphardt *et al.* 2001) can also be followed. The techniques can even be used to record from molecules on the surfaces of intact cells (Sako *et al.* 2000; Benoit *et al.* 2000; Schutz *et al.* 2000) and recordings deep inside cells should soon be possible. Thus

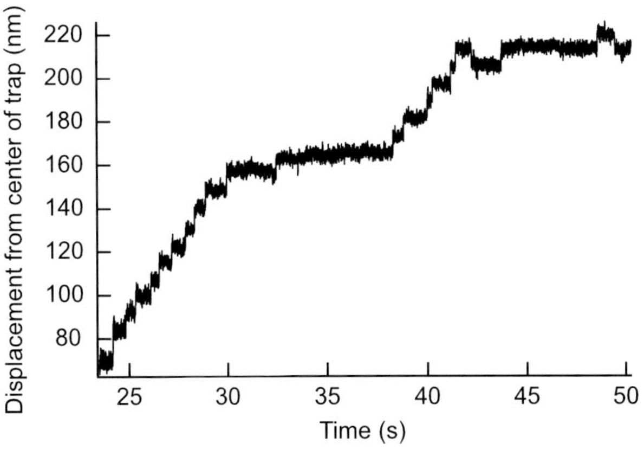

Fig. 5. A rat kinesin molecule taking 8 nm steps along a microtubule at low ATP concentration. (Courtesy of Nick Carter and Rob Cross).

single-molecule techniques are becoming to cell biology what the patch clamp technique and single ion-channel recordings are to neurobiology (Neher & Sakmann 1976; Sakmann & Neher 1995).

5 Atomic structures

The structural and physical basis for motility has been placed on a firm foundation by the solution of the atomic structures of actin (Kabsch *et al.* 1990) and myosin (Rayment *et al.* 1993) and of tubulin (Lowe & Amos 1998; Nogales *et al.* 1998) and kinesin (Kull *et al.* 1996). By fitting the atomic structures into electron micrographs, atomic models of the actin filament (Holmes *et al.* 1990) and the microtubule (Nogales *et al.* 1999) have been built. There are also reasonable guesses for how the motors dock to these filaments (*e.g.* Fig. 6).

The atomic structures have brought many key questions into focus. For example, how do small changes associated with the hydrolysis of ATP (on the order of a few Angströms) lead to protein conformational changes on the order of several nanometers? What determines the directionality of a motor protein? The detailed answers to these problems will require many additional atomic structures: the motors complexed with their filaments, and the motors with different nucleotides bound to them (*e.g.* ATP, ADP and P_i, ADP and no nucleotide). But this will be very difficult, and even when solved, these structures will provide only static pictures, with no kinetic or energetic information: photographing an internal combustion engine at top dead center and at the bottom of the down stroke does not explain how it

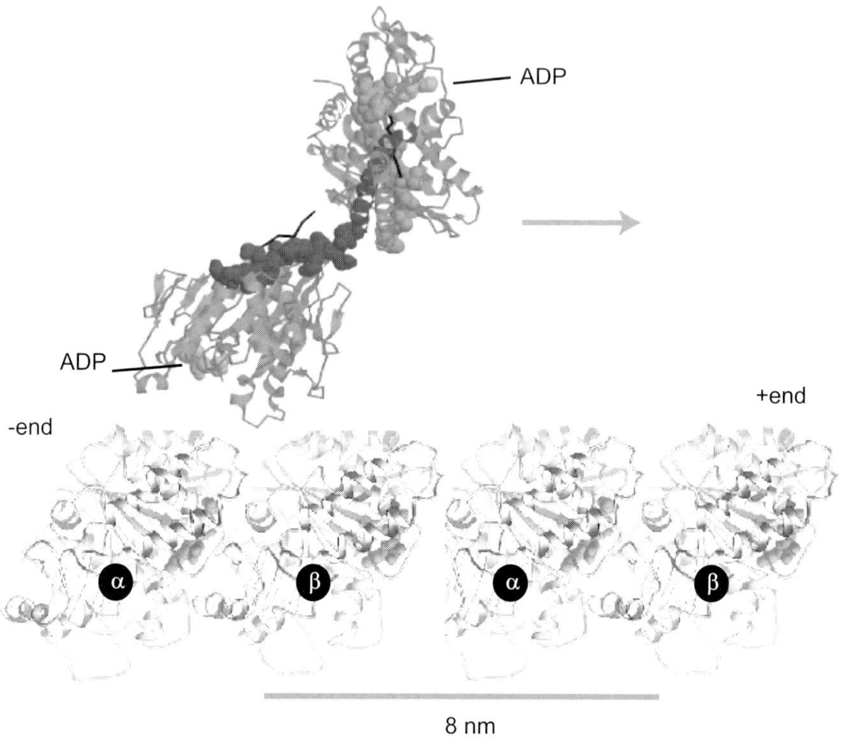

Fig. 6. Kinesin docked to the microtubule. Kinesin has two identical heads joined by the dimerization domain (dark). Each head binds nucleotide, which was ADP in these crystallization conditions. The microtubule is composed of dimers of the closely-related α and β tubulins. The dimers associated "head-to-tail" to form a polar structure that has the β-subunit at the plus end.

works. In the following sections I will address some of physical and thermodynamic questions which are essential to answer in order to understand how conformational changes driven by ATP hydrolysis might generate force and produce directed motion.

6 Proteins as machines

The structural and single-molecule results reinforce the concept of proteins as machines (Alberts 1998). According to this view, a molecular motor is an assembly of mechanical parts–springs, levers, swivels and latches–that move in a coordinated fashion as ATP is hydrolyzed. How does such a molecular machine move in response to a internal and external forces? The answer depends on the Reynolds number, which determines the pattern of fluid flow around a moving object. The Reynolds number is equal to

$$Re = \frac{\rho L \nu}{\eta} \tag{1}$$

where ρ is the density of the liquid, L is the characteristic length of the object (in the direction of the flow), ν is the speed of the movement relative to the fluid, and η is the viscosity. Note that the Reynolds number is dimensionless, and its physical meaning is that it is the ratio of the inertial and the viscous forces. For proteins in aqueous solution $\rho \approx 10^3$ kg/m^3, $L \approx 10$ nm, $\nu \approx 1$ m/s (corresponding to 1 nm per ns, which is on the order of the fastest global conformational changes of proteins) and $\eta \approx 10^{-3}$ Pa.s. This makes $Re \approx 10^{-2}$ (and even less for slower motions). A Reynolds number much less than one means that the inertial forces can be neglected and that the motion is highly overdamped: when subject to a force, a protein will creep into a new conformation rather than undergo oscillations. The time constant of the motion is γ/κ, where γ is the drag coefficient ($\sim \eta L$ according to Stokes law) and κ is the stiffness of the protein.

To get a feeling for how proteins move, imagine that the size of a protein were increased by a factor of 10^7, so that a 10-nm-diameter protein became a mechanical device of diameter 100 mm, fitting nicely in the palm of one's hand (Fig. 7). Now the density and rigidity (Young's modulus) of protein is similar to that of plastic or Plexiglas, so that we can consider that our device is built of plastic (Howard 2001). If the viscosity of the fluid bathing the device is increased by the same factor of 10^7 (by putting it in honey, for example), then the ratio of the inertial to the viscous forces will the same for both the protein and the macroscopic device. The Reynolds number will be unchanged (see Howard 2001 for the detailed argument) and the pattern of fluid flow will be preserved, just scaled in size. However, to deform the plastic device to the same relative extent as the protein will require a much larger force because the device has a much greater cross-sectional area: whereas a force of only 1 pN might be needed to induce a protein conformational change of 1 nm (corresponding to a strain of 10%), a force of 100 N, corresponding to a weight of 10 kg, would be required to produce the same strain in the plastic device. In response to the respective forces, the protein and the mechanical device will move at the same initial speed, but because the protein conformational change is so much smaller, the relaxation of the protein will be complete in much less time: a relaxation that took an almost imperceptible 100 ns for the protein will take a leisurely 1 s for the device. The work done scales with the volume, so the free energy corresponding to the hydrolysis of one molecule of ATP scales to ~100 J.

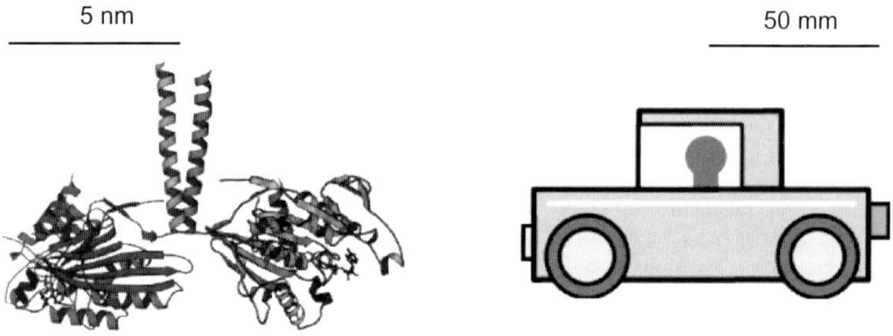

Fig. 7. Comparison of scales between a motor protein (left) and a toy car (right).

Quantity	Motor protein	Macroscopic device
Dimension	10 nm	100 mm
Material (Youngs modulus)	Amino acids (2 GPa)	Plastic (2 GPa)
Solution (viscosity)	Aqueous (1 mPas)	Honey (10 kPas)
Speed	1 m/s	1 m/s
Reynolds number	0.01	0.01
Time constant	100 ns	1 s
Force	1 pN	100 N
Energy	$1\text{–}100 \times 10^{-21}$ J	1–100 J

7 Chemical forces

In addition to mechanical forces, proteins are subject to chemical forces. By chemical forces, we mean the forces that arise from the formation or breakage of intermolecular bonds. For example, consider what happens when a protein first comes in contact with another molecule: as energetically favorable contacts are made, the protein may become stretched or distorted from its equilibrium conformation. Chemical forces could also arise from changes in bound ligands: the cleavage of ATP (Fig. 4) will relieve stresses within the protein that had been built up when the ATP bound. If a protein can adopt two different structures, the binding of a ligand or the change in a bound ligand could preferentially stabilize one of these structures and therefore change the chemical equilibrium between the structures. In this way, we imagine that a chemical change produces a local distortion that in turn pushes the protein into a new low-energy conformation.

To understand how protein machines work, it is essential to understand how proteins move in response to these chemical forces. Just as a chemical force might cause a protein to move in one direction, an external mechanical force might cause the protein to move in the opposite direction. For example, the binding of a ligand might stabilize the closure of a cleft, whereas an external tensile force might stabilize the opening of the cleft; as a result, the mechanical force is expected to oppose the binding of the ligand. Thus mechanical forces can oppose chemical reactions and conversely, chemical reactions can oppose mechanical ones. If the chemical force is strong enough, the chemical reaction will proceed even in the presence of a mechanical load: in this case we say that the reaction generates force.

8 Effect of force on chemical equilibria

The influence of a mechanical force on the chemical equilibrium between two (or more) structural states of a protein can be calculated with the aid of Boltzmann's law. If the difference between two structural states is purely translational–*i.e.* if state M_2 corresponds to a movement through a distance Δx with respect to state M_1 as occurs when a motor moves along a filament against a constant force–then the difference in free energy is $\Delta G = -F \cdot \Delta x$, where F is the magnitude of the force *in the direction* of the translation. If the length of a molecule changes by a distance Δx as a result of a conformational change, then the difference in free energy is

$$\Delta G \cong \Delta G^0 - F\Delta x \tag{2}$$

where F is the tension across the molecule and ΔG^0 is the free energy difference in the absence of tension. The equality is exact if the molecule is composed of rigid domains that undergo relative translation, or if the two structural states have equal stiffness. Application of Boltzmanns law shows that at equilibrium

$$\frac{[\mathrm{M}_2]}{[\mathrm{M}_1]} = \exp\left[-\frac{\Delta G}{kT}\right] \cong \exp\left[-\frac{\Delta G^0 - F\Delta x}{kT}\right] = k_{\mathrm{eq}}^0 \exp\left[\frac{F\Delta x}{kT}\right] \tag{3}$$

where K_{eq}^0 is the equilibrium constant in the absence of the force. *The crucial point is that an external force will couple to a structural change if the structural change is associated with a length change in the direction of the force.* If the change in length of a molecule is 4 nm, then a force of 1 pN will change the free energy by 4 pN.nm $\cong kT$, the unit of thermal energy where k is Boltzmanns constant and T is absolute temperate. According to equation (3), this will lead to an e-fold change in the ratio of the concentrations. Because protein conformational changes are measured in nanometers, and energies range from 1 kT to 25 kT (ATP hydrolysis) (Fig. 4 legend), it is expected that relevant biological forces will be on the scale of piconewtons (10^{-12} N).

An expression analogous to equation (2) holds for the effect of voltage on membrane proteins (Hille 1992). If a structural change of a membrane protein such as an ion channel is associated with the movement of a charge, Δq, across the electric field of the membrane, then the energy difference between the open and closed states will include a term $V\Delta q$, and this makes the opening sensitive to the voltage, V, across the membrane (Fig. 8). The openings of the voltage-dependent Na and K channels that underlie the action potentials in neurons are strongly voltage dependent: classic experiments by Hodgkin and Huxley showed that the ratio of open probability to closed probability increased approximately e-fold per 4 mV (Hodgkin 1964). This indicates that the opening of each channel is associated with the movement of about six electronic charges across the membrane ($\Delta q = kT/V \cong 6e$, where e is the charge on the electron). The predicted movement of these electronic charges has been directly measured as a non-linear capacitance of the membrane (Armstrong & Bezanilla 1974). Protein conformational changes are sensitive to many other "generalized" forces including membrane tension, osmotic pressure, hydrostatic pressure, and temperature. Sensitivity to these forces requires that conjugate structural changes occur in the protein, respectively area, solute accessible volume, water accessible volume and entropy (Howard 2001).

The chemical analogy to force and voltage is chemical potential, a measure of the free energy change associated with a molecular reaction. For example, if a ligand at concentration L preferentially binds to one state of a protein over another, then the difference in free energy between the two states is equal to $kT \ln L \cdot \Delta n$, where $kT \ln L$ is the chemical potential of the ligand and Δn is the number of ligand-binding sites on the protein. If a protein is subject to a combination of mechanical forces, electrical forces and chemical forces, then the free energies will add, allowing one to calculate how physical and chemical forces trade off against each other under equilibrium conditions.

9 Effect of force on the rates of chemical reactions

To understand how force might influence the kinetics of protein reactions is more difficult. Proteins are very complex structures, and a full description of the transition from one structure to another would require following the trajectories of each of the amino acids. The problem is even more difficult because it expected that there are a huge number of different pathways from one structure to another, and so a full description would require the enumeration of all the different pathways, together with their probabilities. This is simply not possible at present given that we do not even understand how a protein folds into even one structure. Thus we need a simple model for the kinetics of protein reactions.

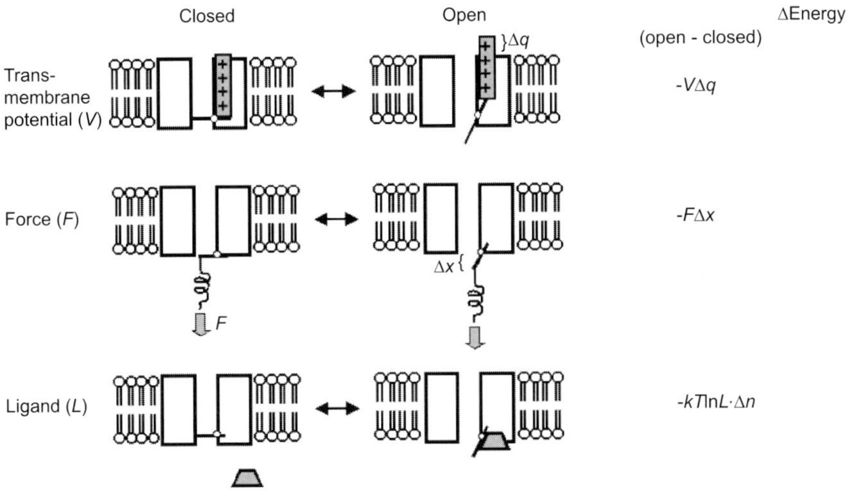

Fig. 8. Effects of generalized forces on protein conformations. In this example an ion channel sitting in a membrane can be either open or closed depending on the position of the gate. If the gate is coupled to a movement of charge (top), then the channels probability of being open will depend on the membrane potential. If it is coupled to a vertical movement (middle), then the probability will depend on a vertical force. And if the opening of the gate is coupled to the binding of a small molecule (bottom), then the probability will depend on that molecules concentration ($\ln L$) and the number of binding sites (Δn).

The simplest model for a chemical change between two reactants M_1 and M_2 is a first order process:

$$M_1 \underset{k_{-1}}{\overset{k_1}{\rightleftarrows}} M_2 \qquad \frac{\mathrm{d}[\mathrm{M}_1]}{\mathrm{d}t} = -k_1[\mathrm{M}_1] + k_{-1}[\mathrm{M}_2]. \tag{4}$$

This reaction is said to obey first-order kinetics because the rate of change depends linearly on the concentration of the species. The constants of proportionality, k_1 and k_{-1} are called rate constants and they have units of s^{-1}. Many protein reactions have successfully been described by one first order processes (*e.g.* McManus *et al.* 1988), though some reactions may not be describable in this way (Austin *et al.* 1975).

Almost inherent within the concept of a first order reaction is that it occurs *via* a high-energy intermediate, called a transition or activated state. If the transition state has a free energy, $G_{\rm a}$, then the rate constant for the transition is equal to

$$k_1 = A\exp\left[-\frac{\Delta G_{\rm a1}}{kT}\right] \qquad \Delta G_{\rm a1} = G_{\rm a} - G_1 \tag{5}$$

where A is a constant called the frequency factor. A similar expression holds for the reverse reaction so that the ratio,

$$\frac{k_1}{k_{-1}} = \exp\left[-\frac{\Delta G}{kT}\right] = \frac{[{\rm M}_2]}{[{\rm M}_1]} \tag{6}$$

accords with Boltzmann's law at equilibrium.

The activated-state concept makes specific predictions of how rate constants depend on external force. If the protein structures are very rigid and the transitions ${\rm M}_1 \rightarrow {\rm M}_{\rm a} \rightarrow {\rm M}_2$ are associated with displacements x_1, x_a, and x_2 *in the direction of the force*, F, then the energies of the states will be decreased by Fx_1, $Fx_{\rm a}$ and Fx_2 respectively. This implies that

$$k_1 \equiv A\exp\left[-\frac{\Delta G_{\rm a1} - F\Delta x_{\rm a1}}{kT}\right] = k_1^0\exp\left[\frac{F\Delta x_{\rm a1}}{kT}\right]$$

$$\Delta G_{\rm a1} = G_{\rm a} - G_1 \qquad \Delta x_{\rm a1} = x_{\rm a} - x_1. \tag{7}$$

An analogous expression holds for k_{-1}. Note that the ratio of the forward and reverse rate constants must give the correct force dependence for the equilibrium (Eq. (3)).

A useful way of thinking about the effect of force on the reaction rates is that it tilts the free energy diagram of the reaction (Fig. 9). If the displacement of the activated state is intermediate between the initial and final states ($x_1 < x_{\rm a} < x_2$), then a negative external force (a load) will slow the reaction, whereas a positive external force (a push) will accelerate the reaction. However, if $x_{\rm a} = x_1$–*i.e.* if the transition state is reactant-like–then force will have has little effect on the forward rate constant. On the other hand, if $x_{\rm a} = x_2$–*i.e.* if the transition state is product-like–then the force will have little effect on the reverse rate constant. If the displacement of the activated state is not intermediate between the initial and final states, it is even possible that a load could increase the forward rate constant (if $x_{\rm a} < x_1$), though in this case the backward rate would be increased even more.

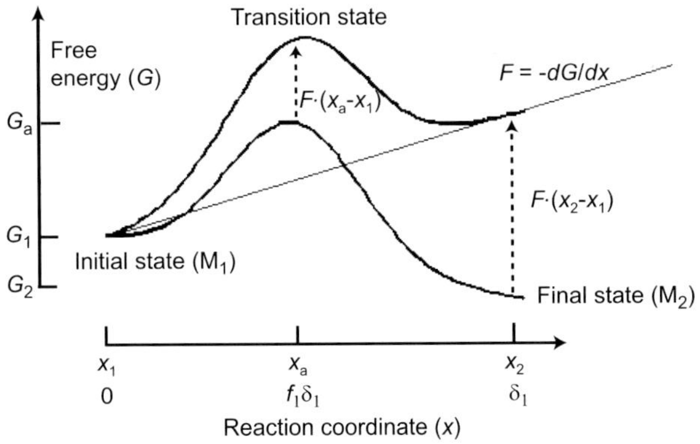

Fig. 9. Transition-state concept. A high energy barrier limits the rate of the transitions between the initial and final states of a chemical reaction.

10 Absolute rate theories

To predict the absolute rate of a biochemical reaction, a more detailed theory is needed in order to estimate the frequency factor A. Two such detailed theories are the Eyring rate theory and the Kramers rate theory. Both require that the reaction coordinate, the parameter that measures the progression of the reaction, be specified. If a protein changes overall length as a result of the $M_1 \rightarrow M_2$ transition, then we could make length the reaction coordinate, though, many other reaction coordinates are possible; indeed, the distance between any two atoms that move relative to one another during the reaction could be used as a reaction coordinate. If the protein is subject to a force, then the natural reaction coordinate is the length of the protein in the direction of the force.

The Eyring rate theory was originally introduced in the 1930s to describe reactions between small molecules such as the bimolecular reaction $2ClO \leftrightarrow Cl_2 + O_2$. The theory assumes that the reaction occurs *via* a transition state which is in equilibrium with the reactants. The transition state breaks down to the products when one of its molecular vibrations becomes a translation (Eyring & Eyring 1963; Moore 1972; Atkins 1986). The rate is

$$k_{\mathrm{E}} = \varepsilon \frac{kT}{h} \exp\left[-\frac{\Delta G_{\mathrm{a1}}}{kT}\right] \qquad \Delta G_{\mathrm{a1}} = G_{\mathrm{a}} - G_1 \tag{8}$$

where $\varepsilon \sim 1$ is an efficiency factor (equal to the probability of making the transition when in the transition state), and the frequency factor, kT/h, is equal to $\sim 6 \times 10^{12}\ \mathrm{s}^{-1}$, where h is the Planck constant, and G_{a} is the free energy of the activated state (ignoring the vibrational degree of freedom that breaks down). According to the Eyring equation, the absolute rate is the frequency factor reduced by the exponential term: assuming that the efficiency is unity, a reaction with a rate constant of 10^3 s^{-1} would have an activation free energy of 22 kT. The Eyring theory proved to be a better model than collision theory to describe the rates of small-molecule reactions in the gas phase. It might also apply to covalent changes of proteins and their ligands. But there is no reason to think that it would apply to global conformational changes of proteins in aqueous solution where there are many degrees of freedom and the motions are expected to be highly overdamped.

Kramers had a different view of the passage through the transition state (Kramers 1940). He assumed that the transition corresponded to a diffusive motion of a particle out of a potential well. This might correspond to the thermal fluctuation of two protein domains held together by a flexible region. Let the (reduced) mass be m, the damping equal γ, and the spring constant of the well be κ. If the motion is underdamped, then the Kramers expression for the rate is

$$k = \frac{\omega_0}{2\pi} \exp\left(-\frac{\Delta G_{\mathrm{a1}}}{kT}\right) \tag{9}$$

where $\omega_0 = \sqrt{\kappa/m}$ is the resonance frequency associated with the potential well. This is a generalization of the Eyring expression which holds for quantum mechanical and non-quantum cases (Haenggi *et al.* 1990). If the motion is overdamped, the rate is

$$k = \frac{1}{2\pi\tau^*} \exp\left(-\frac{\Delta G_{\mathrm{a1}}}{kT}\right) \tag{10}$$

where τ^* is the geometric mean of the damping time constant in the well (γ/κ) and the damping time constant at the top of the barrier (γ/κ^* where κ^* is the negative of the second derivative at the top of the energy barrier). Note that in the overdamped case, the pre-exponential factor depends on the shape of the energy barrier because the efficiency of the reaction (probability of making the transition) depends on the shape. If the peak is precipitous–if there is an absorbing boundary at the top of a harmonic well–then the rate constant in the overdamped case is

$$k_{\mathrm{K}} = \frac{1}{\pi\tau}\sqrt{\frac{\Delta G_{\mathrm{a1}}}{kT}} \exp\left[-\frac{\Delta G_{\mathrm{a1}}}{kT}\right]. \tag{11}$$

Because global conformational changes of proteins are expected to be overdamped (Howard 2001), the overdamped equations are more appropriate, and it is for the overdamped result that the Kramers paper is most widely appreciated (Haenggi *et al.* 1990). According to the overdamped Kramers rate theory, the frequency factor is approximately equal to the inverse of the relaxation time, $\tau = \kappa/\gamma$ where κ is the stiffness of the protein and γ is the damping from the fluid (the other terms are close to unity). This makes intuitive sense: we can think of the protein sampling a different energy level every τ seconds, because τ is the time over which the protein's shape becomes statistically uncorrelated. The protein can react only when it attains the energy of the transition state, and the probability of this occurring is proportional to $\exp(-\Delta G_{a1}/kT)$.

The Eyring and Kramers rate theories represent two extreme views of the mechanism of global conformational changes of proteins. In the Eyring model, the transition state is like the initial state. A sudden, local chemical change (such as the binding of a ligand or the chemical change in a bound ligand) creates a highly strained protein which subsequently relaxes into a new stable conformation. In the overdamped Kramers model, the transition state lies towards the final state on the reaction coordinate: after the protein has diffused into the transition state, the reaction proceeds and locks the conformational change in. The Kramers model is a "foot-in-door" mechanism, in the which the foot plays the role of the chemical change that prevents the opening from being reversed. Obviously these are two extreme cases, and in general the transition state could be anywhere in between the initial and final states (or even outside). The important point of all this is that the various mechanisms can be distinguished: *in the Eyring model the forward rate is independent of force whereas in the Kramers model the forward rate (and perhaps the reverse rate as well) will depend on force.* Thus force dependence offers a way to tell whether a transition is chemical in the sense that it is reached by localized fluctuations, or whether the transition is physical in the sense that it is reached by global spatial fluctuations. The latter, Kramers-type mechanisms have been termed a thermal ratchet mechanism (see below).

11 Role of thermal fluctuations in motor reactions

From a physical viewpoint, we expect thermal fluctuations to play crucial roles in the motor reaction because thermal forces at the molecular scale are large compared to the average directed forces that motor proteins generate. For example, an unconstrained protein will diffuse through a distance equal to its own size, ~10 nm, in tens of microseconds; this is three orders of magnitude faster than the duration of the ATP hydrolysis cycle, which is typically 10 to 100 ms (motors typically hydrolyze 10–100 ATP per second while they are moving, Table 1). Indeed, it is their noisy, diffusive environment that distinguishes molecular machines from the man-made machines of our everyday world. From a chemical viewpoint, we also expect thermal fluctuations to play crucial roles: all chemical reactions require thermal energy in order for a molecule to enter the transition state (see last section), and the chemistry of the hydrolysis reaction is no exception. The challenge is to merge these two views.

An extreme model is that movement of a motor to its next binding site on the filament is purely diffusive, and the role of ATP hydrolysis is to somehow rectify diffusion so that motion in the wrong direction is blocked (Braxton 1988; Braxton & Yount 1989; Vale & Oosawa 1990). This is reminiscent of the pawl and ratchet discussed by Feynman (Feynman *et al.* 1963): if a pawl (the motor) and ratchet (filament) are at different temperatures then they can do work without violating the second law of thermodynamics. But motor proteins cannot literally be heat engines because the diffusion of heat is so rapid over molecular dimensions that thermal gradients will dissipate within picoseconds, much faster than the timescale of the biochemical reactions (Howard 2001). Instead of being driven by temperature differences like heat engines, motors are driven by chemical reactions (ATP hydrolysis) that are out of equilibrium.

These ratchet ideas have inspired a number of papers in the physics literature that explore the fundamental requirements of directed motility (Ajdari & Prost 1992; Magnasco 1993; Rousselet *et al.* 1994; Astumian & Bier 1994; Zhou & Chen 1996). Some of the physical models are very specific, and fail to explain how real motor proteins move. For example, Astumian and Bier proposed a mechanism in which a motor alternates between times when it diffuses along a filament and times when it is subject to an asymmetric energy profile (a ratchet). However, such a purely diffusive, thermal ratchet model fails in two respects. First, the maximum force that it can generate against viscous loads is small, only $2\ kT/\delta$, where δ is the distance between binding sites (Hunt *et al.* 1994). But for kinesin, which has $\delta = 8$ nm, this force is only 1 pN, much less than the measured value of 4 to 5 pN (Hunt *et al.* 1994). And second, because the motor diffuses in the right direction only half the time, it is expected that two molecules of ATP would be hydrolyzed for each forward step. But for kinesin, there is one step per ATP hydrolyzed (Coy *et al.* 1999; Iwatani *et al.* 1999). While they may fail in some specific cases, the physical models have never-the-less been very important because they have clearly defined the two conditions that are necessary for directed motion: spatial asymmetry (a ratchet), which for motor proteins arises from their stereospecific binding to polar filaments; and temporal asymmetry which arises from an out-of-equilibrium chemical reaction (*e.g.* ATP hydrolysis).

A less extreme picture of motility is that only part of the motor diffuses to the next binding site, rather than the whole motor. In other words, the motor is elastic, like a spring, and picks up mechanical energy *via* thermal fluctuations; once it is "cocked" it can then bind to the filament and generate force. This was the picture in the original A.F. Huxley model of muscle contraction (Huxley 1957a), and it forms the basis of other ratchet models (Cordova *et al.* 1992). If the cocking of the spring requires a global conformation change, then the diffusion time will be limited by the stiffness of the spring and the damping from the fluid (as well as possible internal damping within the protein). This a Kramers-type mechanism. The Huxley model was abandoned in the 1970s (Huxley & Simmons 1971; Eisenberg & Hill 1978; Eisenberg *et al.* 1980) in favor of a mechanism in which a local conformational change in the nucleotide-binding pocket drives the subsequent global conformational change. This is an Eyring-type mechanism. The Kramers-type mechanism was abandoned because it was argued that it would take too long for myosin to pick up an appropriate amount of mechanical energy by diffusion; but if a more reasonable mechanical efficiency of muscle is assumed (~50%) then the diffusion time is not prohibitive (Hunt *et al.* 1994).

Thermal energy always plays a crucial role in chemical reactions. In the local, Eyring-type model, thermal fluctuations are still needed to get the molecule into the transition state for the localized structural change. The advantage of the local mechanism is that it is faster: because a short lever is stiffer than a longer one and because the damping on a small domain is less than the damping on a large domain, high-energy states can be reached much more quickly through localized conformational changes. For example, if the lever ratio is 10, then a local conformational change occurs 1000-times faster than a global change (Howard 2001). The extent to which force-generating protein conformational changes are due to diffusive global conformational changes that are locked in by chemical changes, or to localized chemical changes that drive global changes must be determined experimentally by measuring the dependence of the rates on force.

12 A mechanochemical model for kinesin

To illustrate how these concepts apply to motor proteins, I present a model for the chemomechanical cycle of kinesin. Structural, biochemical and biophysical experiments suggest that the two heads of kinesin alternate in their binding to the microtubule, first one head leading and then the other (Fig. 10, Schief & Howard 2001, but see Hua *et al.* 2002). For simplicity, we assume that the ADP and P_i concentrations are zero (as is approximately the case in the *in vitro* assays). This gives the following kinetic equation

$$\mathrm{M} \underset{k_{-1}}{\overset{k_1 T}{\leftrightarrow}} \mathrm{M \cdot D \cdot P} \xrightarrow{k_2} \mathrm{M \cdot P} \xrightarrow{k_3} \mathrm{M} \qquad (12)$$

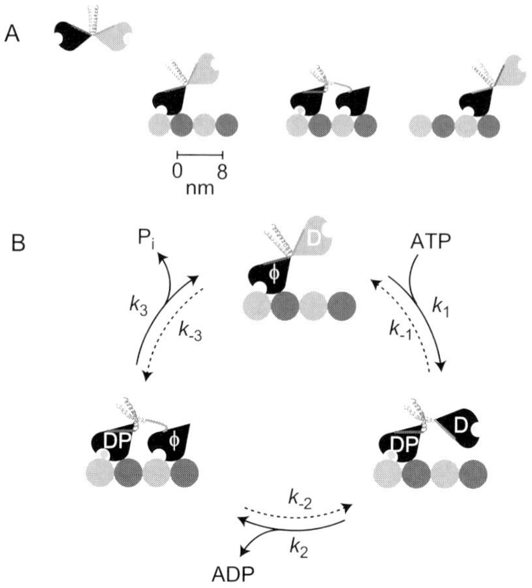

Fig. 10. Hand-over-hand model for kinesin. T = ATP, D = ADP, Pi = phosphate, ϕ = no nucleotide.

where M symbolizes the motor, T the ATP, D the ADP, and P the phosphate ion.

The solution to this equation is

$$\nu = \delta \cdot k_{\mathrm{ATPase}} = \delta \cdot k_{\mathrm{cat}} \frac{[\mathrm{T}]}{K_{\mathrm{M}} + [\mathrm{T}]} \qquad k_{\mathrm{cat}} = \frac{k_2 k_3}{k_2 + k_3} \qquad K_{\mathrm{M}} = \frac{k_3(k_2 + k_{-1})}{k_1(k_2 + k_3)} \tag{13}$$

where ν is the average speed of movement, k_{ATPase} is the ATPase rate, and δ is the distance per ATP (8 nm for kinesin).

In accordance with our discussion above, we model the force-dependence by assuming that the transitions between states are associated with structural changes through distances $\delta_i = x_{i+1} - x_i$, so that the rates depend on a *load*, F, in the opposite direction to δ_i, according to

$$k_{+i} = k_{+i}^0 \exp\left[-f_i \frac{F\delta_i}{kT}\right]$$

$$k_{-i} = k_{-i}^0 \exp\left[(1 - f_i) \frac{F\delta_i}{kT}\right] \tag{14}$$

where f_i is the location of the transition state as a fraction of the distance between the i and $i+1$ states, and k_{+i}^0 and k_{-i}^0 are the rate constants in

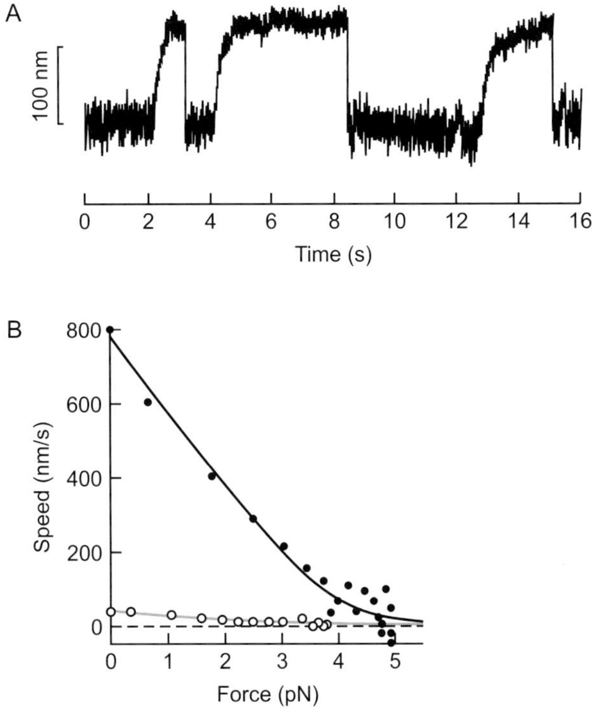

Fig. 11. A. Three examples of a kinesin molecule walking along a microtubule at a high ATP concentration. The molecule is pulling against a flexible glass fiber and stalls when a maximum force of ~5 pN is reached. B. The speed as a function of the force at high (filled circles) and low (open circles) ATP concentration. The curves correspond to the model in the text.

the absence of load (Fig. 10). Note that $\delta_1 + \delta_2 + \delta_3 = 8$ nm, the step size. Normally we think that f_i lies between 0 and 1, but this is not an absolute requirement. For the hand-over-hand model, we make the simple assumption that all $f_i = 1$, meaning that the transition state is displaced all the way towards the final state. This puts all the force sensitivity in the forward step, like a Kramers-type mechanism. The solid curves in Figure 11 are generated with $\delta_1 = 1$ nm, $\delta_2 = 1$ nm, $\delta_3 = 6$ nm; $k_1^0 = 100\ \mu\mathrm{M}^{-1}.\mathrm{s}^{-1}$, $k_{-i}^0 = 3000\ \mathrm{s}^{-1}$, $k_2^0 = 105\ \mathrm{s}^{-1}$, and $k_3^0 = 5000\ \mathrm{s}^{-1}$.

13 Conclusions and outlook

The agreement between the model curve and the experimental data in Figure 11 shows that the ideas presented in this review are plausible: the motor reaction can be described as a small number of first order chemical reactions in which force effects the rate constants through a Boltzmann-type exponential prefactor. However, the picture is very incomplete. What is the reaction coordinate for a molecular motor? How does diffusion and thermal agitation drive a protein's moving parts during the transition from one structural state to another? There are clearly many conceptual questions to be addressed, in addition to the practical ones. It is likely that ideas from protein folding will prove useful, and that experiments using single-molecule techniques will become increasingly useful for probing the motor reaction.

Study of the mechanics of motor proteins has given us a deeper understanding of biological force generation, and in particular how mechanical, chemical, and thermal forces act on proteins. Because conformational changes of proteins, RNA, and DNA are driven by the mutual action of forces, it is likely that the principles learned from the interactions of motors with their filaments should have wide application to other macromolecular machines.

References

[1] A. Ajdari and J. Prost, *C.R. Acad. Sci. Paris II.* **315** (1992) 1635.
[2] B. Alberts, *Cell.* **92** (1998) 291-294.
[3] B. Alberts, A. Johnson, J. Lewis, M. Raff, K. Roberts and P. Walter, *Molecular biology of the cell*, 4th Ed. (Garland Pub., New York, 2002).
[4] R.D. Allen, N.S. Allen and J.L. Travis, *Cell Motil.* **1** (1981) 291-302.
[5] C.M. Armstrong and F. Bezanilla, *J. Gen. Physiol.* **63** (1974) 533-552.
[6] R.D. Astumian and M. Bier, *Phys. Rev. Lett.* **72** (1994) 1766-1769.
[7] P.W. Atkins, Physical chemistry, 3rd Ed. (W.H. Freeman, New York, 1986).
[8] R.H. Austin, K.W. Beeson, L. Eisenstein, H. Frauenfelder and I.C. Gunsalus, *Biochemistry* **14** (1975) 5355-5373.
[9] M. Benoit, D. Gabriel, G. Gerisch and H.E. Gaub, *Nat. Cell Biol.* **2** (2000) 313-317.
[10] S.T. Brady, **317** (1985) 73-75.
[11] S.M. Braxton, Synthesis and use of a novel class of ATP carbamates and a ratchet diffusion model for directed motion in muscle (Ph.D. Washington State University, Pullman, 1988).
[12] S.M. Braxton and R.G. Yount, *Biophys. J.* **55** (1989) 12a.
[13] D. Bray, Cell movements, 2nd Ed. (Garland Pub., New York, 2000).
[14] N.J. Cordova, B. Ermentrout and G.F. Oster, *Proc. Natl. Acad. Sci. U.S.A.* **89** (1992) 339-343.
[15] D.L. Coy, M. Wagenbach and J. Howard, *J. Biol. Chem.* **274** (1999) 3667-3671.
[16] A.A. Deniz, T.A. Laurence, G.S. Beligere, M. Dahan, A.B. Martin, D.S. Chemla, P.E. Dawson, P.G. Schultz and S. Weiss, *Proc. Natl. Acad. Sci. U.S.A.* **97** (2000) 5179-5184.

[17] E. Eisenberg and T.L. Hill, *Prog. Biophys. Mol. Biol.* **33** (1978) 55-82.

[18] E. Eisenberg, T.L. Hill and Y. Chen, *Biophys. J.* **29** (1980) 195-227.

[19] H. Eyring and E.M. Eyring, Modern chemical kinetics (Ed. Reinhold Pub. Corp. New York, 1963).

[20] R.P. Feynman, R.B. Leighton and M.L. Sands, The Feynman lectures on physics (Ed. Addison-Wesley Pub. Co., Reading, Mass., 1963).

[21] T. Funatsu, Y. Harada, M. Tokunaga, K. Saito and T. Yanagida, *Nature* **374** (1995) 555-559.

[22] I.R. Gibbons, *Proc. Natl. Acad. Sci. U.S.A.* **50** (1963) 1002-1010.

[23] P. Haenggi, T. Talkner and B. Borkovec, *Rev. Mod. Phys.* **62** (1990) 251-341.

[24] B. Hille, Ionic channels of excitable membranes, 2nd Ed. Sinauer Associates (Sunderland, Mass, 1992).

[25] A.L. Hodgkin, The conduction of the nervous impulse (Ed. Liverpool University Press, Liverpool, 1964).

[26] K.C. Holmes, D. Popp, W. Gebhard and W. Kabsch, *Nature* **347** (1990) 44-49.

[27] J. Howard, Mechanics of Motor Proteins and the Cytoskeleton (Ed. Sinauer Associates, Sunderland, MA, 2001).

[28] J. Howard, A.J. Hudspeth and R.D. Vale, *Nature* **342** (1989) 154-158.

[29] W. Hua, J. Chung and J. Gelles, *Science* **295** (2002) 844-848.

[30] A.J. Hunt, F. Gittes and J. Howard, *Biophys. J.* **67** (1994) 766-781.

[31] A.F. Huxley, *Prog. Biophys. Biophys. Chem.* **7** (1957a) 255-318.

[32] A.F. Huxley and R. Niedergerke, *Nature* **173** (1954) 971-976.

[33] A.F. Huxley and R.M. Simmons, *Nature* **233** (1971) 533-538.

[34] H.E. Huxley, *J. Biophys. Biochem. Cytol. (J. Cell Biol.)* **3** (1957b) 631-648.

[35] H.E. Huxley and J. Hanson, *Nature* **173** (1954) 973-976.

[36] S. Iwatani, A.H. Iwane, H. Higuchi, Y. Ishii and T. Yanagida, *Biochemistry* **38** (1999) 10318-10323.

[37] W. Kabsch, H.G. Mannherz, D. Suck, E.F. Pai and K.C. Holmes, *Nature* **347** (1990) 37-44.

[38] H.A. Kramers, *Physica* **7** (1940) 284-304.

[39] W. Kühne, Wagenschieber, H. Kühne, C.A. Kofoid and Lees Museum, Untersuchungen über das Protoplasma und die Contractilität (Ed. W. Engelmann, Leipzig, 1864).

[40] F.J. Kull, E.P. Sablin, R. Lau, R.J. Fletterick and R.D. Vale, *Nature* **380** (1996) 550-555.

[41] J. Liphardt, B. Onoa, S.B. Smith, I.J. Tinoco and C. Bustamante, *Science* **292** (2001) 733-737.

[42] J. Lowe and L.A. Amos, *Nature* **391** (1998) 203-206.

[43] K. Luby-Phelps, P.E. Castle, D.L. Taylor and F. Lanni, *Proc. Natl. Acad. Sci. U.S.A.* **84** (1987) 4910-4913.

[44] M.O. Magnasco, *Phys. Rev. Lett.* **71** (1993) 1477-1481.

[45] O.B. McManus, D.S. Weiss, C.E. Spivak, A.L. Blatz and K.L. Magleby, *Biophys. J.* **54** (1988) 859-870.

[46] H. Miki, M. Setou, K. Kaneshiro and N. Hirokawa, *Proc. Natl. Acad. Sci. U.S.A.* **98** (2001) 7004-7011.

[47] W.J. Moore, Physical chemistry, 4th Ed (Prentice-Hall, Englewood Cliffs, N.J., 1972).

[48] H. Nagashima and S. Asakura, *J. Mol. Biol.* **136** (1980) 169-182.

[49] E. Neher and B. Sakmann, *Nature* **260** (1976) 799-802.

[50] E. Nogales, M. Whittaker, R.A. Milligan and K.H. Downing, *Cell.* **96** (1999) 79-88.

[51] E. Nogales, S.G. Wolf and K.H. Downing, [published erratum appears in Nature 1998 May 14; 393 (6681): 191]. *Nature* **391** (1998) 199-203.

[52] H. Noji, R. Yasuda, M. Yoshida and K. Kinosita Jr., *Nature* **386** (1997) 299-302.

[53] I. Rayment, W.R. Rypniewski, K. Schmidt-Base, R. Smith, D.R. Tomchick, M.M. Benning, D.A. Winkelmann, G. Wesenberg and H.M. Holden, *Science* **261** (1993) 50-58.

[54] J. Rousselet, L. Salome, A. Ajdari and J. Prost, *Nature* **370** (1994) 446-448.

[55] B. Sakmann and E. Neher, Single-channel recording, 2nd Ed. (Plenum Press, New York, 1995).

[56] Y. Sako, S. Minoghchi and T. Yanagida, *Nat. Cell. Biol.* **2** (2000) 168-172.

[57] W. R. Schief and J. Howard, *Curr. Opin. Cell. Biol.* **13** (2001) 19-28.

[58] J.M. Scholey, Motility assays for motor proteins. (Ed. Academic Press, San Diego, Calif., 1993).

[59] G.J. Schutz, G. Kada, V.P. Pastushenko and H. Schindler, *Embo. J.* **19** (2000) 892-901.

[60] M.P. Sheetz and J.A. Spudich, *Nature* **303** (1983) 31-35.

[61] J.A. Spudich, S.J. Kron and M.P. Sheetz, *Nature* **315** (1985) 584-586.

[62] F.B. Straub, *Stud. Inst. Med. Chem. Univ. Szeged.* **2** (1941-2) 3-15.

[63] K. Svoboda, C.F. Schmidt, B.J. Schnapp and S.M. Block, *Nature* **365** (1993) 721-727.

[64] A. Szent-Gyorgyi, *Stud. Inst. Med. Chem. of Szeged.* **1** (1941) 17-26.

[65] A. Szent-Gyorgyi, *Stud. Inst. Med. Chem. Univ. Szeged.* **1** (1941-2) 67-71.

[66] R.D. Vale, T. Funatsu, D.W. Pierce, L. Romberg, Y. Harada and T. Yanagida, *Nature* **380** (1996) 451-453.

[67] R.D. Vale and F. Oosawa, *Adv. Biophys.* **26** (1990) 97-134.

[68] R.D. Vale, T.S. Reese and M.P. Sheetz, *Cell.* **42** (1985) 39-50.

[69] M.D. Wang, M.J. Schnitzer, H. Yin, R. Landick, J. Gelles and S.M. Block, *Science* **282** (1998) 902-907.

[70] G.J. Wuite, S.B. Smith, M. Young, D. Keller and C. Bustamante, *Nature* **404** (2000) 103-106.

[71] J. Yajima, M.C. Alonso, R.A. Cross and Y.Y. Toyoshima, *Curr. Biol.* **12** (2002) 301-306.

[72] T. Yanagida, M. Nakase, K. Nishiyama and F. Oosawa, *Nature* **307** (1984) 58-60.

[73] H.X. Zhou and Y.D. Chen, *Phys. Rev. Lett.* **77** (1996) 194-197.

COURSE 3

MODELLING MOTOR PROTEIN SYSTEMS

T. DUKE

Cavendish Laboratory,
Madingley Road,
Cambridge CB3 0HE, UK

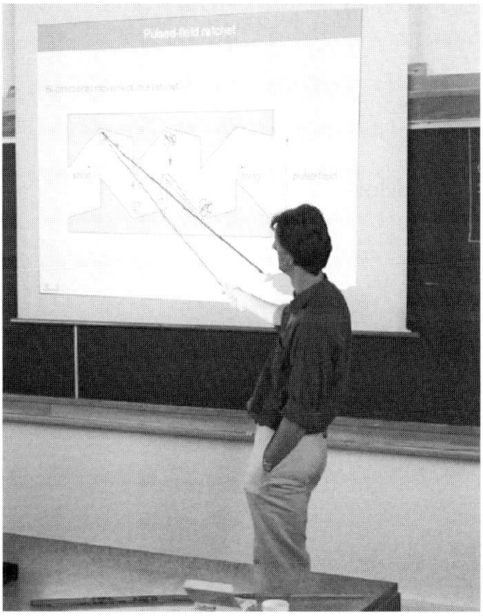

Contents

MODELLING MOTOR PROTEIN SYSTEMS

T. Duke

Abstract

Motor proteins are the cell's workforce. They are specialized molecules which convert chemical energy to mechanical work, thereby generating force and directional motion. In some situations motor proteins act individually, but more commonly they cooperate in large ensembles to accomplish cellular functions. How do molecular motors work? And how do they work *together*? This lecture course recounts theoretical approaches to these questions, which complement the experimental investigations of motor protein systems described in Jonathon Howard's course. The methods of non-equilibrium statistical mechanics permit a general analysis of how chemical energy can most effectively be used to generate the movement of an individual motor. The class of models known as "isothermal ratchets", in particular, is a powerful tool for discussing the principles of energy transduction. More specific kinetic models, such as the "swinging lever-arm" model which is based on the known structure and chemistry of the myosin protein, indicate how these molecules are designed to work efficiently together to drive the contraction of muscle. Both classes of model indicate what types of collective effects can arise when many molecules operate in concert. Cooperative interactions within a team of motor proteins may lead to dynamical instabilities and hysteretic behaviour, which can be exploited to generate oscillations. Physiological processes which may rely on such instabilities include the vibration of insect flight muscle and the undulation of spermatozoid flagella. Motor proteins also play functional roles in sensory systems – hearing being a particularly intriguing example. Hair bundles, which are the mechano-sensors that detect motion in the inner ear, have been found to vibrate spontaneously. Motor proteins appear to play a role either in generating the oscillations, or in maintaining the bundles at the threshold of the oscillatory instability. Poised on the verge of vibrating, they are especially responsive to faint sounds.

1 Making a move: Principles of energy transduction

1.1 Motor proteins and Carnot engines

Molecular motors' principal role in biological cells is the generation of force and motion [1–4]. The basis of their function is thus the transduction of chemical energy to mechanical work. When trying to understand how this is achieved, the most important consideration to bear in mind is the tiny size of these molecular machines. Motor proteins live in the Brownian domain [5–14]. Their operation is based on individual chemical reactions, which are accompanied by energy changes that do not greatly exceed the thermal energy. As a result, they function in a highly stochastic way. The consequences are strange. For example, motor proteins cannot be held still in one place, owing to the continual bombardment by molecules in the surrounding solution. Conversely, molecular motors cannot exert a sustained force without the continual consumption of chemical energy.

In this unfamiliar Brownian realm, our knowledge of how macroscopic engines work, based on the laws of classical thermodynamics, is of little help. Molecular motors are not miniature Carnot engines [8,12]. Nevertheless, the microscopic model of a Carnot engine invented by Feynman [15] serves as a useful starting point, because it introduces the notion of a microscopic *ratchet*. In our everyday experience, ratchets are devices that are used to rectify motion – to ensure that movement in one sense is permitted and motion in the opposite sense is denied. Feynman asked whether such a device, contructed on the microscopic scale, could be used to rectify the fluctuations caused by Brownian motion. His machine consisted of a rotating vein connected to a cogwheel with saw-teeth, whose movement was restricted by a pawl. He demonstrated that the random percussion of molecules on the vein could drive the cog round, but only on one condition: the cog must be enveloped by molecules at a different temperature to those surrounding the vein. Feynman's *thermal ratchet* operates exactly like a Carnot engine. It can convert heat to mechanical work provided that a difference in temperature between cog and vein is maintained. Might a motor protein function the same way? Could a molecular motor use an exothermic chemical reaction to produce heat and raise the temperature of the surroundings, and then use the difference in temperature to drive a microscopic Carnot engine? The answer is no. At the scale of individual molecules, thermal diffusion is far too rapid to permit the maintenance of a temperature gradient. Molecular motors must therefore function *isothermally*. They must directly transduce chemical energy to mechanical work, without passing through the intermediary of heat.

1.2 Simple Brownian ratchet

Can microscopic ratchets be employed to rectify Brownian motion at a single, fixed temperature? To demonstrate that they can, it suffices to provide an illustration. Perhaps the simplest Brownian ratchet consists of a particle moving in one dimension in a periodic but asymmetric potential which is switched on and off at regular intervals of time [16–19]. Consider, for example, the saw-tooth potential in Figure 1, characterized by asymmetry factor γ and periodicity a.

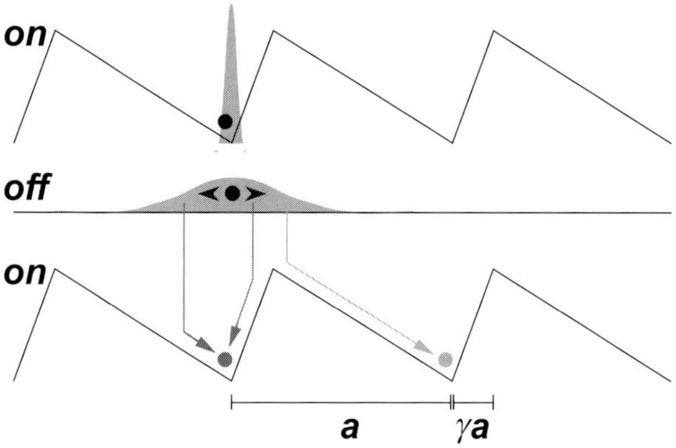

Fig. 1. Movement of a Brownian particle in a ratchet potential which is switched on and off at regular intervals of time. The particle, which is localized when the potential is on, starts to diffuse freely when the potential is turned off and has a Gaussian probability distribution (shown in grey). When the potential is switched back on, the particle slides down the potential gradient. Owing to the asymmetry, there is a higher probability that it hops to the right than to the left.

If the peak of the potential is considerably higher than the thermal energy, the particle tends to be localized at the bottom of one of the wells when the potential is on. But when the potential is switched off, the particle is free to diffuse by Brownian motion. Suppose that the diffusion coefficient is D and that the potential is turned on again after time T. Then if $\gamma^2 < DT/a^2 < (1-\gamma)^2$, there is quite a high probability that the particle will drop into the next potential well to the right, but only a small chance that it will fall to the left. Thus with continued pulsing of the potential at intervals of time T, the particle moves noisily but, on average, it drifts to the right. The Brownian motion has been rectified. Yet no external force has been applied – the potential is globally flat at all times.

Of course, the laws of thermodynamics have not been violated here; in order to generate directional motion, an external agent has had to do work.

There is an input of energy each time that the potential is switched on, raising the potential energy of the particle. This energy is subsequently transferred to the environment by viscous dissipation as the particle drifts towards the bottom of the well. Thus the Brownian motion has been rectified at the cost of transforming some mechanical work to heat.

1.3 Polymerization ratchet

We would like to develop a general formalism to study the movement of a Brownian particle in a ratchet potential. An illustrative example which serves as a useful starting point is the process of polymerization of a filament from a large pool of monomers. This is a stochastic process which is driven in the direction of polymer extension by the negative free energy change, ΔG, which occurs each time a monomer is added. If a mechanical force F is applied to the tip of the polymer, however, the growth can be slowed, or even reversed [20, 21]. Because the insertion of a monomer of size a at the tip requires the energy Fa, the thermodynamic principle of detailed balance imposes a condition on the rates $r^{\pm}$ at which monomers bind and unbind at the tip:

$$r^+/r^- = \mathrm{e}^{-(Fa+\Delta G)/kT}. \tag{1.1}$$

The growth velocity $v = a(r^+ - r^-)$ can be determined if the two rates r^+ and r^- are known independently. These rates, however, depend on the energy barriers which the system must overcome in order to insert or detach a monomer – barriers which depend on microscopic details of the polymerization process and which are not fixed by thermodynamic laws. Suppose that insertion of a monomer at the end requires passage over a potential barrier $U(F)$ which depends on the external force. If $U \gg kT$, polymerization events are thermally activated with the Arrhenius rate $r^+ = r_0\mathrm{e}^{-U/kT}$ and the growth velocity is

$$v = r_0 a\, \mathrm{e}^{-U(F)/kT}\left[1 - \mathrm{e}^{-(Fa+\Delta G)/kT}\right]. \tag{1.2}$$

Two important conclusions can be drawn: Thermodynamic laws only determine the stall force for which the velocity vanishes, $F_0 = -\Delta G/a$; and the form of the force-velocity relationship can vary widely, depending on the unknown function $U(F)$ which determines the strain-dependence of the polymerization reaction.

Kramers [22] introduced the notion that an individual chemical reaction can be modelled as the stochastic motion of a Brownian particle over a potential barrier. Extending his approach, we see that polymerization in the absence of a force is formally equivalent to stochastic motion in an

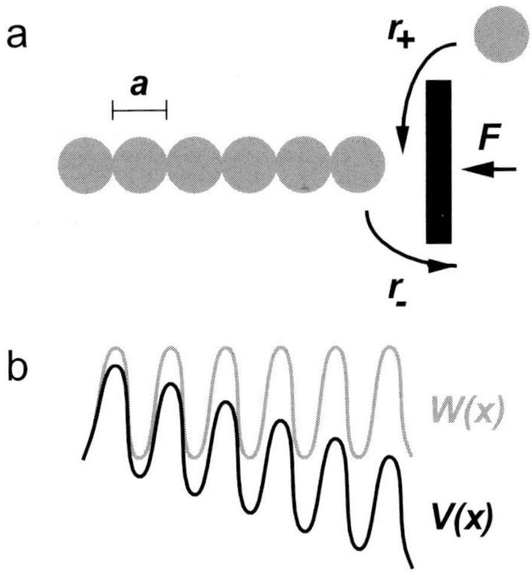

Fig. 2. a) Polymerization of a filament under the constraint of an applied force. **b)** The stochastic polymerization process may be mapped onto the Brownian motion of a particle in a potential $V(x)$, which is the sum of a periodic potential $W(x)$ and a uniform potential gradient proportional to the polymerization energy ΔG.

effective energy landscape

$$V(x) = W(x) - \frac{\Delta G}{a}x, \tag{1.3}$$

where $W(x)$ is a periodic function (which is related to the potential barrier) and x is a position variable along a linear axis (Fig. 2). The polymers of different length correspond to the local minima of $V(x)$. A particle moving stochastically in this potential would drift downhill, corresponding to the direction of polymerization. When an opposing force is applied, the total energy of the corresponding particle changes to $V(x) + Fx$. It can be seen that this becomes a periodic function when $F = F_0 = -\Delta G/a$, at which point there is no net drift of the particle and, correspondingly, polymerization is halted.

The motion of a Brownian particle in a tilted periodic potential, and simulataneously under the influence of an external force F, is most conveniently discussed using the Smoluchowski equation. The probability density $P(x,t)$ of finding the system at position x at time t obeys

$$\begin{aligned} \partial_t P &= -\partial_x J \\ J &= -\frac{1}{\zeta}\left[(\partial_x V + F)P + kT\partial_x P\right], \end{aligned} \tag{1.4}$$

where J is the current. The stationary solution which satisfies $\partial_t P^{\rm stat} = 0$ corresponds to constant current J and is given by

$$P^{\rm stat}(x) = {\rm e}^{-\frac{V(x)+Fx}{kT}} \left[A + \frac{J}{D} \int_0^x {\rm d}y {\rm e}^{\frac{V(y)+Fy}{kT}} \right], \tag{1.5}$$

where A is a constant of integration and $D = kT/\zeta$ is the diffusion coefficient. Normalization (one particle per period) and the condition of periodicity determine both A and the current J. The average velocity $v = Ja$ is

$$v = \frac{Da[1 - {\rm e}^{-(Fa+\Delta G)/kT}]}{\int_0^a {\rm d}x \, {\rm e}^{-\frac{V(x)+Fx}{kT}} \int_x^{x+a} {\rm d}y \, {\rm e}^{\frac{V(y)+Fy}{kT}}}, \tag{1.6}$$

which is the same as equation (1.2), with an effective energy barrier

$$U(F) = kT \ln \left(\int_0^a \frac{{\rm d}x}{a} \, {\rm e}^{-\frac{W(x)+(F-F_0)x}{kT}} \int_x^{x+a} \frac{{\rm d}y}{a} \, {\rm e}^{\frac{W(y)+(F-F_0)y}{kT}} \right) \tag{1.7}$$

and attempt rate $r_0 = D/a^2$. If the form $W(x)$ of the energy barrier is known, then the strain-dependence $U(F)$ can be calculated and the force-velocity relation determined. For example, if $W(x)$ has minima at $x = Na$, and sharply peaked maxima at $x = (N + \gamma)a$, then $U(F) \approx U(0) + \gamma aF$. Force-velocity relations corresponding to different forms of $W(x)$ are shown in Figure 3.

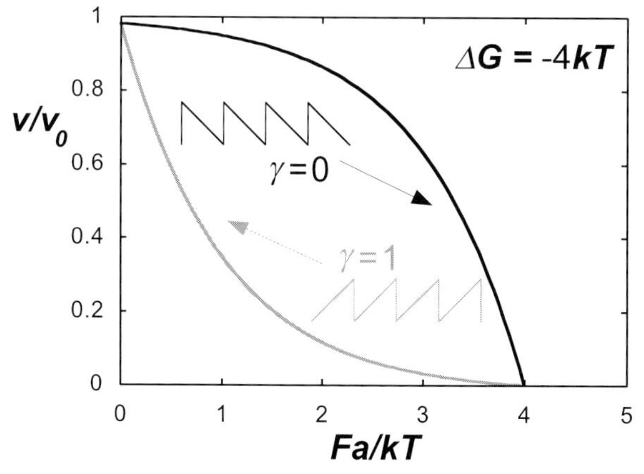

Fig. 3. Growth velocity of a polymer as a function of the restraining force, for two different realizations of the periodic potential $W(x)$, whose form is indicated.

1.4 Isothermal ratchets

The reason for the spontaneous motion of the Brownian particle in the polymerization ratchet is clear; it occurs because of the global tilt of the potential which, in turn, corresponds to the non-equilibrium nature of the polymerization reaction. Yet we have seen, in the preceding example, that directional movement can also occur in a globally flat potential, provided that the potential changes with time. We shall now explore what conditions must be satisfied for such motion to occur. More specifically, we shall focus our attention on a Brownian particle that can exist in a number of different states. In each state, the particle experiences a periodic potential $W_i(x) = W_i(x+a)$; and the particle makes stochastic transitions from state i to state j at rate $\omega_{ij}(x)$. Systems of this type, which generate motion in the absence of an external force, and at a constant temperature, are called *isothermal ratchets* [12].

The dynamics of a particle in an isothermal ratchet may be described by generalized Smoluchowski eqations

$$\begin{aligned} \partial_t P_i &= -\partial_x J_i + \sum_j (\omega_{ji} P_j - \omega_{ij} P_i) \\ J_i &= -\frac{1}{\zeta} \left[(\partial_x W_i + F) P_i + kT \partial_x P_i \right] \end{aligned} \tag{1.8}$$

where $P_i(x,t)$ denotes the probability of finding the particle at time t at position x and in state i. The total probability $P \equiv \sum_i P_i$ and total current $J \equiv \sum_i J_i$ satisfy an ordinary Smoluchowski equation of the type of equation (1.4), but with an effective potential:

$$J = -\frac{1}{\zeta} \left[(\partial_x V_{\text{eff}} + F) P + kT \partial_x P \right]. \tag{1.9}$$

This can be shown by introducing the local fractions of occupation in conformation i

$$\lambda_i(x) \equiv \frac{P_i(x)}{P(x)}, \tag{1.10}$$

which describe how P is divided amongst the different states. Expressing the total current as

$$J = -\sum_i \frac{1}{\zeta} \left[(\lambda_i \partial_x W_i + kT \partial_x \lambda_i + \lambda_i F) P + kT \lambda_i \partial_x P \right] \tag{1.11}$$

we find

$$V_{\text{eff}}(x) = \sum_i \int_0^x \mathrm{d}x' \lambda_i \partial_x W_i. \tag{1.12}$$

The system generates spontaneous motion (*i.e.* motion in the absence of the external force F) if the effective potential is globally tilted. This tilt can be characterized by the effective energy difference per period

$$\Delta G_{\text{eff}} = \sum_i \int_0^a \mathrm{d}x' \lambda_i \partial_x W_i. \tag{1.13}$$

In order for ΔG_{eff} to be nonzero, two conditions must be satisfied: (i) the system as a whole, *i.e.* the potentials $W_i(x)$ and/or the transition rates $\omega_{ij}(x)$, must be *asymmetric* with respect to $x \to -x$; (ii) the system must be *out of thermodynamic equilibrium.*

If the system is completely symmetric, then so are the functions $\lambda_i(x)$ and the integral in equation (1.13) vanishes by symmetry. No spontaneous motion occurs. If the system is in thermodynamic equilibrium, the transition rates satisfy the condition of detailed balance

$$\frac{\omega_{ij}}{\omega_{ji}} = \mathrm{e}^{(W_i - W_j)/kT}. \tag{1.14}$$

In this case, the steady state solution is given by the Boltzmann distribution $P_i^{\text{stat}} \sim \mathrm{e}^{-W_i/kT}$, and the integrand in equation (1.13) is the derivative of the periodic function $-kT \ln(\Sigma \mathrm{e}^{-W_i/kT})$, so $\Delta G_{\text{eff}} = 0$. Again, no motion occurs.

To summarize, the two conditions for spontaneous motion in an isothermal ratchet are *broken spatial symmetry* and *broken detailed balance.*

1.5 Motor proteins as isothermal ratchets

Motor proteins interact with cytoskeletal filaments. Because these are polar polymers, this interaction may be characterized as an asymmetric periodic potential, with the periodicity equal to the monomer size a. Molecular motors also hydrolyse ATP and can therefore exist in a number of distinct chemical states, depending on whether or not nucleotide is bound (and which type of nucleotide is bound). Crucially, the chemical state of the motor modulates the interaction between the motor and the filament. It is therefore natural to model a molecular motor as a particle in an isothermal ratchet. In each state i (identified with the chemical state of the motor) the particle experiences a different potential $W_i(x)$ (identified with the motor-filament interaction). A simple illustration is given in Figure 4.

As the hydrolysis reaction proceeds around the cycle, the laws of thermodynamics state that the transition rates satisfy

$$\frac{\omega_{i\to i+1}}{\omega_{i+1\to i}} = \mathrm{e}^{-(W_i - W_{i+1})/kT} \Theta_{i\to i+1} \tag{1.15}$$

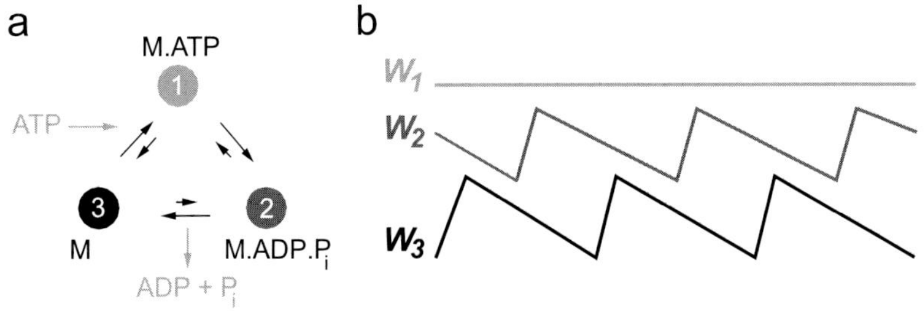

Fig. 4. a) The hydrolysis cycle of a typical molecular motor. The motor passes through a number of distinct chemical states in which a different nucleotide (or no nucleotide) is bound. **b)** The interaction potential between the motor and the cytoskeletal filament depends on the chemical state. In this case the motor in state 1 (with ATP bound) is detached from the filament, and the other two states are bound to the filament.

where

$$\Theta_{i\to i+1} = \mathrm{e}^{-\Delta\mu_{i\to i+1}/kT} \tag{1.16}$$

and $\Delta\mu_{i\to i+1}$ is the change on chemical potential that would accompany the reaction in free solution.

So we see that the two elements required to generate movement are in place. The polarity of the filaments ensures broken spatial symmetry of the ratchet potentials; and the coupling of the motor-filament interaction to the hydrolysis reaction breaks detailed balance by the factors $\Theta_{i\to i+1}$, thereby ensuring that the transitions between the ratchet states are not just thermally activated. Note, however, that if the product of the factors around the cycle $\prod_i \Theta_{i\to i+1}$ were equal to unity, global equilibrium would be preserved and no movement would occur. This is not the case in a biological cell because for each traversal of the cycle, one molecule of ATP has been hydrolysed to ADP and P_i. The total free energy change is therefore $\Delta\mu_{ATP}$ and $\prod_i \Theta_{i\to i+1} = \mathrm{e}^{-\Delta\mu_{ATP}/kT}$. The continual manufacture of ATP in the cell maintains the reaction ATP $\to$ ADP + P_i out of equilibrium, with $\Delta\mu_{ATP} \approx -20\ kT$. It is the nonequilibrium nature of this reaction which ultimately drives the motor.

1.6 Design principles for effective motors

We have seen how to generate motion, but what functional form of the potentials and the transition rates would make for an especially effective motor? Desirable properties of a motor protein such as kinesin, which acts individually to transport substances in the cell, include swiftness of

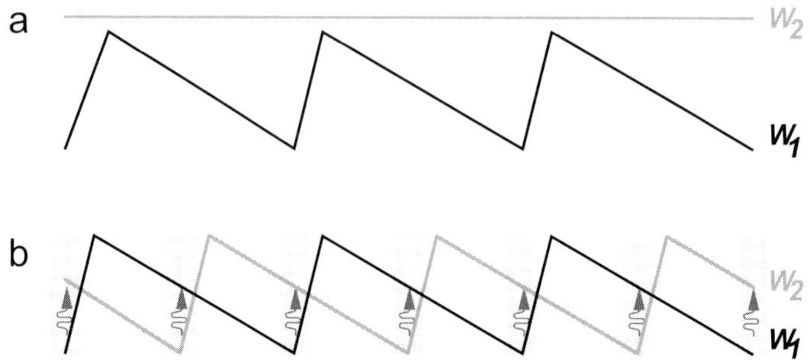

Fig. 5. Examples of two-state isothermal ratchets. In **b)** the stimulated transitions between the states are localized to the shaded regions.

movement, the ability to resist high loads, and the efficient transduction of chemical energy to mechanical work. Can the formalism that we have developed suggest what principles of design might lead to such properties? We noted that the force-velocity relation can vary widely depending on the shape of the effective potential, which in turn depends on the detailed functional form of both the periodic ratchet potentials and the transition rates. Nonetheless, we can gain some insight by examining a simple 2-state model.

Consider therefore the system shown in Figure 5a, where the saw-tooth potential W_1 (with barrier height $W \gg kT$) represents a bound state of the motor, and the flat potential W_2 represents a detached state. We assume that the binding rate is constant, $\omega_{21} = \alpha$, and write the detachment rate as

$$\omega_{12}(x) = \alpha \mathrm{e}^{(W_1 - W_2)/kT} + \Omega \tag{1.17}$$

where the first term on the right hand side corresponds to thermal transitions and the rate Ω, which is assumed to dominate, represents transitions that are stimulated by coupling to a non-equilibrium process. The effective generation of directional movement in this isothermal ratchet requires careful matching of two time scales. First, the particle in the detached state must diffuse a distance greater than γa, but smaller than a, in the typical time α^{-1} that it takes to bind. The optimal situation occurs roughly at the midpoint of this range, when

$$\alpha \approx \frac{kT}{\gamma \zeta a^2} \cdot \tag{1.18}$$

Second, when in the bound state, the particle must have time to drift to the bottom of the potential well before it detaches again. Taking into account that the force acting on the particle in the bound state is approximately

W/a, this condition may be written

$$\Omega < \frac{W}{\zeta a^2} \cdot \tag{1.19}$$

Thus this very simple isothermal ratchet generates spontaneous motion only if the chemical transition rates are adjusted to match the dynamical parameters of the system. Moreover, it cannot support a significant load, despite the high value of the force that is generated in the bound state. This happens because the particle in the detached state slips backwards very readily. Indeed, the motor will stall if it slips a distance of order γa when it is detached, as it will then fail to diffuse beyond the peak of the potential W_1. Taking into account that the slip velocity is F/ζ and the time spent detached is α^{-1}, with α specified by equation (1.18), the stall force is

$$F_0 \approx \frac{kT}{a} \cdot \tag{1.20}$$

The force that the motor can support is determined by the thermal energy and not by the chemical energy available to pump the transitions between states. In summary, this simple 2-state ratchet would be a very poor design for a motor: it is quite awkward to get it to operate at all; and when it does work, it is puny and wasteful of energy.

A system which provides a vastly improved performance [12,23] is shown in Figure 5b. In this case, the potential W_2 has a similar saw-tooth form to W_1, but is *shifted* by half the lattice constant. In addition, both transitions are assumed to be stimulated at a *position-dependent* rate $\Omega(x)$, which is significant only when the particle is close to the bottom of a well. The advantages of this system are clear. The particle in state 1 moves down the long slope of the potential W_1 to the bottom of the well, whereupon it makes a stimulated transition to state 2. It then moves down the slope of W_2, reverting to state 1 when it reaches the bottom. There is now no need to match the kinetic rates with the dynamical parameters to generate effective movement. And because the particle always experiences a potential gradient, the stall force is now determined by the barrier height, $F_0 \approx W/a$, rather than the thermal energy. From equation (1.16), it can be seen that in order to stimulate the transitions from the bottom of the wells, the magnitude of the chemical energy change $\Delta\mu_{ij}$ that accompanies each change of state must exceed the potential energy change, which is equal to half the barrier height. If $|\Delta\mu_{12}| = |\Delta\mu_{21}| \approx W/2$, little energy is dissipated as heat during the transitions and the maximal mechanical efficiency of the system can be close to unity. In this case, the stall force is governed by the chemical potential $\Delta\mu_{\mathrm{ATP}} = \Delta\mu_{12} + \Delta\mu_{21}$ that is available to power the motor,

$$F_0 \approx -\frac{\Delta\mu_{\mathrm{ATP}}}{a} \cdot \tag{1.21}$$

The principles of effective energy transduction outlined above appear to be used by actual molecular motors. It is believed, for example, that kinesin moves along microtubules by advancing its two motor domains in a coordinated, "hand-over-hand" fashion [24,25]. Such cooperation would require a switch-like mechanism, whereby the binding of one domain induces a chemical change in the other, stimulating it to detach. This model corresponds quite closely to the type of isothermal ratchet we have just discussed, where the potential W_1 represents the kinesin-microtubule interaction with one of the motor domains bound, and W_2 the interaction with the other domain bound. Suppose that the domains can only bind at localized binding sites, situated at the minima of the potentials, and that binding causes the stimulated detachment of the other head. Then the transitions between the two potentials are localized, as in Figure 5. The ratchet generates directional movement very effectively and can support a high load. Another molecular motor that makes use of localized transitions to ensure a high efficiency of operation is the F1 component of ATP synthase. In this case, three motor domains work together in a highly coordinated way to generate rotational motion [26]. At any one time, each domain is at a different stage of the hydrolysis cycle: one has ATP bound, one has ADP bound and the third has no nucleotide bound. This motor can be modeled as a three-state ratchet [27] in which the potentials have offset minima, and transitions between states are controlled by a molecular switch so that they take place only at particular angles of rotation.

More generally, position-dependence of the transition rates can be associated with *strain-dependence* of the chemical reactions, and a shift in the position of the potential minima between two states can be associated with a *conformational change* of the motor domain. In the next section we shall explore these concepts in the context of the interaction between myosin and actin.

2 Pulling together: Mechano-chemical model of actomyosin

2.1 Swinging lever-arm model

Isothermal ratchets are a convenient formalism for discussing the generic behaviour of motor proteins. However, for specific motor systems, for which structural and chemical information is available, models based on chemical kinetics can be more appropriate since they may more immediately be related to experimental data. Such is the case for actomyosin. The resulting models [28–34] are termed *crossbridge models*, reflecting the fact that the myosin head forms a transient bridge between the thick and thin filaments in a muscle fibre.

Early electron micrographs of crossbridges bound to the thin filament indicated that there were two, structurally distinct, attached states [35]. Recent crystallographic studies [36], together with micromanipulation experiments [37,38], have refined this view, strongly suggesting that the distal portion of the myosin head acts as a lever arm to amplify small changes in structure occurring at the nucleotide binding site. This has led to the *swinging lever-arm model* [39], which is shown in Figure 6a.

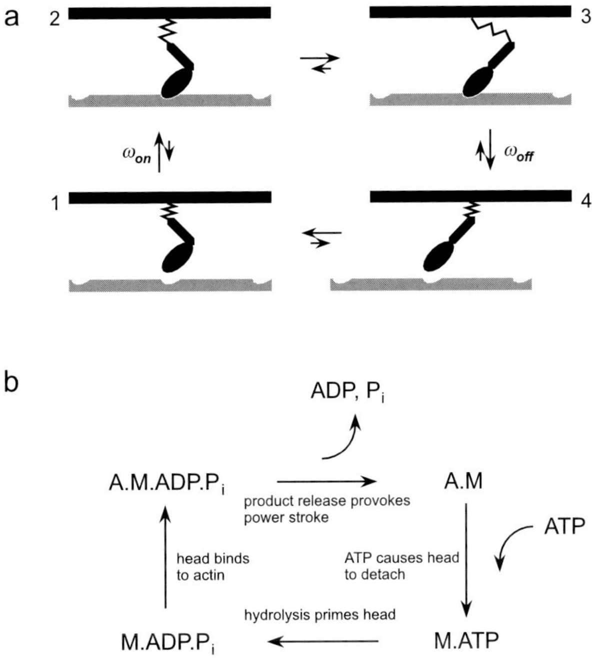

Fig. 6. Swinging-lever arm model of the mechano-chemical cycle of actomyosin.

The basic assumption is that the myosin head can undergo a structural change, provoked by chemical reactions at the nucleotide binding site, which alters the orientation of the lever arm. Biochemical techniques [40] have elucidated the nucleotide reactions which are coupled to the structural change and which modulate the myosin-actin interaction (Fig. 6b). The hydrolysis cycle proceeds in four steps: (1) ATP hydrolysis occurs when the myosin is detached. Energy is not liberated at this stage but is used instead to prime the lever arm. (2) The head binds to the thin filament. (3) The products of hydrolysis, P_i and ADP are released, provoking a significant rotation of the lever arm. The large change in free energy which accompanies product release is not liberated as heat, but is stored in an elastic element which connects the myosin head to the thick filament. Because the head is now in a state that generates a shear force between the thick and the thin filament, this transition between bound states is often referred to as the *power-stroke*. If the force is not compensated by an applied load, the filaments will slide relative to one another. (4) Finally, a new molecule of ATP binds to the myosin head, destabilising its interaction with the actin filament. The myosin detaches, completing the cycle.

In the cell, the concentration of ATP is maintained well above its equilibrium value, so that the actomyosin cycle proceeds only in the forward direction (the free energy of hydrolysis is approximately $20k_{\mathrm{B}}T$, so the probability of cycling backwards is $\mathrm{e}^{-20} \approx 10^{-9}$ in the absence of an external load). As successive hydrolysis reactions occur, the mysoin acts as an "oar" which "rows" along the thin filament.

2.2 Mechano-chemical coupling

Kinetic models of myosin action may be based on this 4-state cycle, but care must be taken to ensure consistency with the laws of thermodynamics. Each state i can be considered to be internally at equilibrium, and may therefore be characterised by a Gibbs free energy G_i which includes contributions from the myosin-actin interaction and the nucleotide-myosin interaction. In contrast to ratchet models, the motor domain is taken to be immobilized when it is bound to the actin binding site. Nevertheless, the crossbridge is assumed to be compliant, so that the thin filament can still move relative to the thick filament. Then the free energy of the actomyosin complex G_i depends on the relative displacement x of the thin and the thick filaments, owing to the mechanical deformation of the myosin molecule. In a given state, the actomyosin undergoes Brownian motion in the potential $G_i(x)$ and the instantaneous force exerted on the thin filament by the myosin head is equal to the gradient of the potential, $F_i = \partial_x G_i$.

For each transition between a pair of states, the principle of detailed balance dictates that the ratio of forward and reverse rates is related to the difference of the free energies:

$$\frac{\omega_{ij}}{\omega_{ji}} = \mathrm{e}^{-(G_j - G_i)/kT}. \tag{2.1}$$

In transitions involving a bound state, $G_j - G_i$ will, in general, depend on the relative displacement of the thick and thin filaments, so either the forward or the reverse transition rate (or both) must be strain-dependent. Put another way, the kinetic rates depend on the force experienced by the myosin molecule. This *mechano-chemical coupling* [30] is the key to understanding motor protein action. Thermodynamics provides strong constraints on theoretical models. The relations between kinetic rates must be associated

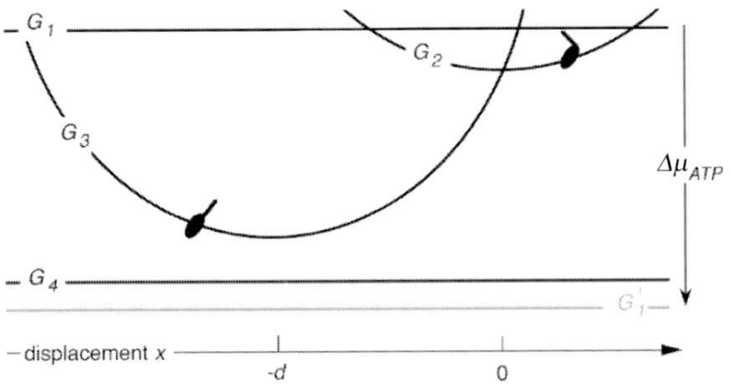

Fig. 7. Free energies of actomyosin states in the swinging lever-arm model.

with the mechanical properties of the molecules. However, there remains considerable freedom to choose the functional form of individual rates.

In most mechano-chemical models of actomyosin, the dependence of free energy on position is simplified and given a physical interpretation by supposing that the myosin molecule contains a linear elastic element. Thus two parameters characterise the mechanical properties of a crossbridge: the distance d that the distal end of the lever moves when the myosin makes a transition between the two attached states; and the spring constant K. The force exerted by the myosin head in each state is then Kx and $K(x+d)$, respectively, where x is the shear displacement between the two filaments. The values of the structural parameters are difficult to measure experimentally, but recent data suggests that $d \approx 5\,\mathrm{nm}$ and $K \approx 1$ pN/nm [41].

The corresponding free energies G_i are indicated in Figure 7. Note that traversal of the cycle in the clockwise direction is tightly coupled to the hydrolysis of one molecule of ATP so, although the actomyosin returns to the same state, the free energy of the whole system has changed by $\Delta\mu_{\mathrm{ATP}}$.

2.3 *Equivalent isothermal ratchet*

Although crossbridge and isothermal ratchet models are not entirely equivalent, it is possible to construct a ratchet model which strongly resembles a crossbridge description. As an example, we outline an isothermal ratchet which would generate approximately the same dynamics as the swinging lever arm model. The ratchet model involves transitions between four different potentials W_i, which represent the actomyosin interaction, as shown in Figure 8. Potentials W_1 and W_4 represent detached states and are therefore flat. The potentials W_2 and W_3, which represent the two bound states,

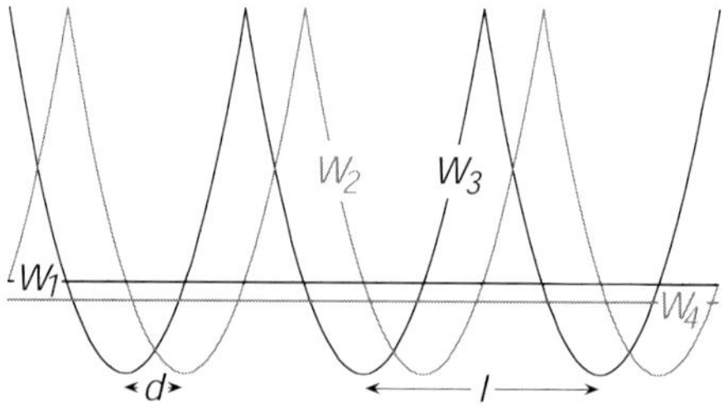

Fig. 8. 4-state ratchet equivalent to the swinging lever-arm model.

are each symmetric and periodic, with period l equal to the distance between the binding sites on the thin filament ($l = 34$ nm for actin). The minima of W_2 are shifted by d relative to those of W_3, and the curvature of both potentials is $\partial_x^2 W = K$. Transitions between these states are stimulated by chemical reactions occurring at the nucleotide binding site. Indeed, the assumption of tight-coupling in the swinging lever-arm model implies that the transition rates are governed by equations (1.15) and (1.16). Although the individual periodic potentials W_i of this ratchet are symmetric, the system as a whole is asymmetric if $d < l/2$, thereby satisfying one of the conditions for movement. The second condition is that the transitions between the potentials W_i violate detailed balance, *i.e.* that the product of the factors Θ_{ij} is not equal to unity. Since $\Pi\Theta_{ij} = \exp(-\Delta\mu_{\rm ATP}/kT)$, this is assured if the nucleotide hydrolysis reaction is maintained out of equilibrium.

This 4-state model differs from the swinging lever-arm model in only one respect: the ratchet potentials W_i corresponding to the bound states are periodic and have multiple minima, whereas the free energies G_i of the bound states in the crossbridge picture are assumed to have a single minimum (whose location varies according to which site on the thin filament the myosin head is bound). Whether this difference is significant is a question of time scale. If transitions between states occur more rapidly than diffusion over the potential barrier, then the two models differ only in insubstantial detail.

2.4 Many motors working together

Many millions of myosin molecules work together to drive muscle contraction. A hundred or so myosin molecules interact with each thin filament, a few thousand filaments make up each sarcomere, and a single myofibril fibre is typically composed of thousands of sarcomeres in series. A single muscle fibre is then formed by a bundle of a few thousand myofibrils. In order to

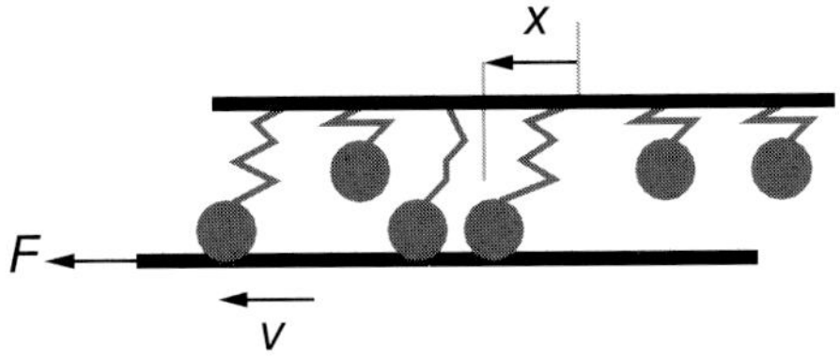

Fig. 9. Protein friction.

understand muscle contraction, therefore, we need to construct a statistical mechanical model, based on the mechano-chemical cycle of an individual myosin molecule that has been outlined above. How do many motors act together to drive a single thin filament? Clearly they can combine their efforts to generate a large force. But they can also get in each other's way and hinder sliding. We shall explore this potential problem first, and see how proteins can create friction [42].

Consider a set of N proteins, each of which independently binds to a filament at rate ω_{on} and detaches at rate ω_{off}, as shown in Figure 9. For the moment, we will assume that these proteins are not motors (and so do not undergo a conformational change when bound) but that, like myosin, they form crossbridges of elasticity K. Suppose that an external force F is applied to the filament, causing it to move with steady velocity v. The mean fraction of proteins that is bound at any one time is $r = \omega_{\text{on}}/(\omega_{\text{on}} + \omega_{\text{off}})$, and the probability that each of these proteins has strain x is $p(x) = (v/\omega_{\text{off}})\exp(-x\omega_{\text{off}}/v)$. When a protein detaches, the elastic energy $\frac{1}{2}Kx^2$ is dissipated as heat. The total dissipation rate due to the proteins is therefore $Nr\omega_{\text{off}}\frac{1}{2}K<x^2> = \zeta_{\text{protein}}v^2$, where

$$\zeta_{\text{protein}} = \frac{NrK}{\omega_{\text{off}}} \tag{2.2}$$

is the effective friction coefficient associated with the proteins. Unless the number of bound proteins is very small, or the detachment rate is very rapid, this *protein friction* is considerably larger than the viscous drag exerted on the filament by the surrounding fluid.

Now suppose that the N proteins are myosin molecules, each of which is executing its mechano-chemical cycle. Let us make a simple assumption about the kinetics, and suppose that the rate-limiting steps are the binding to (2→3) and detachment (4→1) from the filament, and that these occur at *strain-independent* rates ω_{on} and ω_{off}. The fraction r of motors bound to the filament is then fixed. It is convenient to characterize the properties of an individual motor by a unitary velocity $v_{\text{unit}} = \omega_{\text{on}}d$ and a unitary force $F_{\text{unit}} = Kd$. When the filament is loaded with an external force F, its

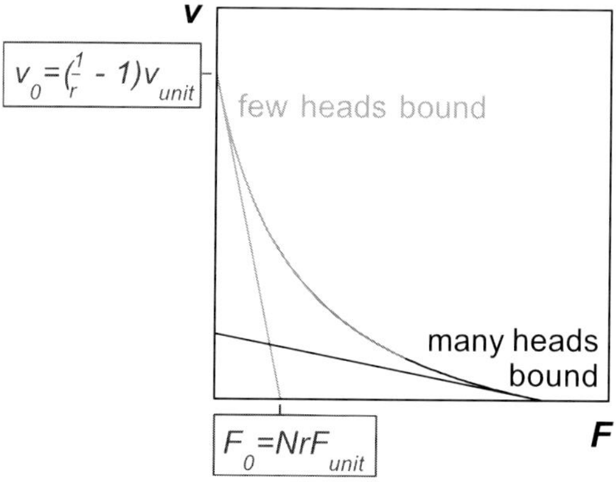

Fig. 10. For strain-independent transition rates, the force-velocity relation is linear. Rapid sliding and a high stall load can be combined if the fraction of bound motors increases with the load (curve).

equation of motion is $F_{\rm mot} = F + \zeta v$, where $F_{\rm mot}$ is the total force generated by the motors. It is clear that when $v = 0$, each bound motor generates force $F_{\rm unit}$, so that the stall force is

$$F_0 = NrF_{\rm unit}. \tag{2.3}$$

Since the system is linear, the force-velocity relation must be a straight line with slope equal to the total effective friction $\zeta_{\rm protein} + \zeta$. Thus, the velocity at zero load is

$$v_0 = \frac{v_{\rm unit}}{r/(1-r) + \tau_{\rm visc}\omega_{\rm on}} \tag{2.4}$$

where $\tau_{\rm visc} = \zeta/NrK$ is the viscous relaxation time of the system. If the viscous relaxation is rapid compared to chemical reaction times, as is the case in a sarcomere,

$$v_0 \approx v_{\rm unit}\left(\frac{1}{r} - 1\right). \tag{2.5}$$

Then, as indicated in Figure 10, a small fraction of bound motors ($r \ll 1$) produces rapid sliding in unloaded conditions, but provides only a low stall force. If, on the other hand, the majority of motors are bound ($r \approx 1$), the sliding is slow, but the stall force is high. Neither situation is ideal for muscle contraction, where a combination of rapidity and strength is required. However, this combination can be achieved by regulating the fraction of bound motors r according to the load. What is required is some kind of control mechanism whereby few motors are bound at zero load, but more are recruited as the load increases.

2.5 Designed to work

How can regulation of the fraction of bound motors be accomplished? The observation that the two products of hydrolysis are not released simultaneously provides a clue. Phosphate is released first, and this provokes the power-stroke of the lever arm. Subsequently ADP is released more slowly, and this makes way for the immediate binding of a new ATP molecule, which precipitates detachment of the head from the actin filament. Thus the detachment step ($4 \rightarrow 1$) is really controlled by the rate of ADP release. Now, we might expect that the departure of any ligand from the nucleotide binding site would result in a small change of conformation there, which gets amplified by the lever arm. Thus ADP release might cause the distal end of the lever arm to shift by a modest amount δ, as indicated in the more complete model of the actomyosin cycle [33] shown in Figure 11. Mechanochemistry would then make the rate of ADP release *strain-dependent*. If a muscle fibre is loaded isometrically so that the filaments are prevented from sliding, each bound motor will be strained, exerting approximately the unitary force $F_{\mathrm{unit}} = Kd$. As a result, the small movement of the lever arm accompanying ADP release will involve an amount of mechanical work equal to $F_{\mathrm{unit}}\delta$. This slows down the reaction by a factor $\exp(-Kd\delta/k_{\mathrm{B}}T)$, compared to the unstrained situation at zero load. The delayed detachment

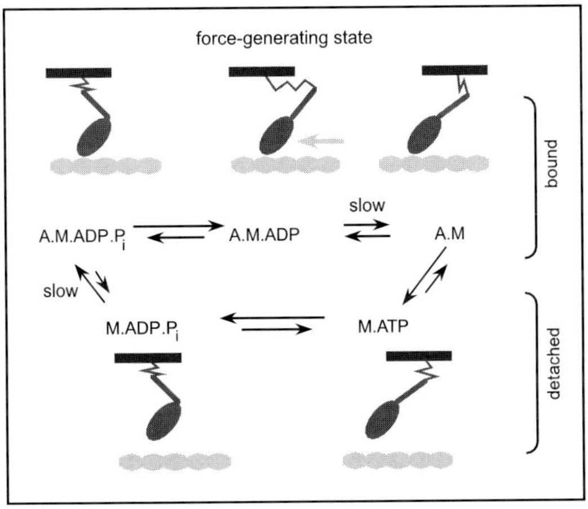

Fig. 11. Model of the actomyosin cycle.

augments the fraction of motors bound to the thin filament, helping to support the high load.

It thus appears that the lever-arm structure of myosin serves a dual function. The large swing of the arm on phosphate release is designed to generate force. And the smaller shift on ADP release is designed to endow muscle with a "continuous automatic transmission", by adjusting the number of motors appropriately to the load. This second shift need not very big ($\delta \approx 1$ nm), which may explain why it has not been observed for skeletal-muscle myosin. Movement of the lever on ADP release has, however, been detected in a number of other myosins [43, 44].

There is a second aspect of the structural design of myosin, which concerns the strength of its elastic element. Clearly this should not be too weak, or a motor would produce only a feeble unitary force F_{unit}. However, it should not be too strong either, since the power-stroke would be inhibited if the mechanical work required to move the lever arm, $\frac{1}{2}Kd^2$, exceeded the free energy change that accompanies phosphate release, $\Delta\mu_{\text{P}_\text{i}}$. In fact, it appears that the molecular design of myosin pushes this limit as far as possible, choosing a value of $\frac{1}{2}Kd^2$ that slightly exceeds $\Delta\mu_{\text{P}_\text{i}}$. This is slightly surprising, because it means that an individual myosin molecule would be ineffective, not having enough energy to power its stroke. But a team of motors working together is actually more efficient in this case, for the following reason: when a myosin head binds to the filament, it cannot immediately execute its power-stroke, but the transition is postponed until filament sliding (due to the action of other motors) has strained the motor and reduced the work accompanying the power-stroke to $\Delta\mu_{\text{P}_\text{i}}$. Consequently, all of the energy of phosphate release gets stored in the elastic element, and none gets wasted as heat.

2.6 Force-velocity relation

Although muscle contraction is clearly a dynamical process, individual pairs of thick and thin filaments may be treated as though they are in "quasi-mechanical equilibrium". This is because the viscous relaxation time is short compared to the time scale of chemical reactions, so that each time an individual motor changes state, the filaments quickly readjust their position as shown in Figure 12.

A simple stochastic simulation algorithm can therefore be used to determine the sliding velocity for a given load [33]. The resulting force-velocity relation is shown in Figure 13. The shape of the curve is concave [45], owing to the regulation of the fraction of bound motors discussed above. However, another interesting feature is apparent. As the load approaches the stall force (or "isometric tension"), the velocity drops abruptly to zero. Precisely this type of behaviour is seen in the experimentally determined

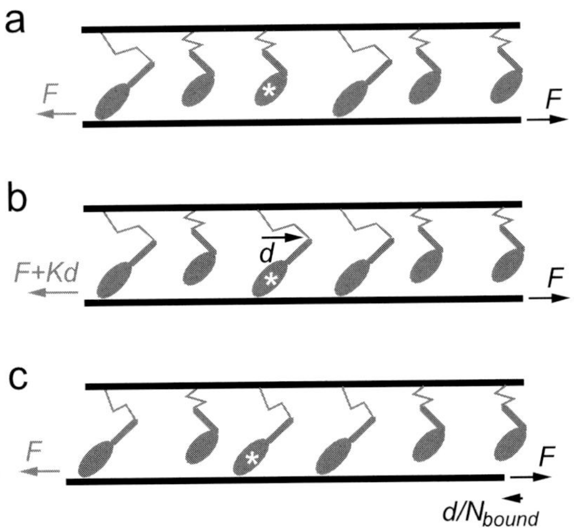

Fig. 12. a) In mechanical equibrium, the sum of the forces exerted by the motors (grey arrow) opposes the applied load (black arrow). **b)** If one of the motors binds and undergoes its power-stroke, the force balance is momentarily upset. **c)** As a consequence, the filaments slide to restore mechanical equilibrium.

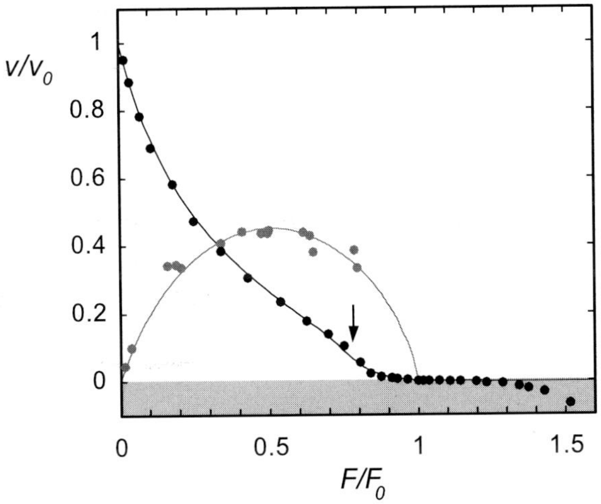

Fig. 13. Force-velocity relation (black) and efficiency (grey) of muscle. Experimental data (points) are compared with model predictions (lines).

force-velocity relation of single muscle fibres [46]. In the model, it is caused by the inhibition of the power-stroke at high loads. As already mentioned, when $\frac{1}{2}Kd^2$ slightly exceeds $\Delta\mu_{\mathrm{P_i}}$, a motor needs the assistance of its team mates to effect its power-stroke. Close to isometric conditions, the other motors do not advance the filament rapidly enough to help, and sliding breaks down. At still higher loads, almost no power-stroke transitions can occur and the myosin heads simply bind and detach reversibly, without releasing the products of hydrolysis. In doing so they create a high protein friction, so the filaments slide backwards (corresponding to muscle extension) only slowly. This suggest another reason why a strong elastic element may be favourable. It permits the muscle to support a variety of high loads while remaining at an almost constant length and without expending much energy.

2.7 Dynamical instability and biochemical synchronization

The abrupt decline of sliding velocity at high loads actually corresponds to a sudden change of dynamical behaviour. As shown in Figure 14, the filaments no longer slide smoothly at a constant velocity, but move in a stepwise fashion [33]. The transition is a direct consequence of the fact that myosin heads must cooperate with one another in order to execute their power-strokes. At zero load, they are able to do so in an uncorrelated way, with individual molecules binding and detaching at random intervals. Those heads which have been bound for the longest period of time get dragged into a position in which they exert negative force. This balances the positive force exerted by the heads that have just executed their power-stroke. Because of the

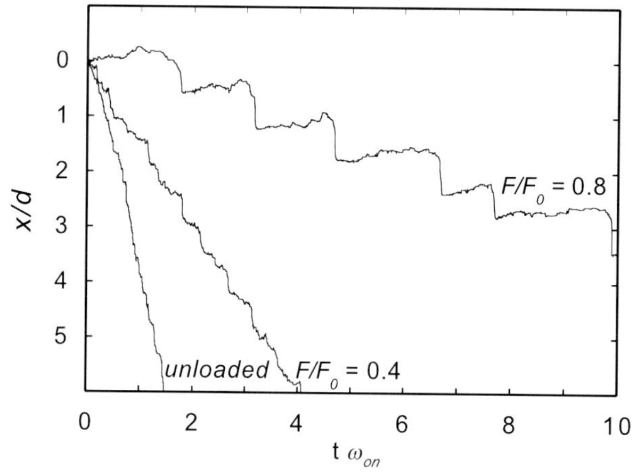

Fig. 14. At high load, the dynamics changes from smooth sliding to stepwise motion.

strain-dependence of ADP release, it is the former subset of heads that has the highest instantaneous detachment rate. Their continual dissociation causes the thin filament to advance sufficiently rapidly to ensure that any newly bound heads are always able to execute their power-stroke. Consequently, the sliding is smooth. As the load is raised and the average sliding velocity diminishes, this situation breaks down. The power-stroke transitions of individual motors begin to fail. Nevertheless, if an individual head does succeed in accomplishing a power-stroke, it causes the thin filament to advance by a small amount and reduces the strain of all the other attached heads, thereby facilitating their own power-stroke transitions. Owing to this cooperativity, a large fraction of the bound heads can stroke almost simultaneously, like a rowing crew, causing the thin filament to lurch forward through a distance approximately equal to the length of the power-stroke. Subsequently, the heads detach stochastically and they must rebind in sufficient numbers to be able to coordinate another cascade of power-stroke transitions. Thus the dynamics is quasi-periodic, with a period defined by the cycle time of an individual molecule. This suggests that the isometric condition of muscle may not be a steady state, but rather one in which some sarcomeres are lengthening and others are shortening at any instant of time.

2.8 *Transient response of muscle*

There is one experiment in which the absence of a steady state in near-isometric conditions would have a clear, observable consequence: the isotonic transient response, in which an isometric muscle fibre is subjected to a sudden decrease in load. The external mechanical stimulus would synchronise the power-stroke transitions of the bound myosin heads throughout the filament. Subsequently, individual sarcomeres would start to shorten in a stepwise fashion but, owing to filament compliance, the coordination of different sarcomeres would gradually be lost. The consequence would be an apparent, damped oscillatory motion of the fibre. This is precisely what is observed in muscle [47, 48].

3 Motors at work: Collective properties of motor proteins

3.1 *Dynamical instabilities*

One of the fundamental lessons of statistical physics has been that even a very simple system, composed of identical elements with pairwise interactions, can display emergent collective properties. An example is the abrupt condensation of a gas as the temperature falls below a critical value. Such phase transitions, in which one type of collective behaviour becomes unstable and another comportment is favoured, are a consequence of cooperativity in the microscopic interactions. A system in which many motor proteins act collectively is considerably more complex than a gas, because the individual entities are not just passive molecules, but active force generators. A collection of motors is a non-equilibrium system, but by analogy with a thermodynamic system we should expect the cooperativity (which is due to mechano-chemical coupling) to induce dynamical instabilities. The transion from smooth sliding to stepwise motion, discussed in the previous section, is one example. In order to explore other examples, we will return to the isothermal ratchet model, whose simple formulation facilitates the analysis of collective effects.

3.2 Bidirectional movement

In Section 2, we saw that an individual motor can move unidirectionally along a polar filament, and that its direction of motion is determined by the form of the interaction potentials and by the position dependence of the transition rates between different motor states. When a number of motors act together to propel a filament, however, the direction of motion is a team property. The filament might move in either sense, and, in some cases, it might be able to move in *both* directions [49]. To see how such bidirectionality can arise, it is easiest to consider the hypothetical case of a non-polar filament. The absence of asymmetry in the potentials means that an individual motor is incapable of moving directionally. But as shown in Figure 15, multiple motors can propel a filament in *either* direction if the reaction rates are strain-dependent, such that detachment can only take place at a localized position near the bottom of a potential well. Suppose that the motors are driving the filament to the left. Then, owing to the detachment of motors which have recently passed through a potential well, there are more bound motors on the right-facing slope of the potential

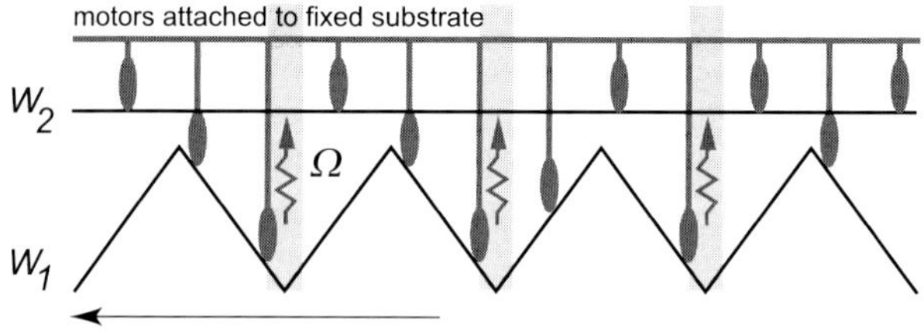

Fig. 15. A team of motors can propel a filament, even if the interaction between an individual motor and the filament is symmetric.

than on the left-facing slope (as shown). Sliding down the slope, these motors produce a force which sustains the filament's leftward movement. An equivalent argument shows that the filament could just as well be driven to the right.

Thus, there are situations in which a team of motors can drive a filament either forwards or backwards. A filament would not travel one way for an indefinite period, however. Stochastic fluctuations in the numbers of motors pushing and pulling would eventually cause an abrupt switch to the alternative steady-state solution [50]. Interestingly, such behaviour has recently been seen in a motility assay, in which the motion of microtubles was observed as they glided over a lawn of Ncd motors absorbed on a surface [51]. Typically, an individual microtubule travelled for several micrometers with its plus end leading, then abruptly reversed direction and travelled for a similar distance in the opposite sense. The speed was approximately the same in each direction and the reversals appeared to occur quite randomly. The particular Ncd molecule used was a mutant with a single amino-acid substitution in the neck region, which links the heads to the coiled-coil tail. The ratchet model suggests that this mutation modifies the effective interaction potential of the motor with the microtubule, so that it is almost symmetric despite the microtubule's polarity.

3.3 Critical behaviour

The collective property of bidirectional motion arises because, when a number of motors are pulling one way, they enhance the likelihood that their teammates will join in and pull in the same direction. This only happens, however, only if the rate of stimulated detachment is faster than a critical value. The system has to be sufficiently far from equilibrium (which means, in practice, that the concentration of ATP has to be above a critical level).

This critical behaviour may be analysed using a simple model 2-state isothermal ratchet model [12, 49], writing the transition rates as

$$\begin{aligned} \omega_{21} &= \alpha \\ \omega_{12}(x) &= \alpha\, \mathrm{e}^{(W_1 - W_2)/kT} + \Omega(x). \end{aligned} \tag{3.1}$$

Here $\Omega(x)$ is the rate of non-thermal transitions from the bound state; it represents detachment stimulated by chemical reactions, and its value characterizes how far the system is from equilibrium. If the motors and the filament are moving with respect to one other at relative velocity v, the probabilities of occupancy of either state satisfy

$$\begin{aligned} \partial_t P_1 + v\partial_x P_1 &= -\omega_{12}(x)P_1 + \omega_{21}(x)P_2 \\ \partial_t P_2 + v\partial_x P_2 &= \omega_{12}(x)P_1 - \omega_{21}(x)P_2 \end{aligned} \tag{3.2}$$

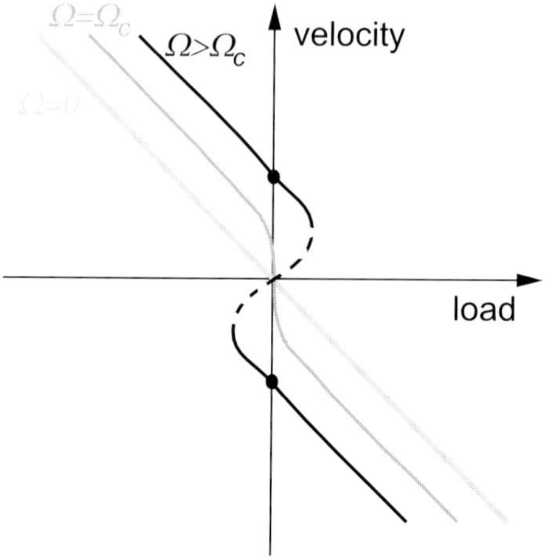

Fig. 16. Force-velocity relation for a team of motors interacting with a symmetric filament.

and $P_1 + P_2 = 1$ gives

$$v\partial_x P_1 = -\left[\omega_{12}(x) + \omega_{21}(x)\right] P_1 + \omega_{21}(x)\ . \tag{3.3}$$

If the motion is opposed by an external load F, the equation for the balance of forces is

$$N \int_0^a \mathrm{d}x P_1 \partial_x W_1 = F + \zeta v \tag{3.4}$$

where the expression on the left hand side is the active force generated by the motors. Equations (3.3) and (3.4) together specify the force-velocity relation. At low velocities the relation may be written as an expansion which, for the case of the symmetric system discussed above, contains only odd terms

$$F = -\zeta v + A(\Omega)v + B(\Omega)v^3 + ... \tag{3.5}$$

The force-velocity relation is shown in Figure 16 for different values of Ω. At equilbrium, $\Omega = 0$, the cubic coefficient vanishes, $B(0) = 0$, and the linear coefficient is negative, representing the protein friction: $-A(0) = \zeta_{\text{protein}}$. Thus the filament simply slides passively backwards, as expected

$$v(\Omega = 0) = -\frac{F}{\zeta + \zeta_{\text{protein}}}\cdot \tag{3.6}$$

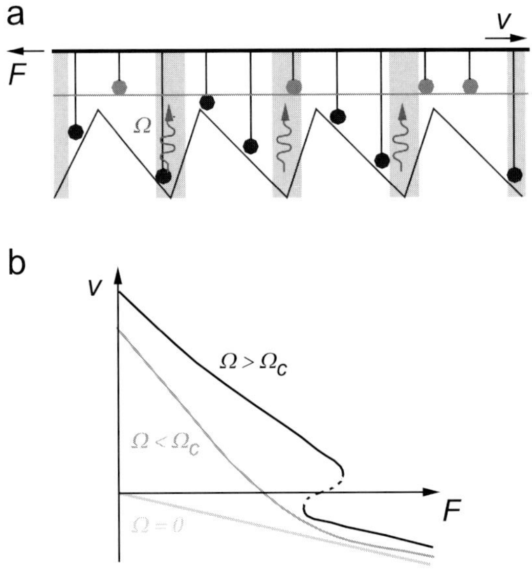

Fig. 17. a) Team of motors interacting with a polar filament. **b)** Possible force-velocity relation.

However, when the system is out of equilibrium, the linear coefficient A can become positive; and above a critical value of the stimulated detachment rate, it can exceed the viscous drag coefficient. Thus, for $\Omega > \Omega_C$, given by $A(\Omega_C) = \zeta$, equation (3.5) can have a non-zero solution for v in the absence of an applied load F. Spontaneous motion occurs with velocity

$$v_{\pm}(\Omega) = \pm \left(\frac{A(\Omega) - \zeta}{B(\Omega)} \right)^{1/2} \sim (\Omega - \Omega_c)^{1/2}. \quad (3.7)$$

In the vicinity of the critical point, the behaviour is generic and the velocity obeys characteristic power-law relations. For example, when a load is applied, the velocity varies as

$$v(\Omega_c) \sim -F^{1/3}. \quad (3.8)$$

Although the occurrence of bidirectionality is especially easy to illustrate for the case of a symmetric ratchet, it is not confined to this situation but is also likely to occur under certain load conditions in asymmetric ratchets [12,49]. A typical case is shown in Figure 17. Crossbridge models can also display regimes in which multiple motors can drive a filament both forwards and backwards [34]. Thus bidirectional motion may be common in naturally occurring motor protein systems, and not just a property of unusual mutants.

Indeed, bidirectional motion was first reported in an actomyosin motility assay [52].

3.4 Oscillations

A region of hysteresis in the force-velocity diagram of the type shown in Figure 17 has an interesting implication: if the filament is connected in series with an elastic element, it will oscillate [53]. The motors first drive the filament forwards and as the spring stretches, the increasing load slows the movement down. But before the filament comes to a halt, the forward motion becomes unstable and the system jumps to the other stable branch, for which the velocity is negative. Consequently the motors start to drive the filament backwards. As they do so, the tension in the spring declines and thus so does the speed. But again, before the filament stops, the motion becomes unstable and the system abruptly jumps back to the original branch. As shown in Figure 18 these motor-driven oscillations have a characteristic saw-tooth form. In general, the amplitude of oscillation depends on the degree of hysteresis, which itself depends on the amount by which Ω exceeds Ω_C.

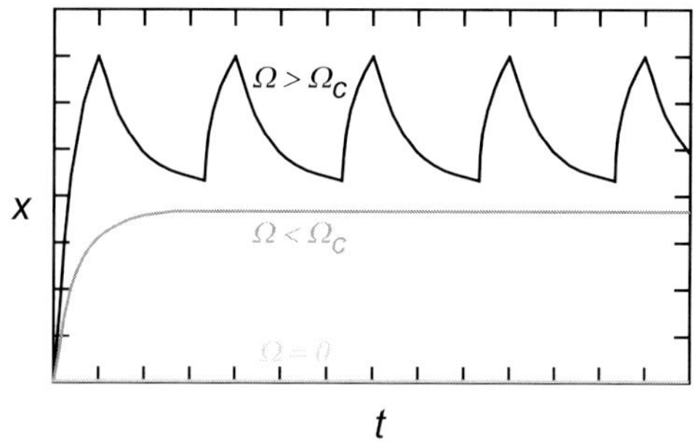

Fig. 18. Oscillations induced by a team of motors in series with a spring.

The critical point $\Omega = \Omega_C$ corresponds to a Hopf bifurcation [53]. It is at this point that sinusoidal oscillations with vanishing amplitude, but a well- defined frequency, set in. Their characteristic frequency

$$\omega_c \sim \left(\alpha \frac{K}{\zeta}\right)^{1/2} \tag{3.9}$$

is the geometric mean of two rates: the typical kinetic rate of the motor α; and the rate of viscous relaxation of the filament $1/\tau_{\text{visc}} = K/\zeta$. Interestingly, the frequency of oscillation can be significantly faster than the cycle time of an individual motor. There is, however, an upper limit to the frequency, since the dynamical instability is eliminated if the external spring is stonger than the total effective elasticity of the motor crossbridges (the elastic constant $k_{\rm xb}$ of a single crossbridge may be estimated in the ratchet model as $k_{\rm xb} = U/l^2$, where U is the height of the potential maximum and l is the period). Thus the maximum attainable frequency is [54]

$$\omega_{\rm c} \sim \left(\alpha \frac{N k_{\rm xb}}{\zeta} \right)^{1/2} . \tag{3.10}$$

There are a number of natural motor systems which oscillate. The vibration of insect flight muscle is too rapid to be controlled by nervous impulses on a cycle-by-cycle basis [55], and is thought to be generated by a dynamical instability of the actomyosin system, which might be based on the type of mechanism described here. Normal striated muscle, in unusual chemical conditions of elevated ADP concentration, also displays large scale oscillations with a characteristic saw-tooth profile [56, 57]. Of special interest, however, is the undulation of spermatazoid flagella, which might be caused by an oscillatory instability of teams of axonemal dynein motors. Before examining this possibility, we shall investigate how the force generated by motor proteins can be used to bend cytoskeletal filaments and generate dynamical patterns.

3.5 Dynamic buckling instability

In a gliding motility assay, fluorescently-stained cytoskeletal filaments are observed as they move across a surface on which motor proteins have been absorbed. The filaments usually glide smoothly, but occasionally they encounter defects, which pin them at their leading end [58]. As shown in Figure 19, the filaments then move in an intriguing way: they either whirl around in a spiral, or they writhe like a snake. These two patterns of behaviour are a consequence of a buckling instability, caused by the force that the motors generate along the contour of the filament. Despite the microscopic size of the system, the dynamics can be treated using continuum elasticity theory [58]. In a standard buckling instability (such as that which occurs when a flag-pole is built too tall) the system comes to a new mechanical equilibrium. In the case of the gliding assay, however, the direction of the applied force changes as the filament bends. As a result, the system evolves to a limit-cycle, instead of a fixed point.

Describing the instantaneous configuration of the filament by the curve $\tilde{r}(s)$, and writing the gradient of the curve $\tilde{u} = \partial_s \tilde{r}$, the equation of motion of the filament is

$$\zeta \partial_t \tilde{r} = f \tilde{u} + \kappa \partial_s^3 \tilde{u} + T \partial_s \tilde{u}. \tag{3.11}$$

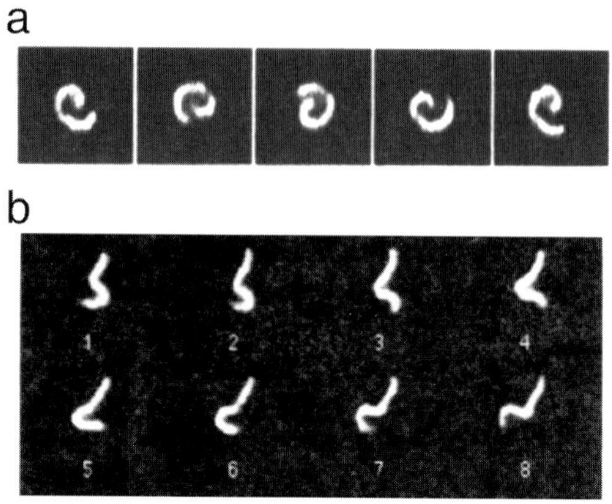

Fig. 19. Dynamical behaviour of cytoskeletal filaments which have been momentarily pinned in motility assays. **a)** Spiral motion. **b)** Undulation reminiscent of a flagellum.

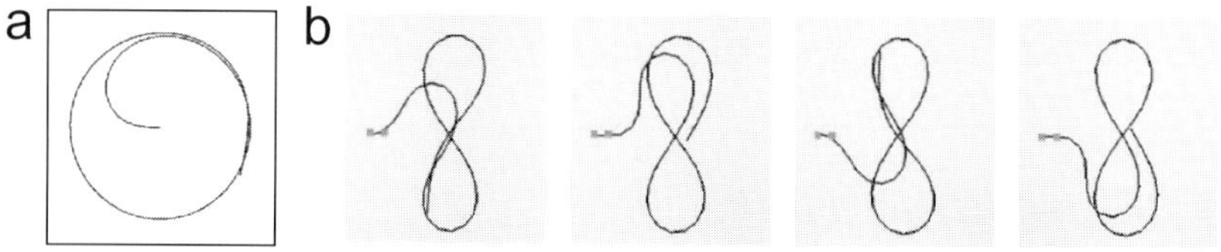

Fig. 20. Dynamic buckling instabilities. **a)** Spiral motion of a filament pinned at one end. **b)** Figure-of-eight motion of a filament pinned at two positions.

Here, f is the force per unit length that the motors exert along the filament contour, and κ is the bending modulus of the filament. The action of the motors additionaly gives rise to an internal compressive thrust T in the filament, which can be determined by imposing the constraint of fixed contour length. This equation must be supplemented with appropriate boundary conditions.

If the position of the leading end is fixed, but the filament is free to swivel about this point, the asymptotic solution is a spinning spiral pattern, shown in Figure 20a. A simple scaling argument reveals how the radius R of the spiral varies with the driving force per unit length. On the limiting circle, the bending moment is $M = \kappa/R$. In the steady state, this must be equal to the torque Γ produced by the external forces (the motor force and the friction). Since the radius of the spiral is the only relevant length scale,

Γ must scale as fR^2. Thus the condition $M \sim \Gamma$ yields the scaling relation

$$R \sim (\kappa f)^{1/3} . \qquad (3.12)$$

Since the bending modulus of cytoskeletal filaments is known, a simple measurement of the size of the spiral provides a direct measure of the motor protein force.

If the filament is pinned at two separate locations near the leading end, both the position and the vector of the end are fixed. In this case, the solution for short filaments is the serpentine motion that is seen experimentally. However, the asymptotic solution for long filaments is the pattern shown in Figure 20b. While the leading end of the filament writhes from side to side, the back end continually traces out a figure of eight – the curve called the "elastica" [59] which can be obtained by pressing on the ends of a thin elastic rod until both ends meet.

3.6 Undulation of flagella

Cilia and spermatzoid flagella share a common structure, called the "axoneme" [60], which is composed of a cylinder of microtubule doublets (Fig. 21a). Dynein motors attached to each doublet interact with the neighbouring doublet around the ring, generating a shear force. Because all of the microtubules are firmly held in place in a basal body (next to the head of a spermatozoan, or at the root of a cilium), the sliding of ajacent microtubule doublets causes to bending of the entire axoneme [61]. Clearly, in order to generate the wavelike undulation of a flagellum, the shear force between a given microtubule doublet must vary with both time and position. Does this require a chemical control mechanism to ensure that the motors on one side of the axoneme work while the motors on the opposite side detach [62]? Or might wave-like motion occur natually as a result of a dynamical instability? The buckling of filaments in a gliding motility assay demonstrates that instabilities can indeed lead to complex periodic movement. In that case, however, the motors exert an external force on the filament. In the axoneme, the motor action is internal. Can a dynamical instability produce undulations in this situation too?

The basic physical properties of the axoneme can be captured by the two-dimensional model shown in Figure 21b, in which a pair of filaments is connected by a set of motor proteins [63]. The rotational symmetry of the axoneme – in which each doublet slides on its neighbour – corresponds to a lack of polarity in the 2-dimensional case. So the motor-mediated interaction between the pair of filaments might be represented as a symmetric isothermal ratchet. As we saw in Section 3.2, this type of system can be bistable, and the total shear force generated between the filaments by the

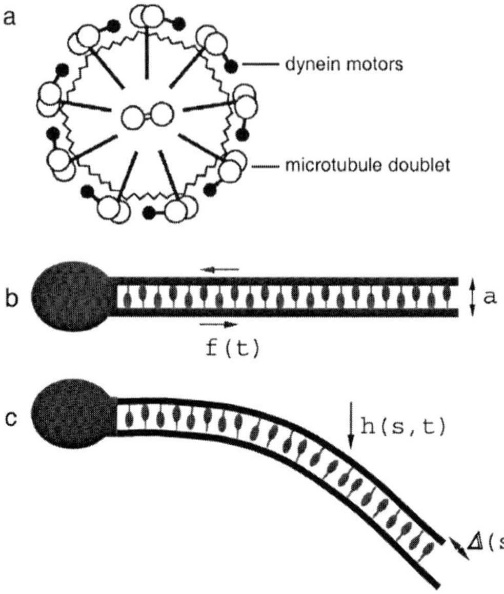

Fig. 21. a) Cross-section through an axoneme. **b)** Model of a flagellum. **c)** Bending caused when the lower filament slides forwards relative to the upper filament.

team of motors can have either sign. The relative sliding of the filaments in response to this force bends the axoneme, as shown in Figure 21c. Now, this movement is countered by a force derived from the elastic bending of the axoneme. To a first approximation, we can consider the motors to be working against a uniform spring. Then, as discussed in Section 3.4, an oscillatory instability can occur, such the shear force generated by the motors varies periodically.

The simplest model [63] of a flagellum, then, supposes that the force per unit length f generated by the motors is independent of the position s along the axoneme, but varies sinusoidally in time, $f = f_0 \cos \omega t$. Writing the perpendicular displacement of the axoneme as $h(s)$, the local sliding displacement is $\Delta(s) = a(\partial_s h(s) - \partial_s h(0))$, where a is the distance between the two filaments. This leads to the following equation of motion,

$$\eta_\perp \partial_t h = -\kappa \partial_s^4 h - a \partial_s f, \tag{3.13}$$

where κ is the bending modulus and $\eta_\perp$ is the friction coefficient per unit length orthogonal to the axoneme axis (which is approximately equal to the viscosity of the surrounding fluid). To solve for the motion, this equation must be supplemented by two boundary conditions at each end. The

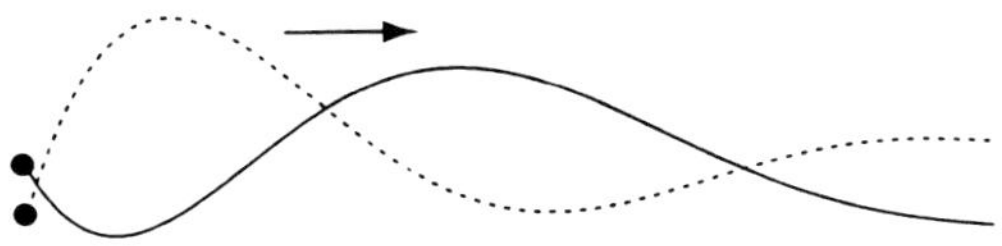

Fig. 22. Undulation generated by oscillatory motor forces within a flagellum.

solution for the situation correponding to a spermatozoan is shown in Figure 22. A bending wave propagates from the head to the tail, if the head experiences a large viscous drag.

A more general treatment [64] recognizes that the elastic force opposing the action of the motors varies along the axoneme, and that the motor force is therefore also a function of position. Solutions may be obtained by considering the form of the response function relating the local motor force to the local sliding velocity, derived for example from a simple two-state isothermal ratchet model. In this case, the stationary solution is stable when the system is close to equilibrium, but a dynamical instability occurs at a critical value of the control parameter Ω. The dynamical mode that is selected, and the characteristic frequency of the movement, both depend upon the length of the axoneme. Long flagella propogate a bending wave, similar to that shown in Figure 22. Shorter axonemes simply flex from side to side, at a frequency which is governed by equation (3.9), with an effective spring constant $K \approx \kappa L^3$ and an effective friction $\zeta \approx \eta_\perp L$. The critical frequency is then

$$\omega_c \approx \left(\frac{\alpha\kappa}{\eta_\perp}\right)^{1/2} \frac{1}{L^2}, \tag{3.14}$$

which increases rapidly as the length of the axoneme decreases.

4 Sense and sensitivity: Mechano-sensation in hearing

4.1 System performance

The performance of our senses is remarkable, and none more so than hearing. The faintest sounds that we can hear impart no more energy, per cycle, than the thermal motion. At the same time, the ear can cope with loud noises that carry more than twelve orders of magnitude more energy. And, of course, it can analyze frequency: two tones that differ by a few percent can be resolved. Just how this astonishing feat is achieved has been a puzzle for over a century, but recent research is beginning to penetrate the mystery.

The key to the ear's acuity is an active system of sound detection [65]. The ear is powered. That this was likely to be the case was first realized by Gold [66], more than fifty years ago. He pointed to a problem with the theory of hearing propounded by Helmholtz [67], who had argued that the ear uses a set of inertial resonators to capture the energy of sound waves. Given that the cochlea is filled with fluid, and the likely size of the resonators is a few microns at most, the strong damping precludes a sharp resonant response. Gold proposed that the ear must work like a regenerative radio receiver, and add in energy at the very frequency it is trying to detect. It was clear to him that such a mechanism must be very delicate, however, as it would require a positive feedback of exactly the right magnitude to cancel the damping. Any less and the ear would be insensitive; any more and it would ring spontaneously.

It was not until the 1970s that the idea of an active cochlea began to be taken seriously. At that time, the first successful measurements on a living ear revealed it to respond far more sensitively than the dead cochlea [68, 69]. Subsequently it was discovered that the ear can spontaneously emit sounds [70]. A microphone placed in the ear cannal usually records a faint hum, but occasionally shrill whistles can be detected as well. Clearly, something within the cochlea is oscillating. In this section, we shall explore how active oscillators can help the ear to hear, and discuss whether motor proteins might be an important component of the active system.

4.2 Mechano-sensors: Hair bundles

A hair bundle is an appendage measuring a few microns high that sticks up above the surface of every hair cell [71] (see Fig. 23). It consists of a number of stereocilia (each composed of a bundle of actin filaments) which slope up against each other to form a pyramidal structure. Each stereocilium is connected to its neighbour by a fine filament, called a "tip link". Shear flow in the cochlea fluid causes the whole bundle to deflect, with each stereocilium pivoting at its base, so that the tip links get stretched. Each tip link connects directly to a tension-gated transduction channel in the cell membrane, which admits potassium ions. So the deflection leads to a change in the ionic current that, it turn alters the cell potential. The hair bundle is therefore a transducer which provides a very direct conversion of a mechanical stimulus to an electrical signal.

In the 1980s it was discovered that hair bundles can oscillate spontaneously [72]. This behaviour had been very difficult to reproduce, however, until recently when it bacame possible to control of the extracellular ionic concentrations, so that they resemble the conditions *in vivo* [73]. In this situation, oscillations occur quite readily. It appears that calcium ions, which are present at low concentration in the fluid surrounding the bundle,

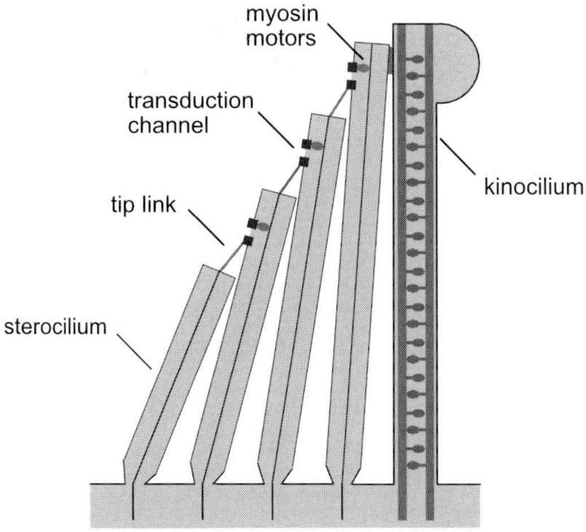

Fig. 23. Hair bundle of a non-mammalian vertebrate.

and which are also admitted by the transduction channel, are crucial to the active process.

Hair bundles also contain molecular motors. Attached to the transduction channel are a number of myosin-1C motors. They are believed to play a role in adapting the hair bundle to varying cellular conditions, by maintaining the proper tension in the tip links [74]. But they might also be more directly involved in generating bundle oscillations. The hair bundles of non-mammalian vertebrates also incorporate a "kinocilium", which has the same architecture as an axoneme. The dynein motors within this structure might also play an active part in the hair bundle response.

4.3 Active amplification

There is considerable evidence, then, that hair bundles contain a force-generating dynamical system that can generate spontaneous oscillations at a characteristic frequency ω_C. In general, the behaviour of such a nonlinear dynamical system may be described by a control parameter C. Above a critical value, $C > C_c$, the system is stable; for $C < C_c$ it oscillates spontaneously. At the critical point $C = C_c$, which is a Hopf bifurcation, the system has remarkable response properties. Because the behaviour close to critical points is generic, we can calculate these properties without a detailed knowledge of the physical process that gives rise to the dynamical instability [75–77].

If a stimulus force $f(t)$ is applied at frequency ω, the hair bundle displacement $x(t)$ has a principal component at that same frequency. So writing the complex Fourier coefficients at frequency ω as $\tilde{f}$ and $\tilde{x}$, the reponse may be expressed as a systematic expansion

$$\tilde{f} = \mathcal{A}\,\tilde{x} + \mathcal{B}\,|\tilde{x}|^2\tilde{x} + ..., \tag{4.1}$$

where $\mathcal{A}(\omega, C)$ and $\mathcal{B}(\omega, C)$ are two complex functions. For a system that undergoes a Hopf bifurcation, the first nonlinear term is cubic. The bifurcation point is characterized by the fact that $\mathcal{A}$ vanishes for the critical frequency, $\mathcal{A}(\omega_{\rm c}, C_{\rm c}) = 0$.

Suppose that the system is poised right at the critical point, $C = C_{\rm c}$. Then the response to a stimulus at the critical frequency, $\omega = \omega_{\rm C}$ has amplitude

$$|\tilde{x}| \approx \frac{|\tilde{f}|^{1/3}}{|\mathcal{B}|^{1/3}}\,. \tag{4.2}$$

This is a highly compressive response, which boosts weak signals much more than strong signals. Indeed, the gain

$$r = \frac{|\tilde{x}|}{|\tilde{f}|} \sim \frac{1}{|\tilde{f}|^{2/3}} \tag{4.3}$$

becomes arbitrarily large for small forces. The critical Hopf oscillator acts as a nonlinear amplifier.

If the stimulus frequency differs from the critical frequency, the linear term in equation (4.1) is non-zero and can be expressed to first order as $\mathcal{A}(\omega, C_{\rm c}) \simeq A_1\ (\omega - \omega_{\rm c})$. When this term exceeds the cubic term in equation (4.1), active amplification is lost and the response becomes linear

$$|\tilde{x}| \simeq \frac{|\tilde{f}|}{|(\omega - \omega_{\rm c})A|}\,. \tag{4.4}$$

The bandwidth of active amplication Δ therefore depends on the level of the stimulus:

$$\Delta \approx \frac{|\mathcal{B}|^{1/3}}{|A_1|}|\tilde{f}|^{2/3}. \tag{4.5}$$

The resonant response of a critical Hopf oscillator is summarized in Figure 24. The active system acts as a sharply tuned high-gain amplifier for weak stimuli, and as a low-gain filter for strong stimuli. These main features are displayed by the response of the basilar membrane in the mammalian cochlea [78].

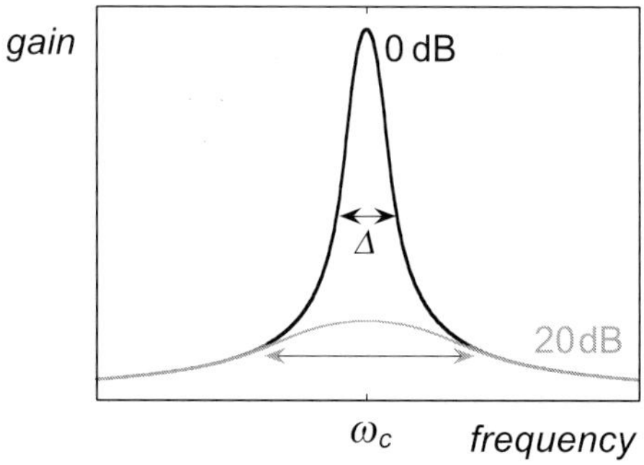

Fig. 24. Hopf resonance. The gain and the sharpness of response are much greater for a weak stimulus (black), than for a stimulus of ten times greater amplitude (grey).

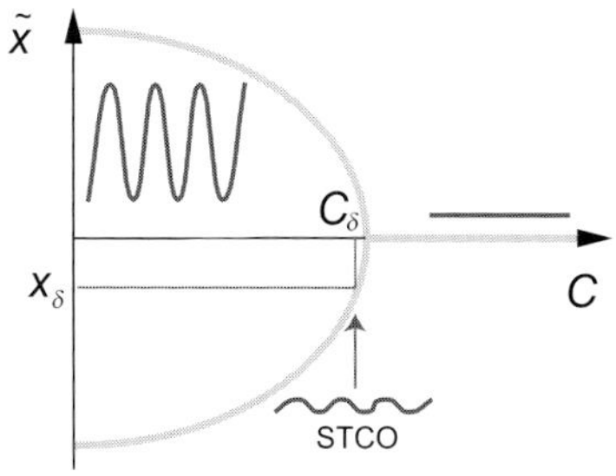

Fig. 25. The working point of the self-tuned system is just on the oscillating side of the bifurcation, yielding self-tuned critical oscillations (STCO).

4.4 Self-tuned criticality

The Hopf resonance is perfectly suited to the ear's needs. It permits frequency descrimination; it boosts faint sounds; and the strongly compressive response provides a huge dynamic range – the twelve orders of magnitude of sound energy that we can comfortably hear give rise to hair bundle displacements that vary by only a factor of one hundred. To profit from the nonlinear amplification, however, each oscillator has to be very close to its critical point. Clearly some kind of regulation mechanism is required to ensure that this is the case.

A feedback mechanism that links the reponse of the system to the control parameter can permit the system to operate *automatically* close to the bifurcation point, whatever its characteristic frequency [76] (Fig. 25). Suppose that some mechanism causes the control parameter to decrease as long as the system does not oscillate. After some time, critical conditions will be reached and spontaneous oscillations will ensue. The onset of oscillations triggers an increase of the control parameter which tends to restore stability. Hence the system converges to an operating point close to the bifurcation point. The following simple feedback, which changes C in response to deflections x, illustrates the general idea:

$$\frac{1}{C}\partial_t C = \frac{1}{\tau}\left(\frac{x^2}{\delta^2} - 1\right) \tag{4.6}$$

where δ is a small amplitude. If no external force is applied, this feedback, after a relaxation time τ, tunes the control parameter to a value C_δ (just less than C_c) for which spontaneous oscillations with $|\tilde{x}| \simeq \delta$ occur. These small-amplitude oscillations are referred to as *self-tuned critical oscillations*. Maintained on the threshold of vibrations by this control mechanism, a hair bundle is exquisitely sensitive to purturbation by periodic stimuli at its characteristic frequency.

4.5 Motor-driven oscillations

What is the physical basis of the dynamical system that generates the oscillations? Because motor proteins are specialized to produce motility, they are an obvious candidate. One possibility is that the kinocilium is the motile element of non-mammalian hair bundles [76]. We have already seen that an axoneme can flex at a frequency that depends on its length, equation (3.9). So in this case, the characteristic frequency of a hair bundle would depend on its architecture. Tall bundles would oscillate at lower frequencies than short bundles. This agrees with what is known about the tonotopic organization of hair cells in the cochlea [79]. It is also plausible that it is the myosin-1C motors joined to the transduction channels, rather than the dynein motors in the kinocilium that cause the oscillations. In either case, the characteristic frequency of the bundle can be significantly faster than the cycle rate of individual motors.

How could self-tuning to the critical point be realized in this system? One possibility is that the influx of calcium ions through the transduction channel down-regulates the motor activity (Fig. 26a) [76]. In this case, the Ca^{2+} concentration plays the role of the control parameter C. Assuming that ion pumps in the cell membrane constantly pump Ca^{2+} out of the cell, C obeys the dynamical equation

$$\partial_t C = -\frac{C}{\tau_{\text{ion}}} + J P_{\text{o}}(x), \tag{4.7}$$

where τ_{ion} is the ionic relaxation time and J is the Ca^{2+} current through an open transduction channel. The probability $P_{\text{o}}(x)$ that a channel is open

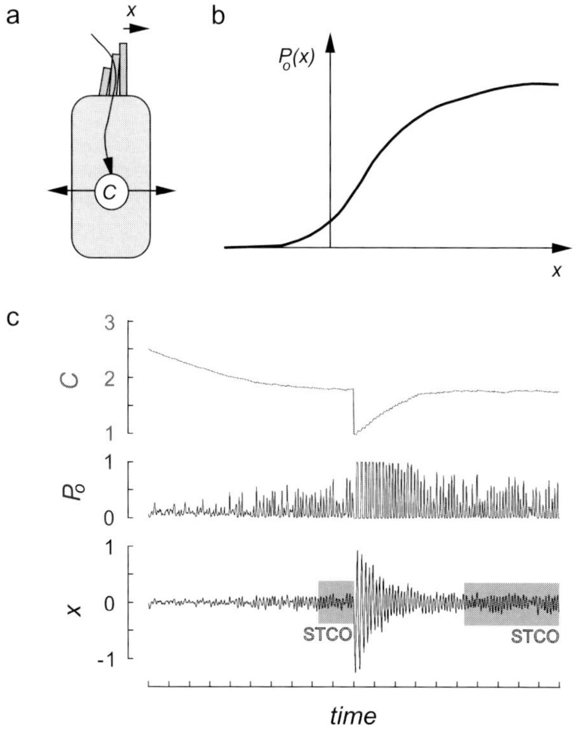

Fig. 26. Self-tuning mechanism. **a)** Calcium flux through the hair cell. **b)** Channel open probability $P(x)$. **c)** The bundle settles down to self-tuned critical oscillations after a relaxation time $\tau_{\rm ion}$, whether the Ca^{2+} concentration C is intially too high or too low.

depends on the hair bundle displacement, as shown in Figure 26b. This is the typical sigmoidal relation, of the form

$$P_{\rm o}(x) = \frac{1}{1 + A\mathrm{e}^{-x/\delta}} \tag{4.8}$$

that is expected if the channel makes rapid stochastic transitions between an open and a closed state, gated by the tension in the tip link. The coefficient A is large, so that when the bundle is still, there is only a slight probability that the channel is open. However, owing to the curvature of $P_{\rm o}(x)$ at $x = 0$, the mean probability of the channel being open increases if the bundle becomes unstable and starts to oscillate. Equation (4.7) then implies that C rises. If the Ca^{2+} ions down-regulate the motors, for example by decreasing the rate Ω at which they detach, the system moves back towards the quiescent regime. This feedback control is a robust way of generating self-tuned critical oscillations, as illustrated in Figure 26c.

4.6 Channel compliance and relaxation oscillations

An alternative mechanism by which oscillations might be generated is suggested by recent micromanipulation experiments on hair bundles [80]. When the tip of a bundle is abruptly displaced by a small amount, the bundle reacts by generating the force in the opposite direction. Indeed, the instantaneous force-displacement relation of the bundle (*i.e.* the relation obtained before adaptation processes mediated by motors or calcium have an affect) displays a region of negative slope (Fig. 27a).

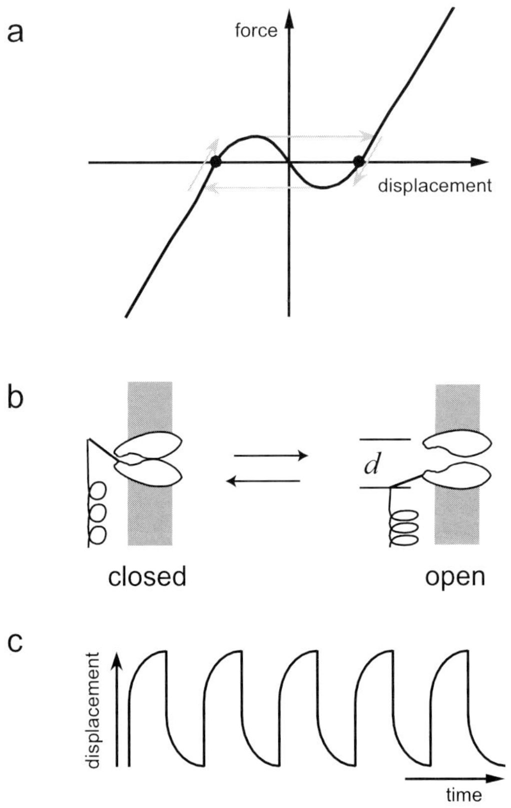

Fig. 27. **a)** Two-state transduction channel with a lever arm. **b)** The force-displacement relation has a region of instability owing to the channel compliance. **c)** The action of the myosin-1C motors can push the system around the arrowed curve, generating relaxation oscillations.

One potential explanation of this behaviour is *channel compliance* [80, 81]. Suppose that the transduction channel has a lever arm, similar to that of a myosin molecule, which amplifies the small movements associated with the opening and closing of the channel (Fig. 27b). Then when the hair bundle is pushed a distance x the positive direction, adjacent stereocilia are sheared by $y = \gamma x$, where γ is a geometric factor that depends on the bundle height. The increased tension T_{tl} in the tiplinks causes the channels to open. Indeed, assuming that the channel kinetics is rapid enough for there to be an equilibrium between open and closed states, the open probability P_{o} is

$$P_{\mathrm{o}} = \frac{1}{1 + A\mathrm{e}^{-K_{\mathrm{tl}}yd/kT}}, \tag{4.9}$$

where K_{tl} is the elastic constant of the tip link, d is the swing of the lever arm and A is a dimensionless constant whose value depends on the free energy difference of the two channel states. The associated movement of the lever arm diminishes the tension in the tip links, so that

$$T_{\mathrm{tl}} = K_{\mathrm{tl}}(y - dP_{\mathrm{o}}). \tag{4.10}$$

The total reactive force of the bundle has contributions from both the tip links and the deformation of the stereocilia pivots,

$$F = N(\gamma T_{\mathrm{tl}} + K_{\mathrm{sp}}x), \tag{4.11}$$

where N is the total number of stereocilia and K_{sp} is the effective Young modulus due to a single sterocilium. The force-displacement relation resulting from equations (4.9–4.11) has a region of negative slope if

$$K_{\mathrm{tl}}d^2 > 4kT. \tag{4.12}$$

In this situation, there is a range of applied forces for which the bundle is *bistable*. If the position of the adaptation motors is fixed, the hair bundle will settle at one or other of the stable positions. However, this state of affairs can be upset if calcium down-regulates the motors, as suggested above. Suppose, for example, that the bundle is at the fixed point with the higher value of x, for which there is a high probability that the transduction channels are open. The Ca^{2+} ions entering through the channel bind to the motors, causing a fraction of them to detach; the diminishing force exerted by the motors causes the tension in the tip links to fall, and the bundle to move backwards. As indicated in Figure 27a, the fixed point vanishes at a critical value of the motor force, and the system then abruptly jumps to the other fixed point. At this lower value of x, the channels are mostly closed. The resulting drop in calcium concentration augments the number of bound motors, increasing the tension in the tip links until the lower fixed

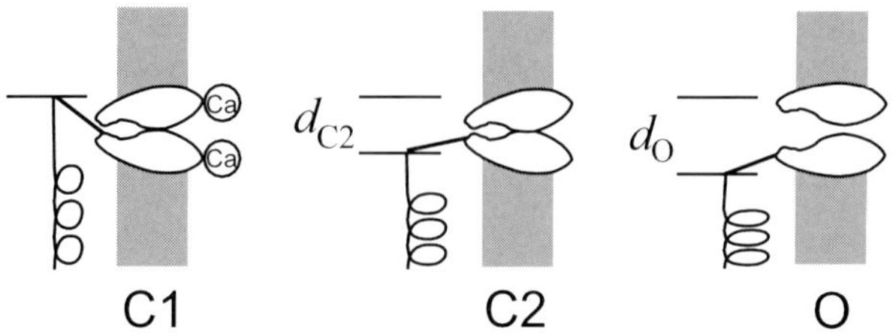

Fig. 28. Three-state channel, in which one of the closed states is stabilized by calcium.

point becomes unstable, whereupon the hair bundle jumps back to its initial position.

This dynamics, whereby the myosin-1C motors shift the bundle from one stable fixed point to another, results in relaxation oscillations with a characteristic form, shown in Figure 27c. Oscillations of this type have been observed in an *in vitro* system [73, 80]. It is not clear, however, whether this mechanism forms the basis of the active amplier. There is no obvious way that the system could be regulated to the critical point at which the oscillations have vanishing amplitude.

4.7 Channel-driven oscillations

There is a second adpatation process that modifies the transduction current, which is much faster than the mechanism mediated by the myosin-1C motors [74, 82]. It depends on the concentration of calcium ions outside the cell and is believed to be caused by Ca^{2+} ions binding to the transduction channels and favouring their closure. Thus, a more appropriate model of the channel might incorporate three states, as shown in Figure 28, in which one of the closed states is stabilized by the binding of Ca^{2+}. The dynamics of calcium, which enters through the transduction channels and is continually pumped out of the cell, as described by equation (4.7), provides a fast feedback. By shifting the channel states and modifying the contribution of the channel compliance, it modifies the force-displacement relation. If the feedback is strong enough, the region of negative stiffness is eliminated and the system has a single fixed point [83]. However, this fixed point may be either stable or unstable and, in the latter case, the bundle executes limit cycle oscillations. Thus the interaction of the calcium with the channel is able to generate a Hopf bifurcation. With their long lever arms, the channels effectively act like molecular motors, driven by calcium rather than by ATP.

The characteristic frequency of oscillation depends on two time scales

$$\omega_c \approx \left(\frac{1}{\tau_{\rm mech}\tau_{\rm ion}} \right)^{1/2}, \tag{4.13}$$

where $\tau_{\rm ion}$ is the relaxation time of the calcium concentration and $\tau_{\rm mech}$ is the viscous relaxation time of the bundle. It scales as

$$\omega_c \sim \left(\frac{N}{L^3 \tau_{\rm ion}} \right)^{1/2}, \tag{4.14}$$

and thus depends strongly on the architecture of the bundle.

Crucially, the stability of the fixed point is determined by the location of the myosin-1C motors. If these motors generate a large force, the hair bundle oscillates; if they generate a small force, the bundle is quiescent. This suggests that the function of the myosin-1C motors is to tune the system to the critical point [83]. Down-regulation of the motors by Ca^{2+} entering through the channels when the bundle oscillates could fulfil this function.

Thus, the two adaptation mechanisms that have been identified for hair cells can together generate self-tuned critical oscillations. Calcium acting on the channels creates a dynamical instability and the molecular motors, acting on a slower time scale, adjust the system to the immediate vicinity of the critical point.

4.8 Hearing at the noise limit

Hair bundles are subject to noise from a number of sources. In addition to the Brownian forces of the molecules in the surrounding fluid, the stochastic nature of the force-generating system adds further randomness to the system. How can the bundle detect a weak signal in the presence of this noise?

As a consequence of the stochasticity, the self-tuned critical oscillations are irregular, as illustrated in Figure 26c. The response of a self-tuned hair bundle to a sinusoidal force with a frequency approximately equal to the bundle's characteristic frequency [76] is illustrated in Figure 29. For weak stimuli, the amplitude of the oscillation does *not* increase with the amplitude of the applied force; this is because the small response to the stimulus is masked by the noisy, spontaneous motion. Instead, the *phase* of the hair-bundle oscillation becomes more regular; as it does so, a peak emerges from the Fourier spectrum at the driving frequency. The height of the peak grows as the cube root of the stimulus amplitude, following the generic behaviour at a Hopf bifurcation specified by equation (4.2).

This suggests how a hair bundle can achieve its remarkable sensitivity to weak stimuli [76]. By profiting from the periodicity of a sinusoidal input,

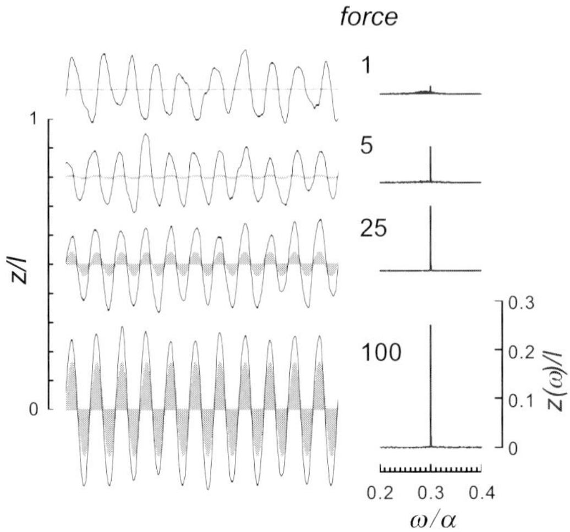

Fig. 29. Detection in the presence of noise. Response to sinusoidal forces of different magnitude (left), and Fourier transform of the response (right). The response of an equivalent passive system in the absence of noise is indicated in grey.

and measuring phase-locking rather than the amplitude of response, the mechano-sensor can detect forces considerably weaker than those exerted by a single molecular motor (if the bundle were a simple, passive structure, its response to such forces would be smaller than its Brownian motion). An important implication of this detection mechanism is that the signal must be encoded by the interval between spikes elicited in the auditory nerve. Paradoxically, the stochastic noise serves a useful purpose. It ensures that the self-tuned critical oscillations of the hair bundle are incoherent, so that the pattern of spontaneous firing in the nerve is irregular. Against this background, the regular response to a periodic stimulus can easily be detected. Another beneficial feature of noise arises from the fact that weak stimuli do not increase the amplitude of oscillation above the spontaneous amplitude. Thus the Ca^{2+} concentration remains constant, the hair bundle stays in the critical regime, and active amplification can be sustained indefinitely.

The active system of detection, in which motor proteins play a key role, explains how the ear can detect sound waves whose energy per cycle is similar to that of the thermal noise.

My understanding of motor protein systems has benefited from conversations with numerous colleagues, including S. Block, L. Bourdieu, S. Camalet, J. Howard, A.J. Hudspeth, A.F. Huxley, S. Leibler, M. Magnasco, P. Martin, and A. Vilfan. I am especially indepted to F. Jülicher and J. Prost, on whose research much of this course is based. The Royal Society provided funding. Last but not least, I thank the organisers and participants of the School who made the summer at Les Houches such an enjoyable experience.

References

[1] T. Kreis and R. Vale, Cytoskeletal Motor Proteins (Oxford University Press, New York, 1993).

[2] B. Alberts, D. Bray, J. Lewis, M. Raff, K. Roberts and J.D. Watson, Molecular Biology of the Cell (Garland, New York, 1994).

[3] D. Bray, Cell Movements (Garland, New York, 2001).

[4] J. Howard, Mechanics of Motor Proteins and the Cytoskeleton (Sinauer, Sunderland Massachusetts, 2001).

[5] M.O. Magnasco, *Phys. Rev. Lett.* **71** (1993) 1477.

[6] J. Prost, J.F. Chauwin, L. Peliti and A. Ajdari, *Phys. Rev. Lett.* **72** (1994) 2652.

[7] R.D. Astumian and M. Bier, *Phys. Rev. Lett.* **72** (1994) 1766.

[8] M.O. Magnasco, *Phys. Rev. Lett.* **72** (1994) 2656.

[9] C.S. Peskin and G.F. Oster, *Biophys. J.* **68** (1995) 202.

[10] T.A.J. Duke and S. Leibler, *Biophys. J.* **71** (1996) 1235.

[11] I. Derènyi and T. Vicsek, *Proc. Natl. Acad. Sci. USA* **93** (1996) 6775.

[12] F. Jülicher, A. Ajdari and J. Prost, *Rev. Mod. Phys.* **69** (1997) 1269.

[13] R.D. Astumian, *Science* **276** (1997) 917.

[14] M.E. Fisher and A.B. Kolomeisky, *Proc. Natl. Acad. Sci. USA* **96** (1999) 6597.

[15] R.P. Feynman, R.B. Leighton and M. Sands, The Feynman lectures on physics, vol. I, chap. 46 (Addison-Wesley, Reading MA, 1966).

[16] A. Ajdari and J. Prost, *C.R. Acad. Sci. Paris II* **315** (1992) 1635.

[17] J. Rousselet, L. Salome, A. Ajdari and J. Prost, *Nature* **370** (1994) 446.

[18] L.P. Faucheux, L.S. Bourdieu, P.D. Kaplan and A.J. Libchaber, *Phys. Rev. Lett.* **74** (1995) 1504.

[19] J.S. Bader, R.W. Hammond, S.A. Henck, M.W. Deem, G.A. McDermott, J.M. Bustillo, J.W. Simpson, G.T. Mulhern and J.M. Rothberg, *Proc. Natl. Acad. Sci. USA* **96** (2000) 13165.

[20] C.S. Peskin, G.M. Odell and G.F. Oster, *Biophys. J.* **65** (1993) 316.

[21] M. Dogterom and B. Yurke, *Science* **278** (1997) 856.

[22] H.A. Kramers, *Physica* **7** (1940) 284.

[23] A. Parmeggiani, F. Jülicher, A. Ajdari and J. Prost, *Phys. Rev. E* **60** (1999) 2127.

[24] W.O. Hancock and J. Howard, *Proc. Natl. Acad. Sci. USA* **96** (1999) 13147.

[25] K. Visscher, M.J. Schnitzer and S.M. Block, *Nature* **400** (1999) 184.

[26] P.D. Boyer, *Annu. Rev. Biochem.* **66** (1997) 717.

[27] H. Wang and G. Oster, *Nature* **396** (1998) 279.

[28] H.E. Huxley, *Science* **164** (1969) 1356.

[29] A.F. Huxley and R.M. Simmons, *Nature* **233** (1971) 533.

[30] T.L. Hill, *Prog. Biophys. Mol. Biol.* **28** (1974) 267.

[31] E. Eisenberg, T. Hill and Y. Chen, *Biophys. J.* **29** (1980) 195.

[32] S. Leibler and D.A. Huse, *J. Cell Biol.* **121** (1993) 1357.

[33] T.A.J. Duke, *Proc. Natl. Acad. Sci. USA* **96** (1999) 2770.

[34] A. Vilfan, E. Frey and F. Schwabl, *Euro. Phys. Lett.* **45** (1999) 283.

[35] M.K. Reedy, K.C. Holmes and R.T. Tregear, *Nature* **328** (1987) 536.

[36] I. Rayment, H.M. Holden, M. Whittaker, C.B. Yohn, M. Lorenz, K.C. Holmes and R.A. Milligan, *Science* **261** (1993) 58.

[37] J.T. Finer, R.M. Simmons and J.A. Spudich, *Nature* **368** (1994) 113.

[38] J.E. Molloy, J.E. Burns, J. Kendrick-Jones, R.T. Tregear and D.C.S. White, *Nature* **378** (1995) 209.

[39] K.C. Holmes, *Curr. Biol.* **7** (1997) R112-R118.

[40] R.W. Lymn and E.W. Taylor, *Biochem.* **10** (1971) 4617.

[41] C. Veigel, M.L. Bartoo, D.C. White, J.C. Sparrow and J.E. Molloy, *Biophys. J.* **75** (1998) 1424.

[42] K. Tawada and K. Sekimoto, *J. Theor. Biol.* **150** (1991) 193.

[43] M. Whittaker, E.M. Wilson-Kubalek, J.E. Smith, L. Faust, R.A. Milligan and H.L. Sweeney, *Nature* **378** (1995) 748.

[44] C. Veigel, L.M. Coluccio, J.D. Jontes, J.C. Sparrow, R.A. Milligan and J.E. Molloy, *Nature* **398** (1999) 530.

[45] A.V. Hill, *Proc. R. Soc. B* **126** (1938) 136.

[46] K.A.P. Edman, *J. Physiol.* **404** (1988) 301.

[47] R.J. Podolsky, *Nature* **188** (1960) 666.

[48] K.A.P. Edman and N.A. Curtin, *J. Physiol.* **534** (2001) 553.

[49] F. Jülicher and J. Prost, *Phys. Rev. Lett.* **75** (1995) 2618.

[50] M. Badoual, F. Jülicher and J. Prost, *Proc. Natl. Acad. Sci. USA* **99** (2002) 6696.

[51] S.A. Endow and H. Higuchi, *Nature* **406** (2000) 913.

[52] D. Riveline, A. Ott, F. Jülicher, O. Cardoso, S. Magnusdottir, J.L. Viovy and J. Prost, *Eur. Biophys. J.* **27** (1998) 403.

[53] F. Jülicher and J. Prost, *Phys. Rev. Lett.* **78** (1997) 4510.

[54] F. Jülicher, *C.R. Acad. Sci. Paris Serie IV* **2** (2001) 849.

[55] J.W.S. Pringle, in R.T. Tregear, *Insect Flight Muscles* (North-Holland, Amsterdam 1977).

[56] K. Yasuda, Y. Shindo and S. Ishiwata, *Biophys. J.* **70** (1996) 1823.

[57] H. Fujita and S. Ishiwata, *Biophys. J.* **75** (1998) 1439.

[58] L. Bourdieu, T.A.J. Duke, M.B. Elowitz, D.A. Winkelmann, S. Leibler and A. Libchaber *Phys. Rev. Lett.* **75** (1995) 176.

[59] A.E.H. Love, *A Treatise on the Mathematical Theory of Elasticity* (Dover, New York, 1944).

[60] I.R. Gibbons, *J. Cell Biol.* **91** (1981) 107s.

[61] C.J. Brokaw, *Science* **178** (1972) 455.

[62] M. Murase, *The Dynamics of Cell Motility* (Wiley, New York, 1992).

[63] S. Camalet, F. Jülicher and J. Prost, *Phys. Rev. Lett.* **82** (1999) 1590.

[64] S. Camalet and F. Jülicher, *New J. Phys.* **2** (2000) 1.

[65] P.J. Dallos, *J. Neurosci.* **12** (1992) 4575.

[66] T. Gold, *Proc. R. Soc. B* **135** (1948) 462.

[67] H.L.F. Helmholtz, *On the Sensations of Tone* (Dover, New York, 1954).

[68] W.S. Rhode, *J. Acoust. Soc. Am.* **49** (1971) 1218.

[69] P.M. Sellick, R. Patuzzi and B.M. Johnstone, *J. Acoust. Soc. Am.* **72** (1982) 131.

[70] D.T. Kemp, *J. Arch. Otorhinolaryngol.* **224** (1979) 37.

[71] A.J. Hudspeth and V.S. Markin, *Physics Today* **2** (1994) 22.

[72] A.C. Crawford and R. Fettiplace, *J. Physiol.* **364** (1985) 359.

[73] P. Martin and A.J. Hudpeth, *Proc. Natl. Acad. Sci. USA* **96** (1999) 14306.

[74] J.R. Holt and D.P. Corey, *Proc. Natl. Acad. Sci. USA* **97** (2000) 11730.

[75] Y. Choe, M.O. Magnasco and A.J. Hudspeth, *Proc. Natl. Acad. Sci. USA* **95** (1998) 15321.

[76] S. Camalet, T.A.J. Duke, F. Jülicher and J. Prost, *Proc. Natl. Acad. Sci. USA* **97** (2000) 3138.

[77] V.M. Eguiluz, M. Ospeck, Y. Choe, A.J. Hudspeth and M.O. Magnasco, *Phys. Rev. Lett.* **84** (2000) 5232.

[78] L. Robles and M.A. Ruggero, *Physiol. Rev.* **81** (2001) 1305.

[79] L.G. Tilney and J.C. Saunders, *J. Cell Biol.* **96** (1983) 807.

[80] P. Martin, A.D. Mehta and A.J. Hudpeth, *Proc. Natl. Acad. Sci. USA* **97** (2000) 12026.

[81] J. Howard and A.J. Hudpeth, *Neuron* **1** (1988) 189.

[82] D.P. Corey and A.J. Hudspeth, *J. Neurosci.* **3** (1893) 962.

[83] A. Vilfan and T. Duke (to be published).

COURSE 4

DYNAMIC FORCE SPECTROSCOPY

E. EVANS* AND P. WILLIAMS**

* Physics and Pathology, University of British Columbia, Vancouver, Canada V6T 2A6; Biomedical Engineering, Boston University, Boston, MA 02215, USA

** Pharmaceutical Sciences, University of Nottingham, Nottingham, UK

Contents

DYNAMIC FORCE SPECTROSCOPY

E. Evans[1] and P. Williams[2]

Part 1: E. Evans and P. Williams

1 Dynamic force spectroscopy. I. Single bonds

1.1 Introduction

Weak-noncovalent interactions govern structural cohesion and mediate most of life's functions from the outer membrane surface to the interior nucleus of a cell. On laboratory time scales, the energy landscape of a weak bond is fully explored by Brownian-thermal excitations, and energy barriers along its dissociation pathway(s) become encoded in a rate of unbonding that can range from $\sim 1/\mu s$ to 1/year. When pulled apart with a ramps of force, the dissociation kinetics become transformed into a dynamic spectrum of unbonding force as a function of the steepness of the force ramps (loading rates). Expressed on a logarithmic scale in loading rate, the spectrum of breakage forces begins first with a crossover from *near equilibrium* to *far from equilibrium* unbonding and then rises through ascending regimes of strength. These regimes expose the prominent energy barriers traversed along the dissociation pathway. Labelled as *dynamic force spectroscopy* [7, 10], this approach is being used to probe the inner world of biomolecular interactions [7, 8, 13, 14, 23, 24, 26, 30] and reveals energy barriers that are difficult or impossible to access by solution assays of near-equilibrium kinetics. These hidden barriers are crucial for specialized dynamic functions of molecules.

In this first chapter of our tutorial, we begin with an outline of the physics needed to understand the impact of force on lifetime of a single bond. Then deriving prescriptions for rate of transition under force, we analyze the stochastic process of unbonding in a probe experiment and

[1]Physics and Pathology, University of British Columbia, Vancouver, Canada V6T 2A6; Biomedical Engineering, Boston University, Boston, MA 02215, USA.
[2]Pharmaceutical Sciences, University of Nottingham, Nottingham, UK.

demonstrate the kinetic origin of the force distribution, the peak of which defines *bond strength*. Finally, we show how these developments come together to establish the method of dynamic force spectroscopy and give examples of single molecule experiments. In the second chapter to follow, we describe how a nanoscale attachment made up of a few bonds fails under force and develop limiting models for use in analysis of probe tests that involve multiply-bonded contacts.

1.1.1 Intrinsic dependence of bond strength on time frame for breakage

Unlike interatomic linkages within nucleic acid – protein – lipid – carbohydrate structures, weak noncovalent bonds between these biomolecules have limited lifetimes and will dissociate under almost any level of force if pulled on for modest periods of time. When close *to equilibrium* in solution, large numbers of molecules continuously bond and dissociate under zero force; thus, application of a field (*e.g.* electrical force or osmotic stress) to the reacting molecules simply alters the ratio of bound-to-free constituents. But at infinite dilution, an isolated molecular complex ("bond") exists *far from equilibrium* and has no strength on time scales longer than the time $t_{\rm off} = 1/K_{\rm off}$ needed for spontaneous dissociation. If pulled apart faster than $t_{\rm off}$, a solitary bond will resist detachment. The unbonding force can range up to – and even exceed – the adiabatic limit $f_\infty \sim |\partial E/\partial x|_{\rm max}$ defined by the steepest gradient in the intermolecular potential $E(x)$ that binds the complex. In other words, if the bond is broken in less time than required for diffusive relaxation ($\sim 10^{-9}$ s), the force must exceed the "brittle" fracture strength of the bond. However, between the extremes in time scale (from a nanosecond to the time for spontaneous dissociation), the force needed to disrupt a weak bond is reduced significantly by thermal activation. Albeit very rarely, Brownian excitations in the liquid environment occasionally contribute large transient impulses of force which, added to a modest external force, exceed the steepest gradient in the intermolecular potential. This enables passage of the confining energy barrier. The physics that governs activated processes in liquids is century old beginning with Einstein's theory of Brownian motion [6] and culminating in Kramers theory for escape from a bound state in liquids [16, 18]. We will use this physics to establish the crucial connection between force – lifetime – and chemistry for a single molecular bond.

1.1.2 Biomolecular complexity and role for dynamic force spectroscopy

What's subtle and daunting about *biomolecular* bonds it that the interactions are usually made up of many atomic scale bonds distributed over diverse regions of large molecules – *i.e.* not localized to a single amino acid

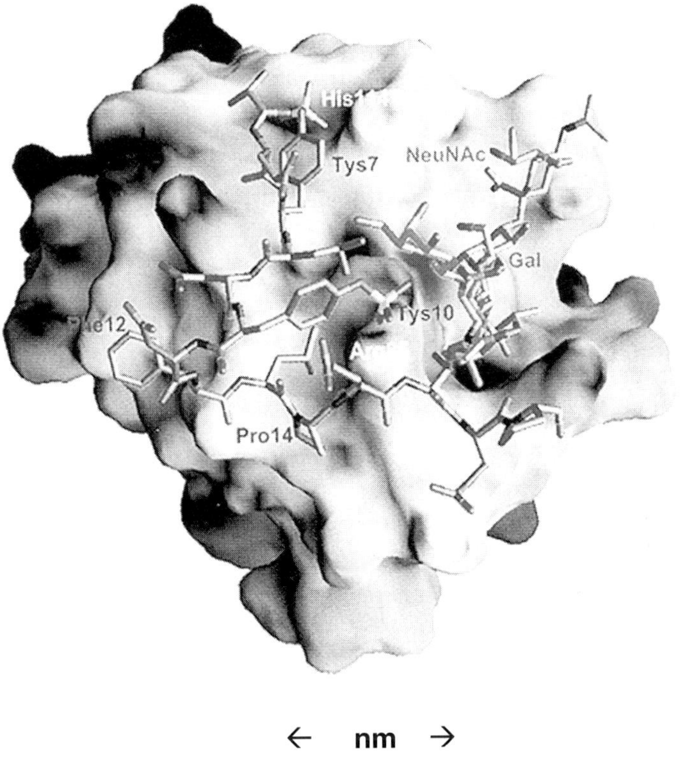

Fig. 1.1. Determined by X-ray diffraction, a vertex (stick) representation shows the important 19 amino acid tip (SGP) of the mucin P-selectin glycoprotein ligand-1 (PSGL-1) in its bound state conformation superposed on the van der Waals surface of the outer lectin domain of the cell membrane receptor P-selectin (taken from Somers *et al.* [29]).

or other small molecular residue. As an illustration, Figure 1.1 shows the structural complex obtained recently for the reactive tip of a glycosylated protein ligand bound to the outer protein domain of its cell surface receptor called P(for platelet)-selectin [29]. Essential in immune function, this interaction enables white blood cells to transiently stick and carry out a rolling patrol of the vascular wall under high shear stress in blood vessels. Referred to as a carbohydrate-protein bond, this ligand-receptor interaction is comprised of several sugar-peptide and sulfopeptide-peptide hydrogen bonds plus a metal-ion coordination bond spread over many residues of both the receptor lectin domain and the tip of the large glycoprotein ligand. Even so, association and dissociation of this ligand-receptor complex in solution seems to exhibit first order kinetics as expected for an ideal "bond", which is modeled as a bound state confined by a single energy barrier. Furthermore, from force probe tests, we will also see that a single barrier dominates the kinetics of dissociation over many orders of magnitude in time scale for this complex interaction. Yet, probe tests of other species of the same class of bonds reveals that a sequence of barriers impede dissociation in different ranges of force. Thus, the landscape of energy barriers in a complex interaction can produce highly specialized dynamic responses in molecular reactions and linkages under stress, which is a principal design requirement for chemistry in living systems. An important step towards understanding these designs lies in probing the relation between force – time – chemistry at the level of single molecules.

In this tutorial, our aim is to show that measuring forces to pull apart single biomolecular complexes over an enormous span of time scales provides a spectroscopic method to explore the energy landscape of barriers which govern dissociation kinetics. By landscape, we mean the free energy profile along a preferential pathway (or pathways) followed most often through configuration space during dissociation; other pathways involve significantly greater energy, which makes their traverse extremely rare. Thus, an energy landscape is viewed to start from a minimum representing the bound state and rise over one or more peaks with intervening valleys to reach the dissociated state as illustrated schematically in Figure 1.2. The peaks are local saddle points in the energy surface and define barriers to kinetics. Because of the thermal (Boltzmann) weighting of the energy barriers, the most prominent barrier is the dominant impedance to kinetics with little retardation to dissociation from passage of lower barriers. When a bond or molecular complex is pulled apart under a ramp of force in a probe test, the barriers diminish in time and thus unbonding force depends on rate of loading (= force/time). As a consequence of diminishing barrier heights, the *most frequent forces* for unbonding plotted on a scale of log(loading rate) yield a dynamic spectrum that images the hierarchy of energy barriers traversed along the force-driven pathway [7, 10]. Thus, the method of *dynamic force spectroscopy* (DFS) probes the inner world of molecular interactions.

1.1.3 Biochemical and mechanical perspectives of bond strength

Given the conceptual energy landscape shown in Figure 1.2, it is useful to compare traditional ways of characterizing the strength of chemical bonds. Starting with biochemistry, the scale for bond strength is usually taken as the free energy difference E_0 between bound and free states – or "binding" energy. In an ideal-dilute solution, the binding energy sets the equilibrium partition of "bound-to-free" constituents, *i.e.* the mole fraction of bound complexes v_{w} [AB] divided by the product of free reactants v_{w} [A] v_{w} [B],

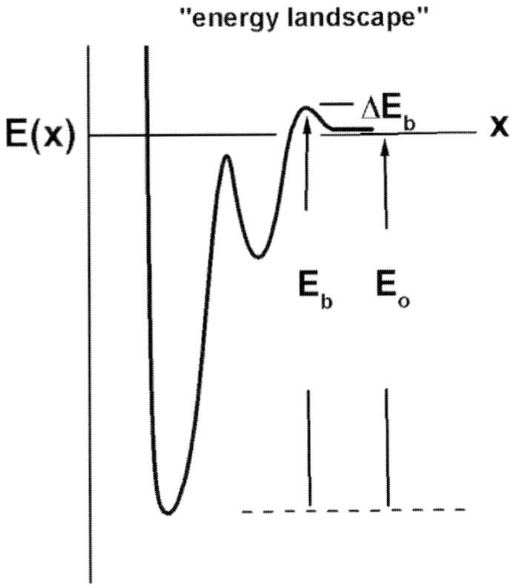

Fig. 1.2. Conceptual energy profile along a scalar reaction coordinate. In this hypothetical example, the energy landscape exhibits a primary bound state and a secondary metastable state punctuated by two energy barriers. From the perspective of biochemistry, only the outer barrier E_b and binding energy $E_0 = E_b - \Delta E_b$ are important in bond formation and dissociation.

where concentrations [number/volume] are converted to a scale of mole fraction by the partial molar volume of water v_w (*e.g.* $\sim$ one liter per 55 Moles). At equilibrium, the ratio $K_{eq} = [AB]/\{v_w[A][B]\}$ expresses the thermodynamic balance, $k_BT\log(K_{eq}) = E_0$, between reduction in (mixing) entropy and gain in free energy from binding. The important dynamical corollary to thermodynamic equilibrium is "detailed balance" where the number of complexes that form per unit time K_{on} [A] [B] must exactly equal the number that dissociate per unit time K_{off} [AB]. The "on" rate K_{on} (M^{-1} time^{-1}) and "off" rate K_{off} (time^{-1}) are empirically-defined parameters.

As such, the connection between equilibrium thermodynamics and phenomenological kinetics is through what's called the dissociation constant $K_D = K_{off}/K_{on}$, which has units of concentration and is inversely related to the equilibrium constant, *i.e.* $1/K_{eq} = v_w K_D$.

Consistent with the label, lowering the concentration of reactants below K_D leads to complex dissociation and increasing concentration promotes complex formation. Introduced by Van't Hoff and Arrhenius in the late 19th century [16], the long-held phenomenological view is that kinetic rates start with primitive-attempt rates driven by molecular excitations but then are discounted dramatically by an inverse exponential (Arrhenius) dependence on a dominant energy barrier. For instance, once the reacting molecules come together rapidly by diffusion, the entrance barrier ΔE_b shown in Figure 1.2 would retard association on approach to the bound state, *i.e.* $K_{on} \sim \exp(-\Delta E_b/k_BT)$. Likewise, given that inner barriers are more than k_BT lower, the height E_b of the paramount-outer barrier relative to the primary minimum would govern the rate of dissociation, *i.e.* $K_{off} \sim \exp(-E_b/k_BT)$. Since the difference $E_b - \Delta E_b$ in energy barriers equals the binding energy E_0, the ratio of kinetic rates is consistent with "detailed balance" at equilibrium. However, what's clearly missing in this biochemical perspective of bond strength is force!

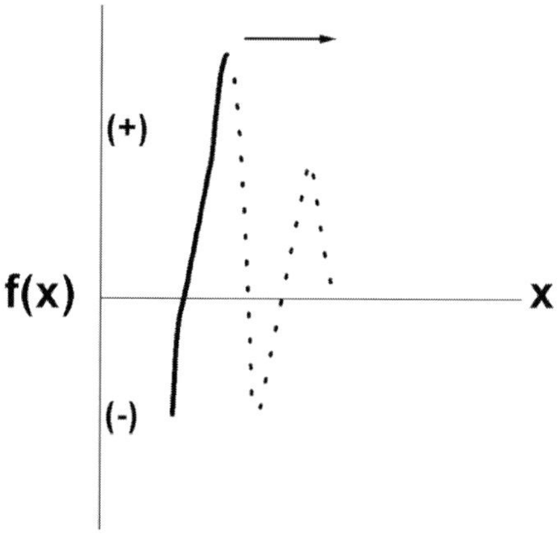

Fig. 1.3. Schematic of the force (solid curve) required to displace molecular components of a bond given the conceptual energy landscape in Figure 1.2. From the viewpoint of mechanics, the complex should become unstable and dissociate from the location of the steepest energy gradient just beyond the primary minimum. The remainder of the energy landscape (dotted curve) would appear not to affect bond strength.

In contrast to the biochemical perspective, classical mechanics is precise in its prescription of bond force. Specifically, the force required to separate interacting molecules is the gradient in energy along the landscape (or interaction potential). The subtlety is that not all positions along the energy profile are accessible as we apply increasing force to pull molecules apart. As sketched in Figure 1.3, only regions of the energy contour with monotonically-increasing gradients would be stable under rising force, which could leave major portions of the landscape as "virtual" or unmapped by a pulling force. Once force exceeds the steepest gradient in energy (just beyond the primary minimum in Fig. 1.2), the molecules would jump apart. Hence, from a mechanical perspective, bond strength is independent of barrier heights or features of the landscape other than the maximum gradient. Here, what's missing is time and temperature!

1.1.4 Relevant scales for length, force, energy, and time

At the beginning, it is important to introduce the relevant length, force, energy, and time scales appropriate to measurements of single bond properties. The increment of length is obviously the size of a small molecule, which is taken as a nanometer (about three water molecules end-to-end). One nanometer is comparable to the mean spacing between molecules at a concentration of ~1 mole/liter and five hundred-fold smaller than the wavelength of green light. Next, the characteristic scale of force needed to speed up dissociation and quickly break weak-noncovalent bonds is a piconewton (pN). One piconewton is about one ten-billionth of a gram weight (10^{-10} gm) or ten thousand-fold smaller than can be measured with an analytical microbalance. The product of length and force scales reveals the appropriate scale for energy – thermal energy $k_{\mathrm{B}}T$ – which is ~4 pN nm at biological temperatures (~300 K), which is better known as ~0.6 Kcal/mole for Avogadro's number ($\sim 6 \times 10^{23}$) of molecules.

Time scale is a much more complicated issue. At the atomic level, the time scale for excitations is typically 10^{-15} s or comparable to the frequency of the light photons emitted or adsorbed in atomic transitions. Much longer, however, kinetics in vacuum or gas phase reactions are theorized to start at an attempt frequency defined by thermal energy over Planck's constant, $k_{\mathrm{B}}T/h \sim 10^{13}$/s. This frequency characterizes thermally-driven transitions in a quantum oscillator model of chemical dissociation as developed by Eyring [16]. However, in condensed liquids or inside compact structures like proteins, kinetics are slowed significantly by dissipative collisions between and within the molecular components as well as with the myriad of other molecules in the surrounding solvent. For example, an instantaneous impulse of momentum from a thermal collision in water will die out on a time scale of $\sim 10^{-12}$ s or less as set by the ratio of damping to molecular inertia. Because of damping, many impulses are needed to separate molecular components over a distance comparable to the scale of bond length even in the absence of a bonding interaction.

Hence, as will be shown next, kinetics in the overdamped world of biomolecular interactions begin on a time scale set by a diffusive relaxation time for the bond. This relaxation time is on the order of 10^{-9} s which means that attempt frequencies for unbonding in liquids are four orders of magnitude slower than predicted by vacuum theory. The corresponding attempt frequency of $\sim 10^{9}$/s is then diminished many orders of magnitude by bond chemistry to reach laboratory kinetic rates of as long as 1/month or more for dissociation of weak biomolecular interactions – *i.e.* a range of more than sixteen orders of magnitude in time scale! Most important, this enormous span in time scale corresponds to breakage forces that range from the maximum gradient in the molecular interaction potential (~nN) under nanosecond detachment to zero force under detachment slower than the spontaneous dissociation rate $1/t_{\rm off}$.

1.2 Brownian kinetics in condensed liquids: Old-time physics

To make the connection between force – time – and chemistry, we need to review the physics that underlies kinetics in a liquid environment. Motivated by Einstein's theory of Brownian motion [6], these well-known developments take advantage of the huge gap in time scale that separates rapid thermal impulses in liquids ($<10^{-12}$ s) from slow processes in laboratory measurements. Three equivalent formulations describe molecular kinetics in an overdamped environment (see for example, N.G. van Kampen: *Stochastic Processes in Physics and Chemistry* [33]). The first is a nanoscopic description where molecules behave as particles with instantaneous positions or states $\mathbf{x}(t)$ governed by an overdamped Langevin equation of motion,

$$\mathrm{d}\mathbf{x}/\mathrm{d}t = [\mathbf{f} + \delta f]/\zeta. \tag{1.1}$$

Changes in state are driven by instantaneous force scaled by the mobility of states or inverse of the damping coefficent ζ. The deterministic part of the force ($\mathbf{f} = -\nabla E + \mathbf{f}_{\rm ext}$) includes the local gradient in molecular interaction potential $E(\mathbf{x})$ plus the applied external force $\mathbf{f}_{\rm ext}$. To this is added a random-uncorrelated force $\delta\mathbf{f}$ that embodies the many body collisions associated with the thermal environment. These random impulses are governed by the *fluctuation-dissipation theorem*, $\langle \delta\mathbf{f}^2\rangle \Delta t \sim k_{\rm B}T\zeta$. (Einstein's great insight was to recognize that the average mechanical energy imparted by thermal impulses had to equal thermal energy, *i.e.* the ensemble-average integral of mechanical power $\langle \int_{\Delta t} \delta\mathbf{f}\cdot\delta\mathbf{v}\,\mathrm{d}t'\rangle = k_{\rm B}T$. The assumption of overdamped motions $\delta\mathbf{v} = \delta\mathbf{f}/\zeta$ then yields the autocorrelation relation that governs force fluctuations.) The nanoscopic view can also be described by a stochastic process, which has become the foundation for an important computational technique – *Brownian dynamics* or *dissipative* Monte-Carlo simulations (referred to by its creators as "smart Monte-Carlo" [27]). In this description, the likelihood $P(\mathbf{x}+\Delta\mathbf{x}, t+\Delta t|\mathbf{x}, t)$ that a state $\mathbf{x}(t)$ will evolve to a new state $\mathbf{x}+\Delta x$ over a time increment Δt is the product of the equilibrium (long-time) Boltzmann weight for the step and a diffusive-Gaussian weight for dynamics,

$$\begin{aligned} P(t \to t+\Delta t) &\sim \exp\{-[\Delta E - \mathbf{f}_{\rm ext}\cdot\Delta\mathbf{x}]/k_{\rm B}T\} \\ &\quad\times \exp\{-|\Delta\mathbf{x} - \mathbf{f}\Delta t/\zeta|^2/(4D\Delta t)\}/(D\Delta t)^{1/2} \end{aligned} \tag{1.2}$$

where the diffusivity of states D is taken here to be a constant given by the Einstein–Stokes relation, $D = k_B T/\zeta$. Finally, on time scales that include many thermal impulses ($\sim 10^{-9}$ s and longer), the overdamped dynamics can be cast in a continuum theory where the density of states $\rho(\mathbf{x}, t)$ at location $\mathbf{x}$ and time t evolves according to the Smoluchowski transport equation,

$$\mathrm{d}\rho/\mathrm{d}t = -\nabla \cdot \mathbf{J} \tag{1.3}$$

with the *flux* of states $\mathbf{J} = (\mathbf{f}_{\text{ext}} - \nabla E)\rho/\zeta - D\nabla\rho$ defined by force-driven convection plus the diffusive gradient. Although each description illuminates different features of kinetics in a dissipative environment, Kramers [16, 18] demonstrated that Smoluchowski transport can be used to predict the rate for thermally-activated escape from a deeply bound state.

1.2.1 Two-state transitions in a liquid

To illustrate the important features of chemical kinetics in liquids and the utility of Kramers approach, we begin by examining two-state transitions with an energy landscape modelled by two deep energy minima separated by an intervening barrier (Fig. 1.4). In this 1-D abstraction, a scalar coordinate "x" is assumed to map the transition pathway over a barrier at energy E_{ts} relative to the deepest minimum. Following Kramers, rates of transition between these two states are approximated by stationary fluxes (constant #/time) between the states under appropriate boundary conditions,

$$J = -D[(\partial E/\partial x)\rho/k_B T + \partial\rho/\partial x] = \text{constant}. \tag{1.4}$$

Again treating the diffusivity or mobility of states as locally constant, integration of the stationary flux relates flux to the end-state densities (ρ_1, ρ_2),

$$J = D\{\rho_1 \exp(E_1/k_B T) - \rho_2 \exp(E_2/k_B T)\} / \left\{ \int_{1\to 2} \mathrm{d}x \exp[E(x)/k_B T] \right\}. \tag{1.5}$$

Directional rates of transition $\nu_{1\to 2}$ and $\nu_{1\leftarrow 2}$ between the two states are then found by starting with all states essentially at "1" or "2" and an adsorbing boundary at the final state "2" or "1" (*i.e.* $\rho_2 = 0$ or $\rho_1 = 0$). This leads to the expressions for forward and reverse rates of transition,

$$\begin{aligned} \nu_{1\to 2} &= D\rho_1 \exp[(E_1 - E_{\text{ts}})/k_B T]/L_{\text{ts}} \\ \nu_{1\leftarrow 2} &= D\rho_2 \exp[(E_2 - E_{\text{ts}})/k_B T]/L_{\text{ts}}. \end{aligned} \tag{1.6}$$

Energy-weighted, the major contribution to the pathway integral in equation (1.5) arises local to the transition state and defines a length scale

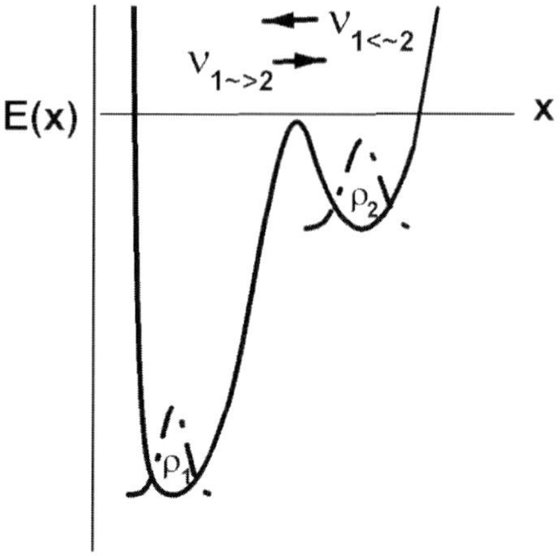

Fig. 1.4. Conceptual energy landscape for a two-state transition.

for barrier width, $L_{\rm ts} \equiv \int_{1\leftrightarrow 2} {\rm d}x \exp[(E(x) - E_{\rm ts})/k_{\rm B}T]$. As expected, "detailed balance" ($\nu_{1\to 2} = \nu_{1\leftarrow 2}$) yields the ratio $(\rho_1/\rho_2)_\infty = \exp[(E_2 - E_1)/k_{\rm B}T]$ required for densities of states at long times.

1.2.2 Kinetics of first-order reactions in solution

Another revealing application of Kramers theory is to bimolecular reactions (A + B $\leftrightarrow$ AB) in solution. Here, we imagine that a 1-D density $\rho_{\rm A}$ (#/length) of reactant (*e.g.* A) exchanges with the bound state (reactant B) at one end of a scalar coordinate x as sketched in Figure 1.5. As such, the reverse rate in equation (1.6) predicts the rate of capture by the attractive potential:

$$\nu_{\rm B\leftarrow A} \approx D\rho_{\rm A} \exp[-\Delta E_{\rm b}/k_{\rm B}T]/L_{\rm ts} \tag{1.7}$$

which may involve passage of an entrance barrier $\Delta E_{\rm b}$ like that sketched in Figure 1.5. To connect with the solution in 3-D beyond the reaction pathway, the 1-D density $\rho_{\rm A}$ is modelled as the product of solution concentration $c_{\rm A}$ (number/volume) and the effective cross section of the reactive site, *i.e.* $\rho_{\rm A} \sim 4\pi x_{\rm b}^2 c_{\rm A}$. In this way, the bimolecular "on" rate in solution is found from the rate of capture per concentration of reactant A, *i.e.* $K_{\rm on} = \nu_{\rm B\leftarrow A}/c_{\rm A}$. Since barrier width $L_{\rm ts}$ will be comparable to barrier location $x_{\rm b}$, "on" rate in solution can be approximated by,

$$K_{\rm on} \sim (4\pi \text{ "}x_{\rm b}D\text{"}) \exp[-\Delta E_{\rm b}/k_{\rm B}T]. \tag{1.8}$$

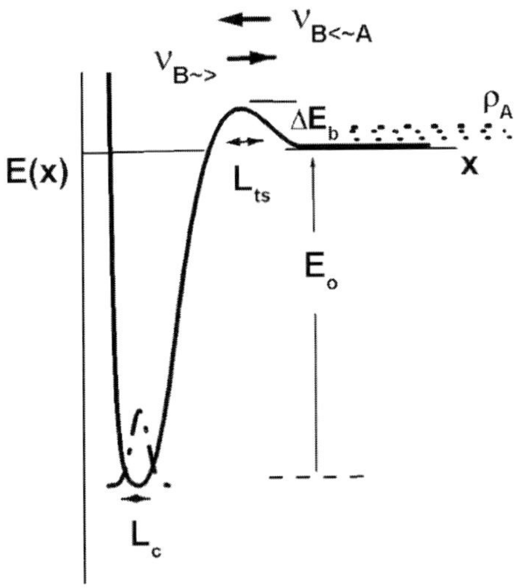

Fig. 1.5. Conceptual energy landscape for capture and release of components in solution.

The prefactor 4π"$x_b D$" in equation (1.8) represents the well-known Smoluchowski rate for diffusion-controlled aggregation which, for two-spheres (A–B), is given by the product of interparticle separation ("x_b" $= r_A + r_B$) at contact and combined particle diffusivity, ("D" $= D_A + D_B$). For nanometer-size molecules in water, the Smoluchowski scale for "on" rate is $\sim 10^9 - 10^{10}$ M^{-1} s^{-1}, whereas typical "on" rates for macromolecular interactions are only $\sim 10^6$ M^{-1} s^{-1}. Thus, formation of macromolecular bonds appears to involve entrance barriers of $\sim 10 k_B T$ in energy.

In examining the "off" rate, the crucial concept is that the constituents in a bound complex are driven to escape by entropy confinement. In 1-D, the relevant entropy gradient is determined by the density of states ρ_1 (#/length) local to the minimum. Under strong confinement, the distribution of states $\rho(x) = \rho_1 \exp[-(E - E_0)/k_B T]$ diminishes rapidly away from the minimum and, hence, the integral $\int_1 \rho(x) \mathrm{d}x \approx 1$ relates the entropy gradient to a confinement length, *i.e.* $\rho_1 = 1/L_c$ and $L_c = \int_1 \exp[-(E - E_0)/k_B T] \mathrm{d}x$. Introducing this length scale into equation (1.6) yields a generic expression for "off" rate ($1/t_{\text{off}} = \nu_{B \to \infty}$),

$$\nu_{B \to \infty} = D \exp(-E_b/k_B T)/L_c L_{ts} \tag{1.9}$$

where kinetics are driven by a diffusive attempt frequency $D/L_c L_{ts}$ and attenuated by the classic Arrhenius dependence on barrier energy E_b. Because of near-thermodynamic equilibration on a local scale, confinement length and barrier width are tied to curvatures ($\kappa = \partial^2 E/\partial x^2$) of the energy landscape at the minimum and transition states, *i.e.* $L_c \approx (2\pi k_B T/\kappa_c)^{1/2}$ and $L_{ts} \approx (2\pi k_B T/\kappa_{ts})^{1/2}$ respectively. (Note: These approximations follow from expansions of energies to quadratic order, $E - E_0 \approx \kappa_c(x - x_0)^2/2$, and, $E - E_b \approx -\kappa_{ts}(x - x_b)^2/2$, in the vicinity of the minimum and transition state respectively.) We now have Kramers' classic prescription [16, 18] for attempt frequency in an overdamped liquid environment, *i.e.* $D/L_c L_{ts} = (\kappa_c \kappa_{ts})^{1/2}/2\pi\zeta$, and the corresponding diffusive relaxation time $t_D = 2\pi\zeta/(\kappa_c\kappa_{ts})^{1/2}$. Although little is known about these molecular-scale properties, damping coefficients of $\sim$2–5 $\times 10^{-8}$ pN-sec/nm are typically deduced from molecular dynamics MD simulations of bonds under force [15, 17, 22]. Assuming a product of length scales $L_c L_{ts}$ on order of 0.01–0.1 nm^2, Kramers theory predicts that kinetics are driven by an attempt frequency of $\sim 10^{10}-10^9$/s, but actual rates of dissociation end up at $\sim$1/s for barrier heights of $\sim 21 k_B T$, or, astonishingly, at $\sim$1/40 years for barrier heights of $\sim 42 k_B T$, and so on!

1.3 Link between force – time – and bond chemistry

1.3.1 Dissociation of a simple bond under force

As illustrated in Figure 1.6, application of a persistent-external force (independent of distance) to a bond contributes a mechanical disjoining potential that deforms the chemical energy landscape. Energy barriers are lowered, displaced inward, and narrowed in ways that significantly affect kinetics. The shapes, levels, and locations of intervening minima are also altered but as we'll see, this has little impact on rate of dissociation provided that there is no switch in location of the primary minimum. Neglecting many subtle features, shifts in locations and changes in widths of barriers merely introduce weak prefactor dependencies on applied force in the unbonding rate (*e.g.* $1/L_{ts} \sim f^{\alpha}$ with $\alpha \sim 1$ or less, see Ref. [10]), which become insignificant for sharp energy barriers and so will be suppressed here. The major impact on kinetics stems from the enormous increase in likelihood of unbonding as barriers fall under applied force.

Force lowers a barrier in proportion to its thermally-averaged projection $x_\beta = \langle x_{ts} \cos\theta_\beta \rangle$ along the pulling direction, *i.e.* $E_b(f) = E_b - f x_\beta$; the angle θ_β accounts for instantaneous deviations of the reaction coordinate from this direction. The reduction in energy under force leads to exponentiation of the rate of barrier passage, *i.e.* $\nu_{\to\infty} = (1/t_D)\exp(-E_b/k_B T)\exp(f x_\beta/k_B T)$, as first hypothesized by Bell [4] over twenty years ago. Consequently, the characteristic scale for force is determined by thermal activation, *i.e.* $f_\beta = k_B T/x_\beta$, which can be surprisingly small since $k_B T \approx$ 4.1 pN nm at room temperature and $x_\beta \sim$ 0.1–1 nm. Thus, because of

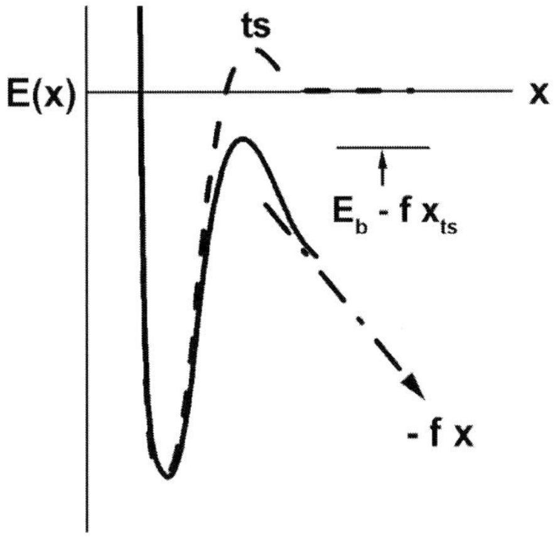

Fig. 1.6. Coupled to the projected reaction coordinate x, a persistent-external force f adds a mechanical potential $-\ f\ x$ that tilts the landscape and lowers the barrier to dissociation. (Note: from here on, f represents an externally applied force f_{ext}.)

thermal activation, pulling on a bond with forces of a few times the thermal force f_β will cause a bond to dissociate a hundred or thousand times faster than it would spontaneously. But to break a bond in less than a nanosecond, it can take 40–50 times the thermal force f_β, as demonstrated in molecular dynamics simulations [15, 17, 21, 22].

1.3.2 Dissociation of a complex bond under force: Stationary rate approximation

Since macromolecular bonds involve widely-distributed atomic-scale interactions, a rough terrain of barriers can exist in the energy landscape even after averaging over fast degrees of freedom before reaching laboratory time scales. When force is applied, outer barriers are driven below inner barriers so that an inner barrier becomes the dominant impedance to unbinding as sketched in Figure 1.7. We will see that switching of the prominent barrier leads to a hierarchy of exponential amplifications in rate of escape under force.

Unlike an ideal single-level transition, analysis of a multilevel transition under changing force is not transparent and usually requires numerical computation or simulation. However, we can develop useful approximations for the effective rate of unbonding over a cascade of N-1 sharp barriers that separate N levels (minima). These approximations follow directly from Kramers stationary-flux theory. Step-wise integration of the flux J from

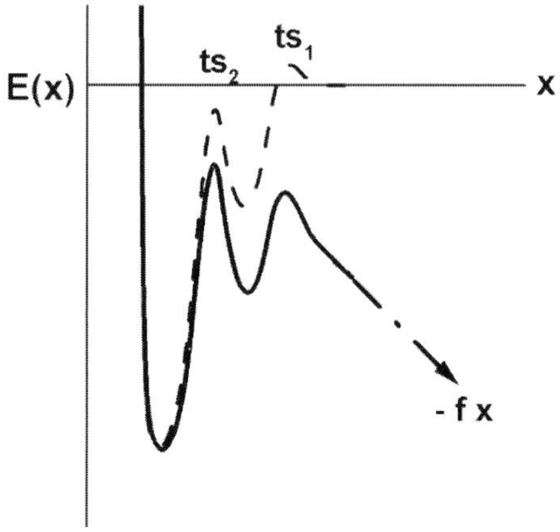

Fig. 1.7. Conceptual switching of dominant energy barriers in complex bond as the outer barrier is driven below an inner barrier by application of force.

one level to the next yields coupled equations that relate the rate of complete transition $\nu_{\rightarrow}(= J$ in 1-D) to the densities of states ρ_n at each energy minimum $E_c(n)$ along the energy contour. With energies defined relative to the first minimum (*i.e.* $E_c(1) = 0$), the initial integration from the first $(n = 1)$ to second minimum $(n = 2)$ gives,

$$\nu_{\rightarrow} L_{ts}(1) \exp[E_{ts}(1)/k_B T]/D = P_1/L_c(1) - P_2 \exp[E_c(2)/k_B T]/L_c(2) \tag{1.10}$$

which is followed by integration through intermediate levels,

$$\begin{aligned}\nu_{\rightarrow} L_{ts}(n) \exp[E_{ts}(n)/k_B T]/D &= P_n \exp[E_c(n)/k_B T]/L_c(n) \ldots \\ &\quad - P_{n+1} \exp[E_c(n+1)/k_B T]/L_c(n+1)\end{aligned} \tag{1.11}$$

all the way to the *unpopulated* Nth level where $P_N = 0$ (*i.e.* an adsorbing state),

$$\begin{aligned}\nu_{\rightarrow} L_{ts}(N-1) \exp[E_{ts}(N-1)/k_B T]/D &= \\ P_{N-1} \exp[E_c(N-1)/k_B T]/L_c(N-1).\end{aligned} \tag{1.12}$$

Here, the density of states ρ_n at each minimum is expressed in terms of the likelihood (probability) P_n of being at that level, scaled by a confinement length $L_c(n)$ as defined earlier, *i.e.*

$$\rho_n = P_n/L_c(n) \qquad \text{and} \qquad L_c(n) = \int_n \mathrm{d}x \exp\{[E(x) - E_c(n)]/k_B T\} \cdot$$

Also, barrier widths $L_{\mathrm{ts}}(n)$ are again introduced to represent the energy-weighted integrals local to each transition state at energy $E_{\mathrm{ts}}(n)$,

$$L_{\mathrm{ts}}(n) = \int_{n \to n+1} \mathrm{d}x \exp\{[E(x) - E_{\mathrm{ts}}(n)]/k_{\mathrm{B}}T\} \cdot$$

For convenience, we idealize the landscape as a sequence of narrow minima and sharp transition states having the same confinement length $L_{\mathrm{c}}(n) = L_0$ and barrier width $L_{\mathrm{ts}}(n) = L_{\mathrm{b}}$. In this way, a common relaxation time $t_{\mathrm{D}} = L_0 L_{\mathrm{b}}/D$ is defined for all transitions, and equations (1.10–1.12) predict Boltzmann-weighted likelihoods for levels, *i.e.*

$$t_{\mathrm{D}}\, \nu_{\rightarrow} \Sigma_{j=1\to N-1}\{\exp[E_{\mathrm{ts}}(N-j)/k_{\mathrm{B}}T]\} = P_1 \tag{1.13}$$

$$\cdot$$

$$t_{\mathrm{D}}\, \nu_{\rightarrow} \Sigma_{j=1\to N-n}\{\exp[E_{\mathrm{ts}}(N-j)/k_{\mathrm{B}}T]\} = P_n \exp[E_{\mathrm{c}}(n)/k_{\mathrm{B}}T] \tag{1.14}$$

$$\cdot$$

$$t_{\mathrm{D}}\, \nu_{\rightarrow} \{\exp[E_{\mathrm{ts}}(N-1)/k_{\mathrm{B}}T]\} = P_{N-1} \exp[E_{\mathrm{c}}(N-1)/k_{\mathrm{B}}T]. \tag{1.15}$$

At equilibrium, we expect each weighted likelihood to be unity, *i.e.* $P_n \to \exp[-E_{\mathrm{c}}(n)/k_{\mathrm{B}}T]$.

However, for the non-equilibrium process of unbonding, the likelihood of being in a particular level will deviate from a Boltzmann distribution. Two particular cases provide limiting initial conditions for the distribution of states in the stationary flux model. The first limit represents a strongly bound complex where the vast majority of states lie in a deep primary minimum, *i.e.* $P_1 \approx 1$. In this case, the approximate rate of unbonding is given by equation (1.13) [7,8],

$$\nu_{\rightarrow} \approx (1/t_{\mathrm{D}})/\Sigma_{n=1\to N-1} \exp[E_{\mathrm{ts}}(n)/k_{\mathrm{B}}T]. \tag{1.16}$$

No surprise, we see that the rate of unbonding is dominated by the largest exponential, which is set by the highest transition state relative to the primary minimum. But for complexes with low-lying secondary minima, partial filling of intermediate levels may affect the unbonding rate in ways that can become important under force. This case is easily treated using the auxiliary requirement that the occupation of inner levels add up to one, *i.e.* $\Sigma_{1\to N-1} P_n = 1$ (as noted by Ajdari *et al.* [3]). In this case, the unbonding rate is found by rearranging equations (1.13–1.15) to obtain the P_n's and then evaluating their sum, *i.e.*

$$\nu_{\rightarrow} \approx (1/t_{\mathrm{D}})/\Sigma_{n=1\to N-1} \exp[-E_{\mathrm{c}}(n)/k_{\mathrm{B}}T]\Sigma_{j\to N-n} \exp[E_{\mathrm{ts}}(N-j)/k_{\mathrm{B}}T]. \tag{1.17}$$

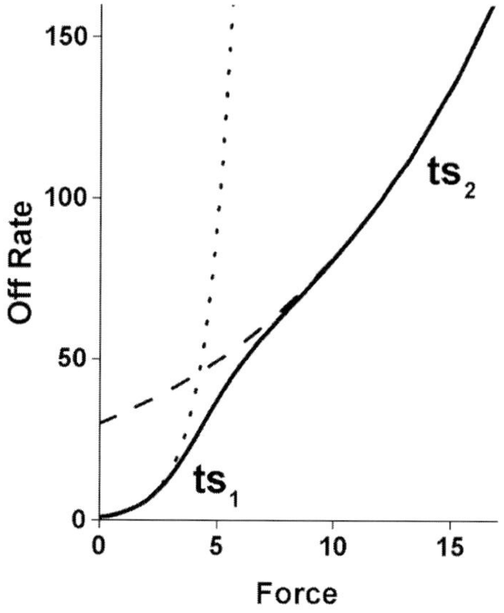

Fig. 1.8. Amplification of unbonding kinetics under persistent force for an energy landscape of two barriers as depicted in Figures 1.3 and 1.7.

As expected, the approximation in equation (1.17) reduces to equation (1.16) when metastable levels are more than a few $k_{\rm B}T$ above the primary minimum.

So how does external force affect the rate of unbonding over a multilevel energy landscape? Given narrow minima bounded by sharp barriers, the energies at these locations will change through proportionalities to force set by projections of the minima $x_{\rm c}(n)$ and barriers $x_\beta(n)$ along the pulling direction, *i.e.* $E_{\rm c}(n) \approx E_0(n) - fx_{\rm c}(n)$ and $E_{\rm ts}(n) \approx E_{\rm b}(n) - fx_\beta(n)$. This introduces exponential dependencies on force into the approximations for unbonding rate, *i.e.*

$$\nu_{\rightarrow} \approx (1/t_{\rm D})/\Sigma_{n\rightarrow N-1} \exp[E_{\rm b}(n)/k_{\rm B}T] \exp[-fx_\beta(n)/k_{\rm B}T] \tag{1.18}$$

$$\nu_{\rightarrow} \approx (1/t_{\rm D})/\Sigma_{n=1\rightarrow N-1}\Sigma_{j\rightarrow N-n} \exp\{[E_{\rm b}(N-j) - E_0(n)]/k_{\rm B}T\} \times \exp\{f[x_{\rm c}(n) - x_\beta(N-j)]/k_{\rm B}T\}\,. \tag{1.19}$$

Hence, consistent with a changing hierarchy of barriers, application of force to a complex bond leads to an sequence of exponential increases in unbonding rate as illustrated in Figure 1.8.

Table 1.1. Master equations for unbonding in an N-level system.

bound
$\mathrm{d}S_1/\mathrm{d}t = -\nu_{1\to 2}S_1(t) + \nu_{1\leftarrow 2}S_2(t)$
.
.
$\mathrm{d}S_n/\mathrm{d}t = -\{\nu_{n\to n+1} + \nu_{n-1\leftarrow n}\}S_n(t) + \nu_{n-1\to n}S_{n-1}(t) + \nu_{n\leftarrow n+1}S_{n+1}(t)$
.
.
unbound
$\mathrm{d}S_N/\mathrm{d}t = -\nu_{N-1\leftarrow N}S_N(t) + \nu_{N-1\to N}S_{N-1}(t)$

1.3.3 Evolution of states in complex bonds

To describe the detailed evolution of states in a complex bond, a hierarchy of "master equations" (Table 1.1) is needed to predict the likelihood $S_n(t)$ of being in the nth level (local minimum) as a function of time. In this Markov sequence, the forward $\nu_{n\to n+1}$ and reverse transition rates $\nu_{n\leftarrow n+1}$ at each barrier depend exponentially on the height of the barrier relative to the adjacent minima. Assuming a simple 1-D topology where $x_\mathrm{c}(n) < x_\beta(n) < x_\mathrm{c}(n+1)$, forward rates of barrier passage would increase under force as,

$$\nu_{n\to n+1} = (1/\tau_{n\to n+1})\exp\{f[x_\beta(n) - x_\mathrm{c}(n)]/k_\mathrm{B}T\} \tag{1.20}$$

and reverse rates would decrease as,

$$\nu_{n\leftarrow n+1} = (1/\tau_{n\leftarrow n+1})\exp\{-f[x_\mathrm{c}(n+1) - x_\beta(n)]/k_\mathrm{B}T\}\cdot \tag{1.21}$$

The prefactors in these expressions are the spontaneous rates of transition defined by the initial height of a barrier $E_\mathrm{b}(n)$ relative to its adjacent minima $E_0(n)$ and $E_0(n+1)$,

$$\begin{aligned}(1/\tau_{n\to n+1}) &= (1/t_\mathrm{D})\exp\{-[E_\mathrm{b}(n) - E_0(n)]/k_\mathrm{B}T\}\\ (1/\tau_{n\leftarrow n+1}) &= (1/t_\mathrm{D})\exp\{-[E_\mathrm{b}(n) - E_0(n+1)]/k_\mathrm{B}T\}\cdot\end{aligned} \tag{1.22}$$

Even with ideal exponential dependencies of transition rates on force, these master equations can only be solved analytically when force is constant. In this case, Laplace transform of the master equations leads to a linear system of algebraic equations in transform space. This set of equations can be diagonalized and inverse transformed to obtain a superposition of decaying exponentials in time that describe the probability of reaching the unbonded state. However, as we will see next, probe tests of bonds and dynamic force spectroscopy in particular involve changing levels of force in time. Under rising force, the dynamics of multilevel transitions can be extremely complex and solution of the master equations will require numerical computation or tedious Monte-Carlo simulation. Fortunately, the strong exponential dependence on force and wide separation of energy barriers in real interactions often allow transitions from bound to free states to be described by a stationary-rate approximation (Eqs. (1.16) or (1.17)) and a single master equation. In this way, we will demonstrate that multilevel unbonding under increasing force leads to a distinct pattern of force *versus* loading rate, which is the basis of dynamic force spectroscopy.

1.4 Testing bond strength and the method of dynamic force spectroscopy

With few exceptions, tests of bond strength with force probes usually follow a common approach. The probe *tip* and substrate are first decorated with reactive molecules using methods that vary from nonspecific adsorption to covalent attachment with heterobifunctional polymer spacers or noncovalent attachment by high-affinity complexes such as biotin-streptavidin. Once prepared, a probe and substrate are repeatedly brought to/from contact by steady-precision movements. If decorated with a very low density of reactive sites, repeated contact between the probe tip and test surface will only result in an occasional bond. Under controlled conditions of contact, a low frequency of attachment provides quantitative verification of the likelihood of rare-single bond events (*e.g.* probability >0.9 when 1 attachment occurs out of 10 touches). When a bond has formed, the tip is held to the substrate and the probe transducer ("spring") is stretched as the surfaces separate. Bond rupture is signalled by rapid recoil of the transducer to its rest position and the maximum transducer extension yields the rupture force. Histories of force over the course of approach – touch – separation with and without formation of a bond are demonstrated in Figure 1.9. After numerous touches, the few detachment forces are cumulated into a histogram (as will be shown later). The peak in this histogram is the *most likely* force for rupture and establishes a statistical definition for *bond strength*. Surprisingly, we'll see later that no matter how carefully measured or precise the technique, bond forces will always be spread in value and the most frequent breakage force will depend on how fast force is applied to the bonds. The crucial feature of the typical probe test is that the force experienced by a bond rises in time. As seen in Figure 1.9, a bond broken under slow loading has a long lifetime but only withstands a small force; whereas, the same type of bond broken under fast loading has a very short lifetime and withstands a large force. We will see that when plotted on a logarithmic scale of loading rate (force/time), measurements of rupture force over an enormous span in rate from very slow to extremely fast image the prominent energy barriers traversed along the dissociation pathway.

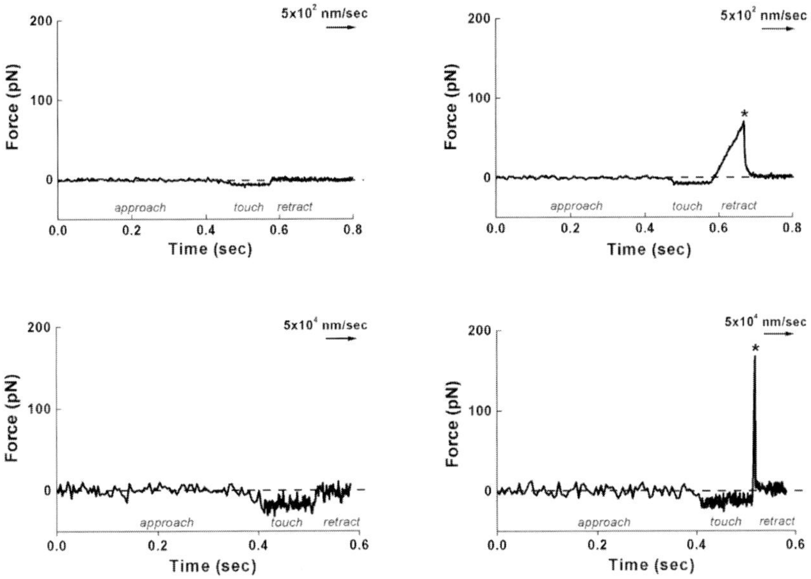

Fig. 1.9. Examples of tests performed at slow (upper plots) and fast (lower plots) speeds with a biomembrane force probe BFP (shown in Fig. 1.10). The force is traced over time in the course of approach–touch–separation with (right) and without (left) formation of a bond.

1.4.1 Probe mechanics and bond loading dynamics

At present, measurements of single bond strength are usually performed with one of three types of apparatus: (i) the atomic force microscope AFM [5] where force is sensed by deflection of a thin silicon nitride cantilever; (ii) the biomembrane force probe BFP [11] where force is sensed by axial displacement of a glass microsphere glued to the pole of a micropipet-pressurized red blood cell (example in Fig. 1.10); and (iii) the laser optical *tweezer* LOT [1,2,19,31] where force is sensed by displacement of a microsphere trapped in a narrowly-focused beam of laser light. Each of these probes acts as a very soft *spring* with a small elastic stiffness κ_f (force Δf per deflection Δx) that ranges from <1 pN/nm to 1 nN/nm. Low values of probe stiffness represent high sensitivity to force for each nm deflection of the spring but also large thermal fluctuations in position ($\delta x^2 \sim k_\mathrm{B}T/\kappa_f$). On the other hand, high probe stiffness represents low sensitivity to force per unit deflection and large thermal fluctuations in force ($\delta f^2 \sim k_\mathrm{B}T \cdot \kappa_f$).

As seen in Figure 1.9, bond strength and lifetime depend critically on how fast force is applied which involves both stiffness of the spring that

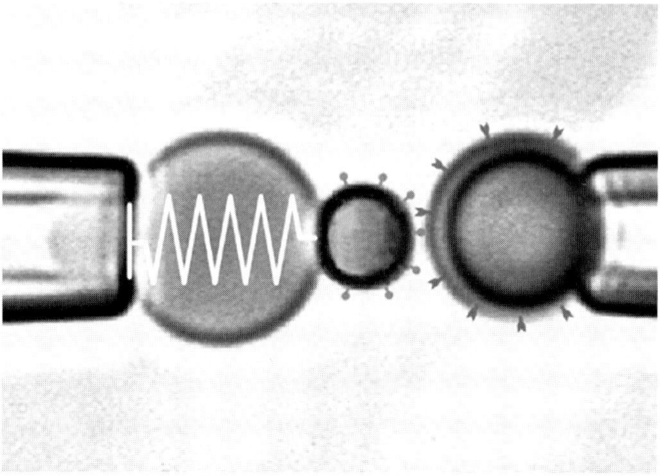

Fig. 1.10. Prepared to test a ligand-receptor interaction, a biomembrane force probe (symbolized by the "spring" – left) with a glass tip decorated by the ligand ("pin heads") is shown on approach to a glass bead (right) decorated by the receptor ("darts"). The pipet suction ΔP applied to the red blood cell transducer (scaled by pipet radius R_p) sets its membrane tension $\tau_\mathrm{m} \sim R_\mathrm{p}\Delta P$ and thereby tunes the elastic stiffness κ_f of the transducer "spring", *i.e.* $\kappa_f \sim \tau_\mathrm{m}$.

pulls on the bond and the pulling speed v. However, even with a known product of probe stiffness and separation speed, the rate of force application to a bond can deviate signficantly from this expected value when soft molecular structures connect the bond and probe. If the bond is linked symmetrically to tip and substrate by components with the same stiffness κ_m, the effective *spring* that pulls on the bond has a compliance given by, $1/\kappa_\mathrm{s} = 2/\kappa_\mathrm{m} + 1/\kappa_f$. Most biomolecules are linked to solid surfaces by highly flexible polymers. These connections have a nonlinear elastic response and a very small thermal scale $k_\mathrm{B}T/(L_\mathrm{p}b)$ for stiffness that depends on contour length L_p and persistence length b of the polymer. Even for relatively short linkers, the stiffness scale is <1 pN/nm. So when connected to a stiff probe (*e.g.* with $\kappa_f > 10$ pN/nm), the soft linker becomes the effective spring that pulls on the bond but applies a nonlinear loading history as seen in Figure 1.11. In addition, bond rupture almost always occurs in the asymptotic regime of loading as the polymer is pulled taut. Here, force diverges $f \approx (k_\mathrm{B}T/cb)/(1 - x/L_\mathrm{p})^\alpha$ as length $x \to L_\mathrm{p}$, with $\alpha = 1$ & $c = 1$ for a freely-jointed polymer and $\alpha = 2$ & $c = 4$ for a worm-like polymer. Under constant pulling speed ($x = vt$), loading increases rapidly with an approximate rate, $r_f(t) \approx (k_\mathrm{B}T/cL_\mathrm{p}b)v/(1 - vt/L_\mathrm{p})^{\alpha+1}$. We will see later

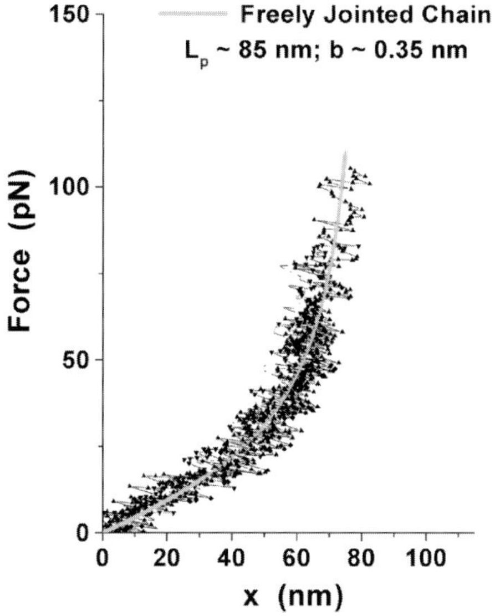

Fig. 1.11. Loading of bonds formed between the biotinylated glass tip of a BFP and an avidinated latex microsphere (superposition of 8 tests). The plot shows that the bonds are anchored to the latex microsphere by long, flexible chains of polystyrene.

that this unsteady loading plays an important role in the dependence of rupture force on detachment speed.

What's subtle with operation of any probe in liquids is that hydrodynamic interactions always accompany relative motion between the probe and substrate. Moreover, each hydrodynamic situation has to be analysed carefully to determine the impact on probe force. But a particularly important effect is the hidden contribution to force Δf that arises when a probe is pulled quickly; the force augmentation is governed by the probe damping coefficient ζ and deflection speed v, *i.e.* $\Delta f = \zeta v$. In general, the damping coefficient is proportional to viscosity η of the liquid environment and a hydrodynamic profile length L_ζ for the probe, *i.e.* $\zeta \approx \eta L_\zeta$. As such, damping coefficients are in the range from $\sim 10^{-5}$ pN-s/nm for a micron-size particle trapped by LOT to $\sim 10^{-4}$ pN-s/nm for a BFP and $\sim 10^{-3}$ pN-s/nm for a long AFM cantilever, which implies viscoelastic response times $t_f \sim \zeta / k_f$ in the range of $\sim 10^{-5}$–10^{-3} s. Unfortunately, damping coefficients are difficult to predict accurately by theory and usually have to be measured. As an example of such measurement, Figure 1.12 shows the recovery of a BFP following release from a deflecting force at three different settings of elastic

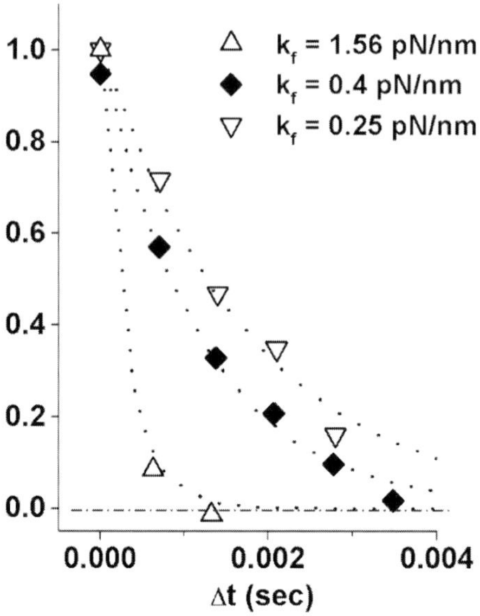

Fig. 1.12. BFP recovery (normalized by initial deflection) following bond rupture events at three elastic stiffnesses k_f. Superposed on each test is the exponential decay (dotted curve) defined by the time scale ζ/k_f with a common damping coefficient $\zeta \sim 5 \times 10^{-4}$ pN-s/nm and each value of stiffness.

stiffness. Exponential fits to each recovery curve with decay times defined by ζ/k_f were used to determine the BFP damping coefficient.

1.4.2 Stochastic process of bond failure under rising force

Because of the enormous gap in time scale between diffusive relaxation ($t_D \sim 10^{-10}$–10^{-9} s) and laboratory experiments ($\sim 10^{-4}$ s to minutes), kinetic rates of bond dissociation can be treated as continuous functions of the instantaneous force on the laboratory time scale. In the limit of large statistics, the master equations listed in Table 1.1 describe the stochastic process of bond failure under dynamic loading. In the idealized context of a simple bond, evolution of a bound complex under force involves the net of forward $\rightarrow$ and reverse $\leftarrow$ transitions in a single master equation,

$$\mathrm{d}S_1/\mathrm{d}t = -\nu_{\rightarrow}(t)S_1(t) + \nu_{\leftarrow}(t)S_2(t) \tag{1.23}$$

$S_1(t)$ is the likelihood of being in the bound state and $S_2(t) \equiv 1 - S_1(t)$ is the likelihood of being detached. When pulled by an elastic linkage $\kappa_s(f)$ at constant speed v, force and time become equivalent statistical variables connected through the dynamical transformation $\mathrm{d}f = [\kappa_\mathrm{S}(f)v]\mathrm{d}t$. Thus, the likelihood of bond survival to a particular time can be described in terms of survival to a particular level of force [7, 10], *i.e.*

$$\mathrm{d}S_1/\mathrm{d}f = -[\nu_\rightarrow(f) + \nu_\leftarrow(f)]S_1(f)/r_f(f) + \nu_\leftarrow(f)/r_f(f) \qquad (1.24)$$

which reveals the intrinsic dependence on loading rate $r_f(f) \equiv \kappa_\mathrm{S}(f)v$. The distributions of bond lifetimes and rupture forces are given by the probability density $p(t) = -\mathrm{d}S_1/\mathrm{d}t$ for rupture times between $t \rightarrow t + \Delta t$ and the probability density $p(f) = -\mathrm{d}S_1/\mathrm{d}f$ for rupture forces between $f \rightarrow f + \Delta f$ respectively.

1.4.3 Distributions of bond lifetime and rupture force

We begin by analyzing the loading condition where the time dependent force does not diminish with separation distance (as in Fig. 1.6). Once force rises above the thermal scale f_β for exponentiation of dissociation rate, molecules that have separated to distances beyond the barrier will continue to drift apart by force faster than diffusion can recombine them. The likelihood of bond survival becomes dominated by rate of dissociation and falls rapidly. So above f_β, the bond is driven *far from equilibrium* and rebinding vanishes ($\nu_\leftarrow S_2 \sim 0$), which leads to simple relations for distributions of lifetimes, $p(t) \approx \nu_\rightarrow(t)\exp\{-\int_{0\rightarrow t}\nu_\rightarrow(y)\mathrm{d}y\}$, and rupture forces, $p(f) \approx [\nu_\rightarrow(f)/r_f(f)]\exp\{-\int_{0\rightarrow f}[\nu_\rightarrow(y)/r_f(y)]\mathrm{d}y\}$. Thus, loaded by a linear spring under a constant pulling speed, the statistics of rupture for a bond confined by a single-sharp barrier become universal on a dimensionless scale $r_f = t_\mathrm{off}\, r_f/f_\beta$ for loading rate [7, 10], *i.e.*

$$p(y) = \exp\{y - [\exp(y) - 1]/r_f\}/r_f \qquad (1.25)$$

where $y \equiv r_f\, t/f_\beta$ or $\equiv f/f_\beta$ and given the expression for rate of passage under force, $\nu_\rightarrow = (1/t_\mathrm{off})\exp(f/f_\beta)]$. For dimensionless rates $r_f \geq 1$, a peak appears in the distribution due to the crossover between exponentiation of the unbonding rate and the precipitous decline in bond survival under increasing force. The distribution peak at $y = y^*$ is defined by $\partial p(y)/\partial y = 0$, which yields the time $t^* = f_\beta\, y^*/r_f$ and force $f^* = f_\beta\, y^*$ for most frequent rupture or *bond lifetime* and *bond strength* respectively. Locating the distribution peak, we see that both lifetime and strength of a bond depend logarithmically on loading rate, *i.e.* $y^* = \log(r_f)$. Hence, under constant speed separation with a linear spring (*i.e.* $r_f = \kappa_\mathrm{S} v$), observation of a straight line regime in a plot of bond strength *versus* log(loading rate) reveals the dominance of a single-sharp energy barrier where the slope $f_\beta = k_\mathrm{B}T/x_\beta$ is set by the barrier projection x_β along the direction of force. Moreover, strength emerges when loading rate surpasses $r_f^0 = f_\beta/t_\mathrm{off}$ and becomes fast enough to contribute $k_B T$ of energy or greater to the bond within the time for spontaneous dissociation t_{off}. Thus, the plot of most frequent rupture force as a function of log(loading rate) can establish both distance scales x_β and spontaneous rates $1/t_{off}$ of passage for barriers traversed in detachment of bonds, which is a spectral image of the energy landscape of a bond [7, 10]. The universal expression in equation (1.25) shows that a distribution of forces is always broadened by kinetics – even with no experimental error – and the spread σ_f in rupture forces is determined by the thermal force scale, $\sigma_f = f_\beta$. Thus, the widths of force distributions should match the slopes of linear regimes observed in a dynamic force spectrum.

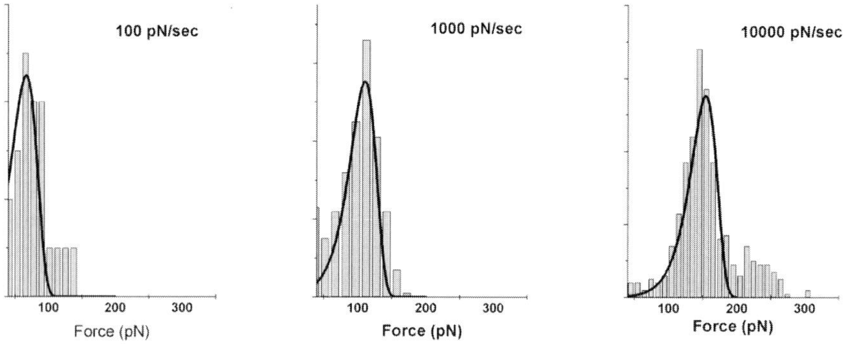

Fig. 1.13. Histograms of detachment forces for bonds between the SGP ligand and P-selectin receptor (Fig. 1.1) obtained at loading rates of 100, 1000, and 10 000 pN/s. Superposed are distributions predicted by the universal distribution for failure of an ideal bond (Eq. (1.25)) rescaled by the slope and intercept of the spectrum of most frequent rupture force *versus* log(loading rate) plotted in Figure 1.14.

To demonstrate that these universal features arise in tests of real bonds, we take results from our recent tests of the carbohydrate-selectin interaction pictured in Figure 1.1, which surprisingly exhibits simple bond-like kinetics under force. Using a BFP, we performed thousands of repeated touches to target microspheres at set loading rates as illustrated in Figure 1.10. From these touches, a few hundred forces were obtained at each rate and collected into histograms, samples of which are shown in Figure 1.13 for a three order of magnitude span in rate. Superposed in Figure 1.13 are the universal probability distributions predicted by equation (1.25) with the thermal force $f_\beta = 19$ pN and spontaneous dissociation rate $1/t_{off} = 0.1\ \mathrm{s}^{-1}$. These force and rate scales were determined from the plot of histogram peaks as a function of log(loading rate), which is the dynamic force spectrum shown in Figure 1.14. From the close match in Figure 1.13, we see that the distributions predicted by the slope and intercept derived from the force spectrum are fully consistent with the histograms of forces measured at each loading rate. This agreement is the requisite control for validity of the dynamic force spectrum. It is important to recognize that each probability distribution is a complete view of the stochastic process of unbonding over all time where the kinetics are imaged through a "shutter" set by loading rate. However, also keep in mind that for complex interactions like the SGP-selectin bond in Figure 1.1, observation of simple bond-like behavior most likely reflects the dominance of one of several energy barriers over a particular – but limited – range of loading rates. This span of dominance is governed by the difference in height of the barrier relative to the next prominent barrier as we will see later.

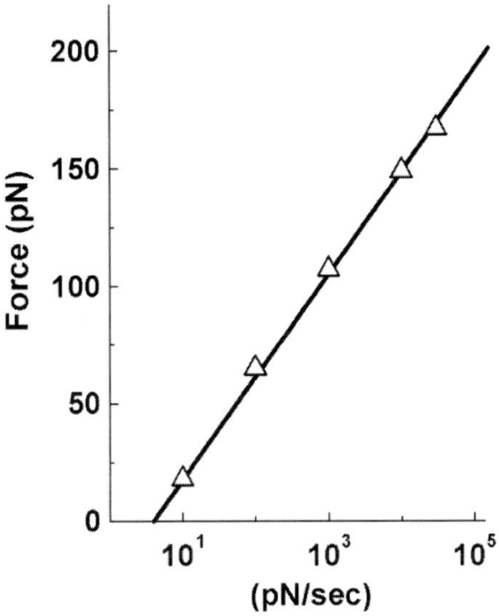

Fig. 1.14. Dynamic spectrum of the most frequent forces in histograms as a function of log(loading rate) for bonds between the SGP ligand and P-selectin receptor (Fig. 1.1).

To examine bond strength f^* in more complicated processes (*e.g.* under nonsteady loading), the location of the maximum, $\partial p(f)/\partial f = 0$, is found from the derivative of the general integral expression for the force distribution,

$$p(f) \approx [\nu_{\rightarrow}(f)/r_f(f)] \exp\left\{-\int_{0\rightarrow f} [\nu_{\rightarrow}(y)/r_f(y)] \mathrm{d}y\right\} . \tag{1.26}$$

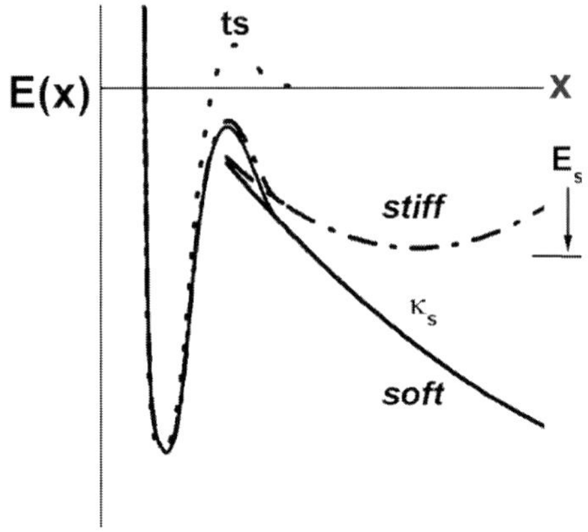

Fig. 1.15. Energy landscapes for a simple bond pulled by a soft and stiff spring.

In this way, the most frequent force is specified by inversion of the following transcendental expression [7, 10]:

$$[\nu_{\rightarrow}]_{f=f^*} = r_f[\partial \log(\nu_{\rightarrow})/\partial f - \partial \log(r_f)/\partial f]_{f=f^*} \tag{1.27}$$

which unfortunately rarely yields a simple expression for force as a function of rate like that for a single-sharp barrier. In the same way, we can directly estimate the spread σ_f in a force distribution using the dimensionless curvature at the peak of the distribution, *i.e.*

$$1/\sigma_f^2 = \left\{[\partial \log(\nu_{\rightarrow}/r_f)/\partial f]^2 - \partial^2 \log(\nu_{\rightarrow}/r_f)/\partial f^2\right\}_{f=f^*}.$$

1.4.4 Crossover from *near equilibrium* to *far from equilibrium* unbonding

Unlike the concept of persistent force with no chance of rebinding after passing the transition state, pulling on a bond with a probe applies a force that diminishes with distance and eventually changes sign to establish a *capture well* as depicted in Figure 1.15. Deepening and moving outward with time, the capture well facilitates rebinding until force becomes large enough to drop the energy level below the bound state minimum. Here, linkage stiffness κ_s is very important since the ratio of rebinding-to-unbinding rates falls in proportion to $\exp(-f^2/2\kappa_s)$. (Note: when possible in the analyses to follow, we will employ a single barrier model for bonds and conveniently express variables in a dimensionless form defined by force $f \equiv f/f_\beta$, unbonding rate $t_{\text{off}}\,\nu_{\rightarrow}$, loading rate $r_f \equiv r_f\,t_{\text{off}}/f_\beta$, and linkage stiffness $\kappa_s \equiv \kappa_s x_\beta^2/k_BT$.)

More subtle, linkage stiffness also modifies the reduction in barrier height under force and introduces a threshold into the dependence of unbonding rate on force, *i.e.* $t_{\rm off}\,\nu_{\rightarrow} = \exp(f - \kappa_{\rm s}/2)$ for $f \geq \kappa_{\rm s}/2$. In other words, force must be large enough to push the capture well beyond the barrier to enable escape. Under rising force, the forward rate of barrier passage will eventually overwhelm the rate of reverse passage and eliminate rebinding, *i.e.* $\nu_{\rightarrow} > \nu_{\leftarrow}$. This condition defines the crossover from *near equilibrium* – to *far from equilibrium* – dissociation.

To determine the impact of force on rebinding, we utilize our earlier analysis of two-state transitions. First, the barrier energy $E_{\rm b}$, confinement length $L_{\rm c}$, barrier width $L_{\rm b}$, and molecular damping $\zeta_{\rm m}$ define the time scale for spontaneous dissociation, *i.e.* $t_{\rm off} = (\zeta_{\rm m} L_{\rm c} L_{\rm b}/k_{\rm B}T)\exp(E_{\rm b}/k_{\rm B}T)$. For simple harmonic loading, the rate of forward transition will increase as an exponential of the force, $t_{\rm off}\,\nu_{\rightarrow} = \exp(f - \kappa_{\rm s}/2)$, given in dimensionless units. Similarly, these properties and the entrance barrier height $\Delta E_{\rm b}$ can be used to define a time scale $t_{\rm on} \equiv (\zeta_{\rm m} L_{\rm c} L_{\rm b}/k_{\rm B}T)\exp(\Delta E_{\rm b}/k_{\rm B}T)$ for spontaneous-inward passage. As expected, the ratio of characteristic times represents the equilibrium constant $K_{\rm eq} = \exp[(E_{\rm b} - \Delta E_{\rm b})/k_{\rm B}T]$ for the interaction, *i.e.* $t_{\rm off}/t_{\rm on} = K_{\rm eq}$. Because of the sharp barrier, we will assume that the height of the entrance barrier is little affected by application of force so the reverse rate of transition is set by the time scale $t_{\rm on}$ and local density of states $\rho_{\rm e}$ at the entrance barrier, *i.e.* $t_{\rm on}\,\nu_{\leftarrow} \approx \rho_{\rm e} L_{\rm c}$. To evaluate the entrance density of states $\rho_{\rm e}$, we use Kramers approach and integrate the rebinding flux from the minimum of the capture well to the entrance barrier, *i.e.*

$$\nu_{\leftarrow}\{\zeta_{\rm s} L_{\rm s}\exp(E_{\rm s}/k_{\rm B}T)/k_{\rm B}T\} = \rho_{\rm s} - \rho_{\rm e}\exp(E_{\rm s}/k_{\rm B}T). \qquad (1.28)$$

Depth of the capture well, $E_{\rm s}/k_{\rm B}T = f^2/2\kappa_{\rm s} - f + \kappa_{\rm s}/2$, and local density of states $\rho_{\rm s} \approx (\kappa_{\rm s}/2\pi k_{\rm B}T)^{1/2} = (\kappa_{\rm s}/2\pi)^{1/2}/x_\beta$ depend on linkage stiffness; the dynamics depend on the linkage damping coefficient $\zeta_{\rm s}$. The Boltzmann-weighted distance $L_{\rm s}$ from the entrance barrier to the minimum of the capture well is dominated by the decay in energy with distance from the barrier and can be approximated by $L_{\rm s} \sim x_\beta/(f+1)$. Matched to the rate of inward passage defined by $t_{\rm on}\,\nu_{\leftarrow} = \rho_{\rm e} L_{\rm c}$, we eliminate the density of states $\rho_{\rm e}$ and establish the dependence of rebinding rate on force, *i.e.*

$$\begin{aligned} t_{\rm on}\,\nu_{\rightarrow} &\approx (\kappa_{\rm s}/2\pi)^{1/2} \\ &\times \exp(-f^2/2\kappa_{\rm s} + f - \kappa_{\rm s}/2)/\{(t_{\rm s}/t_{\rm on})\kappa_{\rm s}/(f+1) + (x_\beta/L_{\rm c})\} \end{aligned} \qquad (1.29)$$

where $t_{\rm s} \equiv \zeta_{\rm s}/\kappa_{\rm s}$ is the characteristic relaxation time of the linkage.

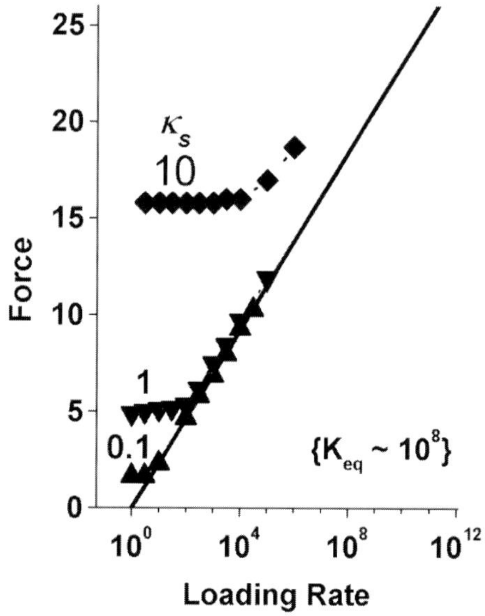

Fig. 1.16. Dynamic spectra of most frequent rupture force as a function of log(loading rate) calculated by numerical solution (closed symbols) to the master equation (1.24) for rupture of a simple bond pulled by a linear spring. The threshold force and crossover to far-from-equilibrium detachment (solid line) depend on the linkage stiffness κ_s and the equilibrium constant $K_{eq} = t_{off}/t_{on}$ of the bond.

Complete unbonding will occur when force rises above the level $f_\otimes$ needed to drive the forward rate of transition above the reverse rate, *i.e.* $\nu_\rightarrow/\nu_\leftarrow > 1$. The crossover force is easily found from the ratio of transition rates [7],

$$\nu_\rightarrow/\nu_\leftarrow \approx (2\pi/\kappa_s)^{1/2} \exp[(f^2/2\kappa_s) - \log(K_{eq})][(t_s/t_{on})/(f+1) + (x_\beta/L_c)]. \tag{1.30}$$

Hence, the force must rise above, $f_\otimes \sim [2\kappa_s k_B T \log(K_{eq})]^{1/2}$ to achieve complete breakage. As shown in Figure 1.16, crossover to far from equilibrium detachment occurs once loading rate is fast enough to produce forces $>f_\otimes$, *i.e.* $r_f > (f_\beta/t_{off})\exp(f_\otimes/f_\beta)$. We see that $f_\otimes$ defines a near equilibrium threshold for strength that depends on mechanical properties of the linkage as well as energetics of the bond [7]. Therefore, rebinding and the threshold for strength are predicted to diminish significantly when bonds are connected to probes by soft linkages like ideal polymers. Without going through the details, a similar ratio of transition rates can be derived for a polymer linker. With a freely jointed polymer of L_p/b

segments and $b \sim x_\beta$, the exponential dependence on depth of the capture well in equation (1.29) is replaced by a power law $(f+1)^{L_\mathrm{p}/b}$, which leads to $f_\otimes \sim (k_\mathrm{B}T/b)[(K_\mathrm{eq})^{b/L_\mathrm{p}} - 1]$. Thus, for long polymer linkers, the near equilibrium threshold drops to very small forces.

1.4.5 Effect of soft-polymer linkages on dynamic strengths of bonds

Polymer linkers are essential for projecting and isolating reactive sites in single molecule experiments. Also, we've just seen that long polymer linkages to bonds help suppress rebinding events. However, polymer linkages alter bond strength under constant speed detachment in ways that are especially important at low speed [12]. Clearly demonstrated in Figure 1.17 with results from the work of Fernandez and coworkers [20], pulling on bonds through polymer connections produces unsteady loading where forces usually increase markedly prior to bond failure. Connected to a stiff probe, a polymer linker dominates the loading rate and under constant pulling speed, loading rate increases essentially as a power law of the force level, $r_f(f) \approx (v/v_\beta) f^{1+1/\alpha}$ as described earlier, where $r_f = r_f t_\mathrm{off}/f_\beta$ and $f = f/f_\beta$. Thus, loading rate becomes parameterized by a characteristic velocity, $v_\beta = (L_\mathrm{p}/\alpha t_\mathrm{off})(x_\beta/cb)^{1/\alpha}$, which represents the speed needed to pull the polymer taut within the time t_off required for spontaneous dissociation. (Recall: $\alpha = 1$ & $c = 1$ for a freely-jointed polymer and $\alpha = 2$ & $c = 4$ for a worm-like polymer). With the exact expression for loading rate $r_f(f) \equiv (v_\mathrm{s}/v_\beta)g(f)$ and assuming far from equilibrium detachment, the distribution of unbonding forces can be predicted by the integral expression in equation (1.26),

$$p(f) \approx (v_\beta/v)\exp\left\{ f - (v_\beta/v)\int_{0\to f} [\exp(y)/g(y)]\mathrm{d}y \right\} /g(f). \tag{1.31}$$

Specified in reference [12], $g(f)$ is a complicated function that approaches $f^{1+1/\alpha}$ once force exceeds f_β (*i.e.* $f > 1$). Since unbonding usually occurs in the asymptotic regime, the most likely force for bond rupture can be easily derived using equation (1.27) and the power law approximation to loading rate,

$$f^* \approx f_\beta \log(v/v_\beta) + f_\beta[\log(f^*/f_\beta - 1 - 1/\alpha) + (1/\alpha)\log(f^*/f_\beta)]. \tag{1.32}$$

Equation (1.32) shows that bond strength deviates from the expected proportionality to log(detachment rate) until loading at high speeds ($v \gg v_\beta$). Moreover, correlation of equation (1.32) to data for rupture force *versus* log(pulling velocity) can only provide values of thermal force f_β and characteristic velocity v_β but not the time scale t_off for unbonding. To determine

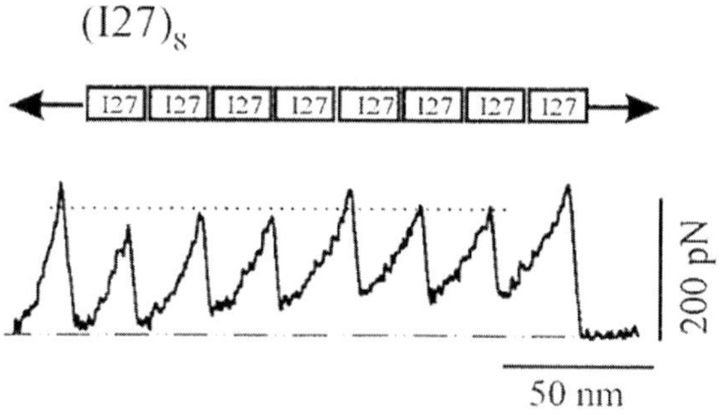

Fig. 1.17. The famous *saw-tooth pattern* of force obtained by pulling on a recombinant construct of eight Ig-I27 domains at constant speed with an AFM (taken from Li *et al.* [20]). Each precipitous drop in force signalled a bond rupture event and unfolding of an I27 domain, which added ~28 nm to the contour length.

the time scale t_{off} from the parameter v_β, we need to know the polymer contour and persistence lengths (L_{p}, b).

Figure 1.17 is a beautiful example of bond rupture under soft polymer loading. Each rupture event led to unfolding of an Ig (I27) domain and added a well defined increment to the total chain length thereby increasing the extensional compliance. Figure 1.18 shows the average forces needed to unfold the I27 domains at different pulling speeds plotted on a logarithmic scale. We can roughly characterize the average force $\langle f^* \rangle$ for unfolding of N domains at a particular speed using equation (1.32) and introducing an effective log(velocity scale) defined by the expression, $(1/N)\Sigma_{j=1\to N} \log[v_\beta(j)] \Rightarrow \log(\langle v_\beta \rangle)$. Superposed on the data in Figure 1.18, the match with equation (1.32) required a thermal force scale of $f_\beta \approx 18$ pN (or $x_\beta \approx 0.225\,\text{nm}$) and an effective velocity scale of $\langle v_\beta \rangle \approx 0.1$ nm/s. At slow pulling speeds, the correlation (solid curve) deviates significantly from the linear behavior expected for steady loading.

To extract single domain properties from the parameters f_β and $\langle v_\beta \rangle$, we need to recognize that there are several (N) equivalent sites for unfolding and that the number of sites diminishes as the length of the chain increases. In particular, the unbonding rate is proportional to the number of folded domains at each step. Hence, the probability distribution for unfolding events and most likely unfolding force change at each step in the process. Given that at each unbonding step $L_{\text{p}}(j) \approx (j+1)\Delta L_{\text{I27}}$ and $t_{\text{off}}\,\nu_{\rightarrow} = (N-j+1)\exp(f)$, the velocity scale for a worm-like chain will vary as $v_\beta(j) \approx (j+1)(N-j+1)(\Delta L_{\text{I27}}/2t_{\text{off}})(x_\beta/4b)^{1/2}$ and result in the following approximation for effective velocity scale: $\langle v_\beta \rangle \approx [N^{2+2/N}/\exp(2-2/N)](\Delta L_{\text{I27}}/2t_{\text{off}})(x_\beta/4b)^{1/2}$. With the value of $x_\beta \approx 0.225$ nm from the

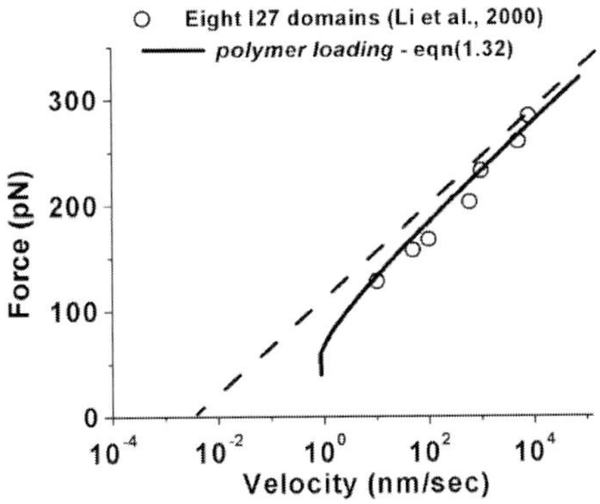

Fig. 1.18. Average forces for unfolding a sequence of eight I27 domains as a function of log(pulling speeds) taken from Li *et al.* [20]. Superposed is the dynamic spectrum predicted by equation (1.32) using a thermal force scale of $f_\beta \approx 18$ pN and an effective velocity scale of $\langle v_\beta \rangle \approx 0.1$ nm/s. The dashed line illustrates the regime of rupture forces that would be measured if the molecular linkage possessed a constant stiffness $\kappa_s \approx 5$ pN/nm.

correlation in Figure 1.18 plus a persistence length $b \sim 0.4$ nm from Li *et al.* [20], this approximation yields $\langle v_\beta \rangle \approx 3.5(\Delta L_{\mathrm{I27}}/t_{\mathrm{off}})$. Thus, with the length increment $\Delta L_{\mathrm{I27}} \sim 28$ nm and $\langle v_\beta \rangle \approx 0.1$ nm/s, we can derive the time for spontaneous unfolding of a single domain, *i.e.*, $t_{\mathrm{off}} \sim 1000$ s. To check our analysis, force distributions were predicted at each unfolding step using equation (1.31) with $f_\beta \approx 18$ pN and $t_{\mathrm{off}} \sim 1000$ s under a pulling speed of 600 nm/s; then the average distribution was compared to the histogram of forces obtained at this speed by Li *et al.* [20]. Plotted in Figure 1.19, the close match verifies that the force events arise from convolution of simple bond-like kinetics with unsteady loading dynamics of a polymer linkage.

1.4.6 Failure of a complex bond and unexpected transitions in strength

Rupture of a complex molecular bond under rising force involves transitions over a changing landscape of multiple barriers. Here, the nontrivial aspect is that the flux of states can vary significantly with time along the reaction coordinate. As pointed out earlier, such details can only be revealed by modeling the evolution of states in a hierarchy of master equations (Table 1.1). Although rupture dynamics under rising force can be quite complicated, we will begin by showing that the most frequent unbonding forces under ramps of force can be reasonably well predicted using the single rate

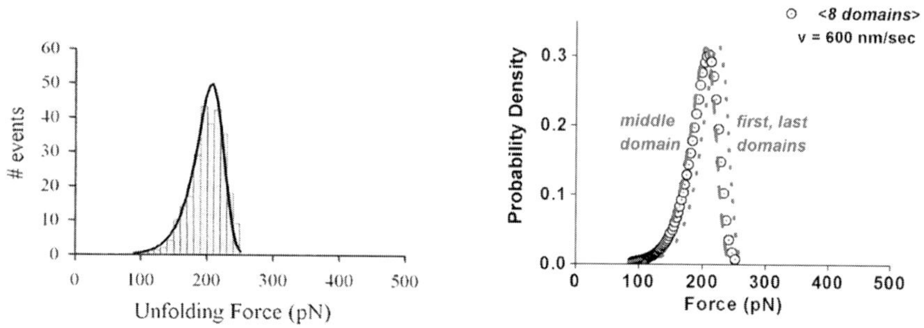

Fig. 1.19. Left: histogram of forces for unfolding eight Ig-I27 domains at a pulling rate of 600 nm/s (taken from Li *et al.* [20]). Superposed is the average probability distribution computed using equation (1.31) plus the parameters $f_\beta \approx 18$ pN and $t_{\mathrm{off}} \sim 1000$ s derived from the correlation in Figure 1.18 as described in the text. Because of changing length with each unfolding step, small shifts in probability distributions occurred as demonstrated at right for the first, last, and middle unfolding events (dashed and dotted curves) relative to the average distribution (open circles).

approximations derived earlier. Under rising force, the single rate approximations can be viewed as representing a dynamic barrier that moves along the reaction coordinate and changes height with time.

With the single rate approximation $\nu_{\rightarrow}$ and a steady ramp of force $r_f = \Delta f/\Delta t$, distributions of unbonding forces far from equilibrium are predicted by the integral expression given in equation (1.26). Although this usually requires numerical computation, the most frequent forces f^* for detachment of a complex bond can be easily derived from the maximum in the distribution, equation (1.27), which locates the peak in the probability distribution. Transforming equation (1.27) to a derivative of the reciprocal rate,

$$1/r_f = -[\partial(1/\nu_{\rightarrow})/\partial f]_{f=f^*} \tag{1.33}$$

we obtain a convenient way to derive the most frequent rupture force f^* from the superposition of transit times that define the single rate approximations. (The spread in the equivalent distribution is given by, $1/\sigma_f^2 = \nu_{N\rightarrow}[\partial^2(1/\nu_{N\rightarrow})/\partial f^2]_{f=f^*}$.) First, applied to the single rate approximation in equation (1.18), the relation between bond strength and loading rate is found to be,

$$1/r_f \approx \Sigma_{n\rightarrow N-1}[t_{\mathrm{off}}(n)/f_\beta(n)] \exp[-f^*/f_\beta(n)]. \tag{1.34}$$

Here, energies $E_b(n)$ and locations $x_\beta(n)$ for each barrier relative to the primary minimum define the forces $f_\beta(n) \equiv k_BT/x_\beta(n)$ and spontaneous passage times $t_{off}(n) \equiv t_D \exp[E_b(n)/k_BT]$ that govern the most frequent rupture force. Next, applied to the single rate approximation in equation (1.19), a similar but more complicated expression is obtained that involves the additional dependence on energies $E_0(n)$ and locations $x_c(n) = k_BT/f_c(n)$ of the intervening levels:

$$1/r_f \approx \Sigma_{n=1\to N-1}\Sigma_{j\to N-n} t_{off}(N-j)[1/f_\beta(N-j) - 1/f_c(n)] \\ \times \exp[-E_0(n)/k_BT]\exp\{f^*[1/f_c(n) - 1/f_\beta(N-j)]\} \cdot \quad (1.35)$$

As for the rate approximations themselves, equation (1.35) reduces to in equation (1.34) when intervening levels remain more than a few k_BT above the primary minimum.

To demonstrate important features of the force spectra approximations, we compare the predictions of equations (1.34) and (1.35) in Figure 1.20 to results for the most frequent rupture forces found by numerical solution to the master equations (Table 1.1) given an energy landscape of two energy barriers. Both approximations cross over smoothly from a linear regime (low slope) for the outer barrier to the linear regime (high slope) for the inner barrier with heights defined relative to the primary minimum. Predicted by equation (1.35) and shown in Figure 1.20 (left), the crossover region spans a broad range of loading rates when the secondary minimum between the barriers is low and differs little from the primary minimum (here $E_0 \sim 1k_BT$). On the other hand, the crossover region diminishes to a narrow range of rates once the intervening level is raised to $E_0 \sim 3k_BT$ as in Figure 1.20 (right) and consistent with the reduced approximation in equation (1.34). The simple geometric construction in Figure 1.20 (left) shows that the width of the crossover region can be anticipated from the intersection between an *intermediate*-linear regime and the limiting linear regimes for the two barriers. Arising from the exponential dependence on force in equation (1.35), this intermediate regime is specified by the differences of location and energy between the intervening minimum and the barrier that follows. As the energy E_0 of the minimum is increased, the intermediate regime shifts to higher loading rates and eventually falls below the crossing of the limiting regimes for the outer and inner barriers. This defines a critical energy E_0^* below which the secondary minimum becomes important.

Superposed on the spectra predicted by the single-rate approximations in Figure 1.20 are the most frequent rupture forces derived from the probability densities of failure computed numerically with the master equations in Table 1.1. As shown by the force distributions for a deep intervening minimum in Figure 1.21 (left), an unexpected jump in bond strength

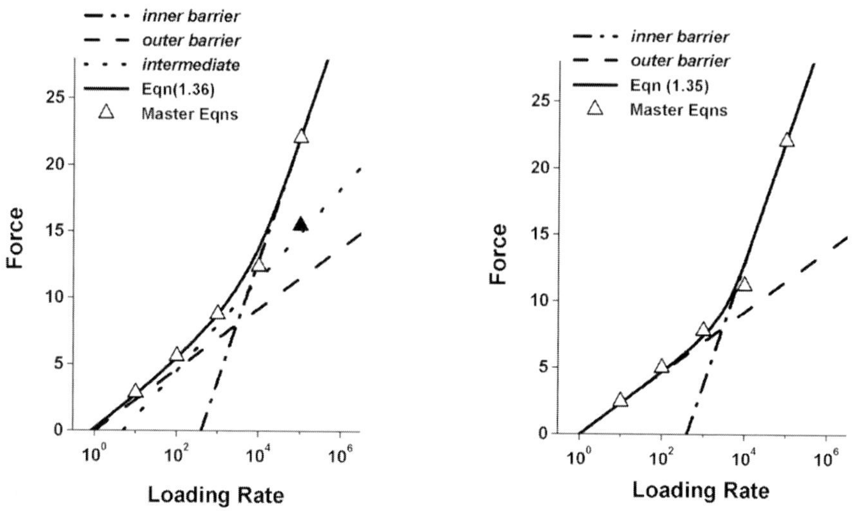

Fig. 1.20. Left: the dynamic force spectrum (solid curve) predicted by the single-rate approximation in equation (1.35) compared to results (open triangles) from numerical solution of the master equations (Table 1.1) for a hierarchy of two barriers. Superposed are linear regimes expected for force ~log(loading rate) defined by the transition rate $1/t_{\text{off}}(n)$ and force scale $f_\beta(n)$ for each barrier relative to the primary minimum. Also plotted is the intermediate-linear regime defined by the differences in energy and location of the outer barrier relative to the secondary minimum. Force, time, and loading rate are scaled by outer barrier properties, which were set equal to one. In these units, $f_\beta(2) = 1$, $t_{\text{off}}(2) = 1$, and $r_f(0) = 1$, properties of the inner barrier and secondary level were specified as, $f_\beta(1) = 4$ and $t_{\text{off}}(1) = 0.01$ plus $f_c(1) = 3$ and $E_0(2) = 1.2k_BT$ respectively. Right: the dynamic force spectrum (solid curve) predicted by the single-rate approximation in equation (1.34) compared to results (open triangles) from numerical solution of the master equations for the same barriers but with $E_0(2) = 3k_BT$.

occurs in the crossover region as a second peak emerges in the distribution under increased loading rate and eventually overtakes the initial peak. On the other hand, when the secondary minimum is raised, only a single peak appears in each distribution that shifts upward and broadens continuously in force with increased loading rate. Thus, energy landscapes with multiple barriers and deep intervening minima can produce dynamical transitions in rupture strength with first order-like coexistence and switching of peaks in distributions of failure events, which is likely to play an important role in dynamic functions of bonds under stress. We see that there are many opportunities for surprises in the highly nonlinear dynamics of bond breakage!

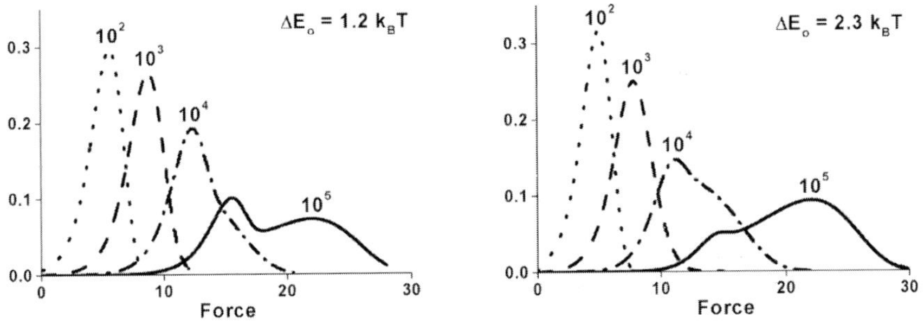

Fig. 1.21. Force distributions predicted by the master equations in Table 1.1 at each decade of loading rate for unbonding impeded by two barriers. The scales of force and spontaneous passage time for the outer barrier have been set equal to one, *i.e.* $f_\beta(2) = 1$, $t_{\text{off}}(2) = 1$, and $r_f(0) \equiv 1$. As for Figure 1.19 above, the inner barrier and secondary level are defined by, $f_\beta(1) = 4$ and $t_{\text{off}}(1) = 0.01$ plus $f_c(1) = 3$ and $E_0(2) = 1.2k_BT$ or $2.3\ k_BT$ respectively.

The power of DFS is the capability to quantify these dynamics in a way that can reveal the hierarchy of barriers in a complex interaction. However, the challenge is to measure forces over many orders of magnitude in loading rates.

Under steady ramps of force in time, the signature of escape over a sharp energy barrier is a straight line in a plot of rupture force *versus* log(loading rate). Although unexpected, tests of the SGP ligand and P-selectin exhibited simple bond-like kinetics under force as seen in Figures 1.13 and 1.14. Here, on the other hand, we will use tests of a far simpler interaction – hydrophobic anchoring of lipids in a membrane – to demonstrate dynamic failure of a complex interaction. To test lipid anchoring strength, we decorated the tip of a BFP with PEG-biotin chains and prepared giant phosphatidyl choline PC lipid vesicles doped with a small fraction of PEG-biotinylated PE lipid (diC14 phosphatidyl ethanolamine). Pre-equilibration of both surfaces with streptavidin SA was then used to block nearly all biotin groups leaving only a few sites for probe-vesicle attachment. Many touches of the BFP tip to vesicles led to a few extraction events as illustrated schematically in Figure 1.22. Sample histograms of the forces are presented in Figure 1.23 for a three thousand-fold range in loading rate.

We see in Figure 1.23 that only small forces were needed to extract diC14 PE lipids from the surface of a C18:0/1 PC lipid bilayer even at fast loading rates. Moreover, we also see that the force distributions begin very narrow at slow rates of loading, then broaden and become bimodal at fast rates of loading. Clearly similar to the complex landscape in the idealized

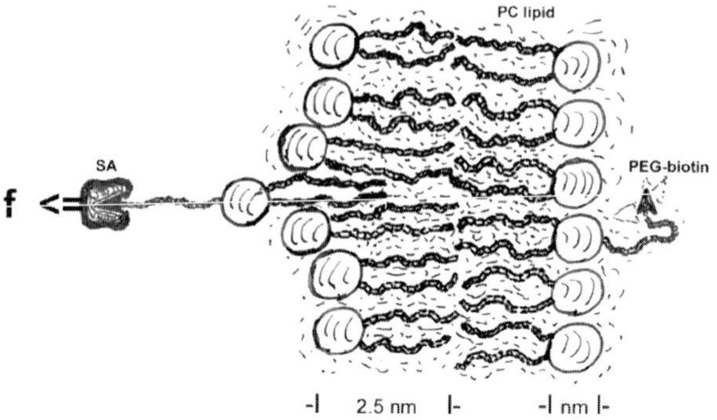

Fig. 1.22. Schematic of PEG-biotin lipid extraction from a lipid bilayer membrane. For typical biomembrane lipids (*e.g.* C18:0/1 phosphatidyl choline), X-ray diffraction yields a mass average thickness of ~4 nm for the full membrane [25]. Augmented by headgroup structure, the distance from the midplane to the headgroup-water boundary is estimated to be ~2.5 nm as sketched above.

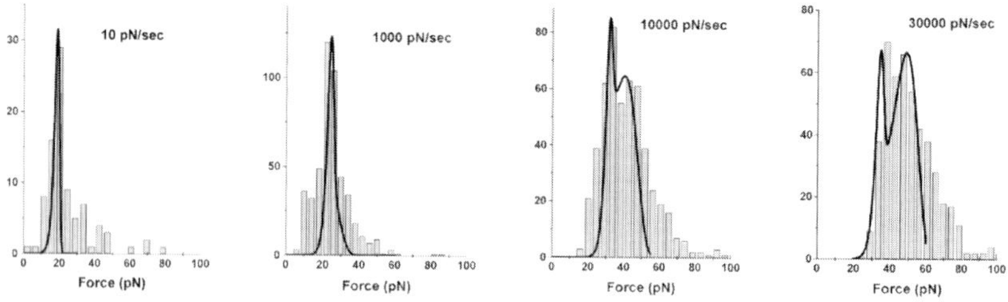

Fig. 1.23. Distributions of forces measured by BFP in extraction of single receptor lipids (biotin-PEG-diC14 PE) from C18:0/1 PC lipid vesicles.

model described above, the histograms in Figure 1.23 correlate well with the probability distributions (superposed) predicted by solution of the master equations for an optimal sequence of two energy barriers punctuated by a deep intervening minimum. Matched to the most frequent forces obtained from experiment, Figure 1.24 shows the dynamic spectrum of anchoring strength predicted by the single-rate approximation equation (1.35) with the locations and energies for the two barriers and intervening level used in the numerical computation of probability densities. As in the model above, the continuous spectrum approaches the limiting linear regime set by the properties of the outer barrier at very slow loading and the limiting linear regime set by properties of the inner barrier at very fast loading.

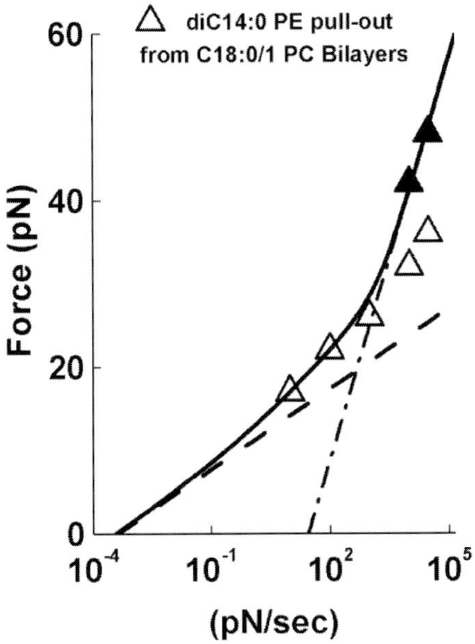

Fig. 1.24. The dynamic force spectrum (solid curve) predicted by the single-rate approximation in equation (1.35) (solid line) compared to locus of most frequent extraction force (open/closed triangles) from the experimental distributions plotted in Figure 1.23. Also shown are the limiting linear regimes for two barriers defined by the transition rate $1/t_{\mathrm{off}}(n)$ and force scale $f_\beta(n)$ for each barrier relative to a primary minimum: $f_\beta(2) = 1.4$ pN and $t_{\mathrm{off}}(2) = 3 \times 10^3$ s for the outer barrier; $f_\beta(1) = 7$ pN and $t_{\mathrm{off}}(1) = 0.25$ s for the inner barrier. Derived from correlation to the evolution of the two peaks in the force distributions with loading rate (Fig. 1.22), the intervening level was found to be characterized by a force scale $f_{\mathrm{c}}(1) = 4$ pN and an energy level $E_0(2) = 2k_{\mathrm{B}}T$ above the primary minimum.

Based on established concepts of hydrophobic interactions [32], we expect the anchoring potential to increase in energy linearly with distance as the lipid is displaced outward along the surface normal. The energy barrier to extraction should scale as the number of $\mathrm{CH_2}$ groups exposed to water with a proportionality of $\sim 1k_{\mathrm{B}}T$ per $\mathrm{CH_2}$ [32]. So taking $t_{\mathrm{D}} \sim 10^{-9}$ s as the diffusive relaxation time for lipid "hopping" motions, the time scale for spontaneous dissociation into aqueous solution would be $t_{\mathrm{off}} \sim (10^{-9}\,\mathrm{s}) \exp(N_{\mathrm{CH_2}})$. Although difficult to measure for diacyl lipids with more chains longer than 12 carbons, this scaling for t_{off} agrees with results for PEG-lipid dissociation from bilayer membranes in solution [28] and even yields a value of $\sim 10^{-9}$ s for the pre-exponential factor. Likewise, analysis of the force distributions in Figure 1.23 and the spectrum in Figure 1.24 yields a time scale $t_{\rm off}(2) = 3 \times 10^3$ s for passage of the outer barrier which is consistent with the spontaneous time for escape of a diC14 receptor lipid under zero force in solution [28]. Another expectation based on this description of hydrophobic interactions is that the thermal force scale f_β should be set by the insertion depth $x_{\rm lipid}$ in the membrane, *i.e.* $f_\beta = k_{\rm B}T/x_{\rm lipid}$. As sketched in Figure 1.22, the distance from the headgroup-water boundary to the midplane of a C18:0/1 phosphatidyl choline bilayer would define a limiting value for insertion depth of $x_{\rm lipid} \sim 2.5$ nm [25], which could be accommodated by full extension of the diC14 PE lipid (*i.e.* 14×0.125 nm for chain length plus $\sim$1 nm for headgroup $\sim$2.8 nm). Interestingly, the distance derived from the thermal force scale for the outer barrier of $x_\beta(2) = k_{\rm B}T/1.4$ pN ≈ 2.9 nm is comparable to the maximum insertion depth of 2.8 nm. However, the slightly longer length derived from force spectroscopy could be due to a small outward deformation of the lipid interface as sketched in Figure 1.22. Because of the large length scale associated with hydrophobic anchoring, we see that the thermal force scale $[f_\beta(2) = 1.4$ pN$]$ is small and the extraction force distributions are very narrow at slow loading rates (Fig. 1.23).

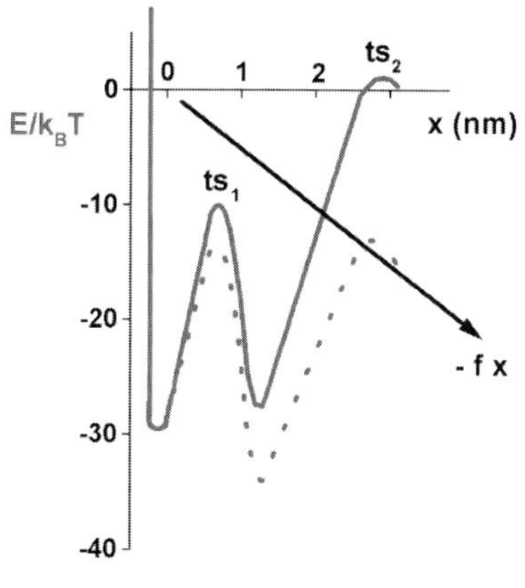

Fig. 1.25. Schematic of the energy landscape for diC14 PE anchoring in a C18:0/1 PC bilayer. Barrier locations and relative energies were obtained from analysis of the extraction force distributions (Fig. 1.23) and force spectrum (Fig. 1.24) as described above. However, to specify the depths of minima, a value has to be specified for the diffusive relaxation time of lipid "hopping" motions, which here was taken to be $t_{\rm D} \sim 10^{-9}$ s.

Consequently, anchoring strength remains weak under slow loading and, even when the time needed for lipid extraction is decreased more than million-fold (*i.e.* from 3000 s to 0.001 s), extraction forces only reach $\sim$20 pN.

Unexpectedly, however, lipid anchoring strengthens abruptly when lipids are forced to dissociate in less than 0.001 s, which requires loading rates 10 000 pN/s and above. Here, a broad second peak appears in the force distributions and the narrow peak wanes. The emergence of a broad peak stems from impedance to dissociation caused by a second barrier characterized by a small length scale, *i.e.* $x_\beta(1) = k_\mathrm{B}T/7$ pN ≈ 0.6 nm. Encountered first along the extraction pathway, this initial barrier probably represents impedance to disruption of the headgroup interface (*e.g.* weak hydrogen bonding as well as hydrophobic exposure) and appears to be followed by a secondary minimum at $\sim$1 nm. Putting together these features, Figure 1.25 demonstrates how dynamic force spectroscopy reveals the inner complexity of hydrophobic anchoring in membranes and significantly extends the test tube picture of the energy landscape for kinetics.

1.5 Summary

We have seen that measuring bond detachment forces and lifetimes under steady loading can provide an intimate view of prominent energy barriers that govern physical strength and limit bond dissociation under stress. Moreover, the distribution of rupture events at each loading speed is a complete picture of the unbonding process in time as barriers fall at a particular rate. Even though dynamic force spectroscopy allows us to look inside molecular interactions, specific experimental requirements must be met in order to achieve a reliable picture of the landscape. Not only do we need an accurate and sensitive force probe, we have to be confident that we are only testing single molecular attachments. This means we have to chemically decorate surfaces with dilute sites for attachments and to also regulate the assembly process (*e.g.* through very soft touch under feedback control). Next, we have to know the compliance properties of the linkage to the bond (*e.g.* the contour and persistence lengths of polymeric connections) and the dynamic response when coupled to the probe. Having met these requirements, it is imperative to measure forces over an enormous range of time scales, or more specifically, loading rates. Having satisfied these criteria, the final-critical step is to compare the histograms of rupture events measured at each loading rate to the probability densities predicted by the thermal forces and spontaneous transition times derived from the dynamic force spectrum.

The authors gratefully acknowledge support from US National Institutes of Health grants HL 65333 & HL31579, Medical Research Council of Canada grant MT7477 (to EE) plus support from the Engineering and Physical Sciences Research Council, the Biological and Biomedical Sciences Research Council, Universitas 21 and the University of Nottingham (to PMW).

References

[1] A. Ashkin, *Biophys. J.* **61** (1992) 569; *ibid, Proc. Natl. Acad. Sci. USA* **94** (1997) 4853.

[2] A. Ashkin, K. Schutze, J.M. Dziedzic, U. Euteneuer and M. Schliwa, *Nature* **348** (1990) 346.

[3] D. Bartolo, I. Derenyi and A. Ajdari (private communication).

[4] G.I. Bell, *Science* **200** (1978) 618.

[5] G. Binning, C.F. Quate and C.H. Gerber, *Phys. Rev. Lett.* **56** (1986) 930.

[6] A. Einstein, *Ann. Physik* **17** (1905) 549; *ibid, Ann. Physik* **19** (1906) 371.

[7] E. Evans, *Faraday Discuss.* **111** (1998) 1; *ibid, Annu. Rev. Biophys. Biomol. Struct.* **30** (2001) 105.

[8] E. Evans, A. Leung, D. Hammer and S. Simon, *Proc. Natl. Acad. Sci. USA* **98** (2000) 3784.

[9] E. Evans and F. Ludwig, *J. Phys. Cond. Matt.* **12** (2000) 315A.

[10] E. Evans and K. Ritchie, *Biophys. J.* **72** (1997) 1541.

[11] E. Evans, K. Ritchie and R. Merkel, *Biophys. J.* **68** (1995) 2580.

[12] E. Evans and K. Ritchie, *Biophys. J.* **76** (1999) 2439.

[13] J. Fritz, A.G. Katopodis, F. Kolbinger and D. Anselmetti, *Proc. Natl. Adad. Sci. USA* **95** (1998) 12283.

[14] M. Grandbois, M. Beyer, M. Rief, H. Clausen–Schaumann and H.E. Gaub, *Science* **283** (1999) 1727.

[15] H. Grubmuller, B. Heymann and P. Tavan, *Science* **271** (1996) 997.

[16] P. Hanggi, P. Talkner and M. Borkovec, *Rev. Mod. Phys.* **62** (1990) 251.

[17] S. Izrailev, S. Stepaniants, M. Balsera, Y. Oono and K. Schulten, *Biophys. J.* **72** (1997) 1568.

[18] H.A. Kramers, *Physica (Utrecht)* **7** (1940) 284.

[19] S.C. Kuo and M.P. Sheetz, *Science* **260** (1993) 232.

[20] H. Li, A.F. Oberhauser, S.B. Fowler, J. Clarke and J.M. Fernandez, *Proc. Natl. Acad. Sci. USA* **97** (2000) 6527.

[21] H. Lu and K. Schulten, *Proteins Struct. Funct. Genet.* **35** (1999) 453.

[22] S.-J. Marrink, O. Berger, P. Tieleman and F. Jahnig, *Biophys. J.* **74** (1998) 931.

[23] P.E. Marszalek, H. Lu, H. Li, M. Carrion-Vazquez, A.F. Oberhauser, K. Schulten and J.M. Fernandez, *Nature* **402** (1999) 100.

[24] R. Merkel, P. Nassoy, A. Leung, K. Ritchie and E. Evans, *Nature* **397** (1999) 50.

[25] W. Rawicz, K. Olbrich, T. McIntosh, D. Needham and E. Evans, *Biophys. J.* **79** (2000) 328.

[26] M. Rief, M. Gautel, F. Osterhelt, J.M. Fernandez and H.E. Gaub, *Science* **276** (1997) 1109.

[27] P.J. Rossky, J.D. Doll and H.L. Friedman, *J. Chem. Phys.* **69** (1978) 4628.

[28] J.R. Silvius and M.J. Zuckermann, *Biochem.* **32** (1993) 3153.

[29] W.S. Somers, J. Tang, G.D. Shaw and R.T. Camphausen, *Cell* **103** (2000) 467.

[30] T. Strunz, K. Oroszlan, R. Shafer and H.-J. Guntherodt, *Proc. Natl. Acad. Sci. USA* **96** (1999) 11277.

[31] K. Svoboda, C.F. Schmidt, B.J. Schnapp and S.M. Block, *Nature* **256** (1993) 721.

[32] C. Tanford, *The Hydrophobic Effect* (Wiley-Interscience, New York, 1973).

[33] N.G. van Kampen, *Stochastic Processes in Physics and Chemistry* (North-Holland Elsevier, Amsterdam, 1992).

Part 2: P. Williams and E. Evans

2 Dynamic force spectroscopy. II. Multiple bonds

2.1 Hidden mechanics in detachment of multiple bonds

Dynamic force spectroscopy (DFS) is a uniquely powerful tool to investigate the physics and chemistry of molecular interactions. It is not surprising, therefore, the extent of research in both the physical and life sciences into the use of DFS for such areas as structural biology, rational drug design and biomaterials science. When attempted, though, it is soon discovered that a major complication in the measurement of bond strength is often the occurrence of multiple bond attachments. Moreover, single attachments between macromolecules often involve bonding interactions distributed over many widely separated groups and behave as multiply-bonded systems. Although careful sample preparation methods and high skills in surface chemistry may mean that the numbers or densities of sites for binding are known, force spectra will always involve some multiply-bonded structures which can be difficult or impossible to interpret because the partition of force and degree of cooperativity amongst binding sites is unknown. Fortunately, examining dynamic rupture strength for a few generic types of multiple bonds can provide useful insights into the nature of hidden interactions in a molecular assembly. We begin by describing simple mechanical scenarios for multiple bonds. The first case is a series or chain of bonds where the force is experienced fully by each bond in the system. By comparison, the second case is a *zipper* which is a series of bonds where the force is applied only to the *lead* bond; once that bond fails, force propagates to the next and so on. Finally, the third case is *parallel* bonds where the force is partitioned equally amongst existing bonds in an attachment. These idealized cases are simple stereotypes of multiple bond attachments. It is important to recognize that structural deformation on the nanoscale can lead to very different conditions of loading for multiple bonds. Still, the series, *zipper*, and *parallel* bond descriptions encompass a broad range of configurations. Thus, we consider two general architectures in which multiple interactions withstand stress; that in which force is shared across the intact connections – the parallel description – and that in which one or more bonds feel all of the force – a serial description. These are illustrated in Figure 2.1. As we see, the *zipper* is a specific case of bonds loaded in series where the bonds connecting two surfaces are broken one after the other. In Part I we have already described a experiment where a series of bonds was ruptured, *i.e.* the unfolding of Ig domains in the muscle protein titin (Sect. 1.4.5 of Part I), which will come up again in this part.

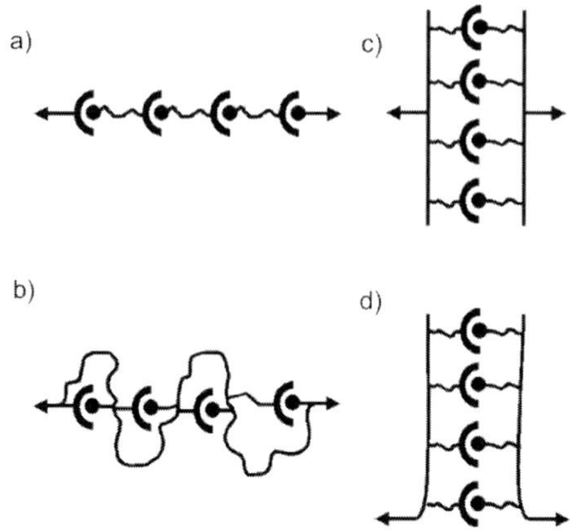

Fig. 2.1. Schematic representations of the prototypes of bonding architecture that we consider here. **a)** Bonds loaded in series. The system fails on breakage of a single bond. **b)** Bonds loaded in series, but breakage of a bond does not lead to failure of the system (analogy of the unfolding of tandem repeats in multidomain proteins). **c)** Bonds loaded in parallel. Force is spread evenly across intact bonds. **d)** A zipper. Bonds are loaded and broken consecutively.

2.2 *Impact of cooperativity*

We first need to consider the profound impact of cooperativity in multiple bond detachment. By this, we mean some type of strong mechanical coupling exists within a array of bonds that only allows the molecules to pass transition states synchronously. Even though atomic scale excitations remain uncorrelated in time, stiff mechanical coupling can create collective modes on long length scales that lead to strong correlations on the time scale for overdamped relaxation of the complex. Such a system behaves like a *macro*-single bond with a barrier given by the sum of the individual barrier energies. But quite different outcomes arise for the *series* and *parallel* loading arrangements shown in Figure 2.1. For a simple-ideal bond, we have seen that the lifetime is set by the exponential magnitude of the energy barrier with a diffusive relaxation time as the prefactor,

$$t_{\text{off}} = t_{\text{D}} \exp(E_{\text{b}}/k_{\text{B}}T). \tag{2.1}$$

With N-identical bonds, cooperative unbinding leads to an enormous increase in the time scale for dissociation as approximated by

$$t_{\text{off}}(N) \approx N t_{\text{off}} \exp[(N-1)E_{\text{b}}/k_{\text{B}}T]. \tag{2.2}$$

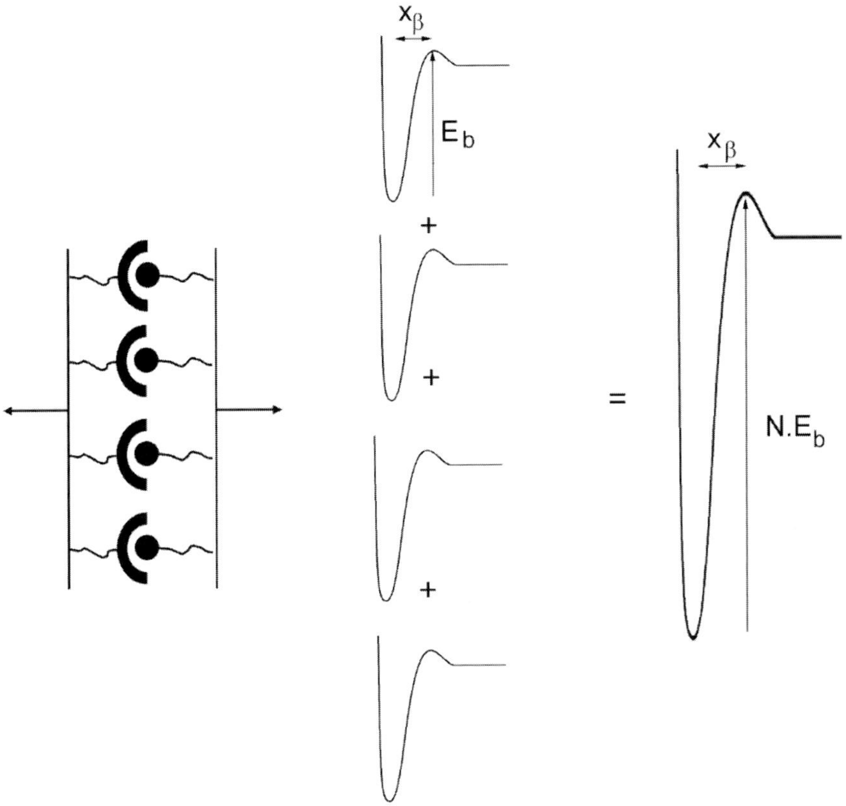

Fig. 2.2. A system of identical attachments that fail cooperatively when loaded in parallel can be considered as a single bond with a N-fold increase in the barrier compared to a single bond. The force scale of the compound bond remains the same as for a single bond interaction.

(Note: The prefactor N is based on the assumption that molecular damping scales with N.) For identical-cooperative bonds in parallel (*cf.* Fig. 2.2), location of the transition state x_β and thermal force scale $f_\beta = k_B T / x_\beta$ remain that for the single bond (see Part I for more details). Hence, the unbinding rate under force f is,

$$\nu_{\rightarrow}(N) \approx [1/t_{\mathrm{off}}(N)] \exp(f/f_\beta). \tag{2.3}$$

For identical bonds in series, each bond contributes an increment in length along the direction of force on unbinding and thus the thermal force scale is lowered N fold, *i.e.* f_β/N. Hence, the unbinding rate for cooperative bond

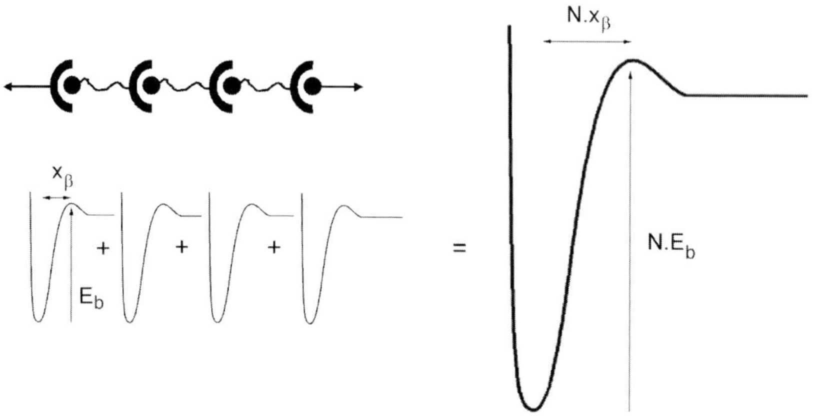

Fig. 2.3. A system of identical attachments that fail cooperatively when loaded in series can be considered as a single bond with both a barrier and force scale N-times that of the single bond. The increase in force scale makes the series connection weaker under force than the parallel system.

in series,

$$\nu_{\rightarrow}(N) \approx [1/t_{\mathrm{off}}(N)] \exp(Nf/f_{\beta}) \tag{2.4}$$

increases exponentially faster with force than for bonds in parallel. Based on the generic relation for most frequent rupture force f^* (defined by, $1/r_f = -[\partial(1/\nu_{\rightarrow})/\partial f]_{f=f^*}$, from Eq. (1.34) in Part I), equations (2.3) and (2.4) show that it takes much less force for cooperative failure of bonds in series,

$$f^* \approx (f_{\beta}/N)[\log(r_f) + 2\log(N) + (N-1)E_{\mathrm{b}}/k_{\mathrm{B}}T] \tag{2.5}$$

than for bonds in parallel, *i.e.*

$$f^* \approx f_{\beta}[\log(r_f) + \log(N) + (N-1)E_{\mathrm{b}}/k_{\mathrm{B}}T] \tag{2.6}$$

when either is broken far from equilibrium.

As shown in Figure 2.4, cooperative failure of multiple bonds has been observed in DFS studies of short strands of DNA (oligonucleotides) using the atomic force microscope (AFM). The DNA double helix is held together by the predominately hydrogen-bonding interaction of nucleotides along the strands. Each interaction accounts for a few $k_{\mathrm{B}}T$ of energy and the extrapolated dissociation rate appears to increase exponentially with the number of base pairs. Hence, increasing the number of base pairs from 10 to 30 leads to an increase in the lifetime of the duplex from several seconds to many years! Interestingly, recent experiments with RNA (to be published) show

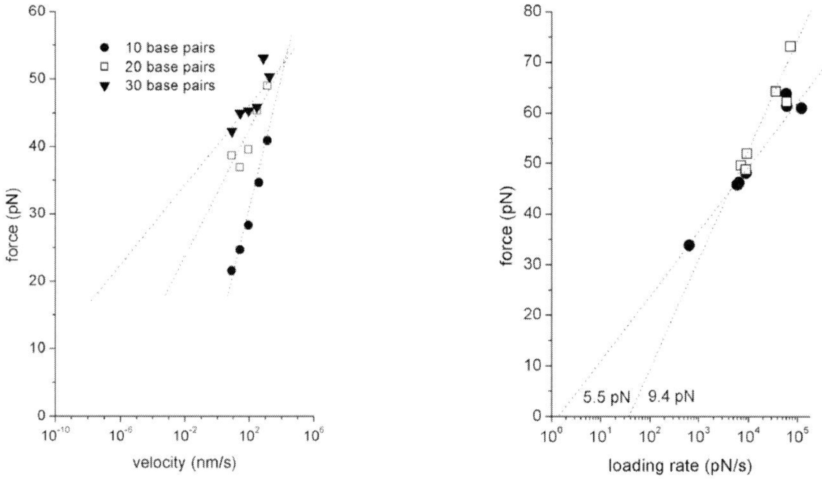

Fig. 2.4. Left: DFS study of the rupture of oligonucleotide strands of 10, 20 and 30 base-pair length behaves like cooperative failure of bonds in series (taken from Strunz *et al.*). The force scale diminishes with increase in length and is approximated by (35/N) pN. The extrapolated off-rate increases exponentially with increasing length. Right: DFS study of RNA duplexes. 12 base pairs produces a force scale of 5.5 pN. It appears that insertion of a 3-base bulge along one of the strands apparently disrupts the cooperativity and effectively doubles the force scale to 9.4 pN.

identical results but addition of a bulge at the mid-point of the duplex disrupts cooperativity and increases of the force scale. Thus, this type of design could be of use to probe the origin of cooperative unbinding.

2.3 Uncorrelated failure of bonds loaded in series

2.3.1 Markov sequence of random failures

Given the soft compliance of most molecular interfaces in liquids, there is no reason *a priori* to expect multiple bonds holding two surfaces together to act cooperatively or pass transition states coincidently in time. For example, several selectin interactions hold a white blood cell to an endothelial wall under strong hydrodynamic loading. Whilst each selectin bond is made up of several widely separated atomic scale interactions that may unbind coincidently, there will be no coincidence between detachment at the different selectin sites even though the rupture event may occur extremely rapidly. Uncorrelated failure of multiply-bonded attachments is distinctly different from that of the cooperative case described in Section 2.2. Like the evolution of states in a complex bond (see Sect. 1.3.3 in Part I), the state of bonding amongst multiple bonds in an attachment is treated as a Markov sequence where a hierarchy of master equations predicts the survival of the attachments as a function of time. In the examples to follow, we will consider multiples of simple bonds with a single transition state but will note the approach to treating multiples of complex interactions with additional-inner transition states. The master equations for unbonding a multiple bonds are identical to those in Table 1.1 of Part I except the "level" in an energy landscape is replaced by the number of bonds that remain during the time course of detachment. (Note: for obvious reasons, we will define the system hierarchy so that subscript "u" represents the unbonded state and that "n" represents the transient n-bonded attachment where $n = 1 \rightarrow N$.) Temporal evolution of the N-bonded state commences with the rate at which N bonds decrease to $N - 1$ diminished by the rate at which $N - 1$ increase to N,

$$\mathrm{d}S_N/\mathrm{d}t = -\nu_{N\rightarrow N-1}S_N(t) + \nu_{N\leftarrow N-1}S_{N-1}(t)$$

and sequentially through intermediate states according to,

$$\mathrm{d}S_n/\mathrm{d}t = -[\nu_{n\rightarrow n-1} + \nu_{n+1\leftarrow n}]S_n(t) + \nu_{n+1\rightarrow n}S_{n+1}(t) + \nu_{n\leftarrow n-1}S_{n-1}(t)$$

ending with complete detachment,

$$\mathrm{d}S_u/\mathrm{d}t = -\nu_{1\leftarrow u}S_u(t) + \nu_{1\rightarrow u}S_1(t).$$

Here unbonding $\nu_{n\rightarrow n-1}$ and rebinding $\nu_{n\leftarrow n-1}$ rates reflect the rates for rupture or formation between populations of "n" and "$n - 1$" bonds, which will depend on the type of architecture and mechanics of loading bonds in the multiple attachment. Again for simplicity, we will base our discussion on the behaviour of identical bonds, each of which has an unbonding rate under force given by $t_{\mathrm{off}}\,\nu_{\rightarrow} = \exp(f/f_\beta)$, and assume that under rapid rise in force, detachment occurs far from equilibrium at each step (*i.e.*, $\nu_{\leftarrow} \sim 0$). Hence, under a force ramp, $r_f = \mathrm{d}f/\mathrm{d}t$, the master equations reduce to,

N bonds

$$\mathrm{d}S_N/\mathrm{d}f \approx -(1/r_f)\,\nu_{N\rightarrow N-1}(N, f)\,S_N(f)$$

$$\cdot$$

$$\cdot$$

$$\mathrm{d}S_n/\mathrm{d}f \approx -(1/r_f)\,\nu_{n\rightarrow n-1}(n, f)\,S_n(f) + (1/r_f)\,\nu_{n+1\rightarrow n}(n+1, f)\,S_{n+1}(f) \tag{2.7}$$

$$\cdot$$

$$\cdot$$

complete detachment

$$\mathrm{d}S_u/\mathrm{d}f \approx (1/r_f)\,\nu_{1\rightarrow u}(1, f)\,S1(f)$$

where, in general, the rate of transition $\nu_{n\to n-1}(n, f)$ at any step of detachment can depend on existing number "n" of bonds as well as force. For instance, the statistical rate of failure of "n" equivalent bonds is n-fold that of a single bond, irrespective of the system architecture and loading dynamics. When breakage of onc bond does not lead to failure of the attachment, it is essential to specify the manner in which the force is redistributed amongst the remaining bonds.

2.3.2 Multiple-complex bonds

As a footnote to Section 2.3.1 above, the situation becomes much more complicated when the attachments are formed by complex bonds with several levels and transition states. For a multiple attachment of N identical complex bonds, each with M levels (local energy minima) and intervening barriers, evolution of the system involves $N \times M$ master equations where survival of the n-bonded state is governed by transport of states between the interior m levels until an unbinding event. Thus, a matrix of functions $S_{n,m}(t)$ is needed to describe the likelihood of being in the mth level of the n-bonded state. Even when each bond rupture occurs with no chance of rebinding, the internal states of complex bonds may still undergo forward and reverse transitions past interior barriers. So for example, steady loading and rupture of N-identical bonds with two transition states are modelled by the following hierarchy,

N bonds

$$
\begin{aligned}
\mathrm{d}S_{N,1}/\mathrm{d}f &= -(1/r_f)\,\nu_{1\to 2}(N,f)\,S_{N,1}(f) + (1/r_f)\,\nu_{1\leftarrow 2}(N,f)\,S_{N,2}(f)\\
\mathrm{d}S_{N,2}/\mathrm{d}f &= -(1/r_f)\,[\nu_{2\to u}(N,f) + \nu_{1\leftarrow 2}(N,f)]\,S_{N,2}(f)\\
&\quad + (1/r_f)\,\nu_{1\to 2}(N,f)\,S_{N,1}(f)
\end{aligned}
$$

$$\cdot$$

$$\cdot$$

$$
\begin{aligned}
\mathrm{d}S_{n,1}/\mathrm{d}f &= -(1/r_f)\,\nu_{1\to 2}(n,f)\,S_{n,1}(f) + (1/r_f)\,[\nu_{2\to u}(n+1,f)\,S_{n+1,2}(f)\\
&\quad + \nu_{1\leftarrow 2}(n,f)\,S_{n,2}(f)]\\
\mathrm{d}S_{n,2}/\mathrm{d}f &= -(1/r_f)\,[\nu_{2\to u}(n,f) + \nu_{1\leftarrow 2}(n,f)]\,S_{n,2}(f)\\
&\quad + (1/r_f)\,\nu_{1\to 2}(n,f)\,S_{n,1}(f)
\end{aligned}
$$

$$\cdot$$

$$\cdot$$

$$
\begin{aligned}
\mathrm{d}S_{1,1}/\mathrm{d}f &= -(1/r_f)\,\nu_{1\to 2}(1,f)\,S_{1,1}(f) + (1/r_f)\,[\nu_{2\to u}(2,f)\,S_{2,2}(f)\\
&\quad + \nu_{1\leftarrow 2}(1,f)\,S_{1,2}(f)]
\end{aligned}
$$

$$\mathrm{d}S_{1,2}/\mathrm{d}f = -(1/r_f)\,[\nu_{2\to u}(1,f)\,S_{1,2}(f) + \nu_{1\leftarrow 2}(1,f)\,S_{1,2}(f)] + (1/r_f)\,\nu_{1\to 2}(1,f)\,S_{1,1}(f)$$

complete detachment

$$\mathrm{d}S_u/\mathrm{d}f = (1/r_f)\,\nu_{2\to u}(1,f)\,S_{1,2}(f).$$

Again at any stage of detachment, the rates of transition can depend on existing number "n" of bonds as well as force and passage of the final transition state $\nu_{2\to u}(n,f)$ is assumed to be terminal for each state of bonding, *i.e.* $\nu_{2\leftarrow u}(n,f) = 0$. Clearly, solving such a system of master equations for many attachments demands tedious numerical computation. But as described Part I, useful analytic approximations (Eqs. (1.18) and (1.19) of Part I) exist for multiple bond attachments which will prescribe effective rates $\nu_{n\to n-1}(f)$ for transition of the n-bonded state to $n-1$ bonds over the course to complete failure. We introduce these approximations where appropriate in our discussion of the different bonding architectures illustrated in Figure 2.1.

2.3.3 Multiple-ideal bonds

Beginning with a chain of identical bonds, each bond experiences the same force but the rate of transition from N to $N-1$ bonds is proportional to the number of bonds in the chain, $t_{\text{off}}\,\nu_{N\to N-1} = N\exp(f/f_\beta)$. If an unbinding event breaks the chain, the probability density for failure or force distribution $p_N(f)$ follows from a single master equation:

$$\mathrm{d}S_N/\mathrm{d}f = -(1/r_f)(N/t_{\text{off}})\exp(f/f_\beta)\,S_N(f) = -p_N(f).$$

This simply rescales the loading rate to give $p_N(f) = N\exp\{f/f_\beta - Nf_\beta[\exp(f/f_\beta)-1]/(r_f t_{\text{off}})\}/(r_f t_{\text{off}})$ and introduces a $\log(N)$ weakening of the most frequent force for failure, $f^* = f_\beta[\log(r_f t_{\text{off}}/f_\beta) - \log(N)]$. By comparison, a series of bonded "knots" (like folded Ig domains of a long titin protein) behave differently in that the chain does not fail with an unbonding event. There is an quick drop in force to a new level which arises from the length of unfolded protein inserted in the chain. The system re-equilibrates and the random process "clock" is reset. Force again increases although from a level which is not necessarily zero. Also, as discussed previously in Section 1.4.5 of Part I, adding a large length of unfolded material into the chain can have an important impact on the loading rate. For unravelling a series of knots, therefore, rupture events become separated in time and produce a "saw-tooth pattern" of force with very nonlinear loading characteristics. Each of these force "spikes" can be considered as a distinct test of the knot-bond strength amongst the n remaining knots under anharmonic loading conditions.

By comparison, there is little drop in force on breakage of a bond in the case of a *zipper* (peeling a sequence of bonds). Nearly the same level of force is transferred to the next bond and so on. Again, there are N independent tests of bond strength but starting at randomly-distributed initial loads. Assuming identical bonds under steady loading, the rate of failure for each bond has the same dependence on force and force is proportional to time. Thus, evolution of a zipper can be modelled by the following master equations,

N bonds

$$\mathrm{d}S_N/\mathrm{d}f = -1/(r_f t_{\mathrm{off}}) \exp(f/f_\beta)\, S_N(f)$$

.

.

$$\mathrm{d}S_n/\mathrm{d}f = -1/(r_f t_{\mathrm{off}}) \exp(f/f_\beta)\, S_n(f) + 1/(r_f t_{\mathrm{off}}) \exp(f/f_\beta)\, S_{n+1}(f)$$

.

.

complete detachment

$$\mathrm{d}S_u/\mathrm{d}f = 1/(r_f t_{\mathrm{off}}) \exp(f/f_\beta)\, S_1(f) = p_N(f).$$

Somewhat more complicated than simultaneously pulling on a series of bonds, the master equations can still be integrated to find the probability density $p_N(f)$ for final detachment of a *zipper* with N identical bonds. A simple transformation of variables, $u \equiv f_\beta[\exp(f/f_\beta) - 1]/(r_f t_{\mathrm{off}})$, leads to the Poisson-like result,

$$p_N(f) \sim u^{N-1} \exp(-u)\ \mathrm{d}u/\mathrm{d}f = \{f_\beta[\exp(f/f_\beta) - 1]/(r_f t_{\mathrm{off}})\}^{N-1}\ p_1(f)$$

where $p_1(f) = \exp\{f/f_\beta - f_\beta[\exp(f/f_\beta) - 1]/(r_f t_{\mathrm{off}})\}/(r_f t_{\mathrm{off}})$ is the probability density for failure of the one bond. As shown below in Figure 2.5, a *zipper* of bonds results in a $\log(N)$ strengthening of the most frequent force for complete detachment.

2.3.4 Equivalent single-bond approximation

Although often necessary, solving the system of master equations for multiple-bond attachments is clearly time consuming and requires detailed definitions of many molecular scale parameters. Thus, as for a complex-single bond described in Section 1.4.6 of Part I, we employ analytic approximations for effective rate of unbinding $\nu_{\rightarrow}$ in a multilevel system to collapse the master equations for complete detachment of multiple bonds to a rate of failure $\nu_{N\rightarrow}$ for an "equivalent single bond". In this way, the most frequent

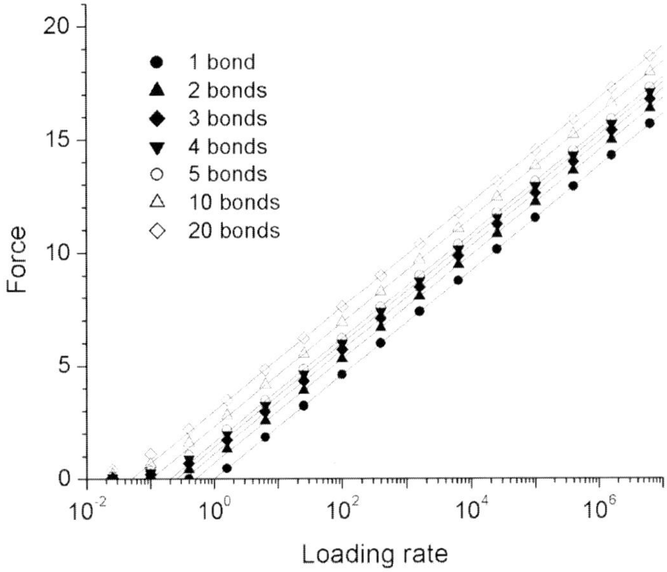

Fig. 2.5. Rupture forces (scaled by f_β) found by solution to the master equations (symbols) for a zipper of between 2 and 20 bonds. (Loading rate is scaled by f_β/t_{off}.) Also shown by solid lines are the forces predicted by the equivalent single attachment model, which closely match the computational results at forces $f/f_\beta > 1$.

rupture force or *strength* of a multibond attachment f^* can be easily estimated using the generic expression for location of the peak (at $\partial p_N/\partial f = 0$) in the probability density for failure of the equivalent single bond (Eq. (1.33) in Part I),

$$1/r_f = -[\partial(1/\nu_{N\rightarrow})/\partial f]_{f=f^*}. \tag{2.8}$$

The case of N-identical simple bonds in series (Fig. 2.1a) is trivial. Here, each bond experiences the same force history and any rupture event leads to failure of the attachment. Thus, the rate of failure events is N-fold faster than the rate for one bond (as any one of the N may break). Because of the N-fold scaling of rate, it follows from equation (2.8) that the attachment is weakened slightly as demonstrated earlier, $f^* = f_\beta[\log(r_f t_{\mathrm{off}}/f_\beta) - \log(N)]$, compared to a single bond at the same loading rate.

Serial linkages of *dissimilar* bonds are, however, more subtle. We expect rupture to occur at the *weakest* bond and naively also expect that *strong versus weak* should follow the scale set by the energy barriers sustaining the bonds. But surprisingly, thermal force scales for exponentiation of unbinding rates are also important factors in the determination of *strong versus weak*. This follows from the failure rate defined by the sum of the unbinding rates for each bond. Taking properties (t_{off} or E_{b}, and f_β) of the bond with the smaller barrier energy for a reference, the combined rate of failure of two bonds in series can be expressed as, $t_{\text{off}}\,\nu_{\rightarrow} \approx \exp(f/f_\beta)\{1 + \exp[-\Delta E_{\text{b}}/k_{\text{B}}T + f\Delta(1/f_\beta)]\}$; $\Delta E_{\text{b}} > 0$ and $\Delta(1/f_\beta)$ are the differences in barrier energy and rate-exponentiation scale of the bond with a higher barrier. Based on the combined rate of failure, the bond with the smaller energy barrier remains the weak bond so long as the inequality, $\Delta E_{\text{b}}/k_{\text{B}}T > f\Delta(1/f_\beta)$ holds. However, if the bond with a higher barrier has a greater amplification of rate under force or smaller force scale (*i.e.*, $\Delta(1/f_\beta) > 0$), a crossover, $\Delta E_{\text{b}}/k_{\text{B}}T < f\Delta(1/f_\beta)$, will occur as the force is increased. Beyond the crossover, the bond with the higher barrier will become the most likely site of failure (*i.e.*, the *weak* bond). Thus, in a DFS spectrum, an abrupt reduction of slope from a linear regime at low loading rates to the next regime at higher rates signals a switch in site of failure amongst bonds in a series linkage.

In zipper-like failure (Fig. 2.1d), bonds break in sequence but at random times from first to last. Hence, far from equilibrium, the failure of the nth bonded state to $n-1$ attachments proceeds at a common rate, $\nu_{n\rightarrow n-1}(f) = (1/t_{\text{off}})\exp(f/f_\beta)$, so given N identical bonds, the approximate rate for breaking all bonds is essentially the reciprocal of the sum of times to break each bond,

$$1/\nu_{N\rightarrow}(f) = t_{\text{off}}\Sigma_{1\rightarrow N}\exp(-f/f_\beta) = Nt_{\text{off}}\exp(-f/f_\beta). \qquad (2.9)$$

Thus, the effective loading rate for the equivalent single-bond attachment is diminished by N fold and equation (2.8) shows that final separation of a *zipper* of identical bonds requires a slightly larger force than one bond (see Fig. 2.5),

$$f^* \approx f_\beta[\log(r_f t_{\text{off}}/f_\beta) + \log(N)]. \qquad (2.10)$$

This approximation is easily verified by calculation of the peak (at $\partial p_N/\partial f = 0$) of the *zipper* probability density described in the previous section. For comparison, imagine that each unbinding event was punctuated by force drops to zero and that the loading rate remained constant. Then, the time needed to *unzip* N bonds would be nearly N-fold longer, *i.e.* $\Sigma_N f^*(1)/r_f = N(f_\beta/r_f)\log(r_f t_{\text{off}}/f_\beta)$. But when force propagates instantly from one bond to the next, the time for complete detachment is only slightly longer than needed to break a single bond, $f^*(1)/r_f + (f_\beta/r_f)\log(N)$.

2.4 *Uncorrelated failure of bonds loaded in parallel*

2.4.1 Markov sequence of random failures

The same series of master equations (2.7) describe the evolution of the n-bonded states for systems loaded in parallel (Fig. 2.1c). However, here force is distributed across the bonds that are intact. As for series loading, the probability of a bond failure increases with the number of intact bonds; but in parallel loading, amplification of unbonding kinetics is reduced due to sharing of the force. Partitioning of force between N bonds is equivalent to an N-fold increase in the thermal force. For bonds in parallel, therefore, the failure rate and effective thermal force scale for each level are $n/t_{\rm off}$ and nf_β respectively, *i.e.* $\nu_{n\to n-1}(f) = (n/t_{\rm off})\exp[f/(nf_\beta)]$. Even with no rebinding at any step, analytical solution of the master equations is not possible in general for N bonds loaded in parallel so we will employ the equivalent single-bond attachment model to derive an analytical approximation for rupture strength and then compare this prediction to results obtained from numerical computations.

2.4.2 Equivalent single-bond approximation

For uncorrelated failure of bonds in parallel, we assume that force is shared equally amongst *existing* bonds in the attachment. Thus, the force experienced by bonds increases with each failure event during detachment, *i.e.* $f/\text{bond} = f/(N-j)$ from $j = 0 \to N-1$. Also, assuming identical bonds, there are $N-j$ possibilities for unbinding at each step, which increases the frequency for bond failure by the factor $(N-j)$. Thus, the effective rate to detach multiple bonds is estimated by the reciprocal of the times to pass from one state of bonding to the next,

$$1/\nu_{N\to}(f) = t_{\rm off}\Sigma_{1\to N}(1/n)\exp[-f/(nf_\beta)]. \qquad (2.11)$$

Using this single-bond approximation, equation (2.8) predicts that the force needed to rupture multiple-bonds loaded in parallel is essentially (but not quite) N-fold larger than for a single bond at the same rate of loading (see Fig. 2.6). Beyond the low force range, the strength of a multiple-bond attachment is predicted to follow a transcendental relationship given by,

$$f^* \approx Nf_\beta[\log(r_f) - \log(f/f_\beta)] \qquad (2.12)$$

which shows that the most probable rupture force is always somewhat less than N-times the single bond value. It is often assumed that multiple attachments will produce exactly N-fold increase in strength or rupture force, which is only approached in the large N limit. This contradicts the naive interpretation invoked in many early AFM studies to rationalize the

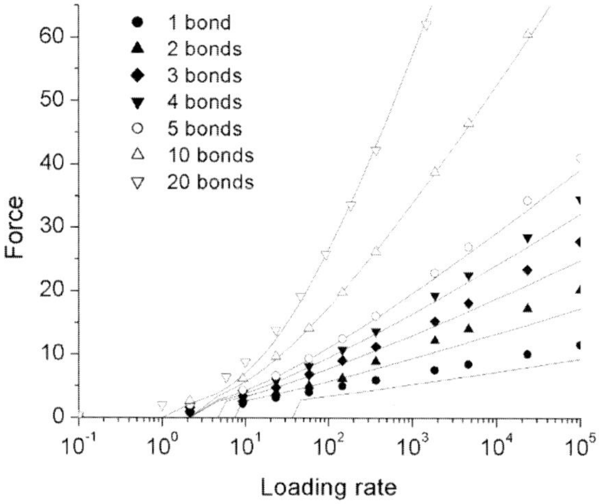

Fig. 2.6. Rupture forces (scaled by f_β) found by solution to the master equations (symbols) for detachment of between 2 and 20 bonds loaded in parallel. (Loading rate is scaled by f_β/t_{off}). The strength of the attachment increases in reasonable proportion to the number of bonds in the attachments. For comparison, the rupture force predicted by the equivalent single-bond attachment approximation (Eq. (2.12)) is plotted as the solid curves, which closely approaches the computational results especially for large numbers of bonds.

apparent presence of a few bonds. Moreover, we will show next that a unitary rupture force per bond cannot be derived from simple analysis of force histograms. The reason is that bond strength is a dynamic property and dependent on the rate of loading. Thus, sharing the load between multiple attachments reduces the single-bond loading rate and hence shifts the force at which bonds break at each step in detachment.

2.5 Poisson statistics and bond formation

The most obvious feature anticipated in tests of multiple bond attachments is that distributions of rupture forces will be very broad. The immediate question that comes to mind is whether or not the spread σ_N in the force distribution and the mean force $\langle f_N \rangle$ are related in some well-defined way to the number of bonds N in the attachments. From our equivalent single-bond attachment model, we would expect (Eq. (2.12)) that the mean rupture force will scale with N at very fast loading rates. Moreover, as noted in Part I, the spread in a force distribution for single bond rupture is set by the slope of force *versus* log(loading rate), *i.e.* $\sigma_f \sim f_\beta$. Hence, we might naively expect the spread in the force distribution for rupture of multiple bonds loaded in parallel to scale as $\sigma_N \sim N f_\beta$ when N is large. This behaviour implied at large N seems consistent with the perspective of "force quanta". As such, we might expect to use Poisson statistics to derive a "unitary force" from moments of force distributions for multiple bond detachment. Indeed, such approaches have been proposed and employed in several studies [15–18]. The rationale begins with the view that if the probability of bond formation on contact is low, the number of bonds present in attachments will be sampled from a Poisson distribution. Thus, as predicted by this distribution, the average number of attachments formed in an infinite number of repeated contacts, $\overline{N}$, equals the variance in number formed, σ_N^2. The next – and seriously flawed – assumption is that all forces scale with the number of bonds in an attachment multiplied by the unitary force f_1 so that both the average rupture force and variance in force would be, $\langle f_N \rangle = \overline{N} f_1$ and $\sigma_N^2 = \overline{N} f_1^2$ respectively. Hence, the unitary rupture force f_1 could be found by dividing the variance of a force distribution by the mean, or more robustly, determining the slope of variance *versus* mean force taken from experiments with a random number of bonds/attachment. However, the assumption that all n-bonded attachments are characterized by a common-unitary force can't be true given the kinetic dependence of rupture force on loading rate. Even though the mean force for rupture of multiple bonds at fixed loading rate is closely proportional to N at large N, the variance in distributions changes nonlinearly at large N as shown below (and demonstrated qualitatively by the analytical expression for an equivalent single bond, $1/\sigma_N^2 = \nu_{N\rightarrow}[\partial^2(1/\nu_{N\rightarrow})/\partial f^2]_{f=f^*}$).

To critically test the Poisson hypothesis, we used the master equations to simulate force distributions $[p_N(f) = \mathrm{d}S_u/\mathrm{d}f]$ for rupture of multiple bond attachments and then determined both the mean and variance. In particular, we focussed on a dimensionless rate, $r_f = r_f t_{\text{off}}/f_\beta = 1000$ because it is comparable to the effective rates employed in recent AFM studies of mulitiple biotin/streptavidin bonds. This dimensionless loading rate is based on actual loading rates r_f of 10^5 to 10^6 pN/s and transition state properties of $f_\beta \sim 40$ pN and $t_{\text{off}} \sim 0.1$ s under fast loading. In performing these computations, the initial populations of bonding states were defined by the Poisson distribution

$$S_n = \exp\left(-\overline{N}\right) \frac{\overline{N}^n}{n!} \tag{2.13}$$

where we varied $\overline{N}$ from 0.1 to 20. As evident in Figure 2.7, variance in force diminishes with increase in the mean force. The derivative of variance with respect to mean force is largest (~4.6) at low force and is always less than the mean force for single bond rupture (~6.9). Hence, the rupture

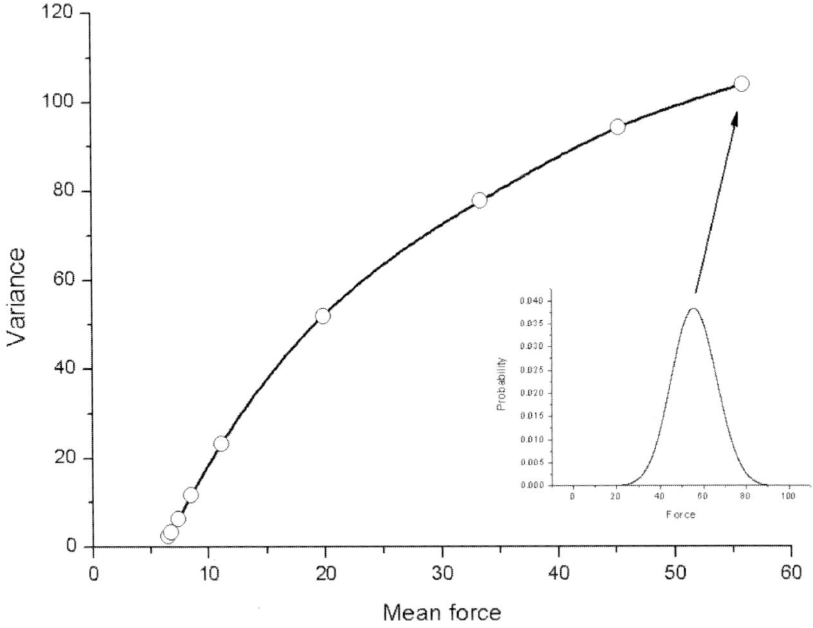

Fig. 2.7. Variance in rupture force distributions plotted against the mean as the average number of bonds formed on contact $\overline{N}$ varies between 0.1 and 20 under the assumption of precise parallel loading. The distributions were obtained by solution to the master equations computed at a dimensionless loading rate of 1000. Clearly, the variance in force does not increase linearly with the mean and demonstrates that a "Poisson-type" analysis cannot be applied to tests of multiple bond attachments. Inset is one of the distributions of rupture forces computed for multiple bonds distributed randomly about an average 20 bonds/attachment where at a rate of 1000, the most probable force for rupture of a single attachment is 6.9.

force for a single bond can only be estimated from extrapolation of the slope of variance *versus* mean force when the probability of attachment is well below 1. Inset in Figure 2.7 is one of the distributions of rupture forces computed for multiple bonds distributed randomly about an average 20 bonds/attachment. Clearly, the rupture forces in this distribution fall off monotonically away from a single peak with no evidence of quantization or hint of multiple peaks. So we see that the ratio of variance – to – mean of force distributions for large numbers of bonds/attachment cannot provide the force for rupture of a single bond.

Single molecule experiments are not easy to achieve in practise. To measure the force between two molecules, the likelihood of bond formation on contact has to be reduced to a "limit of dilution" through control of surface chemistry, molecular density, contact force and time. Even when the contact area and surface chemistry are controlled, multiple interactions can still become significant. Unfortunately, as just discussed, there appears to be no simple way to extract precise-quantitative information about a single bond from measurements of multiple bond detachments. However, in this regard, Poisson statistics can provide an important estimate of the likelihood of multiple interactions amongst a series of tests of bond strength. If a low probability of bond formation exists upon contact and we control the mechanics of contact to be equivalent each time, then the fraction of surface contacts that adhere in repeated trials $N_{\mathrm{f}}/N_{\mathrm{t}}$ (number forces against total number of tests) can be used to estimate the mean number of bonds/attachment,

$$\sum_{n=1}^{\infty} \exp\left(-\overline{N}\right) \frac{\overline{N}^{n}}{n!} = \frac{N_{\mathrm{f}}}{N_{\mathrm{t}}} \tag{2.14}$$

and therefore the mean number per contact is predicted to be,

$$\overline{N} = \ln\left(\frac{N_{\mathrm{t}}}{N_{\mathrm{t}} - N_{\mathrm{f}}}\right) \tag{2.15}$$

which we can relate to the fraction of single-bond events $N_{\mathrm{s}}/N_{\mathrm{t}}$ in the data,

$$\frac{N_{\mathrm{s}}}{N_{\mathrm{t}}} = \frac{N_{\mathrm{t}}}{N_{\mathrm{f}}} \overline{N} \exp\left(\overline{N}\right) = \frac{N_{\mathrm{t}} - N_{\mathrm{f}}}{N_{\mathrm{f}}} \ln\left(\frac{N_{\mathrm{t}}}{N_{\mathrm{t}} - N_{\mathrm{f}}}\right) \cdot \tag{2.16}$$

For example, a 70% frequency of adhesive-attachment implies that the average number of bonds formed on each contact is ~1.2 and, most important, that nearly half of the forces measured will be from multiple interactions. On the other hand, a 20% frequency of attachment implies that nearly 90% of the forces recorded are due to single rupture events. Hence, our aim in experiments should be to achieve a reduced level adhesion events (around 20%) which always involves the frustrating trade off between likelihood of single events vis a vis the duration of data acquisition. To emphasize again our comments from the previous chapter, each rupture event is a particular sample from a distribution of forces and therefore many hundred forces are needed to describe the process at each rate. With adhesion events only occurring one in every ten or so trials, and with the need to test many orders of magnitude in loading rate, many thousands of tests are required to explore the full kinetic process. This introduces significant challenges to instrument design, requires durable sample preparation, and tests the patience of the investigator!

2.6 Summary

We have enlarged the scope for dynamic strength of molecular interactions to include detachment of multiple bonds. We've emphasized that whilst experiments may be designed with the utmost care, contacts in experiments can often lead to multiple bond attachments because of the vagaries of molecular scale association. In this context, we've identified three prototypical mechanisms for detachment of bonds in attachments: *series*, *parallel*, and *zipper* loading. In nearly all situations of adhesive rupture, multiple bond interactions involve uncorrelated nanoscale excitations and thus the kinetics of failure can be treated as independent random processes with only force related to time and the number of surviving bonds in an attachment. However in the exceptional case of mechanically-constrained multiple bonds, cooperative unbinding can arise and thereby the system behaves as a single "compound" bond. As a rare but significant example, experiments on oligonucleotides have revealed the profound impact of cooperativity, which is to extend the natural lifetime of a complex enormously. Focussing principally on uncorrelated failure of bonds, we first treated a series of bonds that break in random sequence, then a zipper of bonds that break in deterministic sequence but at random times, and finally multiple bonds loaded in parallel that again fail in random sequence but which share an equal force that increments with each unbinding event. To follow the detailed evolution of a collection of bonds, we described how a series of master equations is employed to account for both deterministic and random changes force. But even with no rebinding between bond-breakage steps, analytical solution of the master equations is only possible in special cases. So by comparison to numerical computations of the master equations, we demonstrated that an equivalent single-bond model provides a good approximation for strength of multiple bond attachments. To first order, multiple bond attachments loaded in series or peeled apart like a zipper have strengths comparable to that of a single-constituent bond. For a series of chemically distinct bonds, the most likely site of failure is the "weakest" bond but we showed that "weak" *versus* "strong" must be viewed as dynamical characteristics. By comparison to systems dominated by single bond strength, multiple bonds loaded in parallel withstand forces nearly proportional to the number but only in the limit of large numbers of bonds and very large forces. Because of the nontrivial relationship between attachment strength and the number of bonds, there is no reliable way to extract characteristics of single bond dynamics from statistics of multiple-bond force measurements when the numbers of bonds vary. Moreover, there's no reason to expect that detachment of an assembly of many bonds will lead to an identical mechanical loading of each bond. Indeed the opposite should be expected – *i.e.* a subset of bonds will take the load at each step with the remainder only weakly stressed if at all. As the bottom line: the most significant uncertainty in interpreting rupture force for attachments with multiple bonds is knowing which of these scenarios most closely represents the nanoscale mechanics of detachment.

The authors gratefully acknowledge support from US National Institutes of Health HL65333, HL31579, Medical Research Council of Canada grant MT477 to (EE) and the Engineering and Physical Sciences Research Council, the Biological and Biomedical Sciences Research Council, Universitas 21 and the University of Nottingham to (PMW).

References

[1] E. Evans and K. Ritchie, *Biophys. J.* **72** (1997) 1451.
[2] U. Dammer, *et al., Biophys. J.* **70** (1996) 2437.
[3] S. Allen, *et al., FEBS Lett.* **390** (1996) 161.
[4] J.M. Williams, T.J. Han and T.P. Beebe, *Langmuir* **12** (1996) 1291.
[5] E. Evans, *Faraday Discuss. Chem. Soc.* **111** (1998) 1.
[6] R. Merkel, *et al., Nature* **397** (1999) 50.
[7] E. Evans, *et al., Proc. Natl, Acad, Sci, USA* **98** (2001) 3784.
[8] E. Evans, *Annu. Rev. Biophys. Biomol. Struct.* **30** (2001) 105.
[9] T. Strunz, K. Oroszlan, R. Shafer and H.-J. Guntherodt, *Proc. Natl. Acad. Sci. USA* **96** (1999) 11277.
[10] L. H. Pope, *et al., Eur. Biophys. J. Biophys. Lett.* **30** (2001) 53.
[11] M. Rief, *et al., Science* **276** (1997) 1109.
[12] E. Evans and K. Ritchie, *Biophys. J.* **76** (1999) 2439.
[13] C. Gergely, *et al., Proc. Natl. Acad. Sci. USA* **97** (2000) 10802.
[14] H. Skulason and C.D. Frisbie, *J. Amer. Chem. Soc.* **112** (2000) 9750.
[15] F. Stevens, Y.-S. Lo, J.M. Harris and T.P. Beebe, *Lanmguir* **15** (1999) 207.
[16] Y.-S. Lo, *et al., Langmuir* **15** (1999) 1373.
[17] Z.Q. Wei, *et al., Surf. Sci.* **459** (2000) 401.
[18] Y.-S. Lo, Y.J. Zhu and T.P. Beebe, *Langmuir* **17** (2001) 3741.

SEMINAR 1

POLYMERIZATION FORCES

M. DOGTEROM

FOM Institute AMOLF,
Kruislaan 407, 1098 SJ Amsterdam,
The Netherlands

POLYMERIZATION FORCES

M. Dogterom

The mechanical framework (cytoskeleton) of higher order (eukaryotic) cells consists of three types of protein filaments: actin filaments, intermediate filaments, and microtubules [1]. The assembly or polymerization of both actin filaments and microtubules has been implicated in cellular force generation processes. Examples are: the pushing forward of membranes by polymerizing actin filaments in the leading edge of crawling cells [2]; the propulsion of *Listeria* bacteria through their host cell, again by polymerization of actin filaments [3,4]; and the motion of chromosomes by the assembly and disassembly of microtubules during the process of cell division [5]. In this seminar I will describe how force is expected to affect the assembly of cytoskeletal filaments [6–8] and describe experiments designed to measure the forces generated by single growing microtubules.

Let us first consider the assembly of a single linear filament [9]. Above the so-called critical subunit concentration, the (concentration-dependent) rate at which subunits attach (the on-rate) is higher then the (concentration-independent) rate at which they detach (the off-rate). The velocity of polymerization is positive and given by:

$$V = \delta(k_{\rm on} - k_{\rm off}) \tag{1}$$

where δ is the subunit size.

Figure 1A shows schematically the free energy difference ΔG between the on-and off-states of a subunit as well as the activation barrier for transitions between the two. The absolute values of the on- and off-rates depend on the activation energy, but the ratio $k_{\rm on}/k_{\rm off}$ does not. In the absence of load this ratio is simply given by $\exp(\Delta G/k_{\rm B}T)$, where $k_{\rm B}$ is Boltzmann's constant and T is the absolute temperature. When a load is applied, the

This work is part of the research program of the "Stichting voor Fundamenteel Onderzoek der Materie (FOM)", which is financially supported by the "Nederlandse organisatie voor Wetenschappelijk Onderzoek (NWO)".

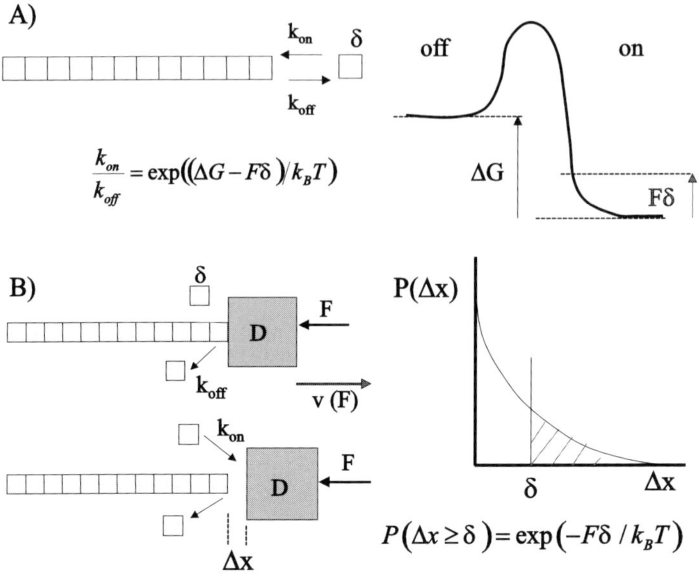

Fig. 1. Polymerization of a single filament in the presence of a load (see also text). **A)** Thermodynamic arguments predict that the ratio between the on- and off-rate decreases exponentially with an applied force. The gain in free energy, ΔG, associated with the assembly reaction is reduced by $F\delta$, the amount of work performed per assembling subunit. **B)** Alternatively, the assembly process can be viewed as a thermal ratchet, where the on-rate is reduced by the probability $P(\Delta x \geq \delta)$ that thermal fluctuations create a gap between the filament and a barrier large enough to insert a new subunit. D is the diffusion constant of the barrier.

gain in free energy due to assembly of a subunit is reduced by the amount of work performed to ΔG-$F\delta$, which gives:

$$\frac{k_{\rm on}(F)}{k_{\rm off}(F)} = {\rm e}^{(\Delta G - F\delta)/k_{\rm B}T} = \frac{k_{\rm on}(0)}{k_{\rm off}(0)}{\rm e}^{-F\delta/k_{\rm B}T}. \tag{2}$$

This formula gives the relative effect of force on the on- and off-rates, but not the absolute effect. To obtain the latter, the activation energy is needed. The growth velocity of the filament as a function of force (the force-velocity curve) is given by:

$$V(F) = \delta\left(k_{\rm on}{\rm e}^{-qF\delta/k_{\rm B}T} - k_{\rm off}{\rm e}^{(1-q)F\delta/k_{\rm B}T}\right) \tag{3}$$

where the value of q depends on how much the off-rate is affected by the force relative to the on-rate. The force needed to stall the assembly of the filament is independent of q and given by

$$F_{\text{stall}} = \frac{\Delta G}{\delta} = k_{\text{B}}T \ln \frac{k_{\text{on}}}{k_{\text{off}}} \,. \tag{4}$$

A more mechanistic view on the force generation process is provided by the thermal ratchet model originally introduced by Peskin *et al.* to describe force generation by actin filaments [6]. In Figure 1B one imagines that assembly takes place in contact with a "barrier" on which a load F is applied. Insertion of new subunits takes place at the contact point between the growing filament and the barrier. In the absence of thermal fluctuations there is no space between these two objects and proteins are physically prevented from attaching to the filament. Thermal fluctuations (*e.g.*, in the form of diffusion of the barrier) create temporary gaps that allow new subunits to add themselves to the end of the filament. After such an addition of another subunit, the barrier can no longer diffuse back to its original location and has thus been "pushed" forward by the assembling filament against the applied load. The off-rate is in this scenario not affected by the force. When diffusion of the barrier over distances of order δ is fast compared to the time between subsequent subunit additions, the rate of assembly is simply the on-rate in the absence of any force or barrier, multiplied by the probability of opening a gap large enough for a new subunit to be added. This probability depends on the energy associated with moving the load F over a distance δ and is given by $\exp(-\delta F/k_{\text{B}}T)$. In this case the force-velocity curve is given by:

$$V(F) = \delta(k_{\text{on}}\mathrm{e}^{-F\delta/k_{\text{B}}T} - k_{\text{off}}) \tag{5}$$

which is the same result as equation (3) for a force-independent off-rate ($q = 1$).

In our experiments we aim at measuring the force-velocity behavior for the assembly of cytoskeletal proteins. We hope to learn (by comparison with available models) how the assembly free energy of proteins is converted into mechanical work. But also to find out how large these forces are compared to the forces that are generated by molecular motors that use cytoskeletal filaments as rails and play an equally important role in cellular motility processes. Of course the assembly of cytoskeletal filaments is not as simple as the situation sketched in Figure 1, and care should be taken in comparing experimental results with the simple predictions above. In particular, microtubules consist of (typically) 13 protofilaments forming a hollow tube, assembled from tubulin protein dimers. For an individual microtubule, periods of assembly randomly alternate with periods of disassembly, a process that depends on the hydrolysis of associated GTP molecules [10].

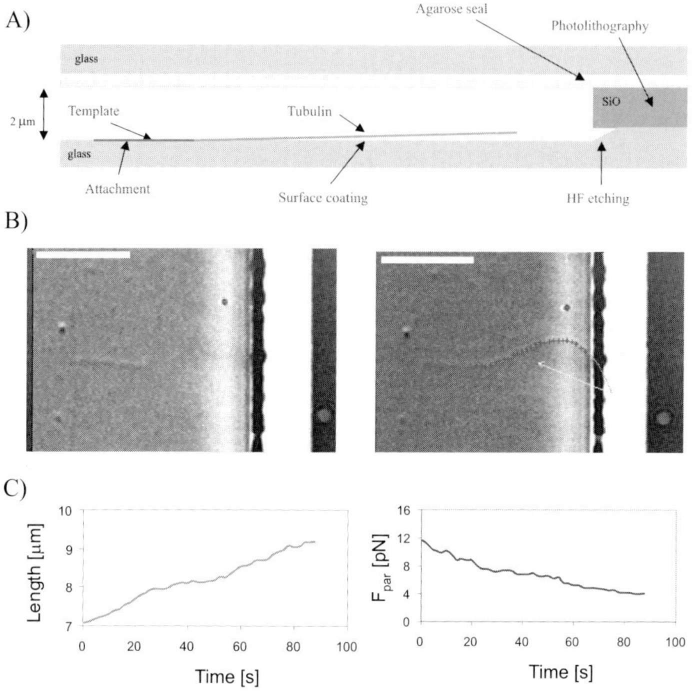

Fig. 2. Experiment to measure the force-velocity curve for a single growing microtubule. **A)** Schematic representation of the experimental set-up (see text). **B)** DIC microscopy images of a growing microtubule before (left) and after (right) encountering the barrier. The crosses indicate the shape of the microtubule as recognized by our image analysis software. The solid line is the result of fitting this shape to the shape of an elastic rod. The arrow gives the direction of the force as obtained from the fit. Scale bar is 5 μm. **C)** Contour length (left) and parallel component of the force (right) as a function of time as obtained from the fits. The initial (zero force) growth velocity of this microtubule was around 2.5 μm/min [12].

Figure 2A shows schematically the experimental set-up that we developed to study, quantitatively, the forces generated by single growing microtubules [11, 12]. Short pieces of stabilized microtubules (templates) are biochemically attached to a glass substrate that has been decorated with barriers for growing microtubules. The barriers are lines (15 μm wide, 2 μm high) of silicon-monoxide that were deposited with lithography techniques. When tubulin proteins are added, microtubules grow from the templates, some of which encounter the barriers. An important technical detail of these barriers is that they were created with a small "undercut" by briefly etching the substrate in hydrofluoric acid. These undercuts force the microtubule ends to stay in the focal plane of the microscope and prevent them from sliding upwards after reaching the barrier. When microtubules hit the barrier they generally continue to grow. To accommodate the increase in length, two things can happen: either the elongating microtubule end slides laterally along the barrier, giving rise to a modest deflection of the microtubule. Or the microtubule end is hindered in this lateral motion (by an encountered irregularity in the barrier profile) which results in a more dramatic buckling of the microtubule as its growing end pivots around a fixed contact point with the barrier; see Figure 2B. In this last case the elastic restoring force of the buckled microtubule puts a significant load on the microtubule, directly affecting the further growth of the filament. Using image analysis to obtain the shape of the growing microtubule as a function of time and fitting the obtained curves to the shape of an elastic rod, both the magnitude and the direction of the force acting on the microtubule end can be determined [13]. Also the increase in microtubule length and thus the growth velocity can be derived from these fits (Fig. 2C). Calibration is provided by an independent measurement of the flexural rigidity and the forces one finds are close to the critical buckling force of a homogeneous elastic rod, given the appropriate boundary conditions.

Note that compared to the situation in Figure 1, it is in this case not the barrier that is fluctuating but the position of the assembling filament end itself. In addition the load is not applied externally, but caused by the elastic deformation of the filament itself. This however does not change the ratchet behavior, as long as the fluctuations in the gap size are sufficiently fast and the applied load is not itself dependent on the gap size.

Figure 3 shows a force-velocity curve for microtubules obtained after averaging many data such as shown in Figure 2. As said before one should take care in comparing these experimental results with the simple Brownian ratchet model described above. A microtubule consists of 13 laterally connected protofilaments forming a hollow tube, instead of a single linear filament, and the geometrical details of the growth process are not well known. (To make matters worse, electron microscopy studies of microtubules have suggested that growing ends consist of sheet-like structures that close into hollow tubes during the assembly process [14].) In addition, there is the possibility that the hydrolysis of GTP, responsible for the occasional switching to a shrinking state, should be taken into account to understand the response of the growth process to force. In this case the stall force may no longer be simply connected to the free energy associated with tubulin assembly.

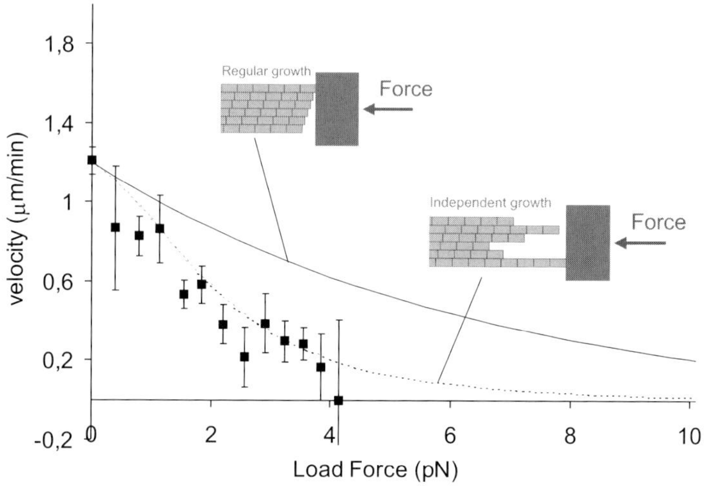

Fig. 3. Force-velocity curve for the growth of single microtubules [11]. Under these conditions the initial (zero force) growth velocity was around 1.2 μm/min. The solid line is predicted by a regular assembly process; the dotted line is predicted by a model that assumes 13 independently growing filaments. For simplicity only 6 protofilaments are shown in these drawings. In both cases the stall force is arbitrarily set to 18.5 piconewton.

Given what *is* known about the structure of microtubules, it is tempting to try to include the geometrical details of a growing microtubule into a thermal ratchet model. A simplistic generalization of the original ratchet model describing the growth process of a polymer consisting of two filaments (as is the case for actin) certainly does not fit the available data [11]. In this case it is assumed that the size of the gap needed to insert each new subunit is equal to the added microtubule length per dimer: δ/n, where n is the number of filaments in the polymer (inset Fig. 3; regular growth). Mogilner and Oster therefore generalized the ratchet model described above in a different way [15]. They assumed 13 laterally connected, independently growing ratchets, initially arranged as a staircase with subsequent shifts equal to one 13th of the subunit size, with the longest filament in contact with the barrier at any time. Growth of any particular filament requires a fluctuation of the barrier large enough for that filament to insert a new subunit, which thus becomes a function of the distance of the end of that filament to the barrier (inset Fig. 3; independent growth). Through numerical solutions as well as simulations of this model the steady state distribution of filament-barrier distances can be determined as a function of applied force and with this the average growth rate can be calculated. The outcome fits the available experimental data surprisingly well. The fit is however very insensitive to the only free parameter in this model: the stall force (*i.e.* the ratio between the bare on- and off-rates, given that their difference is fixed by a measurement of the growth velocity at zero force), and no conclusions about the stall force can be made [16]. Even though this model makes a lot of implicit assumptions and there is no reason to insist on its details, the outcome of the comparison with the experimental data may in fact suggest that the end of a growing microtubule (under these conditions) looks more like a irregular pointed structure than a blunt end. More data, taken at different initial growth velocities, should allow us in the future to test this as well as other hypotheses.

To be able to compare between different possible models (including the ones discussed here) we would also like to be able to measure the stall force directly. The buckling technique does however not allow for such a measurement. Due to the geometry of the experiment the force on the microtubule end never increases after the buckling of the microtubule has started, and in fact decreases during the course of the experiment (due to the strong dependence of the critical buckling force on the filament length: $F_{\rm crit} \sim L^{-2}$). Microtubules that are attached relatively close to the barrier stop growing as soon as they encounter the barrier, apparently because the force needed to overcome their critical buckling force is too large. In these cases we have no direct way of measuring the force applied (although estimates can be made). Currently we are therefore working on a second experimental set-up based on optical tweezers techniques [17]. In this set-up the microtubule template is attached to two micron-sized silica spheres. These beads are each held in an optical trap orienting the growth direction of the microtubule towards a barrier similar to the one used in the previous experiment. In this case the distance to the barrier should ideally be chosen such that the force needed to buckle the microtubule is larger than the expected stall force. In response to the growth of the microtubule the beads will move with respect to the centers of the optical traps, thereby linearly increasing the force on the growing microtubule end until growth stops. An independent calibration of the trap stiffness will then give a direct measure of the force applied [18].

From a biological point of view, the microscopic details of forces generated in contact with an artificial glass barrier may not be all that interesting. What will be important in the future is to repeat these types of experiments with barriers consisting of chromosomes, kinetochore complexes, or simple motor-coated surfaces. Comparison with our current experiments may reveal important hints as to how molecular growth details and the force generating process are affected or even regulated by interaction with these specific barriers.

I would like to thank my current and previous collaborators on this project: Bernie Yurke, Marcel Janson, Sander van Doorn, Catalin Tanase, Bela Mulder, Mahilde de Dood and Jacob Kerssemakers.

References

[1] B. Alberts, A. Johnson, J. Lewis, M. Raff, K. Roberts and P. Walter, *Molecular Biology of the Cell* (Garland Publishing, New York, London, 2002).

[2] D. Bray, *Cell Movements: from Molecules to Motility* (Garland, New York, 2000).

[3] L.A. Cameron, P.A. Giardini, F.S. Soo and J.A. Theriot, *Nature Rev. Mol. Cell Biol.* **1** (2000) 110-119.

[4] D. Pantaloni, C.L. Clainche and M.-F. Carlier, *Science* **292** (2001) 1502-1506.

[5] S. Inoué and E.D. Salmon, *Mol. Biol. Cell* **6** (1995) 1619-1640.

[6] C.S. Peskin, G.M. Odell and G.F. Oster, *Biophys. J.* **65** (1993) 316-324.

[7] J.A. Theriot, *Traffic* **1** (2000) 19-28.

[8] M. Dogterom, M.E. Janson, C. Faivre-Moskalenko, A. van der Horst, J.W.J. Kerssemakers, C. Tanase and B.M. Mulder, *Appl. Phys. A* **75** (2002) 331-336.

[9] T.L. Hill, *Linear Aggregation Theory in Cell Biology* (Springer-Verlag, New York, Berlin, Heidelberg, 1987).

[10] A. Desai and T.J. Mitchison, *Annu. Rev. Cell Dev. Biol.* **13** (1997) 83-117.

[11] M. Dogterom and B. Yurke, *Science* **278** (1997) 856-860.

[12] M.E. Janson and M. Dogterom, in preparation.

[13] F. Gittes, E. Meyhofer, S. Baek and J. Howard, *Biophys. J.* **70** (1996) 418-429.

[14] D. Chrétien, S.D. Fuller and E. Karsenti, *J. Cell Biol.* **129** (1995) 1311-1328.

[15] A. Mogilner and G. Oster, *Eur. Biophys. J.* **28** (1999) 235-242.

[16] G.S. van Doorn, C. Tanase, B.M. Mulder and M. Dogterom, *Eur. Biophys. J.* **29** (2000) 2-6.

[17] J.W.J. Kerssemakers, M.E. Janson, A. van der Horst and M. Dogterom, *Biophys. J.* **82** (2002) 415a.

[18] K. Visscher, S.P. Gross and S.M. Block, *IEEE J. Sel. Topics Quant. Elect.* **2** (1996) 1066-1076.

COURSE 5

THE PHYSICS OF *LISTERIA* PROPULSION

J. PROST

Institut Curie, Section de Recherche,
UMR 168 du CNRS/IC
Physico-Chimie Curie, 26 rue d'Ulm,
75248 Paris Cedex 05, France

Contents

THE PHYSICS OF *LISTERIA* PROPULSION

J. Prost

1 Introduction

Listeria is a pathogenic bacterium, which can be dangerous for immune deficient individuals. It can be found almost everywhere, in particular in food such as soft cheese and smoked salmon. After ingestion, it is able to penetrate in the cellular system where it moves from cell to cell and divides on average every twenty minutes. The reason why it can move from cell to cell is that it develops a comet-like tail (Fig. 1), which pushes the bacterium forward and deforms the plasma membrane until it invades the neighbouring cell. Since it is inside the cellular system it is hard to be detected by the immune system. In order to understand the *Listeria* propulsion mechanism, a particularly intense scientific activity has been developed over these last years [1,2]. Why should one be particularly interested in this problem? The reason is that *Listeria* motility is due to the polymerisation and cross-linking of an actin gel (*i.e.* the comet) just like eukaryotic cell motility is due to the polymerisation of actin in the cell lamellipodium. It is then believed that learning something on *Listeria* is useful for understanding eukaryotic cells as well. Of course studying *Listeria* does not avoid studying eukaryotic cells since there are many more aspects to eukaryotic motility than to *Listeria* motility [3] (like adhesion, molecular motors etc.). Yet this allows us to select one aspect, namely actin polymerisation and cross-linking, in geometrical conditions, which are much simpler than those of eukaryotic cells since the process is exterior to the bacterium. One can in particular use cell extracts to perform *in vitro* experiments. Yet *a priori* simpler than eukaryotic cell motility, *Listeria* motility has its mysteries: in particular a mutant has been observed to move by a succession of jumps followed by periods during which the bacteria is essentially immobile [4] (Fig. 2). During the waiting period the gel grows around the whole bacterium producing some kind of sheath. Eventually the bacterium gets expelled from the sheath, hence the jerky motion. In the following I give a few guidelines for thinking about the physics of the propulsion mechanism, and show that it is essentially a continuum mechanics problem with very unusual boundary conditions.

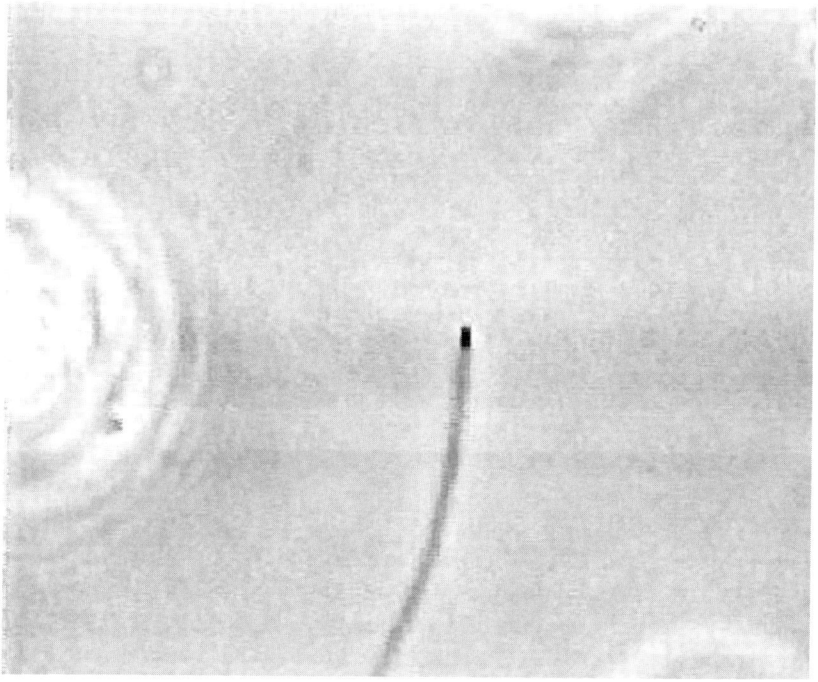

Fig. 1. Example of a wild type *Listeria* and its homogeneous comet.

2 A genuine gel

2.1 A little chemistry

Before getting to the characterisation of the gel, it is necessary to give a few tips concerning the biochemistry of the polymerisation process. First, it is now well established that in order to observe the formation of a comet, a particular enzyme called ActA must be present on the surface of the bacterium. *In vitro* experiments can be made by placing bacteria in cell extracts, or in reconstituted extracts. The presence of ActA is necessary but not sufficient for getting a polymerisation process comparable to the one observed *in vivo*. Actin filaments are polar: they have two fundamentally different extremities. One, called barbed (or plus) end, can polymerise while the other, called pointed (or minus), can simultaneously de-polymerise. Of course this can only happen if energy is constantly fed into the system and this is done by hydrolysing ATP in ADP. Phosphorylated monomers polymerise at the barbed end, while dephosphorylated monomers de-polymerise at the pointed end. In the absence of any other protein the polymerisation

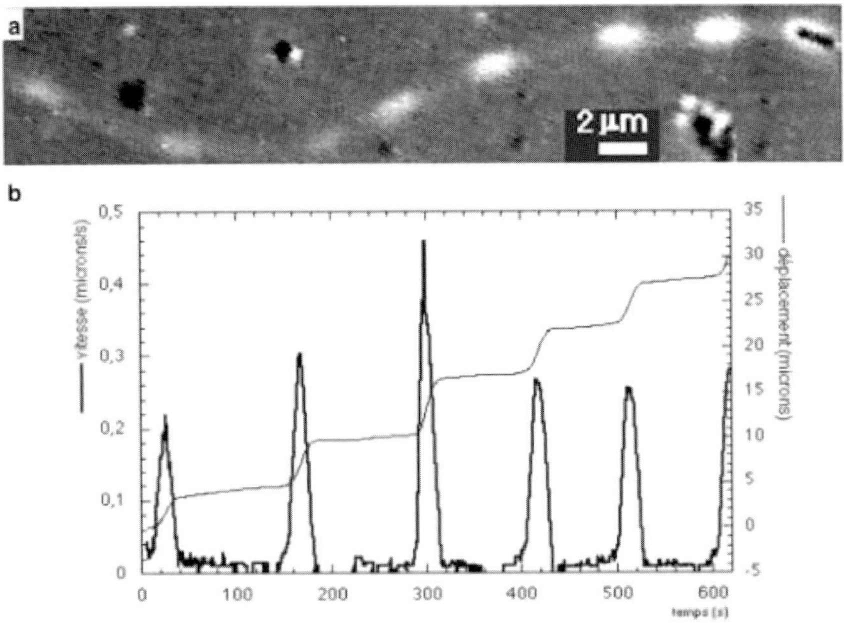

Fig. 2. ActA$_{\Delta 21-97}$ *Listeria* mutant (Courtesy Lasa *et al.* [4]): **a)** DIC view of the modulated comet, **b)** displacement and velocity as a function of time.

rate of actin is about two orders of magnitude slower than *in vivo*. In order to obtain the right values three other types of proteins must be added:

a) cofactor, which speeds up de-polymerisation at the pointed end and thus speeds up turn over and polymerisation at the barbed end;

b) capping protein, which caps free barbed ends and localises polymerisation at the bacterium surface or its immediate neighbourhood;

c) protein complex called Arp 2/3 which provides branching to the network [5,6], and also speeds up the gel formation.

The identification of the minimum number of constituents necessary to reproduce *in vitro* the bacterium motion was an important step towards a quantitative understanding of the process [7]. Another important step is the demonstration that it is possible to replace the bacterium itself by inert beads such as polystyrene beads, on which the enzyme ActA is grafted or adsorbed [8,9]: I will give more details on this aspect in the last two sections of this article.

2.2 Elastic behaviour

In the introduction of this article I have written without further justification that the comet like structure of Figure 1 was a gel, and that polymerisation was taking place at the bacterium surface. Proving that polymerisation is taking place at the bacterium surface was shown by using fluorescently labelled actin [10, 11]. This still does not tell us that the actin gel is a real gel in the sense that it has the mechanical properties of an elastic body. This can be done by cutting pieces of the comet using laser surgery techniques, and measuring the bending modulus of the comet [12, 13]. These experiments show that the comet does have elastic behaviour over time scales of minutes. Elastic moduli are found in the kilo-Pascal range with a large total spread of two orders of magnitude. This spread is not related to a slow de-polymerisation process known to exist in the comet and which can be studied independently. These experiments show that the gel elastic properties must be explicitly taken into account in the physics of the motion.

Next one wants to know about the connection between the bacterium surface and the gel. By using laser tweezers or better electric fields one can exert piconewton forces between bacterium and comet during typically a minute: no relative motion at a micron resolution can be detected [13]. This shows that the bacterium is firmly connected to the gel. More quantitatively, if one describes the bacterium-gel lateral interaction by a friction coefficient, such experiments put a lower limit to the friction coefficient, four orders of magnitudes larger than the hydrodynamic friction coefficient of the bacterium on the surrounding fluid!

3 Hydrodynamics and mechanics

3.1 Motion in the laboratory frame

One can split the problem of *Listeria* motion into two parts. First an external and simple part deciding which of the bacterium or the comet moves with respect to the surrounding fluid, second an internal part describing the motion of the bacterium relative to the comet. The Reynolds numbers in this problem are extremely low (*i.e.* of the order of 10^{-7}) this means that only friction forces should be retained; under such conditions the external dynamics reads:

$$f^{\text{ext}} = \zeta_{\text{b}}\, \nu_{\text{b}} + \zeta_{\text{c}}\, \nu_{\text{c}}.$$

Where f^{ext} is an external force acting on the bacterium/comet system, $\zeta_{\text{b}}, \zeta_{\text{c}}, \nu_{\text{b}}, \nu_{\text{c}}$ are the friction coefficients on the surrounding fluid and velocities of the bacterium and the comet respectively. The polymerisation process itself is responsible for the existence of a relative velocity between bacterium and comet: $\nu_\mathrm{b} - \nu_\mathrm{c} = \nu$. Note that ν is not related simply to the polymerisation rate ν_p, as illustrated in the following. Combination of the two equations allows us to extract the velocity of the bacterium with respect to the surrounding fluid:

$$\nu_\mathrm{b} = \frac{\zeta_\mathrm{c}\,\nu + f^\mathrm{ext}}{\zeta_\mathrm{b} + \zeta_\mathrm{c}}\,.$$

Only in the absence of external force and when $\zeta_\mathrm{c} \gg \zeta_\mathrm{b}$ is $\nu_\mathrm{b} \approx \nu$.

This limit is obtained as soon as the comet length is larger than the bacterium size, and does not depend on the surrounding fluid viscosity since both friction coefficients are proportional to it. Note also that since the friction of the gel on the bacterium surface is at least four orders of magnitude larger than ζ_b, an increase by at least four orders of magnitude of the viscosity is required to influence ν in a significant way, *via* the developed stressed as shown in the following. Such an experiment has been done recently [14]. Yet, for most practical purposes one can ignore the world external to *Listeria* for discussing its motion. Note eventually that external forces cannot be mixed with internal ones in the force conservation equation.

3.2 Propulsion and steady velocity regimes

Since experiments show that the comet is indeed a gel in the continuum mechanics sense, this implies that one has to understand what kind of stresses are generated by the polymerisation-gelation process. If polymerisation was taking place only at the rear part of *Listeria* then life would be simple. ν would be simply the polymerisation velocity ν_p. One could write for instance: $\nu \cong \nu_\mathrm{p} = a(k_+^b c^\mathrm{i} - k_-^b)$, in which $k_+^b, k_-^b, c^\mathrm{i}, a$ are respectively the polymerisation and de-polymerisation rates, an actin monomer concentration and size at the bacterium surface. Most microscopic theories are concerned with the calculation of ν_p [15–18].

However the gel grows not only at the rear of the bacterium. In other bacteria like *Shigella* [19], or in vesicles developing comets [20,21], the comet is hollow which means that there is no rear gel. To understand the propulsion mechanism one has to understand that once a first layer has been polymerised and cross-linked, a subsequent polymerisation can only occur if the first layer is stretched to a new position leaving space for it (Fig. 3). The stretching costs elastic energy and the release of this energy is the driving force for the motion. More precisely, if B is the shear modulus of the gel the stored elastic energy per unit length reads:

$$W_\mathrm{e} \cong B(e/r)^2 2\pi r e L.$$

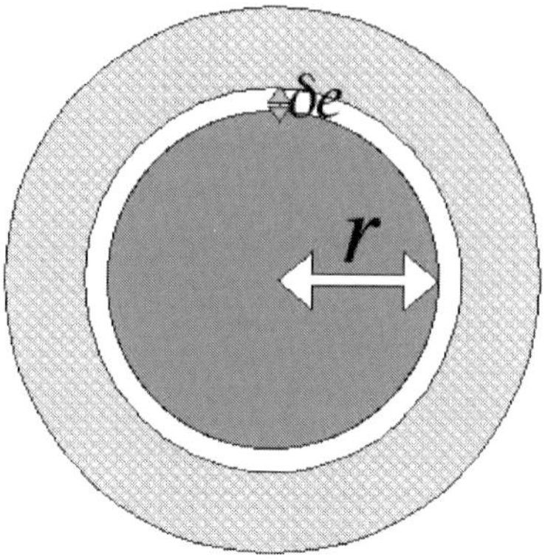

Fig. 3. The grey gel layer, initially with inner radius r must stretch to a new radius $r + \delta e$ to allow for a new gel layer to form.

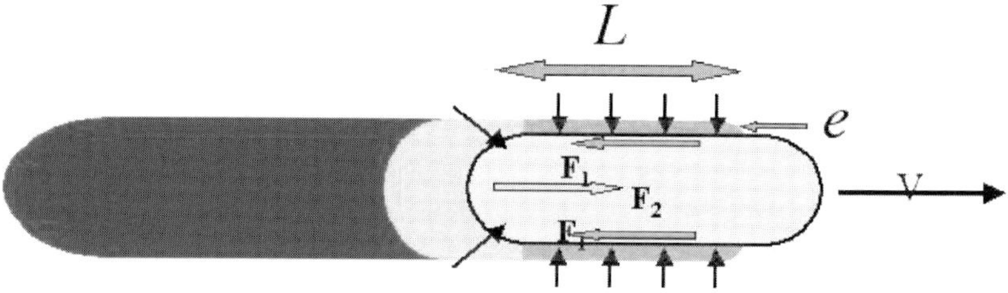

Fig. 4. The elastic force, gives rise to a propulsion force F_1 along the bacterium axis; surface friction equilibrates F_1 with F_2.

Where $e, r, L, (e/r)$ are the gel thickness, the bacterium radius, length and tensile strain, respectively (Fig. 4). Thus the propelling force is:

$$f_{\mathrm{p}} \cong \frac{1}{2\pi} Be^3/r.$$

The gel-bacterium friction balances this force (remember we have shown that external hydrodynamic friction is in most practical situations entirely negligible). We will discuss in more details the notion of gel/bacterium friction, but for the sake of argument let us first describe the friction force by a friction coefficient ξ: $f_{\mathrm{f}} = \xi\nu 2\pi rL$. One obtains immediately an expression for the velocity: $\nu \cong \nu_{\mathrm{i}} e^3/r^2 L$. In which $\nu_{\mathrm{i}} = (\frac{B}{\xi})$ is an intrinsic velocity scale, related only indirectly to the polymerisation process. Two limits merit discussion.

In the first neither the developed stress nor the actin monomer depletion, are large enough to influence the polymerisation process. Then in steady state:

$$e = \left(\frac{\nu_\mathrm{p}}{\nu}\right) L$$

since the time $\frac{e}{\nu_\mathrm{p}}$ needed to generate a thickness e, must equal the time $\frac{L}{\nu}$ to advance one length L, then

$$\nu \cong (\nu_\mathrm{i})^{1/4}\, \nu_\mathrm{p}^{3/4} \left(\frac{L}{r}\right)^{1/2} .$$

Note that in this regime the bacterium velocity relative to the gel ν can be larger than the polymerisation rate ν_p, and that it is not proportional to it. Note also that it depends only weakly on the comet gel properties ν_i: a two orders of magnitude change of the gel elastic modulus (everything else being kept constant), results in a factor of three change of the velocity ν only. This explains why the experimentally observed velocity spread is by no means comparable to the one found for the gel modulus.

In the second regime the thickness saturates to a value e^* controlled either by the developed stresses or by the actin monomer diffusion process (figure). Then:

$$\nu = \nu_\mathrm{i} \frac{e^{*3}}{r^2 L} .$$

We will show that under appropriate circumstances $e^* = e_2 \ln(\nu_\mathrm{p}^0/\nu_\mathrm{dp}^0)$, in which e_2 is a length proportional to the bacterium radius r, and $\nu_\mathrm{p}^0, \nu_\mathrm{dp}^0$ are the polymerisation and de-polymerisation rates in the absence of stress respectively. In this case the dependence of the bacterium velocity on the polymerisation rate is extremely weak!

3.3 Gel/bacterium friction and saltatory behaviour

In the above discussion, we have used the notion of surface friction without further justification. The physical nature of this friction may be understood the following way: during the polymerisation process the actin filaments spend some time τ_c connected to the bacterium surface, and some other time τ_d disconnected to it. When the gel moves with respect to the bacterium the connected filaments gradually distort until they detach. A force results from the distortion, as first understood by Tawada and Sekimoto [22]. The average force per unit area reads: $F_\mathrm{f} = n_\mathrm{c}\Phi$ where n_c is the average number of connected filaments and Φ a typical force per filament. In steady state: $n_\mathrm{c} = n\frac{\tau_\mathrm{c}}{\tau_\mathrm{c}+\tau_\mathrm{d}}$ where n is the number of enzymes per unit area on the bacterium surface. The force Φ is simply given in terms of the product of a filament elastic modulus K multiplied by a typical displacement $\nu\tau_\mathrm{c}$.

A first regime of small velocities is easy to discuss. Both τ_c and τ_d have their intrinsic thermodynamic value τ_c^0 and τ_d^0, and the notion of a friction coefficient with a velocity independent value emerges as anticipated: $\xi = \frac{\tau_\mathrm{c}^{02}}{\tau_\mathrm{d}^0+\tau_\mathrm{d}^0} nK$. If one estimates the gel elastic modulus on dimensional grounds: $B \cong kT\lambda_\mathrm{p}/\lambda^4$, and the filament surface modulus by $K \cong kT\lambda_\mathrm{p}/\lambda^3$, in which λ_p is the actin filament persistence length and λ the average distance between cross-links, then the intrinsic velocity takes the very simple form: $\nu_\mathrm{i} \cong \frac{\tau_\mathrm{c}^0+\tau_\mathrm{d}^0}{\lambda n \tau_\mathrm{c}^{02}}$, which further simplifies to $\nu_\mathrm{i} \cong \frac{\lambda(\tau_\mathrm{c}^0+\tau_\mathrm{d}^0)}{\tau_\mathrm{c}^{02}}$. Although it is possible to have reasonable values of λ, nothing is known on the connected and disconnected times.

A second regime is that of high velocities. The connections are broken in times much shorter than the thermodynamic connection time τ_c^0, such that the work done on the connection with the enzyme is of the order of the potential barrier w_b hindering the escape of the filament from its bound state. This condition requires: $K\nu\tau_\mathrm{c}a_\mathrm{b} \cong w_\mathrm{b}$, in which a_b is a length of order a. The essential result is that now the connected time is inversely proportional to the velocity and the friction force becomes also inversely proportional to it: $F_\mathrm{f} \cong \frac{nw_\mathrm{b}^2}{\tau_\mathrm{c}K\nu}$. The friction force due to this phenomenon tends to zero, simply because all bounds break. The total friction does not vanish though, since there is always a conventional hydrodynamic friction. The total $F_\mathrm{f}(\nu)$ curve plotted in Figure 5 exhibits the typical shape of a solid on solid friction with stick/slip behaviour. Under such circumstances the saltatory mutant is easy to understand. A conventional steady state smooth motion is obtained when the curve characterising the elastic force intersects the friction curve once. The saltatory behaviour is obtained when the friction oscillates between the high and the low friction regimes. Typically the bacterium starts to accumulate a thick gel layer until the elastic force reaches a value such that the unstable regime is reached, that is until surface bonds break. In this phase the bacterium velocity is very small. Then the gel layer is quickly expelled which gives a velocity burst, and leaves the bacterium surface in its initial state; the cycle can start again. A more elaborate description can be found in references [12, 23]. It is interesting to remark that all known phenotypes can be assembled in a single dynamical state diagram, provided one treats at the same level the side gel and the rear gel [23, 24]. This work shows in particular that the rear part of the gel in general does not participate positively in the propulsive force. Only close to stall force does the rear part contribute positively. As a result, the force-velocity relation was predicted to exhibit two regimes [23], as observed experimentally recently [14]. This analysis further shows that,

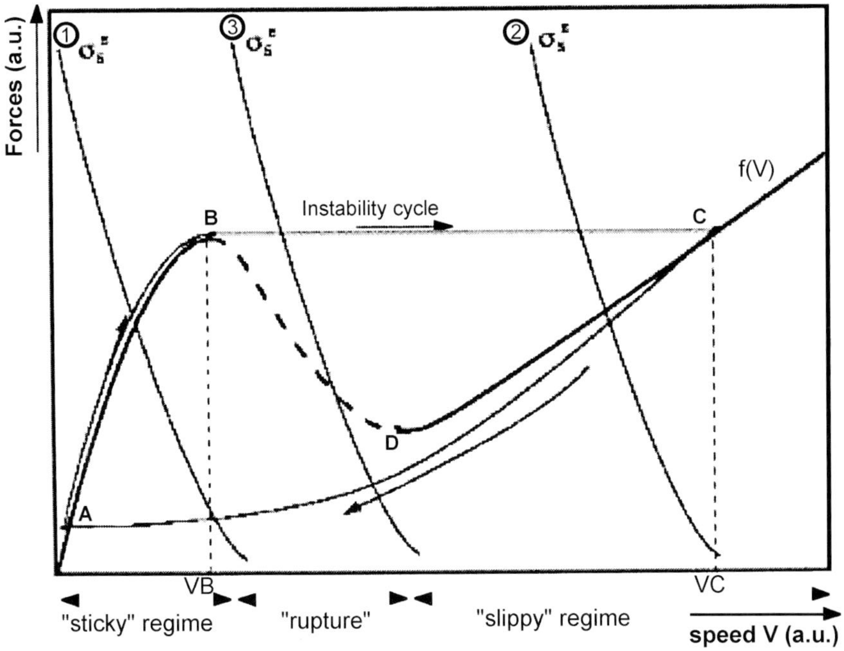

Fig. 5. Typical curve giving the friction as a function of the bacterium/gel differential velocity.

what is called a saltatory mutant, is in fact nothing but the crossing of a boundary line in the state diagram due to the mutation. The crossing, however, might result from many other causes. We illustrate this remark in the following section. Note eventually that although microscopic models are interesting in their own right, a comprehensive analysis cannot ignore the elastic level of interpretation, which naturally provides a correct distinction between internal an external forces.

4 Biomimetic approach

4.1 A spherical *Listeria*

If it is true that *Listeria* needs only to display the enzyme ActA at its surface and for the rest of it steals all the needed compounds to the surrounding cell, it should be possible to replace the bacterium by an inert bead on which this enzyme is grafted. Then, placing the bead in a cell extract or in a reconstituted extract containing all relevant proteins and energy sources,

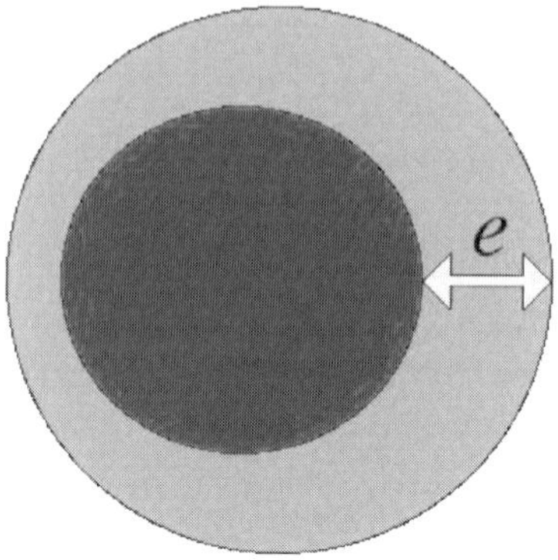

$$e = e_s + \delta e(t) + \varepsilon(t)\cos(\theta)$$

Fig. 6. Example of a symmetry breaking perturbation.

one should be able to observe actin polymerisation and hopefully a comet formation. This has been done in several laboratories [8,9] proving that indeed only ActA was needed at the surface; it was further shown, that human actin polymerisation enzymes could give rise to similar observations [25,26]. Can one learn more with these *in vitro* bio-mimetic assays?

4.2 Spherical symmetry

In many cases, the beads which are used, are spherical and unless a symmetry breaking process develops, the produced actin gel respects the beads symmetry. One observes that the gel growth stops after a given thickness is reached. It is possible to prove that it corresponds to a steady state [27]: polymerisation is still going on at the bead surface, while de-polymerisation takes place at the outer one. The observed thickness is always a fraction of the bead radius, orders of magnitude smaller than the comet length. Why is it so?

The point is that both the polymerisation rate at the inner surface, and the de-polymerisation rate at the outer surface depend on the stresses that develop as we have already explained in the preceding paragraphs. The polymerisation rate at the inner surface decreases under the action of the compressive normal stress while the de-polymerisation rate at the outer surface increases under the action of the tensile stress. When both take on the same value a steady state is reached.

It is possible to be more quantitative by writing standard chemical rate equations:

$$\frac{d\nu_{\rm i}}{dt} = k_+^b c_{\rm i} - k_-^b, \qquad \frac{d\nu_{\rm e}}{dt} = -k_-^p, \qquad \frac{de}{dt} = a\left(\frac{d\nu_{\rm i}}{dt} + \frac{d\nu_{\rm e}}{dt}\right)$$

$\frac{d\nu_{\rm i}}{dt}$, $\frac{d\nu_{\rm e}}{dt}$ are the average number of added monomers per unit time, per filament at the inner and the outer surfaces, k_+^b is the second order rate constant for the addition of a monomer at the inner surface were monomer density is $c_{\rm i}$, and k_-^b, k_-^p are the first order rate constants for removing a monomer from the filament at the inner and outer surfaces respectively. The monomer addition events at the outer surface are rare enough that the corresponding term can be safely neglected in all cases. The superscripts b, p stand for *barbed* and *pointed* and indicate that the polymerisation takes place at the *barbed* end while the de-polymerisation takes place at the *pointed* end. As in the preceding paragraph, e is the gel thickness and a the typical size of a monomer.

The stress dependence of the rates results from the fact that the potential barrier for adding or suppressing a monomer is shifted from its value at zero stress, from a quantity equal to the work given by the force that a given filament exerts on the link of interest. It is thus of the form: $k = k_0 \exp\left(\frac{-fa}{kT}\right)$. The forces are deduced simply from the stresses. We have already noted that the tensile strain was $\frac{e}{r}$ in the geometry of *Listeria* and it is still the case in spherical geometry. Thus the tensile stress, at the outer surface, is $\sigma_{\rm t} \cong B\frac{e}{r}$ and the force per filament is $f_{\rm t} = \sigma_{\rm t} l^2$, where $l = n^{-\frac{1}{2}}$, is the average distance between filaments. The normal stress obeys Laplace's law: $\sigma_{\rm r} = \frac{2T}{r}$ where $T \cong \sigma_{\rm t}.r$ is the total tension across the gel layer. We thus get: $\sigma_{\rm n} \cong 2B\frac{e^2}{r^2}$. With all these remarks we can write:

$$k_+^b = k_+^{b0} \exp\left(-\frac{e^2}{e_0^2}\right), \quad k_-^b = k_-^{b0} \exp\left(\frac{e^2}{e_1^2}\right), \quad k_-^p = k_-^{p0} \exp\left(\frac{e}{e_2}\right).$$

With: $e_i = r(\frac{kT}{a_i l^2 B})^j$ and $j = \frac{1}{2}$ for $i = 0, 1$, $j = 1$ for $i = 2$. In all cases a_i is a length of order a. Note that all e_i scale like r.

4.3 Steady state

In order to discuss the conditions for steady state one still needs to express the monomer concentration at the inner surface $c_{\rm i}$ as a function of its concentration at infinity c_∞. In a first approximation, it is reasonable to assume that the actin monomer concentration obeys a standard diffusion law. Thus in steady state the flux $j = -D\frac{\partial C}{\partial r} = $ const in the gel, and $c = c_\infty$ outside.

Monomer conservation further imposes (for $e \ll r$): $l^2 D \frac{\partial c}{\partial r} = \frac{\mathrm{d}n_\mathrm{i}}{\mathrm{d}t} = \frac{\mathrm{d}n_\mathrm{e}}{\mathrm{d}t}$. These conditions specify entirely the problem. One finds two regimes connected by a smooth crossover. For small radii, the inner concentration is essentially c_∞, and the steady state thickness is solution of the equation:

$$\frac{c_\infty k_+^{0b}}{k_-^{0p}} = \frac{k_-^{0b}}{k_-^{0p}} \exp\left(e^{*2}\left(e_0^{-2} + e_1^{-2}\right)\right) + \exp\left(e^* e_2^{-1} + e^{*2} e_0^{-2}\right).$$

Since all the a_i scale like r, the steady state thickness e^* also scales like r. Actually with numbers relevant to experimental situations one expects:

$$e^* \cong r/10.$$

This is what is observed experimentally for radii smaller than 10 microns [9]. Such numbers imply that the normal stress exerted by the gel on the bead is of the order of one atmosphere!!

Note that if the leading term is provided by the de-polymerisation at the barbed end one finds exactly the expression announced in the second paragraph, that is (with transparent notations): $e^* = e_2 \mathrm{Ln}(\frac{\nu_\mathrm{p}^0}{\nu_\mathrm{dp}^0})$.

The other limit corresponds to what happens on a flat surface. Then no stress is developed, but the thickness is still limited by the monomer depletion due to the need for the monomers to diffuse from outside. Now the steady state condition reads simply:

$$l^2\left(\frac{c_\infty - c_\mathrm{i}}{e^*}\right) = c_\mathrm{i} k_+^b - k_-^b = k_-^p.$$

For all practical purposes k_-^b can be neglected in this stress free situation. It is then easy to infer:

$$e^* = \frac{l^2 D c_\infty}{k_-^p} \cdot$$

For large enough beads, this regime is always obtained. The crossover radius between the two regimes is given by $r_\mathrm{e} \cong \frac{a^{\frac{1}{2}} D l^3 c_\infty^{\frac{3}{2}}}{kT d_-^p}$. Plugging the value of the diffusion constant as measured in solution we estimate the crossover radius in the millimetre range. It turns out that one can clearly observe the two regimes in the 10 microns range, which implies that the monomer diffusion constant is about one thousand times smaller in the gel than in a solution [28]. There may be many reasons for this large difference (temporary fixation sites on the gel, steric hindrance, other objects like the Arp2/3 complex diffusing slowly and limiting the polymerisation rate).

What is more important is that the measured value ($0.02\ \mu\mathrm{m}^2\,\mathrm{s}^{-1}$) is such that diffusion processes cannot be neglected in cells! The crossover length corresponds precisely to typical cell lengths!

4.4 Growth with spherical symmetry

The arguments developed above allow us to write the gel thickness growth as:

$$\frac{\mathrm{d}e}{\mathrm{d}t} = a\left(k_+^{b0}\exp\left(-\frac{e^2}{e_0^2}\right) - k_-^{p0}\exp\left(\frac{e}{e_2}\right)\right)$$

(where we have omitted the k_-^b term for the sake of simplicity). The solution of this equation with initial condition $e = 0$, is a continuous monotonic function, with essentially two regimes:

- at short times the growth is predicted to be linear: $\frac{\mathrm{d}e}{\mathrm{d}t} \cong a(k_+^{b0} - k_-^{p0})$, that is with obvious notations, $e = \nu_\mathrm{p}^0 t$. Such a relation is probably too naive since it ignores the problems of the nucleation of filaments and of their multiplication with Arp2/3, etc.
- at long times, the gel thickness is close to its steady state value e^* and the dynamical equation can be linearised as a function of $\delta e = e - e^*$. The solution is then:

$$\delta e(t) = \delta e(t_1)\exp\left(-\frac{t-t_1}{\tau}\right);$$

$$\tau^{-1} = k_+^{b0}\left(2\frac{e^*a}{e_0^2} + \frac{a}{e_2}\right)\exp\left(-\frac{e_*^2}{e_0^2}\right) = \left(\frac{2e^*}{e_0^2} + \frac{1}{e_2}\right)\nu_\mathrm{p}.$$

The thickness approaches its steady state value exponentially with a time constant growing linearly with the sphere radius r.

4.5 Symmetry breaking

In the following we keep the arguments as simple as possible, and deal only with the thin shell regime. A more elaborate version can be found in reference [29]. As understood first by Sekimoto the important point is that mechanical equilibrium requires that the total integrated tension across the gel thickness must be constant everywhere. If the gel thickness is smaller in some place than in others, since the integrated tension is the same, this means that the tensile stress must be larger. Now, since the de-polymerisation rate at the exterior surface depends exponentially on the tensile stress, this means that the de-polymerisation is faster where the gel is thinner, which amplifies the thinning process. This is a clear sign of instability, which we investigate in the following.

Assume that at some point the gel thickness has picked up a sinusoidal variation of the sort (Fig. 5):

$$e(t) = e_i(t) + \varepsilon(t)\cos(\theta).$$

In which θ is the polar angle in spherical coordinates, e_i and ε the isotropic and anisotropic parts of the gel thickness. We consider the system in the late time regime since in the earlier regime the stresses play essentially no role. Thus we can consider $e_i - e^*$ and ε as small quantities which we need to treat at first order only, for the discussion of stability. The tension being constant, it cannot depend on θ, and can only depend on e_i and ε^2, thus to lowest order:

$$T = B\frac{e_i^2}{r} + 0\left(\varepsilon^2\right) \text{ and, } \sigma_{\mathrm{r}} = \frac{T}{r} = \text{const.}$$

On the contrary:

$$\sigma_\theta = \frac{T}{e(\theta,t)} \cong \frac{T}{e_i}\left(1 - \frac{\varepsilon}{e_i}\cos(\theta)\right) + 0\left(\varepsilon^2\right).$$

As announced the tensile stress is maximal where the thickness is minimal.

If we extend the use of the dynamical equation to this non-homogeneous case, we can write:

$$\frac{\mathrm{d}e(\theta,t)}{\mathrm{d}t} = a\left(k_+^{b0}\exp\left(-\frac{e_i^2}{e_0^2}\right) - k_-^{p0}\exp\left(\frac{e_i}{e_2}\left(1 - \frac{\varepsilon}{e_i}\cos(\theta)\right)\right)\right).$$

Which yields after linearisation:

$$\frac{\mathrm{d}\delta e_i}{\mathrm{d}t} = -\frac{\delta e_i}{\tau}$$
$$\frac{\mathrm{d}\varepsilon}{\mathrm{d}t} = +\frac{\varepsilon}{\tau_0}.$$

As expected, we find that the isotropic part is stable and converges exponentially fast toward the steady state value e^*, and that the symmetry breaking part is always unstable and grows exponentially fast with a time scale $\tau_0 = \tau(1 + 2\frac{e_2 e^*}{e_0^2}) = \frac{e_2}{\nu_{\mathrm{p}}}$. Note that τ_0 grows linearly with the bead radius, just like τ, and is of the same order of magnitude.

Now we know that symmetry-breaking fluctuations are amplified, but where do they come from? Could thermal fluctuations be sufficient?

The typical fluctuation amplitude can be expressed as:

$$\varepsilon_{\mathrm{b}} \cong \left(\frac{kTe^*}{Br^2}\right)^{\frac{1}{2}}.$$

For a 100 nm bead such as used in [8], the corresponding amplitude is in the nanometer range. It might be sufficient to trigger the instability because of its exponential amplification. However, other factors such as enzyme heterogeneous distribution or deviation from pure spherical shape may be more important. For instance if there is a total number N of enzymes at the bead surface, there will always be an imbalance of the order of $N^{\frac{1}{2}}$ between the two sides of the bead. The integrated tension T is still a constant and so is the normal stress $\sigma_{\rm r}$ at the bead surface. However the force acting on filaments now depends on position through the enzyme density dependence. With $l^2(\theta) = l_{\rm i}^2 + 2l_{\rm a}.l_{\rm i}\cos(\theta)$, $l_{\rm i}$, $l_{\rm a}$ being the isotropic an anisotropic parts of the average distance between nucleation enzymes, one obtains a force per filament, $l^2\sigma_{\rm r}$, which also depends on angle and the equation for the anisotropic part of the gel thickness becomes:

$$\frac{{\rm d}\varepsilon}{{\rm d}t} = \frac{\varepsilon}{\tau_0} + \nu_{\rm b}.$$

In which $\nu_{\rm b} = \frac{2l_{\rm a}}{l_{\rm i}}\nu_{\rm p} \cong \frac{2}{\sqrt{N}}\nu_{\rm p}$; $\nu_{\rm p}$ is the polymerisation velocity as defined in the first part of this article. The solution reads:

$$\varepsilon(t) = \nu_{\rm b}\tau_0\left(\exp\left(\frac{t}{\tau_0}\right) - 1\right).$$

In order to assess, which of thermal fluctuations or enzyme heterogeneity, is the leading term in the symmetry breaking source, one should then compare $\nu_{\rm b}\tau_0$, and $\varepsilon_{\rm b}$. Their bead radius dependence is such that at r large enough the enzyme disorder should always win. Plugging reasonable orders of magnitudes, suggests that even for a few nanometers size bead the enzyme disorder is already more important.

An other symmetry-breaking source is the lack of sphericity of the beads. Arguments, very similar to the one we have used for enzymes heterogeneity, can be made. One has essentially to replace $\frac{l_{\rm a}}{l_{\rm i}}$, by $\frac{r_{\rm a}}{r_{\rm i}}$ in which the subscripts refer to the beads radius of curvature variation, with obvious meaning. It is difficult to put numbers on this term, since it depends on the preparation chemistry of the beads. It seems that beads smaller than a micron, have excellent sphericity, controlled by surface tension. In that range one expects enzyme heterogeneity to provide the main symmetry breaking term. It seems more difficult to obtain beads larger than a few microns, in which case spherical aberrations might provide the main symmetry breaking term. It would be interesting to design carefully controlled experiment, for checking quantitatively these predictions. Clever microscopic models have been imagined in order to obtain symmetry-breaking conditions [17,30]. However they do not take into account the very nature of the actin gel.

4.6 Limitations of the approach and possible improvements

In the discussion developed above, we have kept only diagonal stresses. However as soon as the gel thickness is inhomogeneous, some of the elastic energy is released by shear, and a complete analysis should contain the corresponding terms. One can show however that the exposed results are not changed in any significant way [29]. Furthermore, whenever the anisotropy changes with time, the gel redistribution causes friction at the bead/gel interface, a phenomenon that is not included in the present analysis. It is in fact possible to show that it does not modify the structure of the equations but simply renormalizes the onset time τ_0 of the anisotropy.

Indeed, under such circumstances, the tension has an angular dependent part, which must be proportional to the friction coefficient, the velocity of the gel relative to the bead and have the right dimensions:

$$T(\theta) = T_{\rm i} + \kappa\xi r \frac{{\rm d}\varepsilon}{{\rm d}t}\cos(\theta).$$

Where κ is a dimensionless number and ξ is the bead/gel friction coefficient already discussed. The equations are formally unchanged and only the onset time of the modulation is modified to a new value:

$$\tau_0^* = \tau_0 + \frac{r^2}{\nu_{\rm i}}\left(\frac{e_2}{e_0^2} + e^{*-1}\right).$$

The main conclusions are similar to the one obtained before, for instance the symmetry breaking onset time is again proportional to the sphere radius, since e_0, e_2 and e^* are. This expectation is indeed born out by experiment [31]. Furthermore if the friction is very high the spherical steady state can be reached much before symmetry is actually broken. This will in particular happen if the gel is dense, expectation also born out by experiment.

There are several other implicit simplifications in the above presentation: we have considered only one elastic modulus, without specifying whether it corresponds to compression or shear or a combination of them. A proper description is possible by the use of a covariant description of the gel [29]. Actually, the very geometry of the polymerisation/cross-linking process implies that the gel should be anisotropic as well, but keeping this feature adds in complexity without bringing further understanding to the question.

A more important limitation comes from the hidden assumption that the gel density is constant throughout the gel and that all de-polymerisation is located at its external surface. In fact it is known that the gel density decreases exponentially in a *Listeria* comet, over length scales comparable to the comet total length, *i.e.* several tens of microns [13]. One could be tempted to argue that this length scale is much larger than the one we discuss here and forget about this slow bulk de-polymerisation. It would, however, be a wrong argument, since the de-polymerisation mechanism is certainly stress dependent. In fact, the dynamical equation for the gel density ρ should read:

$$\frac{\partial \rho}{\partial t} + \vec{\nu} \cdot \vec{\nabla} \rho = S - k_{\mathrm{d}} \rho.$$

Where, $\vec{\nu}$ is the gel velocity relative to the bead or the bacterium, S is the gel density source localised at or close to the surface, and k_{d} is the de-polymerisation probability per monomer connected to the gel, per unit time. In general, one expects stress dependence similar to the one already described, that is $k_{\mathrm{d}} = k_{\mathrm{d}}^0 \exp(\frac{\sigma_{\mathrm{t}} \phi^2 a_3}{kT})$, in which $\phi^2 = \rho^{-1} a^{-1}$ is the average area spanned by a filament, and a_3 a model dependent length. Far from the surface, in an essentially unstressed comet tail, one does obtain an exponential decrease of the density over a length $L = \frac{\nu}{k_{\mathrm{d}}^0}$. Knowing ν and the comet length, one deduces k_{d}^0 easily. The physical mechanism, behind this de-polymerisation is not obvious: it could be that actin filaments can spontaneously break anywhere, or that reticulation points stabilise the structure and provide the rate limiting step in the de-polymerisation, or that there is a one to one mapping of the pointed end density on the connected monomer density. In all cases a constant average number of monomers should leave the gel for each event. Note that a de-polymerisation from the pointed ends of the filaments cannot in general be represented by such a mathematical structure. For instance if the filaments were parallel on average, all starting from the surface at the barbed end, and with a length distribution, the term would read $-k_{-}^{\mathrm{p}} a\vec{p} \cdot \vec{\nabla} \rho$, in which $\vec{p}$ is the unit vector in the filaments direction. Indeed, under such circumstances $a\vec{p} \cdot \vec{\nabla} \rho$ is a measure of the printed and density.

Because of the exponential dependence of the de-polymerisation coefficient on stress, the length over which the density significantly decreases, may become very short in the presence of such a stress and this mechanism could provide an alternative interpretation of the steady state in spherical symmetry. Under such circumstances, the density decrease occurs essentially over a length such that $\sigma_{\mathrm{t}} = \frac{kT}{\phi^2 a_3}$, or with the scaling laws derived in spherical geometry $e \cong \frac{kT\rho}{Ba_3} r$. The sharpness of the density decrease is controlled by the ρ dependence of B. In all reasonable cases, it is quite pronounced. For instance, if B is proportional to ρ, which is the case whenever the cross-link angular elasticity determines the elastic modulus, then $\rho = \rho_0 \exp(\frac{r_0}{\alpha L}(1 - \exp(\frac{\alpha(r-r_0)}{r_0})))$. In this equation, r_0, ρ_0 denote the radius and density at the bead surface, and α a dimensionless number of order ten. There is a sharp cut off, for $r - r_0 \cong \frac{r_0}{\alpha}$. Further considering that below a threshold density, the gel integrity is totally lost, it is clear that any thickness measurement will give a value very close to $\frac{r_0}{\alpha}$, essentially equivalent to the one derived in Section 4. The drawback of this type of presentation is the added complexity. Its merit is the connection of the comet slow density decrease with the fast one under stress. The same type of result holds for other ρ dependences of B, such as the one used in Section 2.

An observation, which is not accounted for by the present analysis, is the strong density increase of the comet in the saltatory mutant. Simulations, taking into account the detailed chemistry of the polymerisation process, but ignoring the role of the polymerisation enzyme, show a strong density dependence on external stress [16]. A simple phenomenological way of taking this effect into account consists in writing the gel density source s:

$$S = \nu\rho_- \exp\left(\frac{l_{\rm s}}{\nu\tau}\right)\delta_{\rm s}.$$

In which $\delta_{\rm s}$ is the delta function at the surface. $\rho_- = \frac{1}{l^2 a}$, is determined by the enzyme density, the exponential factor expresses the exponential growth due to Arp 2/3, $l_{\rm s}$ the length over which the branching phenomenon can occur and τ a typical capping time. The gel density ρ_+, just outside the proximal domain, reads: $\rho_+ = \rho_- \exp(\frac{l_{\rm s}}{\nu\tau})$. The expression of the growth velocity ν is model dependent. If one assumes that a fraction of order unity of all filaments contributes to the stress, then the natural generalisation of the polymerisation law under stress reads: $\nu = \nu_{\rm p}^0 \exp(-\frac{\sigma_{\rm n}\rho_+^{-1} a_0}{kTa})$. Solving for this set of equations reproduces fairly well the results of the simulation. Indeed, in the large force, slow velocity limit, one finds: $\nu \cong \frac{l_{\rm s}}{\tau {\rm Ln}(\frac{\sigma_{\rm n}}{kT\rho_-})}$, and $\rho = \frac{\sigma_{\rm n}}{kT{\rm Ln}(\frac{\nu_{\rm p}^0}{\nu})}$, in which we have assumed $a_0 \cong a$ for simplicity. In this regime, the velocity is essentially independent of stress and determined by the branching and capping processes. Conversely, the density is practically proportional to the stress. Both features show up clearly in the simulation. For "large" velocities a more conventional regime in which $\rho_+ \cong \rho_-$, and $\nu \cong \nu_{\rm p}^0 \exp(-\frac{\sigma_{\rm n}}{kT\rho_-})$, may exist. Such a formulation should be exploited further.

5 Conclusion

In these lectures we have shown that if one wants to understand the physics of the propulsion mechanism of the bacterium *Listeria*, one must analyse the mesoscopic stress distribution in the actin gel, and the solid on solid friction that the gel exerts on the bacterium surface. Keeping in mind, only the polymerisation process misses so to speak half of the problem. In particular, recognizing the importance of the non-linear friction allows us to understand the saltatory mutant as being a system working in a "stick-slip" regime familiar in solid friction. In general the system gets into the saltatory regime at a well defined threshold called a Hopf bifurcation. In its vicinity measurable quantities such as the velocity and comet density modulations are sinuoïdal functions (of time and space respectively). Away from the bifurcation variations are more abrupt like in Figure 2. It is important to understand that the role of mutation is simply to drive the system in a region of phase space where the behaviour is saltatory, but that many other external perturbations could have the same result without changing the genome. This point is well illustrated in bio-mimetic experiments, where a simple bead diameter change drives the behaviour from regular to saltatory. This example shows that the relation between genotype and phenotype will not be easy to unravel. One will need to fully understand all the complex dynamical diagrams governing biological systems, before being able to fully exploit the formidable genetic data that we are getting now.

It is important to stress that the mesoscopic approach is not antagonistic to microscopic ones, but rather complementary. Microscopic theories can be used as input for writing boundary conditions in mesoscopic theories. They should not only focus on the polymerisation process, but on friction and de-polymerisation as well. An other message is that, even if the *Listeria* propulsion mechanism can be, in some way, representative of the biochemistry involved in more complex Eukaryotic cells, the physics of it will be very different for two simple reasons. Indeed, the stress distribution will be very different because the gel is now produced in the inside of an object topologically spherical, and second because the tangential friction is very different since on a fluid membrane. Yet, what we learn here will be very useful for understanding cell motility. At last, what I find most rewarding, is the potential importance of the bio-mimetic approach for medical applications [32,33].

The ideas developed here result from many discussions with F. Jülicher, K. Sekimoto and C. Sykes. It is a pleasure for me to express my warm thanks. I further thank C. Sykes for her help with the reference list. I am also grateful to M.F. Carlier and D. Pantaloni for introducing me to the biochemistry of actin polymerisation.

References

[1] L.G. Tilney and D.A. Portnoy, *J. Biol. Chem.* **109** (1989) 1597-1608.

[2] F.S. Southwick and D.L. Purich, *New Engl. J. Med.* **334** (1996) 770-776.

[3] D. Bray, *Cell Movements* (New York: Garland, 2001).

[4] I. Lasa, E. Gouin, M. Goethals, K. Vancompernolle, V. David, J. Vandekerkhove and P. Cossart, *EMBO J.* **16** (1997) 1531-1540.

[5] T.M. Svitkina and G.G. Borisy, *J. Cell Biol.* **145** (1999) 1009-1026.

[6] L. Blanchoin, K.J. Amann, H.N. Higgs, J.B. Marchand, D.A. Kaiser and T.D. Pollard, *Nature* **404** (2000) 1007-1011.

[7] T.P. Loisel, R. Boujemaa, D. Pantaloni and M.-F.Carlier, *Nature* **401** (1999) 613-616.

[8] L.A. Cameron, M.J. Footer, A. van Oudenaarden and J.A. Theriot, *Proc. Natl. Sci. USA* **96** (1999) 4908-4913.

[9] V. Noireaux, *et al., Biophys. J.* **78** (2000) 1643-1654.

[10] J.A. Theriot, T.J. Mitchison, L.G. Tilney and D.A. Portnoy, *Nature* **357** (1992) 257-260.

[11] C. Knocks, J.B. Marchand, E. Gouin, H. d'Hauteville, P.J. Sansonetti, M.F. Carlier and M.F. Cossart, *Mol. Microbiol.* **18** (1995) 413.

[12] F. Gerbal, *et al., Pramana* **53** (1999) 155-170.

[13] F. Gerbal, V. Laurent, A. Ott, M.-F. Carlier, P. Chaikin and J. Prost, *Eur. Biophys J.* **29** (2000) 134-140.

[14] J.L. McGrath, N.J. Eungdamrong, C.I. Fischer, F. Peng, L. Mahadevan, T.J. Mitchison and S.C. Kuo, *Science* (2002).

[15] A. Mogilner and G. Oster, *Biophys. J.* **71** (1996) 3030-3045.

[16] A.E. Carlson, *Biophys. J.* **81** (2001) 1907-1923.

[17] A. Mogilner, S. Kuo and G. Oster (preprint).

[18] A. Mogilner and L. Edelstein–Keshet (preprint).

[19] E. Gouin, H. Gantelet, C. Egile, I. Lasa, H. Ohayon, V. Villiers, P. Gounon, P.J. Sansonetti and P. Cossart, *Cell Sci.* **112** (1999) 1697-1708.

[20] C.J. Merrifield, *et al., Nat. Cell Biol.* **1** (1999) 72-74.

[21] J. Taunton, *et al. J. Cell Biol.* **148** (2000) 519-530.

[22] K. Tawadac and K. Sekimoto, *J. Theor. Biol.* **150** (1991) 193-200.

[23] F. Gerbal, P. Chaikin, Y. Rabin and J. Prost, *Biophys. J.* **79** (2000) 2259-2275.

[24] F. Gerbal, Ph.D. Thesis (Universit Paris VII, 1999).

[25] D. Yarar, W. To, A. Abo and M.D. Welch, *Curr. Biol.* **9** (1999) 555-558.

[26] J. Fradelizi, *et al., Nat. Cell Biol.* **3** (2001) 699-707.

[27] V. Noireaux, Ph.D. Thesis (Universit Paris XI, 2000).

[28] J. Plastino and C. Sykes (private communication).

[29] K. Sekimoto and F. Julicher and J. Prost (in preparation).

[30] A. Van Oudenaarden and J.A. Theriot, *Nature Cell Biol.* **1** (1999) 493-499.

[31] A. Bernheim–Groswasser, S. Wiesner, R. Golsteyn, M.-F. Carlier and C. Sykes, *Nature* (2002) (accepted).

[32] E. Friederich, R.M. Golsteyn, D. Louvard, V. Noireaux, J. Prost and C. Sykes, Institut Curie/CNRS, patent No. PCT/FR00/03569 (2000).

[33] J. Fradelizi, E. Friederich, R.M. Golsteyn, D. Louvard, V. Noireaux and C. Sykes, Institut Curie/CNRS patent No. PCT/FR01/00843 (2001).

COURSE 6

PHYSICS OF COMPOSITE CELL MEMBRANE AND ACTIN BASED CYTOSKELETON

E. SACKMANN

Physics Department, Lehrstuhl für Biophysik – E22, Technische Universität München, Munich, Germany

Contents

PHYSICS OF COMPOSITE CELL MEMBRANE AND ACTIN BASED CYTOSKELETON

E. Sackmann, A.R. Bausch and L. Vonna

1 Architecture of composite cell membranes

The composite cell envelope is an impressive example of nature's strategy to design complex materials and machineries with unique and stunning physical properties by self-assembly of hierarchical structures. The most simple prototype of a composite cell membrane is the envelope (often called plasma membrane) of red blood cells.

As illustrated in Figure 1 of the chapter on adhesion it is composed of the central lipid/protein bilayer, the glycocalix exposed to the extracellular space and a thin macromolecular network (the cytoskeleton) coupled to the cytoplasmic leaflet.

1.1 The lipid/protein bilayer is a multicomponent smectic phase with mosaic like architecture

The lipid/protein bilayer forms an electrically insulating barrier. It is a multicomponent two-dimensional smectic liquid crystal composed of roughly 100 different types of pin-like lipid molecules including the amphiphilic steroid cholesteroid and proteins such as ion channels, cell surface receptors or hormone amplifiers. The driving force for the self-assembly is the hydrophobic effect which it is determined by the entropy gain of the water molecules surrounding the lipid. The hydrophobic binding energy of a lipid molecule is roughly proportional to the surface of the hydrocarbon chains. Therefore nature used two-chain lipids (instead of single chain surfactants) the solubility of which is extremely small in water ($<10^{-12}$ M, *cf.* Sackmann 1995). Similarly, integral proteins are anchored in the bilayer by their hydrophobic domains: mostly α-helical segments, the length of which is adopted to the thickness of the hydrophobic center of the bilayer. The matching of the thickness of the bilayer and the hydrophobic domains of the proteins may provide a key driving force for lateral reorganization of the lipids (*cf.* Sackmann 1994, for hydrophobic matching concept). The cell surface receptors are typically embedded in the bilayer with only one α-helical segment, while the bulk of the protein is pointing into the extracellular/intracellular space (*cf.* article on adhesion, Ch. X)[1]. In contrast, ion channels are formed by arrangements of several α-helical segments (*e.g.* thirteen in the case of the anion channel band3, *cf.* Fig. 1).

Despite of the complex composition, small changes in the lipid composition (*e.g.* of the fraction of charged lipids or cholesterol) may have severe consequences for the viability of the cells. It appears that similar to the ensemble of genes (the genom) and proteins (the proteom) each individuum has a well preserved lipid composition (which we could call lipidom).

A large fraction of the biochemical reactions in cells occur at membranes. Prominent examples are (1) the lipid and protein biosynthesis at the ribosomes adsorbed to the endoplasmatic reticulum membrane, (2) the charge separation in photosynthetic membranes, the electron transfer in mitochondrial membranes mediated by a chain of proteins acting alternatingly as electron acceptor and donor or (3) the hormone amplifiers (*cf.* Fig. 13).

To allow for a fast material transport the lipid bilayer is on the average in a fluid state. The two-dimensionality of the bilayer has important consequences for the efficiency of diffusion controlled chemical reactions. The lateral diffusivity depends only logarithmically on the radius of the diffusing particle. Consequently, large integral proteins may diffuse nearly as fast as lipid molecules unless they are coupled to the cytoskeleton. On the other side it becomes more and more evident that the membrane exhibits a mosaic like structure due to local lateral phase separations of the lipids (*cf.* Sackmann 1994). This leads to the (often transient) formation of domains of specific composition such as so-called rafts or small invaginations (buds) called caveoli. These domains appear to contain a high content of cholesterol and sphingomyelins besides proteins[2].

The glycocalix can be considered as a monofilm of macromolecules which

[1]In the following we will denote the lipid protein bilayer of the cell envelope by "plasma membrane" and the complete outer shell of cells as "composite membrane" or "cell envelope", which is composed of the cytoskeleton, the plasma membrane and the glycocalix.

[2]Most natural lipids exhibit hydrocarbon chains with several C-C double bonds and form a randomly mixed fluid phase at physoilogical temperature (*cf.* Sackmann 1995). In contrast bilayers of natural sphingomyelins exhibit fluid-solid coexistence up to 60 °C (*cf.* Döbereiner *et al.* 1993) and it is thus likely that they form gel-like domains in natural membranes, in particular since the outer monolayer of the plasma membrane contains 20% of this lipid. Similarly, the cholesterol content of plasma membranes is about 50 mole % and accordingly to model membrane studies it is expected to form precipitates above 40 mole % (*cf.* Baeyrl & Sackmann 1992). Besides lateral phase separation, domains of specific composition form in fluid membranes by budding which can be induced by adsorption of a protein coat. Thus caveoli are stabilized by the coat protein cavolein.

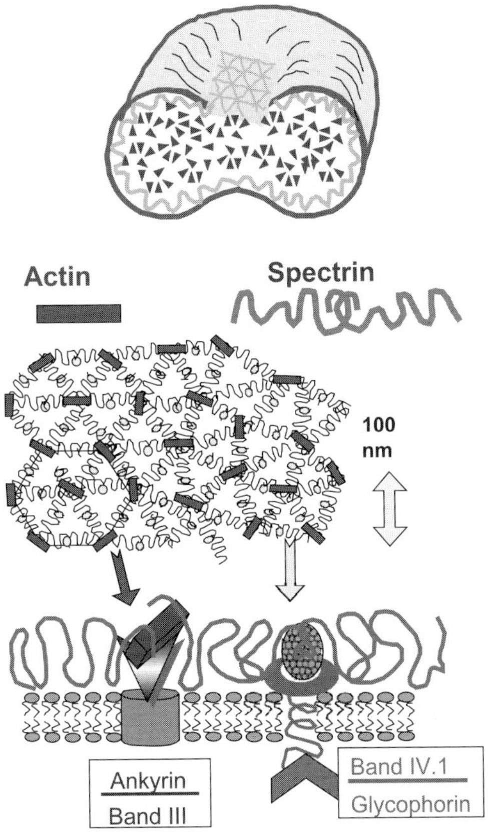

Fig. 1. Composite shell of the red blood cell as most simple prototype of a composite cell membrane. The quasi two-dimensional meshwork is composed of spectrin tetramers and actin oligomers. Spectrin is a highly flexible filament of ~100 nm contour length composed of two interwinded chains. It associates by tail-tail interactions and the tetrameres form the sides of the triangles. The corners consist of actin oligomers (length ~35 nm) which couple to a specific actin binding site at the free end of spectrin (which shares in fact common features with α-actinin, *cf.* Fig. 10). The length of the sides of the triangles is about 80 nm and therefore the spectrin dimers exhibit ellipsoidal rather than rod-like shapes. The network is locally coupled to the bilayer in two ways: the actin oligomer is linked to the band 3 protein (serving simultaneously as anion channel) through a linker called ankyrin; the center of the sides are coupled to the cytoplasmic domain of the glycoprotein glycophorin through the coupling protein band IV.1, which belongs to the family of Ezrin proteins (Bretscher 1999; *cf.* also Sackmann 1995).

is formed mainly by the giant head groups of cell surface receptors (*cf.* article by Bruinsma and Sackmann on Cell Adhesion). It has a typical thickness of 40 to 50 nm, but by the binding of giant macromolecules of the extracellular matrix such as collagen, fibronectin or hyaluronic acid (a giant polysaccharide) to their respective receptors it can be much thicker. The main task of the glycocalix is the communication with the environment but it is assumed to serve also as protective film by repelling bacteria and parasites. Most helpful in this context is the high negative charge which is mostly due to sialic acid-rich oligosaccharides.

1.2 The spectrin/actin cytoskeleton as hyperelastic cell stabilizer

How can a fluid leaflet which is thousand times thinner than paper form stable shells of 10 μm dimension. Although the cohesion between lipid molecules due to the hydrophobic effect is very strong, the red cell envelope would decay into small vesicle if it was not stabilized by the membrane associated fishernet-like meshwork. It is essentially composed of spectrin (a flexible filamentous protein) and rod-like oligomers of actin and these components assemble into a roughly triangular network as shown in Figure 1. The network is coupled to the membrane through association of the linker protein ankyrin and band IV.1 with membrane proteins. The coupling strength is, however, most likely regulated in a dynamic way by phosphorylation and dephosphorylation of the coupling proteins which requires ATP (*cf.* Bennet 1990). It is supposed that for this reason ATP-depletion leads to stiffening of the membrane.

The human body contains about 2.5×10^{12} copies of erythrocyte and the production rate of these cells is about 2.5×10^{6} per second. They are formed by detachment from giant mother cells (megacytes) and contain initially a nucleus together with other cellular compartments. This excess material is expelled by exocytosis resulting in the well defined biconcave sack filled with mainly hemoglobin (*cf.* Sackmann 1995).

The inventiveness of nature becomes really evident if we consider the elastic properties of the shell. Despite of their complex composition, erythrocytes are perfect elastic shells as becomes evident if we deform cells in high frequency electric field (*cf.* Engelhardt & Sackmann 1988). The cell can be stretched by a factor of two and relaxes to its original shape within a period of a second. The elastic deformation can be well described by the three classical modes of deformation: isotropic extension, shearing and bending characterized by the corresponding elastic moduli. Various methods have been developed to measure the elastic moduli (*cf.* Evans & Needham 1986; Engelhardt & Sackmann 1988). The values are compared in Figure 2 with the moduli of a hypothetical plastic shell of the same size and shape made of plastic material (*e.g.* polyethylene). It is seen that the

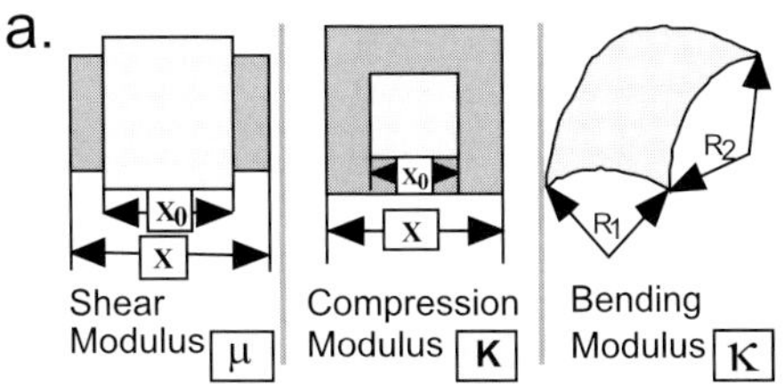

Unique Combination of Elastic Moduli

Material	Shear [dyne/cm]	Bending [erg]	Compression [dyne/cm]
Polyethylene	$3 * 10^{+2}$	$2 * 10^{-10}$	$5 * 10^{+3}$
Erythrocyte	$6 * 10^{-3}$	$5 * 10^{-13}$	$1 * 10^{+3}$

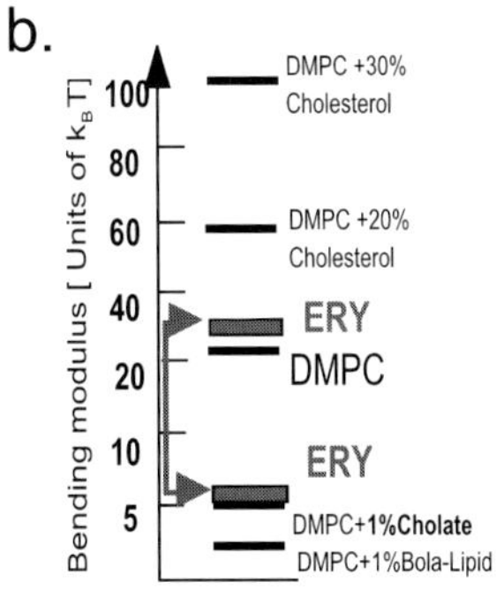

Fig. 2. a) Top: representation of the three classical modes of deformation of an elastic shell. Bottom: comparison of elastic moduli of erythrocyte membrane (RBC) with corresponding parameters of a hypothetical shell of the same shape and size made of technical material (*e.g.* polyethylene). Note the enormous softness of RBC with respect to shearing and bending. **b)** Typical values of bending moduli of membranes presented in terms of thermal energy. Note that cholesterol stiffens while small surfactants soften the membrane. The different κ-values for erythrocytes correspond to different measuring techniques.

biological shell is orders of magnitude softer than the technical one with respect to shearing and bending but is as resistant as soft metal towards area changes. This combination of elastic properties is absolutely essential for the survival of the shell during its 400 km long travel through the blood vessel system consisting to a large part of very narrow capillaries. The softness with respect to bending and shearing prevents elastic instabilities due to buckling (*cf.* Evans 1972) while the lateral incompressibility is important to minimize the loss of ions. This minimizes the ATP-consumption required to maintain the osmotic equilibrium (*cf.* Sackmann 1995). A most spectacular feature of the red cell membrane and giant non-spherical vesicles is the dynamic surface roughness caused by the pronounced thermally excite bending undulations (*cf.* Brochard & Lennon 1975; Strey *et al.* 1995). This so-called membrane flickering has important consequences for the elastic properties of membranes. It gives rise to an entropic lateral membrane tension associated with the entropic free energy required to stretch out the dynamic wrinkles (*cf.* Evans & Rawicz 1990). The most important consequence is the generation of entropy-determined repulsive undulation forces which play a key role for the swelling of multilamellar stacks of membranes in water (*cf.* Helfrich & Harich 1984). It impedes the adhesion of the erythrocytes for instance to tissue surfaces or white blood cells and may thus help to prolong the life time of the cells in the body.

The thermal excitation of erythrocytes or vesicles can be described in terms of a superposition of overdamped spherical harmonic modes (*cf.* Brochard & Lennon 1975; Millner & Safran 1987). If one considers only a small area L^2 (with L small compared to the radius of the shell) the membrane excitations may be represented by planar modes and the dynamic deflection of the membrane at position $\vec{r}$ is

$$u\left(r,t\right) = \sum_{\mathrm{q}} u_{\mathrm{q}}\left(t\right) \exp\left\{i\vec{q}\cdot\vec{r}\right\} \tag{1.1}$$

(where $\vec{r}$ is the position and $\vec{q}$ the wavevector in the plane of the membrane) The amplitudes u_{q} can be related to the bending modulus and the membrane tension σ by the equipartition theorem as

$$L^2\left\langle u_{\mathrm{q}}^2\right\rangle = \frac{k_{\mathrm{B}}T}{\kappa q^4 + \sigma q^2}\cdot \tag{1.2}$$

By measuring the mean square amplitudes $\left\langle u_{\mathrm{q}}^2\right\rangle$ as a function of the wave vector q the elastic modulus and the membrane tension can be measured with high precision for free (*cf.* Brochard & Leuno 1990; Duwe & Sackmann 1990) and adhering vesicles (*cf.* Rädler *et al.* 1995). The bending energies presented in the diagram of Figure 2b have been determined in this way. The most intriguing result of such measurements is the surprisingly low κ value (*cf.* Fig. 2b) found for the erythrocyte membrane. Since the membrane contains 50 mole % of cholesterol one would expect a ten fold higher value of κ_{c}. This is even more surprising since the cell membrane exhibits also a shear elasticity μ due to the membrane-cytoskeleton coupling. The mean square amplitude is expected to be reduced (according to Monte Carlo calculations by Lipowsky & Girardet 1990) to

$$L^2\left\langle u_{\mathrm{q}}^2\right\rangle \approx \frac{k_{\mathrm{B}}T}{2c(k_{\mathrm{B}}T\cdot\mu)^{1/2}q^3 + \kappa q^4 + \gamma q^2} \tag{1.3}$$

where c is a numerical constant $c \approx 1.3$. With the value of $\mu = 5\times10^{-5}$ J/m^2 (*cf.* Fig. 2) the undulations should be suppressed at wavelengths larger than 0.5 μm.

One likely and interesting explanation is that the membrane undulations are mainly excited by fluctuating chemical forces. Such excitations could for instance be caused by the dynamic phosphorylation and dephosphorylation of the proteins (ankyrin, band IV.1) mediating the coupling of the cytoskeleton to the bilayer membrane proteins. They appear to be related to the activity of Mg-ATP-ases (*cf.* Tuvia *et al.* 1999). The local decoupling of spectrin-membrane binding sites is also supposed to play a role for the great softness of red cell membranes.

In normal cells the flickering is suppressed since the bilayer membrane is coupled to the stiff actin cortex. It has, however, been postulated that membrane undulations could control the kinetics of the formation of pseudopods (of *Dictyostelia Discoideum*, *cf.* Oster 1988). The idea is that free volume is created by flickering at the front of the cells mediated by local decoupling of the bilayer from the actin cortex and that this process mediates unidirectional growth of the cortex (*cf.* Oster) in a Brownian ratchet process.

1.3 The actin cortex: Architecture and function

The envelope of nucleated eukaryotic cells is a much more complex multi purpose machinery than the erythrocyte membrane. The inner wall of the outer membrane is stabilized by a roughly 0.5 μm thick crosslinked actin network (called actin cortex) which forms part of the chemomechanical machinery mediating the locomotion of cells (*cf.* Fig. 4 for the case of *Dictyostelia Discoideum*cells). The adaptive mechanical properties of the composite shell are outstanding and enable cells to undergo a stunning manifold of shape changes and mechanical processes.

As an example we consider the function of the endothelial cells which line the inner wall of blood vessels (Fig. 3). The closed cell monolayer forms a selective filter between blood and tissue or between blood and brain (blood-brain-barrier) and have to be permeable for white blood cells in a controllable way. Thus, lymphocytes have to patrol the body continuously for foreign antigens by recirculating from blood through tissue into the lymph and back to the blood (*cf.* Springer 1994). Granulocytes have to penetrate into the tissue to destroy invaders causing inflammations. The selective filter function of the endothelium is associated with two fundamental processes: cell adhesion and cellular shape changes resulting in the formation of gaps within endothelial cell monolayers.

Let us consider first cell adhesion which comes into play several times (*cf.* also Chapter on Adhesion). First, the lateral connection between cells is mediated by the mutual interaction of homophilic (self recognizing) receptors of the cadherin family. Secondly, the endothelial cells are grafted to the basal membrane by binding of integrins ($\alpha_i\beta_i$) to adhesion epitopes of collagen IV ($\alpha_2\beta_1$) or laminin (*e.g.* $\alpha_6\beta_1$). Thirdly, the white blood cells adhere transiently to the endothelial surfaces by binding of receptors of the selectin family (selectin E) to oligosaccharide rich cell adhesion molecules (so-called Gly-CAMs, *cf.* article on adhesion) of the white blood cells surface where selectin recognizes the specific oligosaccharide Sialyl Lewis X. The selectin receptors are enriched on the cell surface during

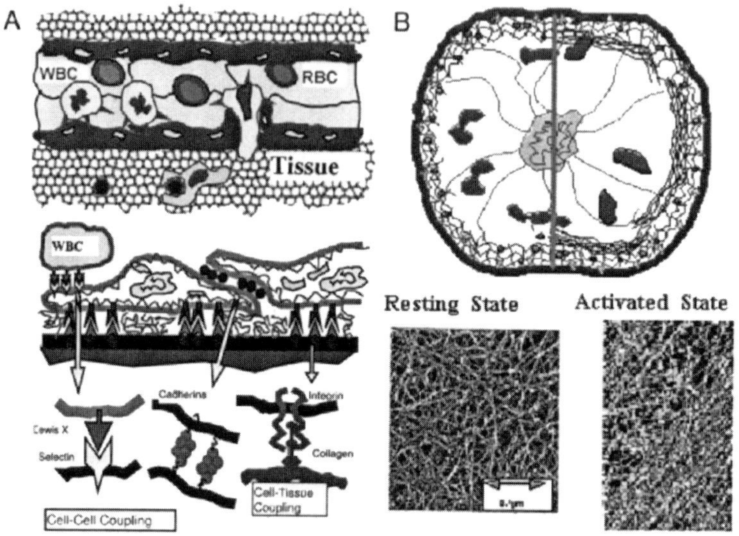

Fig. 3. a) Top: schematic view of endothelial cell monolayer covering the inner wall of blood vessels. Also shown are white blood cells (WBC) which roll along the surface of the endothelium and a WBC penetrating the gap in the cell monolayer formed by local contraction of the endothelium. Bottom: representation of three scenarios of adhesion. Right: coupling of endothelial cell to the wall of a blood vessel through binding of integrin to tissue proteins (collagen IV and laminin). Middle: formation of tight cell-cell contact by homophilic receptor of cadherin family. Left: weak coupling of WBC to endothelial cells by coupling of the endothelial cell surface receptor selectin E and oligosugars of Lewis X-type which are attached to oligosugar-rich molecules (CAMs; *cf.* article on adhesion). Note that the inner surface of the blood vessel (the basal membrane) is composed of a network of collagen (type IV) which is associated with other macromolecules of the extracellular matrix such as laminin and percelan into a multifunctional network (*cf.* Lodish *et al.* 1995). **b)** Left: Schematic view of radnom network of actin in cortex of a resting cell. Right: Schematic view of bundle formation in actin cortex induced by activation resulting in the centripetal contraction of endothelial cells. Note that stress fibers are formed within seconds. Bottom: Freeze fracture electron micrograph of random actin network (left); and the bundled network structure.

inflammation. Close to the site of infection the white blood cell WBC (granulocyte) adheres strongly to the endothelial cell by switching on the interactions between integrins ($\alpha_L\beta_i$) on the WBC with cell surface proteins (E CAMs) on the endothelial cells. This adhesion process is enforced by

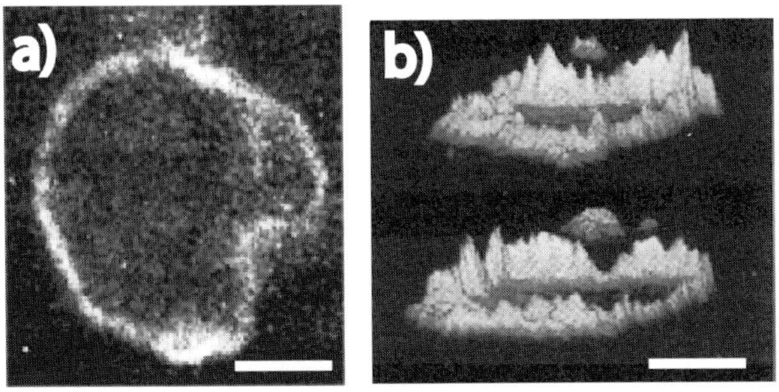

Fig. 4. a) Distribution of actin in cortex of *Dictyostelia* cell visualized by microfluorescence. **b)** Distribution of myosin in *Dictyostelia* cells. Scale bar on both images is 20 μm.

chemoattractant molecules formed close to the site of infection which bind to specific receptors on the WBC membrane and induce the accumulation of integrins ($\alpha_M\beta_2$) on the WBC-cell surface. This accumulation occurs by fusion of intracellular vesicles enriched in the integrins with the plasma membrane or by de novo synthesis of integrins (*cf.* Springer 1994). This shows, firstly, that adhesion is a dynamic process involving genetic expression processes and secondly, that the composite cell membrane is an open system. It is interesting to note that the receptor for the chemoattractant resembles hormone receptors. It spans the membrane with seven hydrophobic α-helices and exhibits a loop which activates $G_{\alpha\beta\gamma}$-proteins (*cf.* Ch. 3.2) after binding of chemoattractants.

The second basic process, the formation of gaps within the endothelium cell monolayer, is associated with a dramatic change of the structure of the actin cortex. The most dramatic and immediate effect occurring within a few seconds after stimulation is the formation of actin bundles (stress fibers) within the actin cortex. Similar actin bundle formation is also observed during the stimulation of blood platelets (thrombocytes) where it leads to the formation of tentacle-like membrane protrusions (so-called filipodia) which are filled by actin fibers. Before we discuss this process in detail we describe in the following the tricks by which nature controls the architecture of the actin cortex and show how rapid structural changes may be controlled through phase transitions within heterogeneous gels.

Actin is the most abundant protein in many eucaryotic cells; *e.g.* blood platelets contain about 0.5 mM or 22 mg/ml of actin. Only about 50% of this actin is polymerized forming filaments with lengths varying from a fraction of a μm to a few μm. The filaments are most likely partially crosslinked (*cf.* Podolski & Steck 1990). Under many conditions (*e.g.* endothelial cells in the resting state or *Dictyostelia* cells in the vegetative h_0-state; *cf.* Jungbluth *et al.* 1994) the crosslinked actin network forms a thin shell of about 0.2–0.3 μm thickness (*cf.* Figs. 3b, 4) which is locally anchored to membrane proteins of the bilayer membrane in a still mostly unknown way. In thrombocytes for instance the inner membrane leaflet is supposed to be coupled to a spectrin-actin network similar to erythrocytes (*cf.* Hartwick 1991). In other cases (such as for instance *Dictyostelia* cells) the membrane coupling of actin is accomplished by various families of coupling proteins to cytoplasmatic domains of cell membrane receptors. Examples are

- Talin, α-actinin and vinculin mediating the binding of actin filaments to the inner domain of integrins (β chain); often these actin binding proteins form part of the focal adhesion complexes (*cf.* Fig. 2 in chapter "Adhesion");

- Cadenin coupling actin to inner domains of cadherins; a connection which is particularly important for the lateral stabilization of endothelial cell layers;

- The family of ezrins such as the band IV anchoring protein of red blood cells which couple actin filaments tangentially to membranes (*cf.* Fig. 2 in chapter "Adhesion");

- Dystrophin, a spectrin-like molecule coupling actin to a glycoprotein complexes of muscle cells and the lack of which causes a severe form of muscle dystrophy.

In most cases the outer domains of the membrane receptors couple to macromolecules of the extracellular matrix, thus providing a pathway of signal transmission between the tissue and the cytoplasm of cells. It should be noted that in vitro experiments show that many actin-membrane coupling proteins can mediate direct binding of actin to lipid bilayers in particular in presence of charged lipids. This holds for talin, vinculin and also for the actin crosslinker filamin (*cf.* Tempel *et al.* 1994). Finally, actin membrane coupling may also be mediated by motor proteins such as myosin I and myosin V. For this purpose, these proteins exhibit membrane binding domains which bind strongly to lipid bilayers in the presence of charged lipids such as phosphatidylserine. Interestingly, the inner leaflet of the plasma membrane and the inner leaflet of intracellular compartments contain about 20% of such acidic lipids and the electrostatic actin-membrane coupling may thus play an important role even under physiological conditions.

2 Physics of the actin based cytoskeleton

2.1 *Actin is a living semiflexible polymer*

As illustrated in Figure 5 actin is a living polymer forming double stranded filaments of several 10 μm length which coexist with monomers at a concentration of about 0.1 μM. In appropriate buffer (containing Ca^{++} or Mg^{++} and ATP) a random network is formed. The average distance between the filaments (the mesh size ξ) decays with monomer (G-actin) concentration according to $\xi \sim c_{\mathrm{A}}^{-1/2}$. The actin filaments are typically about 20 μm long and the contour length L is thus large compared to the mesh size. The length of the filaments can be adapted by so called capping molecules which bind to the fast growing end (*cf.* Ch. 3.1).

Actin is a semiflexible macromolecule. This is demonstrated by pronounced conformational fluctuations which can be directly observed by microscopic observation of fluorescence labelled filaments which are for instance embedded in a network of non-labelled filaments. Another technique for the direct visualization of single filament dynamics will be described below (Fig. 9). The flexibility of the filaments is characterized by the persistence length L_{p}. It is defined as the contour length, s, over which the local orientations (characterized by the tangent vector $\vec{t}$) are correlated (*cf.* Käs *et al.* 1996). The correlation of the orientation of the local tangent to the contour position $s = 0$ and s' decays exponentially (*cf.* Doi & Edwards Sect. 8.8)

$$\langle \vec{t}(0)\vec{t}(s) \rangle = \exp\left(-\frac{s}{L_{\mathrm{p}}}\right) \cdot \tag{2.1}$$

Since the manifold of conformations is a consequence of thermally excited bending fluctuations the persistence length can be related to the filament bending elastic modulus, B, by

$$B = k_{\mathrm{B}}T \cdot L_{\mathrm{p}} \tag{2.2}$$

B is defined through the bending energy function

$$\Delta G_{\mathrm{ela}} = \frac{1}{2} B \int \left(\frac{\partial^2 u}{\partial s^2}\right)^2 \mathrm{d}s \tag{2.3}$$

where u is the deflection of the filament at the contour position s and $\partial^2 u/\partial s^2$ is the local curvature. The bending modulus can be measured most directly by Fourier analysis of the bending fluctuations of fluorescent labelled filaments (*cf.* Käs *et al.* 1996) similar to the procedure for the

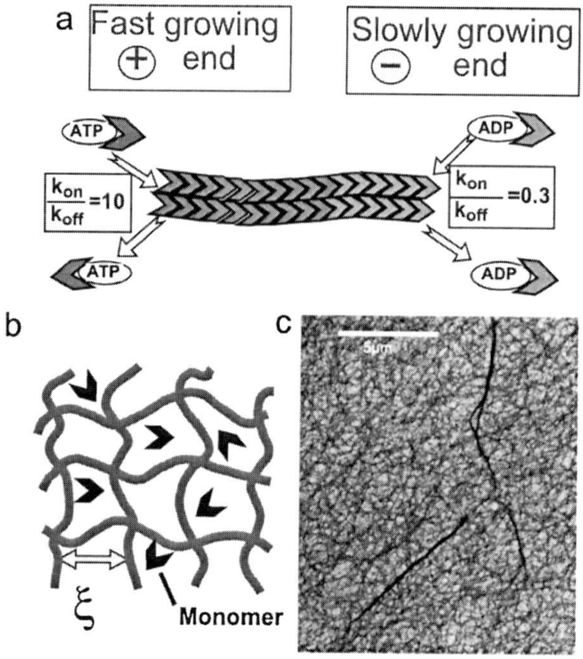

Fig. 5. a) Schematic view of actin as living macromolecule. The filaments are double stranded and exist in a dynamic equilibrium with monomers. They exhibit a fast growing end (also called barbed or plus end) where the rate of monomer association is ten fold higher than the dissociation rate ($k_{\mathrm{on}}/k_{\mathrm{off}} \sim 10$) while at the opposite end (pointed or minus end) the rate of monomer dissociation is higher ($k_{\mathrm{on}}/k_{\mathrm{off}} \sim 0.3$). **b)** At stationary equilibrium the growth rate at the barbed end equals the dissociation rate at the pointed end (a situation which is called treadmilling), resulting in a random network of filaments coexisting with monomeric actin (G-actin, where "G" stands for "globular"). There exists a threshold concentration for polymerization which is of the order of 0.1 μM. **c)** Electron micrograph (negative staining) of network of F-actin (of meshsize $\sim$1 μm). In addition the crosslinker filamin has been added at a molar ratio actin to filamin of 100:1 which leads to the formation of bundles in the network (*cf.* Sect. 3.1).

measurement of the membrane bending modulus. The momentary deflection is

$$u(s,t) = \sum_{\mathrm{q}} u_{\mathrm{q}}(t) \exp\{iqs\} \cdot \tag{2.4}$$

The total bending energy of a filament of length L is $\Delta G_{\mathrm{ela}} = \frac{1}{2} BL \sum_{\mathrm{q}} q^4 u_{\mathrm{q}}^2$

and the equipartition theorem (stating that the average energy per mode is $\frac{1}{2}k_{\rm B}T$) provides a relationship between $u_{\rm q}$ and B:

$$Lu_{\rm q}^2 = \frac{k_{\rm B}T}{Bq^4} \cdot \tag{2.5}$$

Measurements of $u_{\rm q}$ as function of the wavelength of the mode $\Lambda = \pi/q$ allow to measure B rather precisely (*cf.* Käs *et al.* 1996). The bending stiffness of an actin filament is $B \approx 6 \times 10^{-26}$ J·m and thus the persistence length $L_{\rm p} \sim 15$ μm. The persistence length can be modified by a factor of two through binding of actin regulation proteins such as tropomyosin (*cf.* Götter *et al.* 1996). An interesting (still unsolved) question is wether filaments may also be rendered more flexible by local defects (*cf.* Piekenbrock & Sackmann 1992). In this context it is interesting to note that the parallel coupling of the single strands is weak and that filaments unwind locally resulting in pronounced local fluctuations of the torsional angle (*cf.* Bremer *et al.* 1991). Such fluctuations are expected to cause restricted torsional motions of the filaments.

Further evidence for the role of defects is provided by the finding that binding of the sequestering molecule cofilin to actin filaments results in a shortening of the twist of the actin filament by 75%. The effect is mediated by the cooperative binding of cofilin between the two strands of a filament. Local binding of cofilins is expected to change the torsional angle of the filament.

An important quantity characterizing the flexibility of semiflexible filaments is the mean square amplitude of the fluctuations $\langle u^2 \rangle$, also called roughness. Integration over all modes defined in equation (2.5) yields the following scaling relation between the roughness and the filament length

$$\langle u^2 \rangle = \frac{k_{\rm B}T}{B} L^3. \tag{2.6}$$

The roughness has important consequences for the behaviour of filaments in networks where the excitation of the conformational degrees of freedom is impeded. Following the strategy of classical polymer physics (*cf.* Doi & Edwards 1986) the constraints imposed on a single (*e.g.* fluorescence labelled) test chain by the surrounding network can be accounted for by assuming that the filament is surrounded by a tube of constant diameter ξ which is about equal to the mesh size. As illustrated in Figure 6 the filament exhibits restricted wiggling motions within the tube. It is obvious that the confinement by the tube truncates the long wavelength excitations. According to equation (2.6) the longest wavelength unaffected is of the order

$$\Lambda_{\rm e} \approx (L_{\rm p} \cdot \xi^2)^{1/3}. \tag{2.7}$$

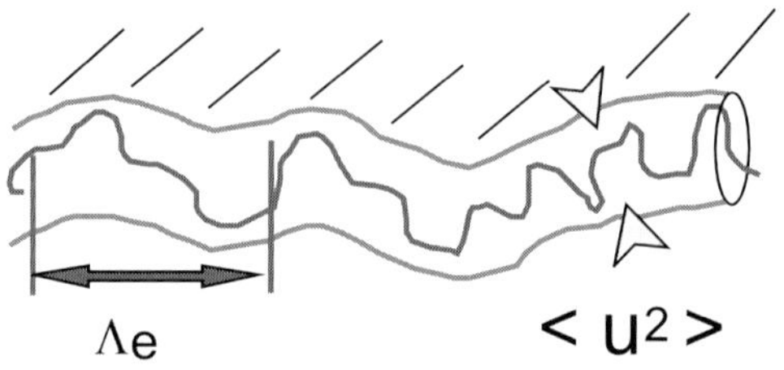

Fig. 6. Restricted local motion of single test filament embedded in macromolecular network. The constraints imposed by the network are accounted for by a tube of constant diameter. The network thus forms an effective medium for the test filament determining $\langle u^2 \rangle$. Note that due to the constraint a maximum bending excitation wavelength Λ_e can be defined which characterizes the bending mode with the smallest wave vector not affected by the wall of the tube.

This follows by noting that $\langle u^2 \rangle \approx (\xi^2)$ and equation (2.2) and defines a new contour length scale: $L_e \approx \Lambda_e$, called "deflection length" or (Odijk) "entanglement length" which determines the crossover between two types of behaviour of the filaments (*cf.* Odijk 1983). Over contour lengths $L < L_e$ the bending fluctuations are determined by the filament bending energy (as free filaments) while for $L > L_e$ (or wave vectors $q \gg L_e^{-1}$) the bending fluctuations are determined by the fluctuating forces generated by the tube walls. The entanglement length plays a crucial role for the frequency dependence of the viscoelastic impedance of actin networks (*cf.* Isambert & Maggs 1996; Hinner *et al.* 1998; Morse 1998).

A special but important feature of semiflexible filaments is that (in contrast to flexible filaments) their elastic behaviour depends on the length, L, of the filament (or of a chain segment considered).

If $L \gg L_p$ the chain behaves as an entropic spring. However, the tension associated with stretching of the filament is not an universal quantity as for flexible polymers but depends sensitively on the bending stiffness. The tension τ associated with an extension δL (in the direction parallel to the long filament axis) is

$$\tau \sim \frac{B^2}{k_B T L^4} \delta L \tag{2.8}$$

and thus depends strongly on the bending stiffness and the filament length. This relationship follows from the following consideration. The elastic energy of a filament under tension is (similar to the situation of membranes forming semiflexible shells)

$$\Delta G_{\mathrm{ela}} = \frac{1}{2} \int \left\{ B \left(\frac{\partial^2 u}{\partial s^2} \right)^2 + \tau \left(\frac{\partial u}{\partial s} \right)^2 \right\} \mathrm{d}s. \tag{2.9}$$

Similar to (1.2), the mean square amplitude of the thermally excited mode of wavelength q is

$$L \left\langle u_{\mathrm{q}}^2 \right\rangle = \frac{k_{\mathrm{B}}T}{Bq^4 + \tau q^2} \cdot \tag{2.10}$$

(Note that τ has dimension $[\tau] = N$).

We consider now the change in length $\delta L = (L - L_\circ)$ and remember that we can express it as $L - L_\circ \simeq 1/2 \int (\nabla u)^2 \, \mathrm{d}s$. The total mean square amplitude is obtained by replacing the sum over all modes $\langle u \rangle \propto \sum u_{\mathrm{q}}$ by an integral $L_\circ/(\pi) \int u_{\mathrm{q}} \, \mathrm{d}q$ (with the limits π/L and π/a, where a is the filament diameter and $L_\circ$ the contour length). It follows

$$\delta L \approx L_\circ \frac{k_{\mathrm{B}}T}{\sqrt{B\tau}} \arctan \left(\frac{L}{\pi} \sqrt{\frac{\tau}{B}} \right) \cdot \tag{2.11}$$

For small tensions (which are relevant for the present lecture) one can expand δL in terms of τ/B and obtains equation (2.8). It is very important to realize that the extension-*versus*-force relationship depends critically on the orientation of the force with respect to the average filament axis (*cf.* Kroy & Frey 1996).

For $L \leq L_{\mathrm{p}}$ the deformation is no longer temperature dependent and the deformation is highly anisotropic (*cf.* Frey and Wilhelm for a discussion of the tangential elasticity in terms of the Euler model of rigid rods). For forces perpendicular to the filament axis the deflection $U_{\max}$ is proportional to B^{-1}: $U_{\max} \propto fL^3/B$ (*cf.* Landau & Lifshitz, Sect. 20).

2.2 Actin network as viscoelastic body

A physical property of actin networks of uttermost importance is its viscoelasticity. It controls the dynamics of many cellular processes such as cell locomotion, centripetal endothelial cell contractions (*cf.* Sect. 5) or cellular shape changes (*e.g.* under shear flow in blood vessel). From the point of view of biological physics even more important is that measurements of viscoelastic properties yield insights into the structural organization of biomacromolecular networks or into the correlation between macroscopic viscoelasticity and the conformational dynamics of single molecules as will be discussed below. Microrheological techniques provide a powerful tool to study the control of cellular processes by viscoelastic properties of the

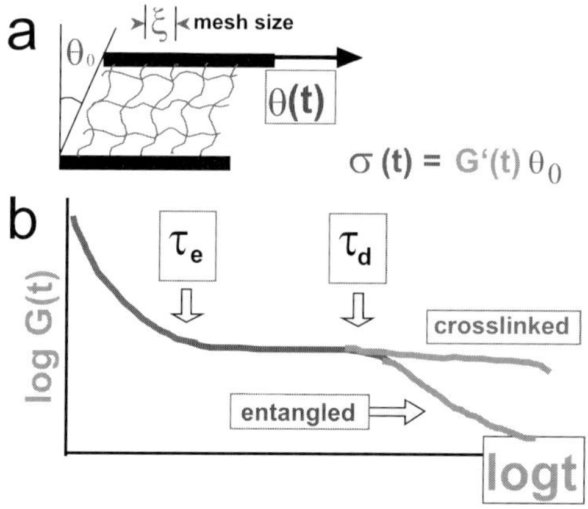

Fig. 7. a) The definition of viscoelastic behaviour of body subjected to sudden shear strain by angle γ. **b)** Time dependence of elastic shear modulus, $G(t)$, of entangled actin network. Note the composition of $G(t)$ of three regimes: an initial **relaxation regime** at $0 \leq t \leq \tau_{\mathrm{e}}$, a **plateau regime** at $\tau_{\mathrm{e}} < t < \tau_{\mathrm{d}}$ and a **terminal regime** at $t \geq \tau_{\mathrm{d}}$.

cell membrane and cytoskeleton through simultaneous systematic studies of model systems and biomaterials (*cf.* Bausch *et al.* 2000; Palmer *et al.* 1999).

To introduce the concept of viscoelasticity we make the following Gedankenexperiment (*cf.* Fig. 7). We confine an actin network between two parallel plates and deflect the top plate tangentially by a distance Δx corresponding to a sudden shearing of the network by an angle Θ_{c}. Observing the effect of the shear onto a single test filament, we recognize that the test filament within the networks would feel a time dependent stress. In the case of small deformations the elastic response is linear and Hooks law holds for this time dependent elastic stress

$$\sigma(t) = G(t)\,\Theta_{\mathrm{c}} \tag{2.12}$$

$G(t)$ is a time dependent elastic modulus called the relaxation modulus. For entangled actin networks we find three time regimes. At short times the networks behave as an elastic body but the elastic constant decreases rapidly with time. After a relaxation time τ_{e} (typically $\sim 10^{-1}$ s), $G(t)$ remains nearly constant over a time regime $\tau_{\mathrm{e}} \leq t \leq \tau_{\mathrm{d}}$ (called plateau regime). At long times $t > \tau_\mathrm{d}$ (called the terminal regime) the entangled network starts to flow like a liquid[3].

Viscoelasticity is determined by two sets of viscoelastic parameters. There are two strategies to measure viscoelastic moduli: the creep response experiment and the oscillatory experiment (*cf.* Doi & Edwards 1983; Tempel *et al.* 1996). In the first case a sudden strain (or stress pulse) is applied and the stress (or strain) response is measured as a function of time. The response is characterized by a time dependent frictional coefficient $\zeta(t)$ (measured in terms of Pa·s) and an elastic modulus $G(t)$ also called relaxation modulus (or an elastic compliance $J(t)$ also called response compliance). $G(t)$ and $\zeta(t)$ are interrelated through the response time τ

$$\tau = \frac{\zeta(t)}{G(t)} \cdot \tag{2.14}$$

In the creep experiments one of the moduli ($G(t)$ or $J(t)$) and τ are measured. However, in general the relaxation behaviour is determined by a whole spectrum of relaxation times. We will ignore this aspect in the present review and refer the reader to the monograph by Ferry (1980) for a rigorous discussion of this point.

In the oscillatory experiment an oscillatory stress (or strain) is applied and the response is analyzed according to the classical method known from the treatment of damped oscillators. The viscoelasticity is characterized by a complex impedance according to

$$\sigma(\omega) = G^*(\omega) \cdot \gamma(\omega) \tag{2.15}$$

with $\gamma(\omega) = \gamma_\circ \exp\{i\omega t\}$ and

$$G^*(\omega) = G'(\omega) + iG''(\omega). \tag{2.16}$$

The real part, $G'(\omega)$, (called the "storage modulus") determines the response in phase with the excitation and is the frequency dependent shear elastic modulus. The imaginary part $G''(\omega)$ represents the out of phase component and is called "loss modulus" since it characterizes the energy dissipation during one cycle (*cf.* Eq. (2.19)). $G''(\omega)$ is related to the shear

[3]For more general time dependent shear strains one has to consider that the momentaneous stress depends on the pre-history of deformation and equation (2.12) has to be replaced by

$$\sigma(t) = \int G(t - t') \frac{\mathrm{d}\alpha(t')}{\mathrm{d}t'} \mathrm{d}t' \tag{2.13}$$

where $\mathrm{d}\alpha/\mathrm{d}t$ is the shear rate.

viscosity by

$$G''(\omega) = \omega \cdot \eta(\omega). \tag{2.17}$$

Various methods for the high precision measurements of $G^*(\omega)$ of soft actin networks are available. These include (i) torsional rheometry enabling measurements of the elastic impedance of macroscopic networks (*cf.* Müller *et al.*; Janmey 1995), (ii) magnetic bead microrheometry (*cf.* Ziemann *et al.* 1994; Amblard *et al.* 1996) shown in Figure 8 and (iii) a force free technique suitable for soft materials is based on the Fourier analysis of the random motion of colloidal probes embedded in the networks (*cf.* Crocker *et al.* 2000).

As mentionend, the moduli have a simple physical meaning: G' is a measure for the energy stored during one cycle of the deformation which is

$$W_{\rm stored} = \frac{1}{4} G'(\omega) \gamma_0^2 \tag{2.18}$$

for a half cycle of the oscillatory excitation. G'' is a measure for the energy dissipated during a half cycle

$$W_{\rm dissip} = \pi G''(\omega) \gamma_0^2. \tag{2.19}$$

It is often helpful to measure the phase shift between both response functions which is defined as

$$\tan\phi = \frac{G''(\omega)}{G'(\omega)} \propto \frac{W_{\rm dissip}}{W_{\rm stored}} \cdot \tag{2.20}$$

An impedance spectrum of the entangled network is shown in Figure 8. As expected $G'(\omega)$ is the mirror image of the relaxation modulus $G(t)$ exhibiting again the three characteristic regimes. It is noteworthy that G'' has a minimum in the plateau regime of $G'(\omega)$ which can be used to define the center of the plateau.

It is also important and helpful to realize that the loss modulus $G''(\omega)$ exhibits maxima in the frequency regimes associated with relaxation processes. Examples are the stress relaxation by selfdiffusion of the chains within the confinement tube leading to the low frequency band of $G''(\omega)$ or the relaxation of the filament bending modes leading to the sharp rise of $G''(\omega)$ at $\omega > 1$ rad/s in Figure 8c.

The viscoelasticity of the network is characterized by two elastic moduli corresponding to the two frequency regimes. The major parameter characterizing the elasticity at low frequencies ($\sim 10^{-1} - 1$ Hz) is the value of $G'(\omega)$ in the plateau regime II (also called rubber plateau regime since it is also a characteristic feature of the viscoelastic behaviour of rubber).

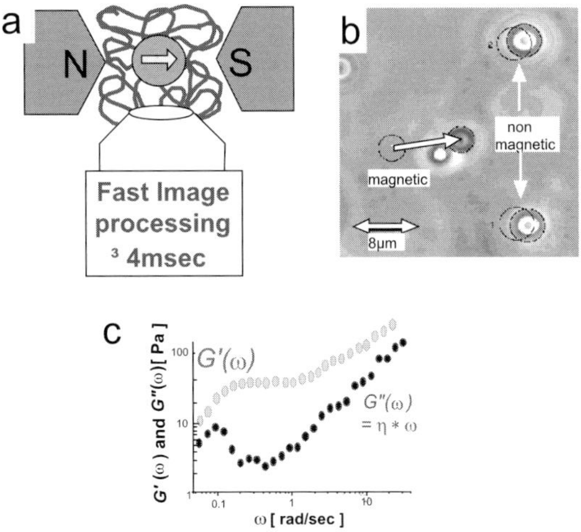

Fig. 8. a) Magnetic bead microrheometry: a powerful tool to study local viscoelastic parameters of soft macromolecular networks. Magnetic beads (called magnetic tweezers) are embedded in the network **a)** and the bead deflection induced by an oscillatory magnetic field or a force pulse is analyzed by fast image processing. By analyzing the bead deflection with ultramicroscopy deflection amplitudes of 5 nm may be observed and the time resolution is 10^{-2} s. Forces from femto-Newton to nano-Newton may be applied. The maximum stress achieved with iron oxide beads of 4 μm diameter is about 500 Pa. By embedding also non-magnetic beads **b)** the strain field induced by local forces can be determined simultaneously. **c)** Viscoelastic moduli $G'(\omega)$ and $G''(\omega)$ for entangled actin network of mesh size $\xi \sim 1$ μm ($c_\mathrm{A} = 300$ μg/ml). Note that $G''(\omega)$ is shifted in the vertical direction by an order of magnitude to facilitate better distinction between and that G' and G'' cross at $\omega \approx 2$ rad/s.

The plateau shear modulus $G'_\circ$ is related to the mesh size ξ and the chain length through the following power law

$$G'_\circ \sim k_\mathrm{B}TL_\mathrm{p}^{-1/5}\xi^{-14/5} \sim k_\mathrm{B}TL_\mathrm{p}^{-1/5}c_\mathrm{A}^{7/5} \tag{2.21}$$

where we made use of the relationship $\xi \propto c_\mathrm{A}^{-1/2}$ (*cf.* Hinner *et al.* 1998).

The high frequency regime ($\nu > 10$ Hz) is determined by the tension of the filaments. The theoretical prediction for this relaxation modulus in this regime is (*cf.* Morse 2001; Gittes & MacKintosh 1998)

$$G'(t) \sim \rho\frac{k_\mathrm{B}TL}{L_\mathrm{p}}\left\{\frac{\zeta L_\mathrm{p}^3}{k_\mathrm{B}T}t\right\}^{3/4} \tag{2.22}$$

where ρ is the polymer density (number of filaments of length L per unit area) and ζ is the frictional coefficient of the filament in the tube. A measure for the cross-over frequency between the two regimes is the reciprocal relaxation time of the mode of wavelength Λ_e

$$\tau_e \sim \frac{\zeta \Lambda_e^4}{B} \cdot \tag{2.23}$$

The power laws have been verified experimentally (*cf.* Isambert & Maggs 1996; Hinner *et al.* 1998) for a large range of concentrations. This scaling law is very helpful (i) to compare the data measured for different concentrations, (ii) to check the purity of actin preparations with respect to crosslinkers or (iii) to estimate the elastic modulus of the actin cortex of cells if the mesh size is known.

2.3 Correlation between macroscopic viscoelasticity and molecular motional processes

The three regimes can be related to distinct molecular processes by studying the conformational dynamics and diffusion of simple test filaments labelled with fluorescent markers (*cf.* Käs *et al.* 1996; Amblard *et al.* 1996) or colloidal probes (*cf.* Dichtl & Sackmann 1999; *cf.* also Fig. 9). In the latter case, the local motion of the test filament is analyzed by measuring the mean square displacements of the colloidal bead as a function of time in the direction parallel $\langle |U_{\|}(t) - U_{\|}(0)|^2 \rangle$ and perpendicular $\langle |U_{\perp}(t) - U_{\perp}(0)|^2 \rangle$ to the long axis of the filament (which defines also the local axis of the tube). By using confocal ultramicroscopy the position of the bead (in the image plane) can be determined with an accuracy of ± 5 nm.

The mean square displacements exhibit a short time regime $t < 0.5$ s where the local segments move isotropically and exhibit a time dependence

$$\langle |U_{\|}(t) - U_{\|}(0)|^2 \rangle = \langle |U_{\perp}(t) - U_{\perp}(0)|^2 \rangle \sim t^{0.75} \tag{2.24}$$

and a long time behaviour where $\langle |U_{\perp}(t) - U_{\perp}(0)|^2 \rangle$ saturates $(\langle |U_{\perp}(t) - U_{\perp}(0)|^2 \rangle = U_{\perp 0}^2)$ while the parallel component increases linearly with time $\langle |U_{\|}(t) - U_{\|}(0)|^2 \rangle \sim t$. The short time motional behaviour for both directions is determined by the local wiggling motion of a semiflexible filament which is confined within a tube and therefore subjected to a line tension (*cf.* Gittes & MacKintosh 1998; Amblard *et al.* 1996).

The different long time behaviour for the two directions yields the following important information:

1. From the distribution $P\left(U_{\perp 0}^2\right)$ of the saturation values $U_{\perp 0}$ of the transverse bead motion one can determine the local interaction potential $V_{\perp}\left(U_{\perp}\right)$ characterizing the confinement of the filament by

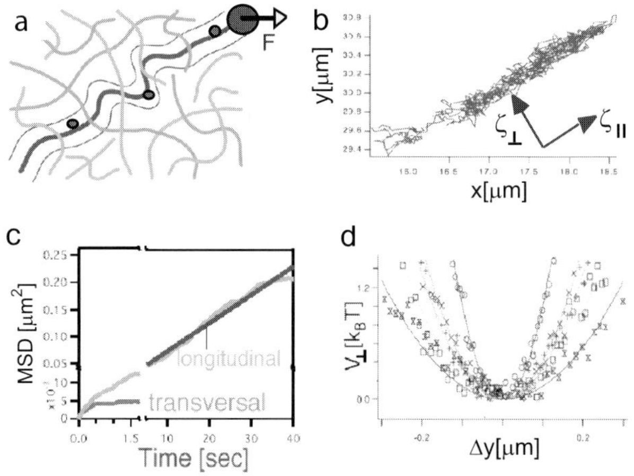

Fig. 9. a) Labelling of test filaments with colloidal beads and a magnetic force probe (=tweezer) at one chain end. The bead can be transiently pulled through the network by application of force pulses and by analyzing the bead deflection. If the bead diameter is small compared to the mesh size this experiment corresponds to microrheometry with linear force probes (*cf.* M. Dichtl 2001). **b)** Random walk of colloidal probe which reflects local motion of filament. By decoupling into component parallel and perpendicular to filament axis the mean square displacements can be measured yielding the reptation (self-) diffusion coefficient and the local potential of interaction between the filament and the surrounding network. **c)** Time dependence of mean square displacement of colloidal bead coupled to filament in direction parallel ($MSD_{\|}$) and perpendicular ($MSD_{\perp}$) to the tube direction. Note that the $MSD_{\perp}$ saturates after a time $\tau_s \sim 0.5$ s while $MSD_{\|}$ increases linearly with time after τ_s. The slope yields the selfdiffusion coefficient of the filaments along the tube direction (often called reptation). **d)** The effective interaction potential determined from the distribution of the perpendicular motion $P\left(U_{\perp 0}^2\right)$.

the network. This is achieved by application of Boltzmann's law: $V_{\perp}\left(U_{\perp}\right) \sim k_{\mathrm{B}}T \ln P\left(U_{\perp 0}^2\right)$ (*cf.* Dichtl & Sackmann 1999). A remarkable finding is that the width of the potential and thus the local tube diameter varies by up to a factor of two along a single filament;

2. The longitudinal mean square displacement yields the selfdiffusion coefficient (the reptation diffusivity $D_{\|}$) of the chain along the tube axis according to

$$D_{\|} \propto \frac{\left\langle |U_{\|}(t) - U_{\|}(0)|^2 \right\rangle}{t} \,. \tag{2.25}$$

This reptation diffusion coefficient determines the terminal relaxation time τ_d of the viscoelastic impedance spectra since τ_d is determined by the time required by the filaments to escape the confinement tube by selfdiffusion. For a chain of $L \sim 20$ μm we determined $D_{\|} \approx 10^{-16}$ m^2/s which corresponds to a value of $\tau_d = 10^{-5}$ s, in good agreement with experimental data (*cf.* Dichtl & Sackmann 1999). The previous microfluorescence analysis yielded larger values of $D_{\|}$ ($D_{\|} \sim 10^{-14}$ m^2/s) (*cf.* Käs *et al.* 1996).

3 Heterogeneous actin gels in cells and biological function

3.1 Manipulation of actin gels

In order to manipulate the structural organization of actin networks for the multiple functions of the composite biological membranes nature uses manifold of actin regulating (or actin binding) proteins. These include:

- Severing proteins (examples: gelsolin, severin) which can cleave filaments and bind to the fast growing end, thus controlling the filament length (which varies from 0.1 to 20 μm in *Dictyostelia* cells; *cf.* Podolski & Steck 1990);

- Sequestering proteins (example: profilin) which bind actin monomers and can thus control the concentration of polymerizable actin monomers and thus the mesh size ξ;

- A manifold of crosslinking proteins which link actin filaments noncovalently and which can induce the formation of randomly crosslinked gels (α-actinin), tree-like branched gels or bundles;

- Actin membrane couplers (examples: vinculin, talin, ankyrin) which can anchor actin filaments to various membrane proteins (*cf.* Fig. 1).

Another key feature of actin manipulation is that the function of the manipulating proteins in cells is regulated by second messenger molecules as illustrated in Figure 10 for the severing protein gelsolin. This second messenger is activated by Ca^{++} and mediates the formation of soft or sol-like networks. The protein is inhibited by the phospholipid phosphatidylinositol-diphosphate (PIP_2) which thus promotes gel formation. Similarly, PIP_2 inhibits the activity of the sequestering protein profilin and thus decreases the mesh size while Ca^{++} is supposed to activate the sequestering proteins thus diluting the network (*cf.* Janmey 1995).

Through the second messengers the structure of the actin cortex is connected to the manifold of cell signalling processes. As will be shown below (for the case of thrombin) this provides a powerful and versatile tool to respond to external signals by structural reorganization. In this context it is

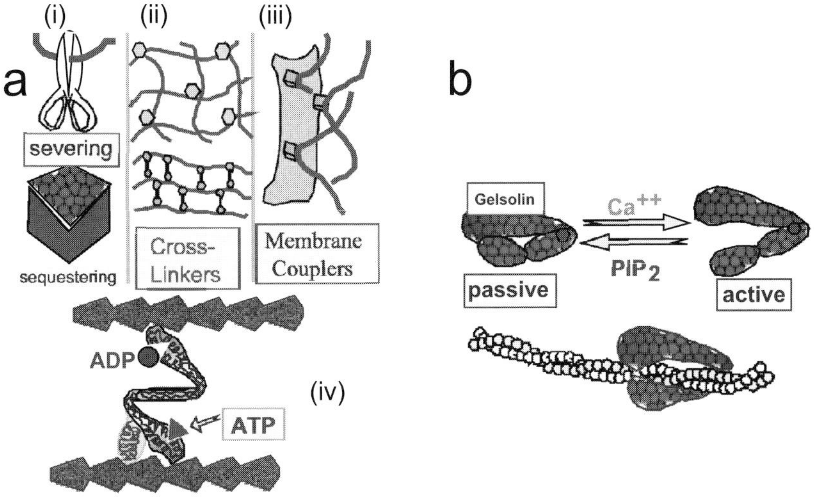

Fig. 10. a) Summary of different classes of actin manipulating proteins enabling the control (i) of the filament length through severing proteins; (ii) of the mesh size through sequestering proteins controlling the fraction of polymerizable G-actin; (iii) of the membrane coupling of the actin cortex and (iv) of the connectivity of the actin filaments through various cross-linkers including myosins shown at the bottom. Note that cross linkers can generate gels of distinct topology such as random networks, tree-like branched networks and bundles. **b)** Control of activity of actin manipulation proteins by second messengers, for example of the severing protein gelsolin. The protein is activated by Ca^{++} which is thus a sol-former and inhibited by binding of phosphatidyl-inositol-diphosphate (PIP_2) which is thus a gel-former (*cf.* Janmey 1995).

highly important to note that actin filaments bind messenger-RNA and can influence its transport and translation (*cf.* G. Bassell & R.H. Singer 1997), which may provide a link between actin structure and genetic expression.

A central property of the cellular cytoskeleton is its ability of rapid reorganisation as demonstrated by the fast generation of actin bundles. One important example is the formation of actin stress fibers during the activation of blood platelets. In this case actin bundling leads to the formation of tentacle-like protrusions which mediates the rapid closing of wounds by self-assembly of the spider-like cells over openings of blood vessels thus impeding bleeding. A second example is the formation of gaps in endothelial cell monolayers by centripetal contraction discussed in Section 3.2.

How can such transition of actin networks be induced in the fraction of seconds or faster? One intriguing possibility is through phase transitions

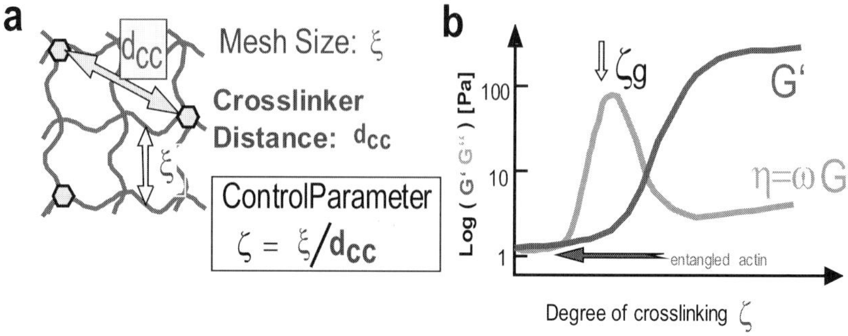

Fig. 11. a) The state of crosslinked actin networks is controlled by two length scales: the mesh size ξ and the average distance between the cross-linkers d_{cc}. The ratio $\zeta = \xi/d_{\mathrm{cc}}$ is an essential parameter characterizing the degree of cross-linking. **b)** Variation of shear modulus $G'(\zeta)$ and loss modulus $G''(\zeta)$ of actin-α-actinin networks with fraction ρ_{act} of active crosslinkers. Note first, that according to equation (3.2) ρ_{act} and thus ζ increases monotonically with decreasing temperature but that the absolute value of ρ_{act} can not be determined and the quantitative relationship between ρ_{act} and ζ is not know. Note secondly, that the storage modulus $G'(\zeta)$ increases abruptly by two orders of magnitude above a threshold-temperature corresponding roughly to the value of $\zeta_{\mathrm{g}} \approx 1$ (called the gel point). The loss-modulus (or the loss angle $\tan\phi = G''(\zeta)/G'(\zeta)$) diverges and exhibits a maximum at the critical value ζ_{g} while $G'(\zeta)$ starts to increase sharply at $\zeta \geq \zeta_{\mathrm{g}}$. This behaviour is typical for a percolation transition into a gel-like state.

within actin networks as described below for actin-α-actinin networks. It is obvious that the state of cross linked actin networks is determined by two length scales: the mesh size ξ and the average distance between the crosslinkers d_{cc} (*cf.* Fig. 11a). The ration $\zeta = \xi/d_{\mathrm{cc}}$ can thus be considered as a control parameter which determines the state of the actin network. For $\zeta > 1$ the distance d_{cc} is larger then the average distance between points of entanglement within the network and crosslinking occurs predominantly at these sites without changing the structure. For $\zeta > 1$ crosslinking induces local contractions of the gel as shown in Figure 12a (thin line $\zeta \sim 1$).

Some crosslinkers bind weakly to actin resulting in a dynamic association-dissociation equilibrium under physiological conditions. One example is myosin II which forms an active and dynamic crosslinker in the presence of ATP (but acts as rigid connector in the absence of ATP). Another prominent example is α-actinin which is also supposed to be involved in the coupling of actin to membranes, the formation of focal complexes and the generation of filipodia and lamellipodia (*cf.* Lodish *et al.* Sect. 22).

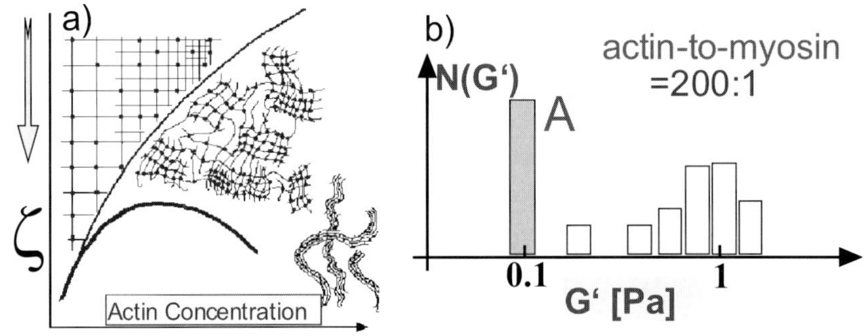

Fig. 12. a) Equivalent phase diagram of actin gels according to Tempel *et al.* (1996). The state of the network is determined by two parameters, the fraction of active cross-linkers $\rho_{\rm act}$ (or ζ) and the mesh size ξ; ξ is related to the actin concentration $c_{\rm A}$ by $\xi \propto c_{\rm A}^{-1/2}$. The thin line defines the phase boundary of the percolation transition and the thick line defines the coexistence of bundles (or random networks) with a dilute solution. Note that the bundle state may also coexist with the heterogeneous network. Electron micrographs of the heterogeneous states can be found in Tempel *et al.* (1996). **b)** Distribution of plateau values of storage modulus of an entangled actin network (bar A) and of a network crosslinked by myosin in the rigor state (ATP depletion). The network exhibits the same microgel state as actin/α-actinin networks.

The association-dissociation equilibrium of the actin/α-actinin system

$$A + \alpha A \overset{K_0}{\rightleftharpoons} C \tag{3.1}$$

is nearly balanced under physiology-like conditions. The association dissociation equilibrium can be shifted in either direction by small temperature variations since the binding energy of the α-actinin-actin complexes is of the order of $k_{\rm B}T$ at 300 K (*cf.* Tempel *et al.* 1996). According to the chemical equilibrium theory the fraction of active crosslinkers is related to the actin concentration by

$$\rho_{\rm c} = \frac{[A]\, K_0 \exp\{\Delta H/k_{\rm B}T\}}{1 + [A]\, K_0 \exp\{\Delta H/k_{\rm B}T\}} \tag{3.2}$$

where K_0 is the prefactor of Van't Hoff's law of the temperature dependence of the equilibrium constant $K = K_0 \exp\{\Delta H/k_{\rm B}T\}$. Therefore the degree of crosslinking (expressed by ζ) can be changed reversibly by variation of the temperature around 20 °C.

As shown in Figure 11b the viscoelastic moduli change in a characteristic way with ζ (or T). The loss modulus $G''(\zeta)$ (expressed in terms of the loss angle $\tan\vartheta = G''/G'$) exhibits a broad band whereas $G'(\zeta)$ starts to increase sharply by to two orders of magnitude at the distinct value of $\zeta = \zeta_{\rm g} \approx 1$ (or $T = T_{\rm g}$) where the broad band of $G''(\zeta)$ exhibits a maximum. This behaviour is distinct typical for percolation transitions (*cf.* de Gennes 1979). This analogy is further supported by the finding that the decaying branch of $G''(\zeta)$ and the divergence of $G'(\zeta)$ exhibit power laws

$$G'(\zeta) \sim (\zeta - \zeta_{\rm g})^{\rm t}, G''(\zeta) \sim (\zeta_{\rm g} - \zeta)^{-{\rm s}}, \tag{3.3}$$

and the exponents agree roughly with theoretical predictions (*cf.* Tempel *et al.* 1996).

The viscoelastic and structural studies by electron microscopy suggest that the behaviour of the actin-α-actinin network can be described by the phase diagram of Figure 12a. If the average distance $d_{\rm cc}$ between cross-linkers is considerably larger then the mesh size a homogeneous gel is formed and the shear modulus is only slightly larger than the plateau modulus of the merely entangled actin network. At $\zeta \geq 1$ a heterogeneous gel starts to form. It consists of domains of tightly crosslinked filaments (with mesh size $\xi' < \xi$) which are interconnected by regions of low polymer density. At further increasing ζ well defined bundles of actin are formed which can coexist with the randomly cross-linked state. Further decreasing ζ the bundles may form an interconnected network which can have a smaller viscoelastic impedance than the heterogeneous gel.

Magnetic bead microrheometry studies of cells show that the local viscoelastic moduli of the cell cytoskeleton vary by more than an order of magnitude (*cf.* Bausch *et al.* 1999) suggesting that the intracellular scaffolds form highly heterogeneous gels such as the coexistence of randomly organized networks and bundles. The heterogel state exhibits unique features. It combines high mechanical stability with the existence of soft voids or pathways for the embeddement of intracellular compartments or the local transport by active forces in the 10 pN range (*cf.* Sect. 4). Another unique property of the heterogel state is that the local elasticity can be changed drastically by the formation or cleavage of a few interconnections between dense clusters or bundles – enabling effective local switching between gel- and sol-like states.

The state of intracellular actin networks can also be controlled by myosin which can act either as highly dynamic linker or as stable crosslinker depending on the presence or absence of ATP, respectively. Evidence for this has been provided by studies of mutants of *Dictyostelia* cells lacking myosin. It has been shown that the effective cytoplasmic viscosity $G''(\omega)$ at low frequencies is much higher in the absence of myosin II than in the presence of this motor protein.

In vitro studies show that in the absence of ATP myosin II behaves similar to α-actinin and forms a heterogel consisting of starlike assemblies which are embedded within a randomly organized network. In contrast, a homogeneous network is formed in the presence of ATP. If ATP is depleted gradually in actomyosin-networks one can observe a gradual transition from a soft network to a heterogeneous gel (M. Keller and E. Sackmann, unpublished data).

A characteristic feature of the heterogeneous gel state (also called microgel state) is that it combines high mechanical stability with local softness. Measurements with magnetic tweezers show that the viscoelastic moduli vary by an order of magnitude between the dense clusters and the soft voids (*cf.* Fig. 12b). Strikingly similar behaviour has been observed for cytoplasms of cells such as macrophages or *Dictyostelia* cells (*cf.* Sect. 6). Thus the shear modulus in macrophages varies between 10 Pa and 600 Pa and the viscosity between 10 Pa·s and 500 Pa·s (*cf.* Bausch *et al.* 1999) strongly suggesting that the cytoplasm exhibits a similar heterogel state as the actin/α-actinin system although all three subsystems of the cytoskeleton are involved. The heterogeneity may be even more pronounced due to the composition of the cytoskeleton from the three subsystems actin, intermediate filaments and microtubuli.

In summary, the heterogel state appears to be ideally suited for the design of cells. It confers upon them high global mechanical stability while the large soft voids between densely crosslinked clusters enable the embeddement of intracellular compartments. When the hard regions form a percolated meshwork the soft regions may form a continuous networks of soft channels enabling the transport of the compartments by weak forces in the ten piconewton range generated typically by Kinesin or myosin motors. The local elasticity may be varied drastically by breakage or formation of a few bundles mediated by activation of severing or crosslinking proteins through second messengers. As will be discussed below (Sect. 6) the long range transport of compartments by relatively weak forces is further made possible by the viscoplastic effect.

3.2 Control of organization and function of actin cortex by cell signalling

The manipulation of the structure and viscoelasticity of actin networks by the actin regulating proteins and the control of the activity of these molecules by second messengers provides the basis for the regulation of the function of the actin cortex by extrinsic signals (*e.g.* hormones, growth factors, inflammational agents) through intracellular signalling (*cf.* Fig. 13b). A simplified scheme of the signal pathways induced by thrombin which mediates the activation of myosin II by phosphorylation of chains of myosin II. This myosin light chain phosphorylation is activated by two major processes. A fast signalling pathway is based on the intracellular increase of the Ca^{++}-level by opening of Ca^{++}-storage vesicles through

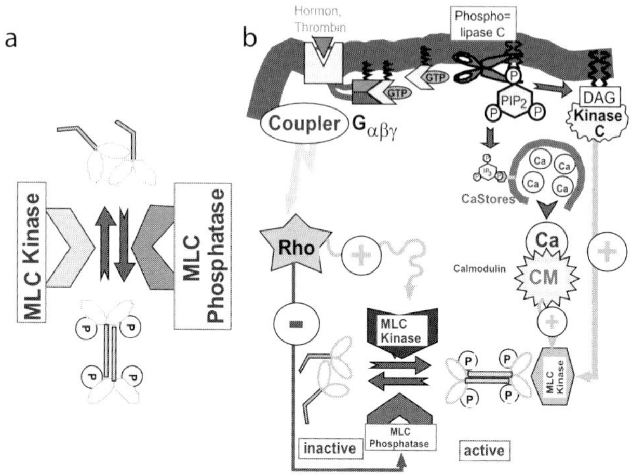

Fig. 13. a) Model of activation of myosin II by phosphorylation of its light chain (MLC) through MLC-kinase. This enzyme activates the myosin head group (S1). The kinase works in tandem with the light chain phosphorylase which dephosphorylates and thus deactivates the MLC. The activation may be accompanied by formation of microbundles (*cf.* Pasternak *et al.* 1989). **b)** Simplified scheme of regulation of myosin activity by various cell signalling pathways induced by the inflammational agent thrombin. There are two major pathways. A fast process (acting in the second time scale) is mediated by Ca^{++} which activates the MLC-kinase through calmodulin. The increase of the Ca^{++}-level is initiated by activation of G-protein switches of the $G_{\alpha\beta\gamma}$-family. The $G_{\alpha\beta\gamma}$-protein binds to the inner loop of the receptor (after thrombin binding) resulting in the exchange of GDP for GTP. The α-domain is thus activated and diffuses in the membrane to the phospholipase C (symbolized as scissors) which is activated and cleaves the inositol-triphosphate (IP_3) form the phosphatidyl-inositol-diphosphate. IP_3 induces the libration of Ca^{++} from storage vesicles by binding to IP_3 receptors of Ca^{++} storage vesicles. A slower pathway (response times in the order of minutes) is mediated by a small G-protein switch of the rho-family: rhoA. The protein (activated again by GDP-GTP exchange) activates the MLC-kinase.

inositol-triphosphate (IP_3). The MLC-kinase can be activated either through calmodulin-Ca^{++}-complexes or through kinase C which is activated by diacylglycerol (DAG). Both DAG and IP_3 are generated from phosphatidylinositol diphosphate (PIP_2) by phospholipase C which is activated by the extrinsic signal.

The thrombin receptor shares two common features with other hormone receptors (*e.g.* adrenalin) and the photon receptor rhodopsin of the visual membrane. It penetrates the membrane with seven hydrophobic α-helixes and exhibits an intracellular loop which contains a binding site for the $G_{\alpha\beta\gamma}$-protein. Binding of the agent to the receptor induces coupling of the G-protein to the loop which is followed by exchange of GDP for GTP in the α-domain of the $G_{\alpha\beta\gamma}$ complex. The GTP-α-domain is detached and activates the enzyme associated with the pathway (phospholipase C in the case of thrombin). The receptor can be deactivated again by phosphorylation of the loop or by coverage by a capping proteins (arrestin in the case of rhodopsin). In this way the activity of the receptor may be switched off and the action of hormones may be downregulated.

4 Micromechanics and microrheometry of cells

Mechanics and hydrodynamics play a key role for numerous biological processes comprising the control of animal motion or the adaptive growth of plants or bones. Mechanical forces control also numerous cellular functions such as cell locomotion, phagocytosis or cell adhesion. An enlightening example is the regulation of the blood pressure through the hydrodynamic shear force exerted by the blood flow or endothelial cells. The hydrodynamic shear stress induces the generation of NO-molecules in a stress dependent manner which control the state of contraction of the smooth muscle surrounding the blood vessels (a process again activated by cell signalling pathways). The production of NO increases with increasing blood pressure (for instance induced by a decreasing diameter of the blood vessel) and this small molecule causes the relaxation of the muscles resulting in the widening of the blood vessel. The study of continuum-mechanical properties of biomaterials from macroscopic to nanoscopic scales is therefore expected to become a challenging field of future biophysical research. One example concerns the viscoelasticity of cells. Measurements of local viscoelastic parameters can yield useful and new insights into the microscopic architecture of cell envelopes or the cytoplasm and into the stimulation of cell signalling pathways by mechanical stresses. A powerful tool to study microscopic mechanical and hydrodynamical properties of cells is magnetic bead microrheometry (*cf.* Fig. 14). Magnetic colloidal beads are coupled to specific sites of cells such as the cell membrane, the actin based cytoskeleton or microtubuli through specific ligands, cell receptors or molecular motors. The magnetic beads are deflected by sequences of short pulses of inhomogeneous magnetic fields and the trajectories of the beads are analyzed as a function of time by particle tracking using confocal laser scanning or phase contrast microscopy combined with fast image processing (*cf.* Keller & Schilling 2001). Viscoelastic parameters (spring constants and viscosities) are obtained by analyzing the viscoelastic response curves (also called creep functions or

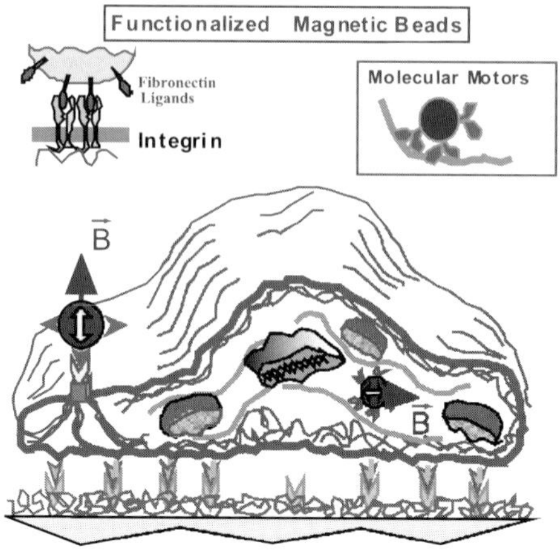

Fig. 14. a) Microrheometry of cell membranes and cell cytoskeleton. Magnetic beads are coupled to specific sites of the cell (cell plasma membrane, cytoskeleton, microtubuli) by functionalization with specific ligands. Viscoelastic response curves are taken by application of force pulses in the nano-Newton range and by recording the deflection as a function of time by particle tracking with fast image processing techniques (*cf.* Fig. 17). The advantage of the method is that pulses may be applied repeatedly to test the reproducibility of the measurement or the linearity of the response and to explore active responses of the cell to external mechanical stimuli.

retardation functions) in terms of mechanical equivalence circuits. A typical creep response curve is shown in Figure 15. It consists of three regimes: an instantaneous elastic deflection (I), a relaxation regime (II) and a region where the bead exhibits continuous flow (III).

The first step is to evaluate the viscoelastic response in terms of a simple mechanical equivalent circuit. It consists of a dashpot (friction $\gamma_\circ$) in series with a so-called Voigt body. The latter consists of a parallel arrangement of a spring (spring constant $k_\circ$) and a series connection of a second spring (k_1) and a dashpot (γ_1). The same model was introduced by Zener to account for the viscoelasticity of solids and is thus also called the standard solid model. The deflection is given by (*cf.* Bausch *et al.* 1998)

$$\frac{x\,(t)}{F} = \frac{1}{k_\circ}\left(1 - \frac{k_1}{k_\circ + k_1}\exp\left\{-\frac{t}{\tau}\right\}\right) + \frac{t}{\gamma_\circ}\,\cdot \tag{4.1}$$

The last term accounts for the flow and characterizes the long time behaviour. For $t \to 0$: $x\,(t) = F/(k_\circ + k_1)$ and the effective spring constant of the membrane is thus $k_{\rm eff} = k_\circ + k_1$. The relaxation time is $\tau = \gamma_1 k_{\rm eff}/k_\circ k_1$

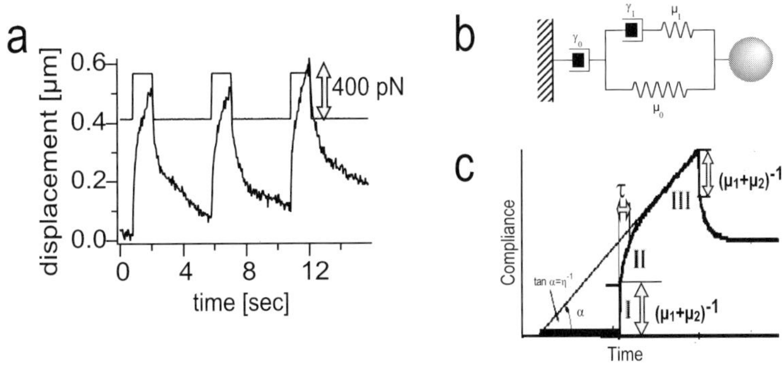

Fig. 15. a) Typical viscoelastic response curve of magnetic bead in cytoplasm of macrophage. Obtained for 1 μm diameter magnetic bead and $f \approx 400$ pN. Note that $J(t)$ is composed of three regimes: a fast elastic response (I), a relaxation regime (II) and a viscous flow (III). **b)** Mechanical equivalent circuit enabling formal representation of creep response curve of cytoplasm of macrophages. The dashpot η_0 accounts for the viscous flow due to the viscoplasticity of cell. The effective total spring constant is $\mu = \mu_1 + \mu_2$. **c)** Typical response function calculated for the Zener model as given in equation (4.1).

and is a measure for the time interval within which the local bonds yield resulting in the flow of the bead.

In reality the relaxation process is determined by a distribution of relaxation times and a more general expression for the deflection $x\,(t)$ would be

$$x\,(t) = F \sum_{\mathrm{i}} J_{\mathrm{i}} \left(1 - \exp\left\{ -\frac{t}{\tau_{\mathrm{i}}} \right\} \right) + \frac{t}{\eta} \tag{4.2}$$

where $J_{\mathrm{i}} = k_{\mathrm{i}}^{-1}$ are elastic compliances characterizing the softness of internal bonds.

The viscoelastic parameters of the mechanical model are relative measures for the membrane stiffness and viscosity. In order to obtain true elastic constants and viscosities of the cell surface one has to analyze the viscoelastic response in terms of the theory of the elasticity of shells. This is an unsolvable problem for large heterogeneously structured shells. Fortunately, measurements of the deformation field evoked within the membrane by a local tangential force showed that the membrane deformation is a local quantity. This deformation field mapping technique is illustrated in Figure 16. It is based on the analysis of the deflection of an assembly of small non-magnetic beads which are anchored to the cell surface together with the magnetic tweezers. It is seen (*cf.* Fig. 15b) that the deflection of

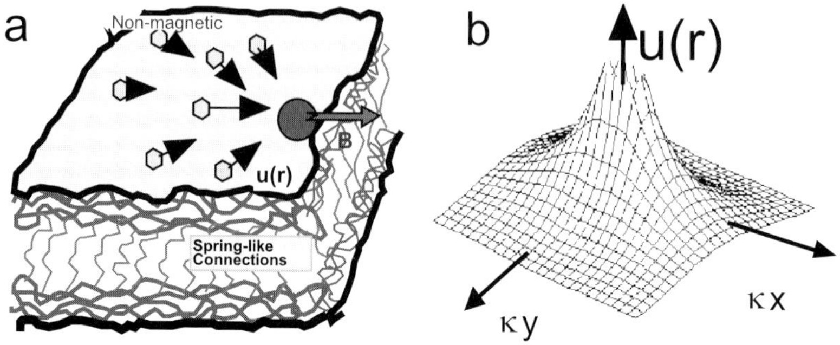

Fig. 16. a) Deformation field mapping technique. For the measurement of the range of the elastic strain field $u(r)$ generated by a tangential force a magnetic bead is coupled to integrin receptors $\alpha_V\beta_3$ through fibronectin and non-magnetic beads are coupled to the cell surface by nonspecific (electrostatic) forces or fibronectin. The tangential deflection $u(r)$ of these colloidal probes induced by lateral deflection of the magnetic tweezer is measured as a function of distance of the probe from the center of the applied force f_0. Mechanical model of thin lobe of cell which corresponds to the situation of spread fibroblasts or endothelial cells. Each of the two juxtaposed membranes are coupled to dense actin network and these are interconnected by inner cytoskeleton. **b)** Plot of lateral displacement $u_x(\vec{r})$ in the direction (x) of the force as a function of the distance from the force as predicted by the theory of deformation of a thin plate coupled to an underlying network which is fixed to the surface. The range of the decay field and the elastic shear modulus is obtained by fitting the theoretical curve $u(r)$ to the experimental data.

the non-magnetic beads, $u(r)$, decays rather rapidly with the distance from the source of the force. The range of the deformation is characterized by a decay length of $\lambda^{-1} \sim 5{-}10$ μm, depending on the cell type. This is a very important result since it shows that the elasticity of composite cell membranes is a local quantity. A detailed theoretical analysis shows that the decay of the deformation field (induced by a point force) with the distance can be explained in terms of the coupling of the lipid/protein bilayer to the intracellular scaffold. The deformation of a thin plate coupled to a network (fixed to a substrate) is determined by the following differential equation

$$\Delta u + \frac{1+\sigma}{1-\sigma}\nabla\nabla u - \lambda^2 u = -\frac{F}{\mu^*}, \tag{4.3}$$

where μ^* is the two-dimensional shear modulus which is related to the bulk modulus μ by $\mu^* = \mu h$, where h is the membrane thickness. The parameter λ has the dimension of a reciprocal length and λ^{-1} is thus a measure for the screening length. It is determined by the coupling strength of the bilayer membrane to the underlying scaffold. This coupling strength can be expressed in terms of a spring constant per unit area ($\chi = J/m^4$) and since $\lambda^{-2} = \mu^*/\chi$, measurements of the screening length yield insight into the coupling strength of the clustered integrin receptors to the actin cortex. This is still a rough model and more refined microscopic theories of composite cell membranes are required to derive reliable viscoelastic moduli from microrheological studies. In particular it is essential to separate passive and active viscoelastic responses as shown in Section 6.

5 Activation of endothelial cells: On the possibility of formation of stress fibers as phase transition of actin-network triggered by cell signalling pathways

The centripetal contraction of endothelial cells required for the formation of gaps within the confluent endothelial cell larger lining the inner wall of blood vessels involves two processes. First the formation of actin stress fibers within or close to the initially random network of the cortex (*cf.* also Fig. 3) and second, the aggregation and activation of myosin II to form acto-myosin motors.

The microscopic details of this process are still largely obscure, but there is ample evidence that the structural change in the actin cortex and the activation of the myosin motor is mediated by intracellular signal transduction involving GTP-hydrolyzing proteins so called *G*-proteins. Several signal pathways are known which regulate the organization and functionality of the actin scaffold on different time scales.

For biophysical studies the process of endothelial cell contraction can be induced by the inflammation agents thrombin and histamine which allow to mimick the gap opening under controlled conditions. Monolayers of endothelial cells from veins of human naval cord are grown on solid supports. Addition of above mentioned thrombin or histamine induces dramatic effects. Firstly, it causes the formation of numerous actin bundles within the actin cortex within a few seconds, which can be visualized by labelling actin with fluorescent derivatives of phalloidin (*cf.* Fig. 17b). Secondly, the stimulation causes centripetal contraction of the cells by disrupture of cadherin-cadherin bonds resulting in the formation of gaps within the initially confluent cell monolayer. This gap formation can be detected with very high sensitivity through the reduction of the electrical impedance measured by electrical impedance spectroscopy (*cf.* Wegener *et al.* 1999; Hillebrandt *et al.* 2001). Thirdly, the shear elastic modulus of the composite membrane increases dramatically. As demonstrated in Figure 17a this stiffening effect can be followed and analyzed in real time by repeated

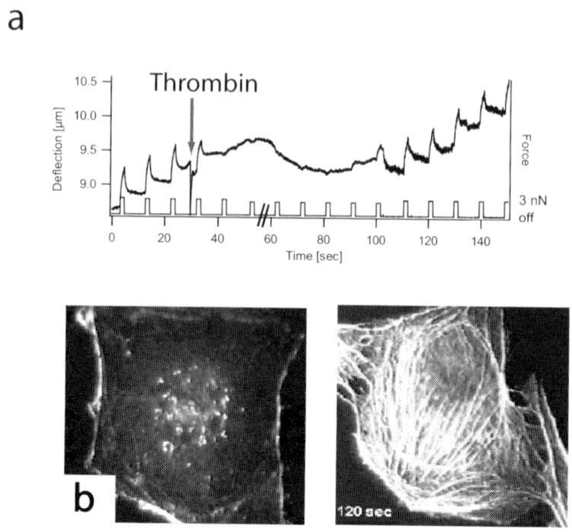

Fig. 17. A real time demonstration of rapid formation of stress fibers associated with dramatic stiffening of composite membranes by magnetic tweezer force spectroscopy. A sequence of viscoelastic response curves evoked by force pulses (amplitude $f_\circ = 2$ nN) is recorded to establish the reproducibility of the measurement. The response curves exhibit a fast response, a relaxation regime and finally flow-like behaviour similar to Figure 15 and are fully reversible. Thrombin is added at the time indicated by an arrow. The deflection amplitude decays within 2–5 s after addition of thrombin, but recovers after a few minutes, although thrombin is still present. As demonstrated by the fluorescence micrographs at the bottom the stiffening is accompanied by formation of stress fibers. Note that the actin forms a randomly organized network before addition of thrombin.

observation of local deflections of the magnetic tweezers coupled to integrins $\alpha_V\beta_1$ through fibronectin. Even forces of up to 10 nN are too weak to evoke appreciable deflections a few seconds after addition of thrombin. The stiffening effect relaxes after about 1 hour despite of the continuous presence of thrombin which is most likely due to the downregulation of the thrombin receptor. The close correlation between actin bundling and membrane stiffening strongly suggests that the stiffening is due to the coupling of the stress-fibers to the integrin clusters.

Following Garcia & Schaphorst (1995) the formation of stress fibers, the membrane stiffening and the cell contraction is associated with the three following biochemical processes: (i) the F-actin concentration increases by 60% (most likely due to inhibition of sequestering molecules), (ii) the intracellular Ca^{++} level increases also by about 50% and (iii) the light chains of myosin II are phosphorylated. Interestingly, in the above described stiffening response a different biochemical pathway is involved: the specific

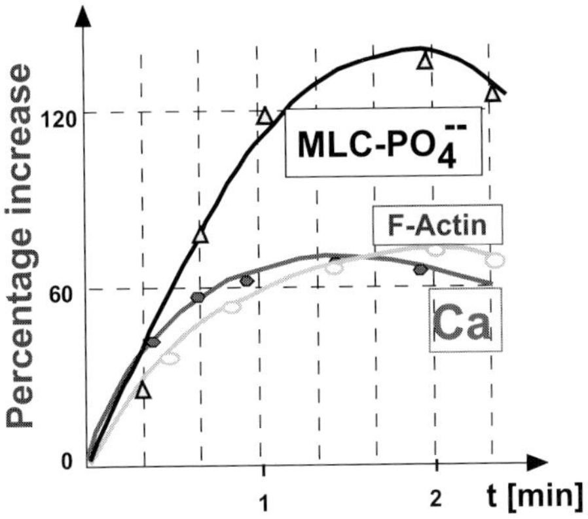

Fig. 18. Biochemical changes induced by thrombin in endothelial cells according to Garcia *et al.* (1995). Concerted time dependent increase of intracellular Ca^{++}-concentration, F-actin content and amount of myosin II activated by phosphorylation of the light chain.

inhibition of parts of the described pathway did not result in an inhibition of the stiffening response. This demonstrates that the quantification of the viscoelastic properties is also a powerful tool to study in real time the biochemical process in living cells.

The growth rate of actin filaments is typically $\sim 1\ \mu\mathrm{m\,s}^{-1}$ and it is hard to imagine that the rapid formation of rather regular arrays of stress fibers is caused by concerted growth of single filaments. In view of the above results, the spontaneous and rapid formation of stress fibers could also be explained in terms of a phase transition of the actin-network if it exists in a state close to the percolation transition. According to the phase diagram of Figure 12a transition from a nearly homogeneous network to a state where the network coexists with actin bundles can be induced both by increasing the F-actin content and/or by activation of the cross-linkers.

It is not known yet whether actin cross-linkers are also activated by thrombin or wether bundling is solely due to the increase of F-actin. In fact, microbundles of myosin could be involved in the actin bundling process as demonstrated in the authors laboratory (M. Keller, authors laboratory, unpublished data). The activation of myosin II by MLC-phosphorylation could thus simultaneously serve the bundling of actin and the formation of contractile fibers. The formation of myosin II microbundles by phosphorylation has been demonstrated for *Dictyostelia* cells. However in this case the phosphorylation site is located at the heavy long arm (*cf.* Pasternak *et al.* 1998; Jungbluth *et al.* 1994).

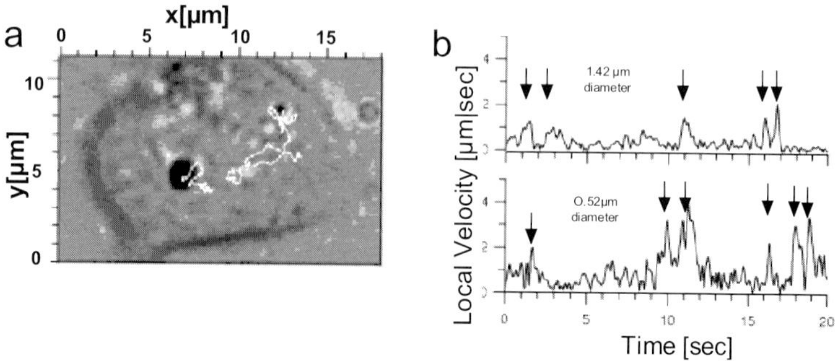

Fig. 19. a) Phase contrast micrograph of intracellular active motion of magnetic beads of 1.4 μm and 0.5 μm diameter in *Dictyostelia Discoideum* cells. The trajectories consist of nearly straight paths interrupted by local random walks. Note that small beads move first towards the center and then back to the periphery. Since the path is several μm long it occurs most likely along microtubuli. **b)** Time sequence of velocity of large (top) and small phagosomes (bottom). The peak velocities indicated by arrows ($\sim$2 μm/s for the large bead and $\sim$4 μm/s for the small bead) correspond to motions along straight paths. The velocities $v \leq 1$ μm/s are mostly found for the local motions.

6 On cells as adaptive viscoplastic bodies

As pointed out above, the cytoplasm is a highly heterogeneous viscoelastic body which must combine high mechanical stability with local softness to enable effective transport of compartments along the microtubuli and actin filaments by forces of the order of ten piconewton. As demonstrated above such behaviour can be realised by heterogeneous gels consisting of entangeled bundles or clusters of tightly crosslinked fibrous macromolecules interconnected by a few thin bundles in such a way that a fully percolated network results. The magnetic bead microrheometry provides a non-invasive technique to measure the distribution of viscoelastic moduli in cells (*cf.* Bausch 1999). A convenient and most natural way to incorporate magnetic beads is uptake by phagocytosis. The engulfed beads are wrapped by the plasma membrane with the inner leaflet of the membrane exposed to the cytoplasm. These "magnetosomes" behave then similar to intracellular particles. Two types of studies can be performed: firstly, measurements of intracellular transport forces and secondly, determinations of spatial distributions of the intracellular viscoelastic parameters (shear moduli and friction coefficients) by creep experiments. As an instructive example we consider measurements in *Dictyostelium Discoideum* cells (*cf.* Feneberg *et al.* 2001). Figure 19a shows that the magnetosomes are transported over large distances within the cell. This transport occurs often repeatedly from the periphery to the center and back and the bead is finally expelled again. Obviously, the trajectories of the beads consist of nearly straight stretches and local random walks. The local velocities along the quasi straight pathes depend on the bead size and are 4 μm/s for the 0.5 μm and about 2 μm/s for the 1.4 μm bead. The velocity of the local random motion varies between 0.1 μm/s and 0.5 μm/s and does not depend appreciably on the size of the beads.

These findings show that the cytoplasm consists of alternating soft and hard regions. In the soft regions the particles can be transported by the molecular motors over long distances (most likely along microtubuli by kinesin or dynein) while they are trapped in the hard regions. Since the velocity does not depend on the bead size the nearly random local walks are most likely due to actively driven local motions of the cytoplasm which is also directly visible by local motion of intracellular particles (*e.g.* vacuolea) using phase contrast microscopy. A very remarkable result is that for cells in which myosin II has been knocked out by mutation the velocity distribution of the small local motion is shifted to lower values. This interesting result suggests that the motor protein acts as softener of the cytoskeleton (if ATP is present).

The local viscoelastic response of the cytoplasm of *Dictyostelia* cells exhibits very complex often non-linear behaviour. Two characteristic response curves are shown in Figure 19. Under the action of a strong shear stress ($\sigma \sim 100$ Pa) the bead is deflected with rapid initial velocity and saturates after a relaxation regime. After turning off the stress it relaxes, however, only partially showing that the bead exhibited viscous flow. For a weak stress ($\sigma \sim 50$ Pa) the bead does not respond appreciably for the first three seconds and is then suddenly deflected by $\sim$2 μm before the deformation saturates. Again, after the the application of the shear stress stops, the bead relaxes only partially, indicating a viscous loss. For higher stresses ($\sigma > 100$ Pa) the bead motion does not saturate and it moves over long distances with force-dependent velocity. Analysis of the numerous response curves showed that the yield strength of the cytoplasm varies between 10 and 20 Pa and the apparent cytoplasmic viscosity varies between 10 and 350 Pa·s. This strongly force-dependent behaviour suggests that the cytoplasm behaves as viscoplastic body exhibiting highly heterogeneous yield forces.

It further suggests tat the bead deflection with constant velocity $\langle v \rangle$ is a consequence of the breakage of internal bonds rather that the internal friction associated with momentum transfer between the bead and the fluid cytoplasm. It is thus more realistic to characterize the constant bead motion in terms of a force-dependent local mobility $\mu(f_{\rm ex})$ according to:

$$\langle v\rangle = \mu(f_{\rm ex})\, f_{\rm ex}. \tag{6.1}$$

Each bond breaks with a rate

$$k = k_0 \exp\left\{\frac{\Delta g - f_{\rm ex}\xi}{k_{\rm B}T}\right\}, \tag{6.2}$$

where Δg is the activation energy of the bond breakage, $f_{\rm ex}\xi$ is the work associated with the bond breakage and ξ is of the order of the local mesh size. The mobility is then related to the local shear stress σ by

$$\mu(f_{\rm ex}) = \mu_0 \exp\left\{\frac{\Delta G - \xi^2\sigma_{\rm ex}}{k_{\rm B}T}\right\}, \tag{6.3}$$

where ΔG is the activation energy of all bonds which must be broken to advance the bead by a distance ξ. For weak external stresses ($\sigma \leq 50$ Pa, Fig. 20, bottom) the deflection scenario depends sensitively on the local yield strength of the cytoplasm. If the latter is smaller than the externally applied stress $\sigma_{\rm ex}$ one observes a similar response curve as for the high force. If the bead is initially embedded in a region of higher yield stress ($\sigma_{\rm yield} > \sigma_{\rm ex}$) only a small deflection is initially observed. It escapes, however, in general from the pinched state after prolonged application of the force and moves fast again, once a softer region is reached.

Note that within the scope of the viscoplastic model the viscosity is replaced by a reciprocal mobility. Despite of the complex viscoelastic behaviour one can measure distributions of the shear elastic moduli. The range of shear elastic moduli depends on the type of cells. For *Dictyostelia Discoideum* cells the values vary between 10 Pa $\leq \mu \leq$ 100 Pa while for macrophages values range from $\mu = 50$ Pa to 600 Pa.

The viscoplastic behaviour is expected to play an important role for the intracellular transport of particles by relatively weak active forces as will be argued in the following. Magnetic tweezers can be applied as force transducers to measure these active transport forces in cells. For this purpose the local motion of magnetosomes is perturbed by application of a force pulse and the active force is obtained by analysing the change of the velocity. The initial velocity is $\vec{v}_{\rm act} = f_{\rm act}/\zeta_{\rm loc}$ where $\zeta_{\rm loc}$ is the local and unknown frictional coefficient. The changed velocity is

$$\vec{v} = \vec{v}_{\rm act} + \vec{f}_{\rm ex}/\zeta_{\rm loc} \tag{6.4}$$

and the active force is thus obtained from the resulting velocity $\vec{v}$ during the pulse according to

$$f_{\rm act} = \zeta_{\rm loc}\vec{v} - f_{\rm ex}. \tag{6.5}$$

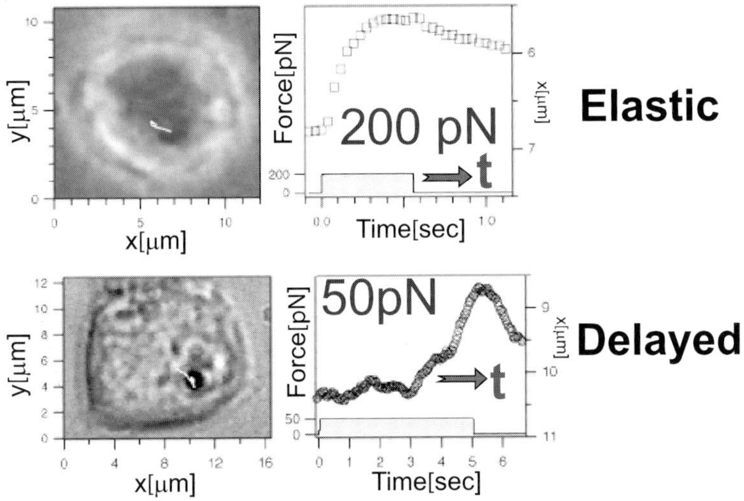

Fig. 20. Nonlinearity of intracellular viscoelastic response: two characteristic examples of viscoelastic response curves induced by force pulses of 200 pN and 50 pN, respectively. Note that the response induced by the large pulse is immediate but does not relax after switching off the pulse while for the small pulse the response is drastically delayed but is partially reversible.

The active force in *Dictyostelia* cells obtained in this way is about $f_{\mathrm{act}} \approx$ 30 pN which would correspond to the concerted action of seven kinesin motors. Active forces of the order of 30 pN are too weak to transport intracellular compartments such as mitochondria or vesicles through the cytoplasmic regions of high mechanical strength exhibiting typically yield forces of 300 pN. However, this problem is overcome by the viscoplastic behaviour of the cytoplasm. As demonstrated in Figure 20 the compartments can penetrate the high elasticity regions by extended application (for a few seconds) of forces considerably smaller then the yield force. After reaching the soft regions exhibiting yield strengths of $\sim$10 Pa rapid transport is possible again. The transport could be further accelerated by active softening of the intracellular regions of high mechanical strength mediated through intracellular signalling.

The situation is further obscured by the active adaptive response of the cells to mechanical stimuli and it is essential to develop further concepts and methods enabling the distinction between passive and active mechanical processes.

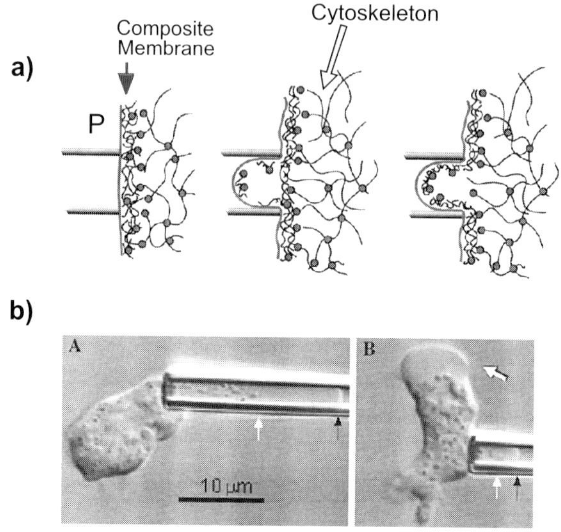

Fig. 21. a) Schematic view of micropipette aspiration technique. Note that aspiration occurs above a threshold pressure $p_{\rm th1}$. **b)** Phase contrast micrograph of *Dictyostelia Discoideum* cell aspired by a micropipette (inner diameter ~3.8 μm) with constant suction pressure at the leading edge (left) and the trailing end of the cell (right). At the cell front, where a new pseudopod is forming, the aspiration length is significantly larger than at the trailing end of the cell. Black arrows denote the end of the aspirated cell lobe. Note that in both cases A and B the aspirated lobe exhibits a clear hyaline cap, free of internal cell compartments. The hyaline cap extends from the tip of the aspirated cell lobe to the beginning of the organelle-containing region (thin white arrow). The bold white arrow in B marks the pseudopod at the leading front of the cell.

7 Controll of cellular protrusions controlled by actin/myosin cortex

Another powerful micromechanical technique to study adaptive mechanical properties of cell envelopes and to gain insight into the coupling strength of the outer shell with the intracellular scaffold is the micropipette aspiration technique (*cf.* Evans & Yeung 1989; Zhelev & Hochmuth 1995). An instructive example is shown in Figure 21. The cell envelope of *Dictyostelia* cells is aspirated by cylindrical pipettes with defined suction pressures (*cf.* Simson *et al.* 2000). Above a threshold pressure $p_{\rm th1}$ the cell envelope disrupts from the cell body and a tube-like protrusion moves into the pipette with nearly constant velocity until a finite protrusion length l_c is reached (*cf.* Fig. 21a). After a recovery phase (typically 10–20 s) the protrusion is actively pulled back provided the applied pressure is kept below a second threshold pressure $p_{\mathrm{th}2}$.

Initially only the lipid/protein bilayer together with a thin layer of actin is pulled into the pipette while the intracellular compartments are hold back by the bulk of the actin shell (*cf.* Fig. 21b). This fraction of the actin cortex forms a closed shell. It holds back the intracellular compartments which are aspired only at a later time if the suction pressure is high.

Dictyostelia cells exhibit several advantages for micromechanical studies of the cell cytoskeleton and the composite shell (*cf.* Schleicher & Noegel 1992). The cytoskeleton can be modified in a controlled way by removing one or several of the actin-binding proteins by mutations. Moreover, distinct cytoskeletal proteins can be fluorescent labelled by fusion with green fluorescent proteins through transfection. The shape of the cells can be controlled by starving the cells for distinct periods of time (*cf.* Schleicher & Noegel 1992). Moving cells exhibit transient polarities and the momentary front and rear of the cells may be distinguished as follows: the locomotion of amoeba like (rounded) *Dictyostelia* cells is associated with three distinct shape changes (*cf.* Schindl *et al.* 1995). First a thin lobe of the cell spreads over the surface for about 10 s forming a flat protrusion (pseudopod) of about 5 μm length. After formation of a new actin cortex in the protruded lobe the trailing end retracts (at fixed position of the leading edge of the pseudopod) and the cell assumes a buldged shape resulting in a strong reduction of the contact area. For several tens of seconds the cell persists in this resting state (typically for 100 s) before a new pseudopod forms in a new direction. In this way the leading front and the trailing end may be distinguished to study the micromechanical properties at sites of distinct activity within the cells. The micropipette experiments show (*cf.* Simson *et al.* 2001)

- The threshold pressure $p_{\mathrm{th}1}$ and thus the coupling strength of the membrane to the actin cortex is remarkably smaller (by 50%) at the leading edge than at the trailing end. The length l_c for a given pressure is about twice as large at the leading front;

- The maximum protrusion length increases roughly linearly with the applied suction pressure;

- The velocity of the advancing protrusion increases linearly with the suction pressure enabling estimates of the viscosity of the composite cell envelope;

- The aspired cell lobe is retraced again after a recovery time of $\sim$20 s, however only against suction pressures smaller than $p_{\mathrm{th}2} \approx 1000$ Pa;

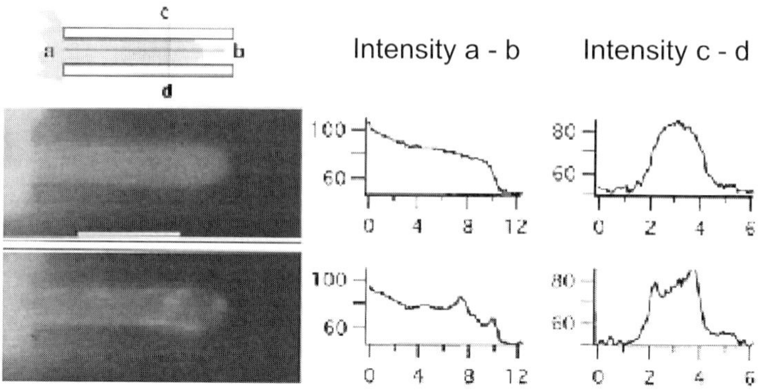

Fig. 22. Demonstration of the reorganization of actin/myosin cortex beneath the membrane (envelope) of aspirated cell protrusion by microfluorescence of transfected cells containing myosin II with attached green fluorescent proteins. Top row: fluorescence micrograph of freshly aspirated protrusion exhibits nearly isotropic distribution of myosin II within the cytoplasm of the aspired cell lobe. Bottom row: myosin II distribution after 20 s (along sections a and b indicated in the figure at the top) showing accumulation of motor proteins close to the wall and the mouth of the pipette.

- During the recovery period the actin/myosin cortex of the aspired cell lobe is re-established again. This follows from microfluorescence studies of the redistribution of myosin II which is labelled with green fluorescent protein by transfection. As shown in Figure 22 myosin II is isotropically distributed in the direction perpendicular to the axis of the pipette and is only slightly accumulated towards the pipette entrance. After several (typically 20) seconds the motor protein accumulates close to the inner wall of the cell membrane (along the whole length of the protrusion). Moreover it is further accumulated close to the entrance of the micropipette. The protrusion can only retract after completion of this reorganization process. A second prerequisite for the retraction is that the cell adheres with the free cell body to the substrate because the protrusions are not retracted again if the cell is freely suspended in the buffer;

- The necessity of myosin II for the retraction process is also demonstrated by the fact that myosin II deficient cells are not able to retract the protrusions.

The micromechanical experiments provide also useful tools to study the effect of mutations of the cytoskeleton in a quantitative way. Removal of myosin II increases the threshold pressure required for the aspiration of the protrusions by a factor of five that is from $p_{\mathrm{th}} = 210$ Pa for wild type cells to $p_{\mathrm{th}} = 1000$ Pa for the mutant. In agreement with the results obtained with magnetic bead microrheometry (*cf.* Sect. 6) this observation shows that the motor protein acts as softener of the actin-based cytoskeleton.

Conclusion: the envelopes of eucaryontic cells are soft, trilamellar shells composed of the central lipid protein bilayer (the biomembrane), the glycocalix, covering the outer surface and the cortex associated with the inner leaflet of the membrane. This composite membrane mediates local structural changes of the glycocalix (for the membrane) and the membrane-associated macromolecular cortex, which plays an essential role for the signal transduction between extracellular space during numerous cellular processes. In this lecture we discussed the physical properties of lipid membranes, the viscoelastic properties of the ultrasoft quasi-twodimensional networks of red blood cells and of pure cytoskeletal networks such as the semiflexible polymer actin and the regulation of the viscoelastic properties of composite membranes of eucaryontic cells.

The red blood cell exhibits outstanding elastic properties unmatched by technical materials: it is extremely soft with respect to bending and shearing but hard as solids towards extensions which is essential for the survival of the cells during their several hundred kilometer long journey through the capillary system. These unique properties are due to the ultrasoft quasi-twodimensional triangular network (mesh size 80 nm) with the sides formed by flexible spectrin filaments and corners formed by actin oligomers.

Pure Actin networks are also of great interest from the point of view of polymer physics. Since the contour lengths (20 m) and persistence lengths (10 m) are easily accessible to the optical microscopy one can visualize and analyze the conformational dynamics, the local and long range diffusivity and the flexibility of single filaments by microflueorescence or the analysis of the Brownian motion of filament coupled colloidal gold. Together with novel colloidal bead microrheometry techniques such studies open the possibility to relate molecular motional properties of the local structural fluctuations to the macroscopic viscoelastic impedance.

We discussed the self-assembly and function of the endothelial cell monolayers lining the inner wall of blood vessels which implies two fundamental and vital processes, cell adhesion (treated in the accompanying article by Sackmann and Bruinsma) and numerous chemo-mechanical processes. The manifold functions of the actin based cytoskeleton (*e.g.* during numerous chemomechanical processes) is determined by the outstanding viscoelastic properties of this prototype of a semiflexible macromolecule. A Phallanx of actin manipulating proteins (including motor proteins of the myosin family) together with the regulation of their activity by second messengers enables cells to switch rapidly between different local structural organizations of the cytoskeleton. Heterogeneous gels can be rapidly formed or reorganized (in sub-second time scales) through phase transitions such as the formation of heterogels composed of densely packed domains (microgels) interconnected by single filaments or bundles from homogeneous networks by switching on the activity of cross-linkers (*e.g.* a-actinin). Such percolated networks combine astonishingly high stifnesses with lage soft voids enabling the uptake and effective transpot of particles. They can transform further into states of bundle-network coexistence.

Magnetic bead microrheometry studies demonstrate that the cytoplasm exhibits typical viscoelastic features of such heterogeneous networks with two fundamental differences. The heteogels are highly dynamic and thus behave as soft viscoelastic bodies and the active transport of intracellular phagosomes is thus determined by a force dependent mobility. The cells can respond to extrinsic forces by adaptive active reorganizations of the cytoskeleton thus facilitating the intracellular transport.

These examples demonstrate the continuous need for new physical techniques, which allow the local manipulation of biomaterials to understand their unique physical properties. The study of their complexity and their heterongenety are not only of fundamental physical interest but also allow the quantification and thus fundamental understanding of important biological processes.

References

[1] F. Amblard, A.C. Maggs, B. Yurke, A.N. Pargellis and S. Yeibler, *Phys. Rev. Lett.* **77** (1996) 4470.

[2] G. Bassell and R.H. Singer, *Curr. Biol.* **9** (1997) 109.

[3] A. Bausch, W. Möller and E. Sackmann, *Biophys. J.* **76** (1999) 573.

[4] T.M. Bayerl and E. Sackmann, *Neutron scattering as a tool to probe the effect of cholesterol on phase separation, structure formation and lipid protein interaction in model membranes.* Cholesterol in Membranes, edited by L. Finegold (CRC Press, Boca Raton, USA, 1992).

[5] V. Bennet, *Phys. Rev.* **70** (1990).

[6] A. Bremer, R.C. Millonig, R. Sütterlin, A. Engel, T.D. Polard and U. Aebi, *J. Cell Biol.* **115** (1991) 689.

[7] A. Bretscher, *Curr. Op. Cell Biol.* **11** (1999) 109.

[8] F. Brochard and J.F. Lennon, *J. Phys.* **11** (1975) 1035.

[9] P. de Gennes, *Scaling concepts in polymer physics* (Cornell University Press, Ithaca, 1979).

[10] M. Dichtl and E. Sackmann, *New J. Phys.* **1** (1999) 181.

[11] H.G. Döbereiner, J. Käs, D. Noppl, I. Sprenger and E. Sackmann, *Biophys. J.* **65** (1993) 1396.

[12] M. Doi and S.F. Edwards, *The theory of polymer dynamics* (Clarendon Press, Oxford, 1986).

[13] H.P. Duwe and E. Sackmann, *Physica* **163** (1990) 410.

[14] H. Engelhardt and E. Sackmann, *Biophys. J.* **54** (1988) 267.

[15] J.C. Crocker *et al.*, *Phys. Rev. Lett.* **85** (2000) 888.

[16] E.A. Evans, *Biophys. J.* **14** (1972) 923.

[17] E.A. Evans and D. Needham, *Faraday Discussion Chem. Soc.* **81** (1986) 267.

[18] E.A. Evans and W. Rawicz, *Phys. Rev. Lett.* **64** (1990) 2094.

[19] E.A. Evans and A. Yeung, *Biophys. J.* **56** (1989) 151.

[20] W. Feneberg, M. Westphal and E. Sackmann, *Eur. Biophys. J.* **30** (2001) 284.

[21] J.D. Ferry, *Viscoelastic properties of polymers* (Wiley, London, 1980).

[22] E. Frey and J. Wilhelm, *Phys. Rev. Lett.* **77** (1996) 2581.

[23] J.G.N. Garcia and K.L. Schaphorst, *J. Inv. Med.* **43** (1995) 117.

[24] F. Gittes and F. MacKintosh, *Phys. Rev. E* **58** (1998) 1241.

[25] R. Götter, K. Kroy, E. Frey, M. Bärmann and E. Sackmann, *Macromol.* **29** (1996) 30.

[26] W. Häckl, U. Seifert and E. Sackmann, *J. Phys. II* **7** (1997) 1141.

[27] W. Helfrich and R.M. Servuss, *Nuovo Cimento D* **3D** (1984) 137.

[28] H. Hillebrandt, A. Abdelghani, C. Abdelghani-Jacquin, M. Aepfelbacher and E. Sackmann, *Appl. Phys. A* **73** (2001) 539.

[29] B. Hinner, M. Tempel, E. Sackmann, K. Kroy and E. Frey, *Phys. Rev. Lett.* **12** (1998) 2614.

[30] H. Isambert and A. C. Maggs, *Macromol.* **29** (1996) 1036.

[31] G. Isenberg, *J. Mus. Res. Cell Mot.* **12** (1991) 136.

[32] P. Janmey, *Cell membranes and the cytoskeleton*, volume 1B of *Handbook of biological physics*, edited by R. Lipowsky and E. Sackmann (Elsevier, Amsterdam, 1995), Chapter 11.

[33] A. Jungbluth, V. von Arnim, E. Biegelmann, B. Humbel, A. Schweiger and G. Gerish, *J. Cell Sc.* **107** (1994) 117.

[34] J. Käs, H. Strey, J.X. Tang, D. Finger, R. Ezzel, E. Sackmann and P. Janmey, *Biophys. J.* **70** (1996) 609.

[35] M. Keller, J. Schilling, and E. Sackmann, *J. Sci. Instrum.* **72** (2001) 3626.

[36] E. Korn, M. Cartier and D. Pantaloni, *Science* **238** (1987) 638.

[37] K. Kroy and E. Frey, *Phys. Rev. Lett.* **77** (1996) 306.

[38] L. Landau and E. Lifshitz, *Theory of elasticity* (Pergamon Press, 2nd edition, 1970).

[39] R. Lipowsky, *Generic interactions of flexible membranes*, volume 1B of *Handbook of biological physics*, edited by R. Lipowsky and E. Sackmann (Elsevier, Amsterdam, 1995), Chapter 11.

[40] R. Lipowsky and M. Girardet, *Phys. Rev. Lett.* **65** (1990) 2893.

[41] H. Lodish, *Molecular cell biology* (W.H. Freeman Company, New York, 1995).

[42] L. Machesky and K. Gould, *Curr. Op. Cell Biol.* **11** (1999) 117.

[43] F.C. MacKintosh, J. Käs and P.A. Janmey, *Phys. Rev. Lett.* **75** (1995) 4425.

[44] A. McGough, B. Pope, W. Chiu and H. Weeds, *J. Cell Biol.* **138** (1997) 771.

[45] R. Merkel, R. Simson, D.A. Simson, M. Hohenadl, A. Boulbitch, E. Wallraff and E. Sackmann, *Biophys. J.* **79** (2000) 707.

[46] S.T. Millner and S.A. Safran, *Phys. Rev. A* **36** (1987) 4371.

[47] D. Morse, *Macromol.* **31** (1998) 7030.

[48] D. Morse, *Phys. Rev. E* **63** (2001).

[49] O. Müller, H.E. Gaub, M. Bärmann and E. Sackmann, *Macromol.* **24** (1991) 3111.

[50] T. Odijk, *Macromol.* **16** (1983) 1340.
[51] G. Oster, *Cell Motil. Cytoskel.* **10** (1988) 164.
[52] A. Palmer, T.G. Mason, J. Xu, S.C. Kuo and D. Wirtz, *Biophys. J.* **76** (1999) 1063.
[53] C. Pasternak, P. Flicker, S. Ravid and J.A. Spudich, *J. Cell Biol.* **109** (1998) 203.
[54] T. Piekenbrock and E. Sackmann, *Biopolymers* **32** (1992) 1471.
[55] J.L. Podolski and T. Steck, *J. Biol. Chem.* **103** (1986) 2747.
[56] A. Radhakrishnan and H.M. McConnell, *Biophys. J.* **77** (1999) 1507.
[57] J. Rädler, T. Feder, H. Strey and E. Sackmann, *Phys. Rev. E* **51** (1995) 4526.
[58] E. Sackmann, *FEBS Lett.* **346** (1994) 3.
[59] E. Sackmann, *Biological membranes: Architecture and Functions*, volume 1A of *Handbook of biological physics*, edited by R. Lipowsky and E. Sackmann (Elsevier, Amsterdam, 1995), Chapter 1.
[60] M. Schindl, E. Wallraff, B. Deubzer, W. Witke, G. Gerish and E. Sacmann, *Biophys. J.* **68** (1995) 1177.
[61] M. Schleicher and A. Noegel, *The New Biologist* **4** (1992) 461.
[62] U. Seifert, *Phys. Rev. Lett.* **74** (1995) 5060.
[63] T.A. Springer, *Cell* **76** (1994) 301.
[64] H. Strey, M. Peterson and E. Sackmann, *Biophys. J.* **69**, (1995) 478.
[65] M. Tempel, W.H. Goldmann, C. Dietrich, V. Niggli, T. Weber, E. Sackmann and G. Isenberg, *Biochemistry* **33** (1994) 12565.
[66] M. Tempel, G. Isenberg and E. Sackmann, *Phys. Rev. E* **54** (1996) 1802.
[67] S. Tuvia, A. Moses, N. Gulayev, S. Levin and R. Korenstein, *J. Phys.* **516** (1999) 781.
[68] J. Wegener, S. Zink, D. Rösen and H.J. Galla, *Eur. J. Physiol.* **437** (1999) 925.
[69] D.V. Zhelev and R. Hochmut, *Biophys. J.* **68** (1995) 2004.
[70] F. Ziemann, J. Rädler and E. Sackmann, *Biophys. J.* **66** (1994) 2210.
[71] A. Zilker, M. Ziegler and E. Sackmann, *Phys. Rev. A* **46** (1992) 7998.

COURSE 7

CELL ADHESION AS WETTING TRANSITION?

E. SACKMANN* AND R. BRUINSMA**

** Physik Department E22,
James-Franck-Straße, D-85748
Garching b. München, Germany
** Universiteit Leiden,
Instituut-Lorentz voor Theoretische
Natuurkunde, Postbus 9506, NL-2300
RA Leiden, The Netherlands*

Contents

CELL ADHESION AS WETTING TRANSITION?

E. Sackmann[1] and R. Bruinsma[2]

Abstract

Cell adhesion is controlled by a complex interplay of short range (lock-and-key) forces mediated by cell surface receptors, a phalanx of (short and long range) non-specific (generic) interactions and, last but not least, membrane elasticity. The physical basis of cell adhesion is explored by the design of simplified model systems mimicking cell and tissue surfaces enabling local measurements of cellular shape changes and adhesion forces by microinterferometry. Cell adhesion can be understood as first order dewetting transition resulting in the formation of adhesion plaques such as focal adhesion sites of cells which allows cell adhesion at astonishingly low receptor densities. The repeller molecules of the glycocalix play a key role for the control of the adhesion transition and the mechanical stability of the adhering cells by relaxing the strength of the binding forces. Stress fibers are postulated to be essential for the stabilization of adhesion domains against leverage through bending moments enforced by hydrodynamic shear forces.

1 Introduction

Cell adhesion is a fascinating albeit very complex biological process which controls many life processes. Examples are the formation of specific organs during the magic process of the transformation of fertilized eggs into embryos or the assembly of endothelial cells into tubular structures during the formation of blood vessels or of the blood-brain barrier [1]. The specificity of cell adhesion is controlled by genetic expression of receptors at the cell surface which bind specifically to adhesion molecules of target cells or tissue by the lock-and-key force principle. However, as illustrated in Figure 1, cell adhesion is also controlled by a manifold of generic forces and, most importantly, by membrane elasticity. The situation is further obscured by often

[1]Physik Department E22, James-Franck-Straße, D-85748 Garching b. München, Germany.

[2]Universiteit Leiden, Instituut-Lorentz voor Theoretische Natuurkunde, Postbus 9506, NL-2300 RA Leiden, The Netherlands.

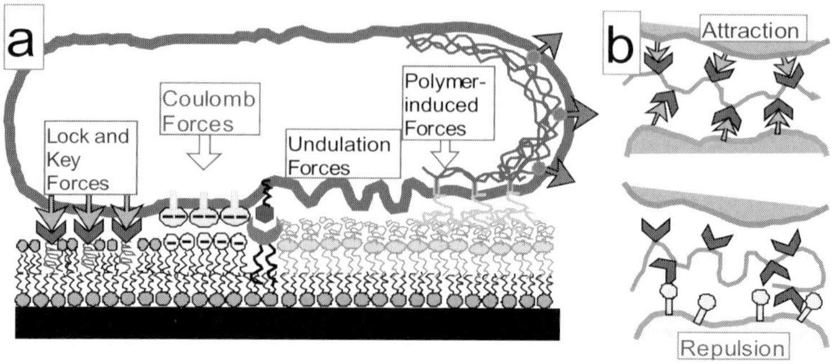

Fig. 1. Control of cell adhesion by interplay of receptor-mediated specific forces, a manifold of generic interfacial forces and membrane elasticity. The cartoon shows the interaction of a cell with a biofunctional surface mimicking target cells or tissue. Non specific forces include: (i) electrostatic interactions which can be mediated by cell surface receptors exhibiting large numbers of acidic sialic acid residues (*cf.* GlyCAM, Fig. 3); (ii) attractive Van der Waals interactions, (iii) repulsive undulation forces due to thermally excited bending undulations (flickering) of lipid protein bilayers; (iv) a manifold of polymer induced forces. The latter may include steric repulsion forces between molecules of the glycocalix such as cell adhesion molecules exhibiting long extracellular chains (*cf.* Fig. 3) but also interactions mediated by extracellular matrix proteins such as hyaluronic acid and fibronectin which are bound to their respective receptors CD44 and integrin, respectively. These interactions can be attractive if the opposing cells (or cells and tissue-surfaces) exhibit receptors for the macromolecules but will be repulsive in the absence of such receptors. Note that fibronectin has also a binding domain for collagen IV a major component of the basal membrane forming the top layer of endothelial cells.

drastic reorganizations of the membrane-coupled, actin-based cytoskeleton induced by receptor-ligand recognition. A prominent example is the formation of focal adhesion complexes formed by a phalanx of actin binding proteins mediating the formation of actin bundles (stress fibers) and their coupling to the cytoplasmic domains of membrane spanning cell surface receptors [1].

Studies of adhesion processes are further hampered by the fact that the formation of stable cell-cell and cell-tissue contacts is a dynamic process involving a sequence of steps [2].

Firstly, as shown in Figure 2, the receptors and ligands are often hidden within the several ten nm thick glycocalix. It is formed mainly by membrane anchored cell surface molecules with very large head groups such as cell adhesion molecules (CAMs) exhibiting head groups which may extend by up

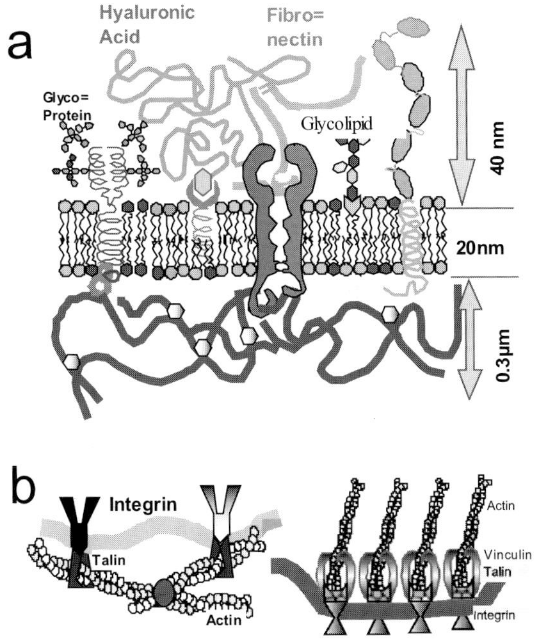

Fig. 2. a) Cartoon of three-layered composite cell membrane. The central leaflet is composed of a multicomponent lipid/protein bilayer. It is coupled at the cytoplasmic side to the actin-based cytoskeleton (the actin cortex); a several hundred nm thick meshwork of actin filaments which are partially interconnected by cross-linkers (including myosin) and which is locally connected to the bilayer membrane. Note that the actin filaments are in fact about twice as thick (8 nm) as the bilayer (4 nm). The outside is covered by the (several ten nm thick) glycocalix which is made up of the head groups of cell receptors (carrying mostly several oligosaccharide chains) of glycolipids but also of huge macromolecules of the extracellular matrix (such as hyaluronic acid and fibronectin) which are transiently coupled to their respective receptors. **b)** Coarse grained models of local coupling of actin filaments to bilayers resulting in the formation of a thin shell of partially cross-linked actin (left) or of bundles (right) that can act as stress fibers and may serve the enforcement of local cell adhesion.

to 40 nm into the extracellular space (*cf.* Fig. 3). Many of these molecules may act as receptors or repellers depending on the surface of the target cell or tissue. In addition the effect of the glycocalix may be enhanced by huge macromolecules of the extracellular matrix such as fibronectins (~70 nm long) or hyaluronic acid (= hyaluran, a linear polysaccharide of up to 25 000 monomers) attached to their respective receptors. The formation

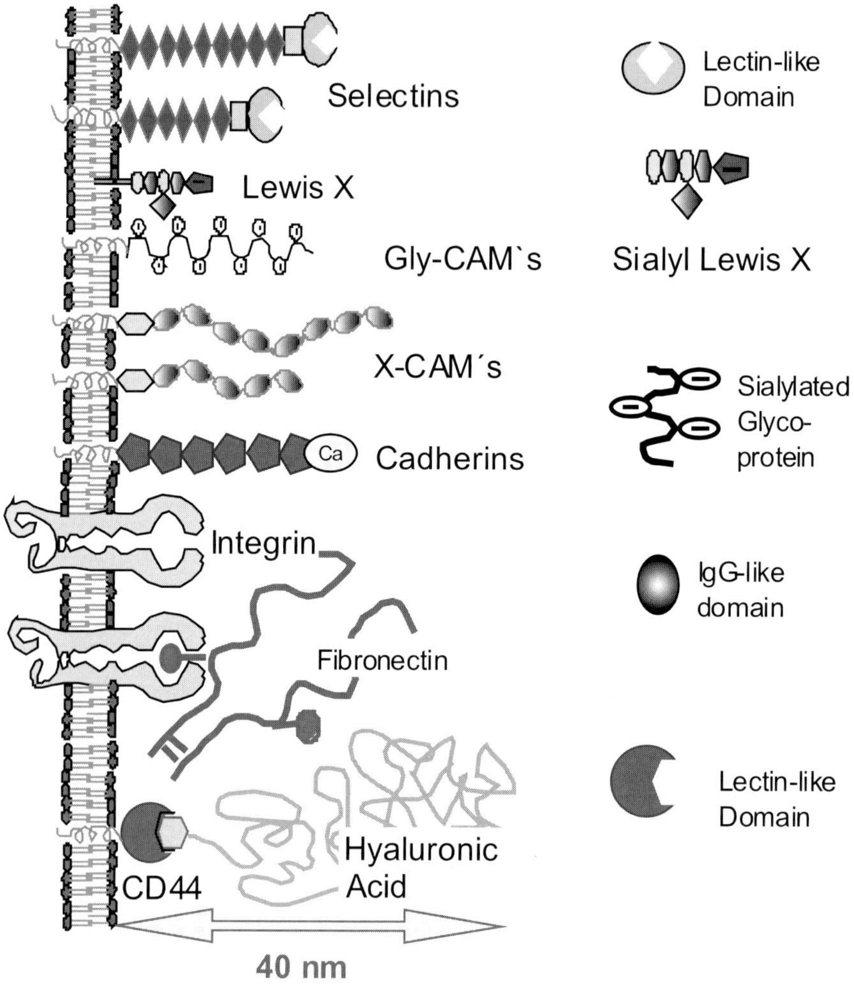

Fig. 3. Text see on page 291.

of tight cell-cell contacts thus requires the expulsion of these repellers from the adhesion zone.

Secondly, the cell adhesion molecules are randomly distributed within the plasma membrane and the formation of tight adhesion zones by receptor segregation is a diffusion controlled process [3].

Thirdly, long term modulations of adhesion associated with adhesion induced cell signaling have to be considered. Examples are the transfer of receptors from cytoplasmic storage compartments to the plasma membrane

Caption to Figure 3: Summary of families of cell adhesion molecules (involved in the process of migration of white blood cells through endothelial cell layers lining the inner wall of blood vessels [2]) stressing structure of extracellular domains only. The lengths of the head groups are approximately drawn to scale. The head groups of the family of **selectin** receptors are composed of chains of identical polypeptides and exhibit a lectin-like outer domain which recognizes glycolipids and oligosaccharide side chains of glycoproteins (Gly-CAMs) with acidic sialic acid residues (such as so-called sialylated Lewis X factors). The different members of the family are distinguished by the number of repeat units. The longest representative with eight repeat units (P-selectin) may extend up to 30 nm into the extracellular space. The superfamily of so-called **cell adhesion molecules (CAM)** comprises two families: X-CAMs and Gly-CAMs. The extracellular domains of the family of X-CAMs are composed of chains of immunoglobulin(IgG)-like polypeptides (composed of about 100 amino acids) which are separated from the α-helical membrane anchor by a domain (also called III-domain) which is also found in fibronectins. The second superfamily of CAM-receptors (also called Gly-CAMs or "muccin"-like receptors) are composed of long polypeptide chains, rich in serin and threonin, to which sialylated oligosaccharide chains are coupled. Gly-CAMs are found on white blood cells (leucocytes). The head group may exhibit up to 80 oligosugar chains and may extend up to 40 nm into the extracellular space. These molecules act both as repeller and receptor [2,3] depending on whether the target cell exhibits a receptor or not. The superfamily of **cadherins** belongs to the group of homophilic (or self-recognizing) receptors which can mediate the tight mutual coupling of cells within cell monolayers (such as the epithelial cell layers of skin or the endothelial cell layers lining the inner wall of blood vessel). The cadherin family also plays a central role for the control of the formation of specific tissue during embryonal development where epithelial cell layers enclosing specific developing organs (*e.g.* the neural tube or the precursor of muscle cells) are interconnected by a specific class of cadherins which are genetically expressed at the cell surface. **Integrin** is the most versatile type of receptor found hitherto. It is a heterodimer composed of an α- and a β-chain with 15 nm long headgroups. Since there exist a number of different types of α- and β-chains a matrix of integrin receptors $\alpha_i\beta_j$ can be formed. Many integrins bind giant macromolecules of the extracellular matrix such as fibronectin (which are about 50 nm long) and these may act as repulsive spacers between cells which compete transiently with receptor-ligand pair formation. A cell receptor which is only involved in cell-matrix interaction is the CD44. It recognizes hyaluronic acid a giant polysaccharide which may exhibit Flory radii of up to $\sim$500 nm and which can thus also contribute to long range cell-cell repulsions.

or the *de novo* synthesis of these receptors. A prominent and fascinating example of the regulation of adhesion by cell signaling is the activation of strong adhesion of leucocytes on endothelial cells by chemoattractants which triggers the emigration of the cells through the endothelial cell layer of blood vessels [2,3]. This process is assumed to be initiated by activation of the G-proteins through the receptor for the chemoattractants which span the membrane with seven helices (thus resembling hormone receptors) and results eventually in the opening of gaps in the monolayer by centrifugal contraction of the endothelial cells [4].

The enormous interest in cell adhesion is exemplified by the more than 20 000 publications on this topic in 6 years. One group of studies concentrate on the molecular basis of specific cell-cell and cell-tissue-interaction and resulted in the discovery of an appallingly large number of receptors and conjugate ligands. Fortunately, these can be classified in terms of a relatively small number of sub groups which share the same structural motifs (*cf.* Ref. [2] and Fig. 3 for a summary). Moreover, concerning the interaction mechanisms the receptors may be divided into two distinct groups: (1) homophilic receptors which interact by non-covalent association with an anti-parallel oriented receptor of the same type in the opposing membrane mostly of the same type of cell and (2) heterophilic receptors which recognize distinct ligands (called adhesion epitopes) of opposing cell membranes or tissue (called "lock-and-key bonds" or "links", below).

A second major group of studies deals with the cell biological basis of cell-cell interaction processes associated with immune responses, such as the above mentioned inflammation-induced selective interaction of white blood cells (leucocytes) with the endothelial cell layers of blood vessels [2]. These studies showed impressively that cell adhesion is a multistep process (as pointed out above) which involves the whole composite cell plasma membrane and that adhesion processes are intimately linked (i) to diffusive lateral reorganizations within the lipid/protein membrane, (ii) to reorganizations of the actin cortex, (iii) to the exchange of material between cytoplasm and plasma membrane through endocytosis and exocytosis and (iv) to genetic expression of new receptors.

2 Mimicking cell adhesion

To study the physical basis of cell adhesion model systems were designed which exhibit key elements of the cell surface involved in adhesion and which enable the simultaneous measurement of free energies of adhesion and adhesion forces with high precision (for literature *cf.* Refs. [5,6]). Test cells are mimicked by giant vesicles with reconstituted homophilic receptors (such as natural lipid coupled homophilic cell surface receptors [5]) or lipid-coupled

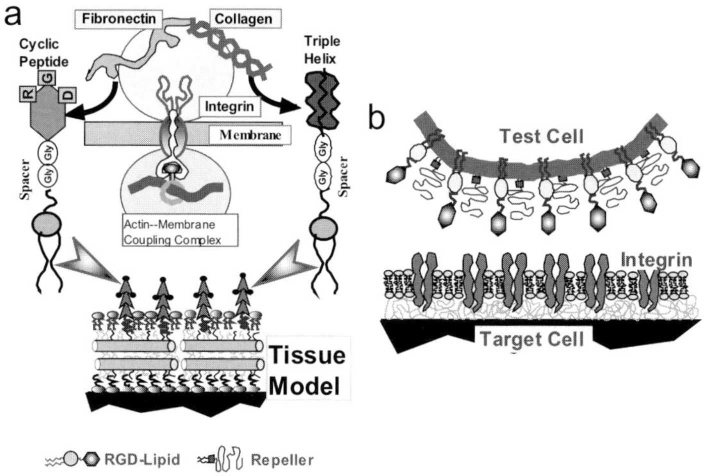

Fig. 4. a) Design of mimetics of tissue surfaces. Fractions of giant extracellular matrix proteins or synthetic peptides (called "adhesion epitopes") which are specifically recognized by receptors (*e.g.* of the integrin family) are coupled to lipids (*via* spacers) which are reconstituted into supported lipid monolayers or bilayers. To avoid denaturing of the proteins, the lipid layers must be separated from solid supports (*e.g.* semiconductors) by ultrathin polymer cushions. In the example shown, these films are made by deposition of multilayers of cellulose filaments exposing alkylchains (called hairy) [5]. On the left side the cartoon shows adhesion epitopes of fibronectin composed of cyclic hexapeptide with arginin-glycin-aspartate (RGD-) tripeptide sequence which are specifically recognized by integrin $\alpha_v\beta_3$ and the right side shows collagen epitopes composed of a recombinant triple helix of collagen [L. Moroder, private communication]. **b)** Typical model system mimicking cell adhesion. Giant vesicle with reconstituted lipid-coupled RGD-ligands shown in **a)** serves as test cell and supported membrane with reconstituted integrin receptors of blood platelets ($\alpha_{2b}\beta_3$) acts as target cell or target tissue. To mimick the repulsion by the glycocalix, lipids with macromolecular head groups (polyethyleneglycol) are reconstituted in one or both membranes [6, 7].

ligands (such as cyclic peptides with arginin-glycin-aspartate tripeptide motifs (so-called RGD-ligands) which are recognized by receptors of the integrin family; *cf.* Fig. 4). The generic long range repulsion is modeled by embedding lipopolymers such as phospholipids with polyethyleneglycol (= PEG) head groups (abbreviated as PEG-lipids or "repellers" in the following). Target cells are modelled by supported membranes with reconstituted receptors (such as integrins recognizing RGD-ligands) or lipid-coupled adhesion epitopes while target tissue is mimicked by coupling of adhesion epitopes to ultrathin hydrophilic polymer cushions anchored on solids [5].

3 Microinterferometry: A versatile tool to evaluate adhesion strength and forces

Microinterferometry is a most versatile technique (1) to visualize cell adhesion and (2) to determine local free energies of adhesion or to measure unbinding forces between membrane-bound receptor-ligand pairs [6]. The technique (called reflection interference contrast microscopy RICM) is summarized in Figure 5. The image is formed by interference of light reflected from the surface of the adhering shells and from the biofunctionalized substrate, respectively. It is essentially a Fourier transformation of the height distribution $h(x, y)$ of the adhering shell hovering over the substrate. The surface profile of the shell can be reconstructed by inverse cosine transform of the intensity distribution $I(x, y)$ with a relative height resolution of about 1 nm and a lateral resolution of about 0.3 μm [5]. The maximum height accessible is $\sim$5 μm, the time resolution is $\sim$10 ms. Several procedures for the improvement of RICM image analysis have been developed [6].

4 Soft shell adhesion is controlled by a double well interfacial potential

The non-specific potential of interaction $V(h)$ between an adherent soft shell and the target cell or tissue is composed of several contributions (*cf.* Appendix A). These include the (attractive) van der Waals potential V_{vdw}, electrostatic interactions V_{el}, repulsive undulation forces V_{und} and the short range repulsions exerted firstly by the supported membrane and secondly by the repeller molecules of the glycocalix. For practical purposes it is useful to fill the model cell with a higher density fluid which attributes a gravitational potential $V_{\text{g}}(h)$ and which corresponds to real biological situations.

The undulation force (discovered by W. Helfrich [8]) is a unique feature of soft membranes. It is a consequence of thermal excitations of bending undulation of the highly flexible bilayers (bending modulus $\kappa \approx 25\ k_{\text{B}}T$). As illustrated in Figure 6, it results in a pronounced dynamic surface roughness of the soft shell membrane. It counteracts adhesion since in order to approach the membrane to the solid surface the Brownian motion has to be frozen in and this corresponds to a decrease in entropy. The resulting disjoining pressure is analogous to the pressure generated by a compressed ideal gas. In biology, undulation forces may play a key role for the control of adhesion of erythrocytes since these cells exhibit very strong undulatory excitations. The undulations are suppressed in normal cell membranes. At small membrane tensions ($\sigma < 10^{-6}$ mN/m) the repulsion decays with

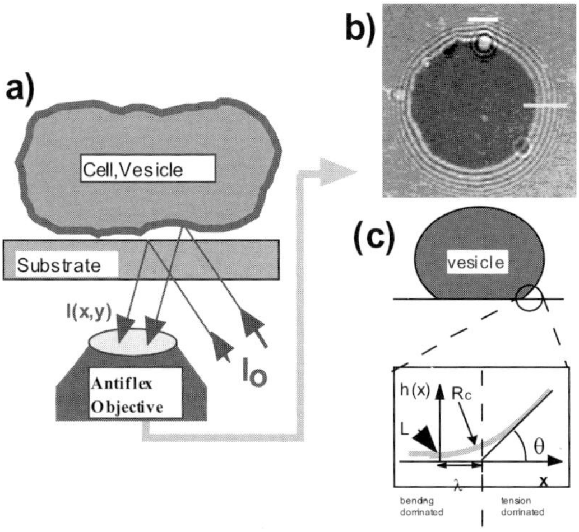

Fig. 5. **a)** Schematic view of image formation by interference of light reflected from the surface of the soft shell and from the substrate surface, respectively. **b)** Interferogram of adhering vesicle on solid support with adhesion zone defined by dark disc. The Newtonian fringes define the contour of the vesicle close to the substrate. The horizontal line defines the direction of contour reconstruction **b)**. **c)** Schematic view of typical surface profile of soft shell vesicle adhering moderate strength. Note that the profile exhibits a straight regime with the slope θ defining a contact angle (very similar to a partially wetting fluid droplet on a solid). However, the finite bending stiffness of the membrane leads to a smooth transition from the free to the adherent membrane at the rim of the adhesion disc defining a contour curvature R_c. The distance between the rim of the adhesion disc (defined by the contact line L) and the intersection of the tangent to the straight profile with the abscissa defines the capillary length $\lambda = \sqrt{\kappa/\sigma}$ (κ is the membrane bending modulus and σ is the membrane tension). λ is a measure for the range of the membrane deformation dominated by bending elasticity that is at $x > \lambda$ the deformation is dominated by lateral membrane tension and at $x < \lambda$ by bending elasticity.

distance d, as $V_{\mathrm{und}} = (k_{\mathrm{B}}T)^2/\kappa d^2$ while it is reduced with increasing tension [8].

Even more important is the repulsion potential due to the repeller molecules which plays the key role for the control of adhesion. It is not only determined by the interaction potential due to the steric repulsion between the head groups but also by the chemical potential of the repellers in the non-adherent part of the shell [6] and determines the depth of the strong attraction potential [5,6].

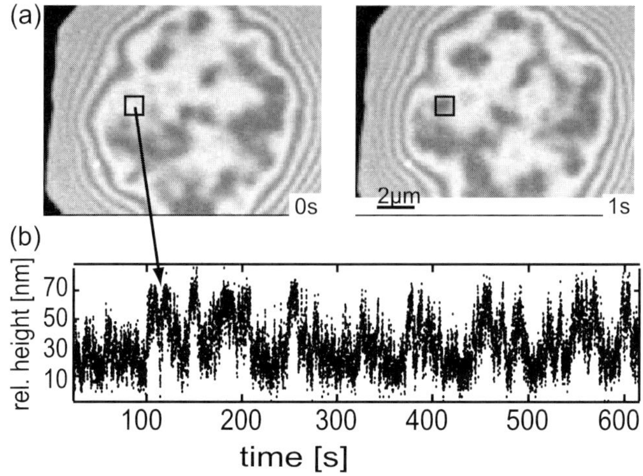

Fig. 6. a) Demonstration of dynamic surface roughness of vesicles caused by thermally excited bending undulations of lipid bilayer as visualized by temporal change of local contrast of leopard-like pattern of adhesion disc (*cf.* left and right image). The distance between dark and bright areas is about 250 nm. **b)** Time fluctuation of membrane-substrate distance (at encircled area of **a)**) demonstrating the oscillatory transition of membrane between state of strong and weak adhesion (courtesy of Stefanie Marx and Jörg Schilling).

Some of the interactions decrease hyperbolically with distance (such as $V_{\rm vdw}$ and $V_{\rm und}$ which decay with the square of the inter-membrane distance) while some decay exponentially (such as the electrostatic and steric forces). It is obvious that the superposition yields a general potential $V(h)$ consisting of an attractive and a repulsive branch but that the different interaction potentials cannot be determined separately. For moderate interaction potentials, $V(h)$ may be approximated by a harmonic potential [10, 11]

$$2V(h) = \partial^2 V/\partial h^2 (h - h_0)^2 = V_0''(h - h_0)^2. \tag{4.1}$$

Under this condition one always finds two limiting situations: at small generic attractions the soft shells hover over the surface at a distance of about 30 nm exhibiting strong flickering (since the adhesion induced tension is small). At increasing attraction the shells start to adhere to the surface forming a circular adhesion disc (as shown in Fig. 5b). The reason for this transition to the adherent state is that the undulations are suppressed by the adhesion-induced tension (which is in fact the case in Fig. 5b). In fact, the "tension-induced" transition from the free to the bound state has typical features of a first order phase transition. As demonstrated by the time series of distance fluctuations in Figure 6, one finds an intermediate situation where the membrane switches between a state of weak and strong adhesion, suggesting that the adhesion is determined by a double well potential with minima at a short ($h_o \sim 10$ nm) and a long distance ($h_1 \sim 40$ nm) [5,6].

The reason for the appearance of a double well potential is the chemical potential ($\mu_R = k_B T \cdot \ln c_R$) of the repeller molecules (of concentration c_R) in the non-adherent part of the membrane which controls the relative height of the two minima. At high repeller content the van der Waals minimum $V_{vdw}(h_o)$ is higher than the shallow minimum and the vesicle hovers 30–40 nm over the surface. With decreasing c_R the Van der Waals minimum is lowered and eventually becomes the absolute minimum resulting in the transition from the weakly to the strongly bound state. The analysis of the local interaction potential by the above flicker spectroscopy suggests the following adhesion laws:

- Since the two minima are separated by an activation barrier this adhesion transition is of first order and is a typical nucleation and growth process. Therefore the initial phase of the adhesion process consists in the formation of adhesion domains which can slowly merge into a single adhesion disc;
- The repeller chemical potential μ_R corresponds to the osmotic pressure π_R exerted by the reservoir of repellers in the bulk (non-adhering part) of the membrane. There exists a state of coexistence of weak and strong adhesion when the two minima exhibit the same depth. This occurs when the osmotic pressure becomes equal to the Van der Waals adhesion energy $\pi_R = W_{vdw}$ [5,12];
- The osmotic pressures of the repellers but also of the receptors in the non-adherent part of the membrane [12] relax the strong attraction forces by orders of magnitude as demonstrated in a model membrane study by Nardi *et al.* [14] where oppositely charged lipids were used as toy receptor-ligand pairs. This is absolutely necessary to maintain the mechanical equilibrium between the intrinsic membrane tension and the adhesion-induced tensions (which is determined by the Young equation) at the contact line.

5 How is adhesion controlled by membrane elasticity?

The physical laws controlling the state of adhesion and the shape of soft shells and of fluid droplets (*e.g.* of water on glass), respectively, share some common features but differ in one aspect: the surface elasticity. The free energy ΔG_{ad} of a partially wetting fluid droplet (that is the work gained by the partial spreading of a droplet) is equal to the gain in adhesion energy

$\Delta G_{\rm ad} = w \cdot A_{\rm c}$ (where w is the specific adhesion energy and $A_{\rm c}$ the contact area; *cf.* Fig. 5) minus the energy cost, $\sigma\Delta O$, associated with the increase, ΔO, of the surface area caused by the deformation of the initially spherical droplet (σ is the surface tension):

$$\Delta G = -wA_{\rm c} + \sigma \cdot \Delta O. \tag{5.1}$$

The work of adhesion of soft shells is also determined by these contributions but, in addition, one has to consider the elastic energy $\Delta G_{\rm b}$ associated with the bending of the fluid shell. The bending deformation of a thin (non-spherical) shell is determined by the mean curvature $H = 1/R_1 + 1/R_2$ (where R_1 and R_2 are the principal radii of curvature of the surface). If the thickness of the shell ($\Delta_{\rm s} \sim 4$ nm for bilayers) is small compared to its radius ($R \sim 10^3$ μm) the bending elastic energy associated with the deformation of the initially flat plate into the adhering shell is, according to Hook's law, $\Delta G_{\rm bend} = 1/2\kappa \int\int H^2 {\rm d}O$. κ is the two-dimensional bending elastic modulus of the membrane which is measured in units of energy. It is typically of the order of 25 $k_{\rm B}T$ for lipid layers at 37 °C.

The total free energy (work) of adhesion is thus

$$\Delta G_{\rm adh} = -wA_{\rm c} + \sigma\Delta O + 1/2\kappa \int\int H^2 {\rm d}O. \tag{5.2}$$

The state of adhesion and the shape of the adhering shell can in principle be calculated by application of the variation calculus methods [13]. This is a very complex problem which can only be solved analytically for the very simple situation of a single-component bilayer vesicle adhering weakly on a homogeneous smooth surface. The adhesion energy can then be accounted for by a harmonic interfacial interaction potential (introduced above). The problem has been solved by Lipowsky & Seifert [10] who showed that the state of adhesion and the vesicle shape depends in a subtle way on two parameters: the normalized adhesion energy $w^* = wR^2/\kappa$ and the reduced volume $v = V/(4\pi R^3/3)$ (where v is the ratio of the actual volume to the volume of a sphere with the same membrane area $A = 4\pi R^2$).

The balance between the gain in adhesion energy and the cost in bending energy has surprising consequences. For instance by monotonous deflation (decrease of v) a vesicle may undergo a transition from the free to the bound state which can, however, be followed by an unbinding transition due to local membrane budding. A second result is that the suppression of the membrane bending excitation (and thus of the undulation force) by adhesion induced membrane tension can induce a binding transition (tension-induced switching).

In the case of mixed membranes (such as the system of Fig. 4) the variational problem can only be solved numerically. Fortunately, the adhesion energy and even the force exerted by receptor-ligand pairs can be measured by interferometric analysis of the contour of soft shells close to the surface on the basis of elastic boundary condition as shown below.

6 Measurement of adhesion strength by interferometric contour analysis

The contour of adhering soft shells close to the contact line (which defines the circumference of the contact disc) exhibits the general shape shown in Figure 5c. The membrane is slightly bent at the contact line and goes smoothly over into a straight line region and the contour is thus completely defined by a contact curvature R_c and a contact angle θ_c. The straight line region is a consequence of the membrane tension induced by the adhesion. In analogy to the well-known Young law on the balance of surface tensions (at the triple line of partially wetting fluid droplets) the contact angle is related to the adhesion strength, W, according to

$$W = \sigma(1 - \cos\theta_c) \tag{6.1}$$

W has the dimension of a surface pressure (force per unit length) and is thus often called spreading pressure. Similarly, the contact curvature is determined by the balance of bending moments which provides a second relationship between the free adhesion energy and the contact curvature $W = \kappa/R_c^2$ [10, 15].

The two boundary conditions provide a powerful tool to determine the physical parameters W and σ through measurements of the geometric parameters θ_c and R_c (if the bending stiffness is measured in a separate experiment *e.g.* by flicker analysis [8]).

A more rigorous analysis of the membrane deformation close to the contact zone in terms of the boundary conditions provides a more general relationship for the contour in the direction, s, perpendicular to the contact line [6]

$$H(s) = \theta_c (s - \lambda) + \theta_c \lambda \exp\left\{-\frac{s}{\lambda}\right\}. \tag{6.2}$$

Where λ is the "capillary length" $\lambda = \sqrt{\kappa/\sigma}$, which is related to R_c as $R_c = \theta_c \cdot \lambda$. By fitting this curve to the contour of the adhering shell (as determined by the microinterferometric technique illustrated in Fig. 5) R_c and θ_c and thus W and σ may be determined with high precision. Note that λ has a simple meaning. It is the length over which the contour is determined by the bending elasticity before it becomes tension dominated and is typically $\lambda \sim 1\ \mu$m.

7 Switching on specific forces: Adhesion as localized dewetting process

The adhesion scenario changes dramatically when specific forces are switched-on. At low receptor densities of the order of $c_{\rm R} \leq 10\ \mu{\rm m}^{-2}$ (which are comparable to the situation in real cells) the adhesion area decomposes into domains of tight adhesion (formed by 2D assemblies of receptor ligand pairs) which are separated by areas of weak adhesions which may exhibit strong flickering. This phenomenon is reminiscent of the formation of adhesion plaques or focal adhesion contact sites of adhering cells and it is in both cases a consequence of adhesion-induced lateral receptor segregation. The adhesion is driven by the lowering of the short range (van der Waals) minimum of the double well potential by switching on the receptor-ligand interaction potential $U_{\rm oo}$ (where $U_{\rm oo}$ is of the order of 10 $k_{\rm B}T$ for integrin-mediated lock-and-key forces corresponding to an energy density of about 10^{-4} J/m^2). It can thus be understood in terms of a first order dewetting transition of a two-dimensional fluid.

It is important to note that the adhesion domain formation is a transient (nucleation and growth) process and that the domains merge in the time course of hours. The coarsening is driven by the line tension arising due to the bending deformation at the rim of the domains exhibiting a width $\xi = (\kappa/V'')^{1/4}$. V'' is defined in equation (4.1). ξ, which is defined in Figure 7a, is called the correlation length [6, 11] and is of the order of $\xi \sim 10$ nm (that is much smaller than the capillary length λ). As is well known from the dynamics of nucleation and growth, the merging of domains with time t is very slow and follows a $t^{1/3}$-law. In biomembranes the merging of adhesion plaques is expected to be further slowed down by the attraction between receptors, their coupling to focal adhesion complexes and actin bundles. The domain structure is thus a quasi-static state on biological time scales.

8 Measurement of unbinding forces, receptor-ligand leverage and a new role for stress fibers

The thermodynamic concept described above breaks down if the binding energy between receptor-ligand pairs becomes considerably larger than the thermal energy ($w \geq 10\ k_{\rm B}T$). This becomes most clearly evident if we consider a situation of soft shells adhering through a small number of pinning centers. These can be exposed and studied if one pulls the shell at the top by a force $F_\perp$ (applied through magnetic tweezers glued to the top pole of the shell; *cf.* Fig. 8 and [6]). The contact line exhibits relatively sharp protrusions of nearly triangular shape. A closer inspection also shows that the tip of the protrusion has rounded edges exhibiting a radius of

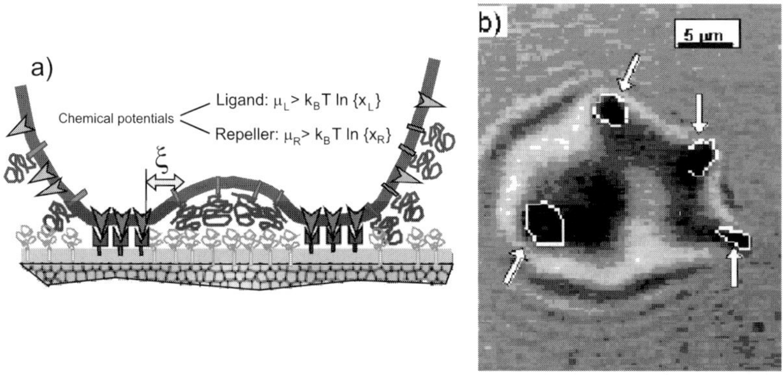

Fig. 7. a) Adhesion leads invariably to lateral segregation of receptors and ligands resulting in the formation of tightly bound adhesion plaques (distance $d_o \sim 10$ nm) separated by weakly adhering often flickering regions (with distance $d_1 \sim 30$ nm) determined by repeller repulsion and undulation forces. The adhesion may be described as a localized dewetting transition. Note that membrane forms a dimple near adhesion plaques extending over a correlation length of $\xi \sim 20$ nm. μ_L and μ_P are the chemical potentials of the repeller (x_i molar fractions). **b)** RICM image of cell model adhering through tight adhesion plaques (encircled and indicated by arrows).

curvature ρ_c [17]. These edges can only exist if they are stabilized by a line force **f** mediated by the membrane tension (*cf.* [15], Sect. 12). The vertical component $f_\perp$ which balances the force generated by the receptor-ligand pairs is related to the contour $\zeta(h)$ of the shell above the substrate (defined in Fig. 8a)

$$f_\perp = \kappa \left(\zeta''' + \rho_c^{-1}\zeta''\right) \tag{8.1}$$

(with $\zeta' = \partial\zeta/\partial h$). This equation follows from the boundary condition for the mechanical stability of an adhering plate with a curved contour and is derived in reference [15] (Sect. 12). It is a second order boundary condition since it depends on the third derivative of the vertical deflection of the shell and $\kappa \cdot \zeta'''$ is the gradient of the bending moment. However, for small radii of curvatures of the contour (ρ_c) the force $f_\perp$ becomes large and the second order boundary condition becomes important. The contour at distances $r > \lambda$ from the edge of the pinning center is no longer a straight line as in the regions of weak adhesion but the slope decreases with the inter-membrane distance h as $\zeta' = F\lambda/2\pi\sigma r$. The components of the line force $f_\perp$ perpendicular to the substrate can be determined by the following trick (illustrated in Fig. 8b). Since the membrane tension σ of fluid membranes is

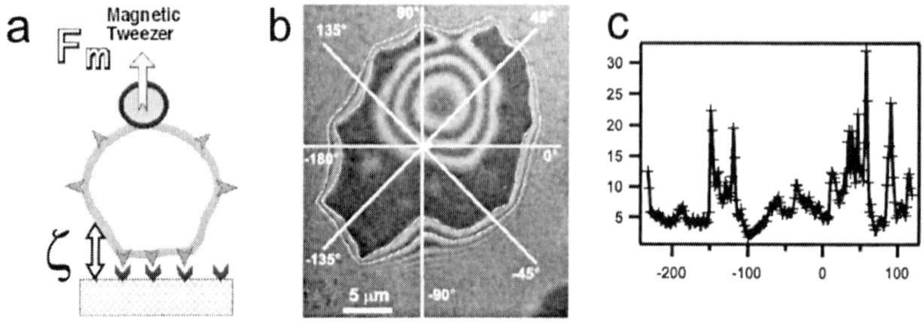

Fig. 8. a) Magnetic tweezer technique to expose pinning centers by pulling vesicle in vertical direction with force F_{m}. ζ is the distance between the substrate and the membrane. **b)** Exposure of pinning center along contact line by application of force at top vesicle with magnetic tweezer. The white lines parallel to the contour line mark contours of equal height above surface yielding local contact angle. Note that the contact line of the pinning center exhibits a triangular shape forming an angle with round tips of radius ρ_{c}. **c)** Plot of line force (force per unit length) along contact line from which receptor-ligand unbinding force can be measured.

isotropic it can be measured by analyzing the contour at any region of weak adhesion along the contact line. If $\theta_{\mathrm{c}}(s)$ is measured along the contours the vertical line force is:

$$f_{\perp} = \sigma \sin \theta_{\mathrm{c}}(s) \tag{8.2}$$

and is related to the applied force as $F_{\perp} = \sigma \int \sin \theta_{\mathrm{c}}(s)\, \mathrm{d}s$. $\theta_{\mathrm{c}}(s)$ can often be measured by the distance $\Delta d(s)$ between the first and second interference fringe yielding $f_{\perp}(s)$. Figure 8b shows a plot of $f_{\perp}(s)$ along the contour and the sharp peaks clearly define the pinning centers. By simultaneously measuring $f_{\perp}$ and the radius of the pinning centers while $F_{\perp}$ is increased, one can measure the line force where the bonds break and thus obtain the binding force per link if the receptor density is known.

A very surprising result is that bonds break at very weak forces (~1 pN). It has been postulated [6] that this is caused by the leverage of the receptor-ligand bond due to an amplification of the unbinding force by the bending moment exerted by the external force on the site of the receptor-ligand pairs. This finding attributes a new role to the stabilization of sites of cell adhesion by focal adhesion complexes associated with stress fibers. By this trick, nature prevents the disrupture of cells subjected to large hydrodynamic shear stresses such as endothelial cells lining in the blood vessels.

9 An application: Modification of cellular adhesion strength by cytoskeletal mutations

The microoptical method opens new possibilities to quantify the cell adhesion strength in order to explore the effect of mutations or deceases on cell-adhesion. Very often cytoskeletal mutations do not affect the phenomenological behavior of cells substantially, and quantitative methods of characterization are necessary [18]. On the other side, specific mutations provide a valuable tool to induce distinct structural alterations of the composite membrane for systematic studies of the correlations between membrane structure and adhesion. Thus, the effect of the coupling strength of the actin cortex to the lipid/protein bilayer has been studied by removal of the actin-membrane coupling protein talin (*cf.* Fig. 2a).

In order to measure the work of adhesion, the membrane tension and the bending modulus simultaneously, the changes of the surface profile by hydrodynamic shear fields were measured. As shown in Table 1 the membrane bending stiffness of the talin-deficient mutant is decreased by a factor of 20 and the adhesion strength W by a factor of 5. In fact, the bending stiffness of the mutant agrees well with that of the pure bilayer containing 50 mole% cholesterol. Two major messages of this experiment are: (i) talin is indeed essential for the coupling of the actin cortex to the membrane and (ii) the adhesion strength is closely related to membrane bending stiffness.

Table 1. Effect of knock-out of talin on the bending stiffness of the cellular envelope and on adhesion strength (according to Simson *et al.* [18]).

Cell type	κ $[k_\mathrm{B}T]$	W_ad $\left[\mu\mathrm{J/m}^2\right]$
wild type	1000	20
mutant	50	5

10 Conclusions

Insight into the physical basis of cell adhesion is gained by interferometric studies of vesicles or cells adhering on substrates which mimick the role of cell or tissue surfaces and adhesion forces can be measured locally by analysis of the contour of adhering shell in terms of the equilibrium of elastic stresses close to the adhesion zone.

The adhesion process is controlled by interplay of short range specific forces between repeller-ligand pairs and long range repulsion between cell surface molecules forming the glycocalix. This competition can lead to spontaneous segregation of receptors resulting in the formation of tight adhesion domains similar to the generation of focal contact sites of adhering cells. The adhesion process shares common features with a first order dewetting transition which explains why a rather small number of receptors can drive cell adhesion. The merging of the adhesion plaques is very slow (typical for coarsening processes) and could be further slowed down by the coupling of cell receptors to the actin cortex. The adhesion strength is closely related to the membrane elasticity and thus to the receptor-cytoskeleton coupling and its measurement can yield valuable insight into modification of the actin cortex by mutations or cell damaging agents.

Our model membrane studies attribute an important role to the glycocalix, namely the maintenance of mechanical equilibrium of adhering cells. Mobile repeller molecules (but also non-adherent receptors) in the non-adhering part of adhering soft shells exerts a two-dimensional osmotic pressure on the adhering membrane fraction which relaxes strong receptor-ligand forces. Surprisingly weak forces on adhering shells (*e.g.* hydrodynamic shear forces) can disrupt receptor-ligand pairs due to leverage. This could be the reason for the strategy of cells to stabilize adhesion plaques by stress fibers.

We greatly appreciate the help of Nikita Ter-Oganessian with the preparation of the manuscript.

A Appendix: Generic interfacial forces

Below we summarize the essential generic interfacial forces governing the adhesion of soft shells. The modification of these forces by steric repulsion associated with membrane undulations (or the dynamic surface roughness) are discussed at the end. The dominant forces are the ubiquitous Van der Waals and electrostatic forces, short range steric or solvation induced forces and steric repulsion mediated by repellers of the glycocalix and/or of the target tissue.

The Van der Waals attraction between a membrane of thickness δ ($\sim$2 nm) and planar surface can be expressed as

$$V_{\mathrm{VdW}} \approx -\frac{H}{12\pi}\left(\frac{1}{h^2} - \frac{1}{(h+\delta)^2}\right) \tag{A.1}$$

where H is the Hamaer constant and is of the order of $k_{\mathrm{B}}T$. The adhesion energy for a distance of $h \approx 10$ nm (the distance enforced typically by repellers) is $V_{\mathrm{VdW}} \sim 10^{-6}$ J/m^2. It is by an order of magnitude smaller for two interacting bilayers. For bare membranes d is about 2 nm (*cf.* Evans *et al.*) and $V_{\mathrm{VdW}} \sim 10^{-5}$ J/m^2. The electrostatic interaction between two differently charged membranes in the presence of electrolyte exhibiting charge densities (per unit area) σ_1 and σ_2, respectively, is best expressed in terms of the disjoining pressure $P = \partial V/\partial h$ [19].

$$P(h) = \frac{8\pi}{\varepsilon} \frac{\left(\sigma_1^2 + \sigma_2^2 - 2|\sigma_1|\sigma_2 \cos \kappa h\right)}{(2 \sin \kappa h)^2} \propto \mathrm{e}^{-\kappa h} \tag{A.2}$$

where ε is the dielectric constant ($\varepsilon \approx 80$) and κ is the Debye screening length which depends on the salt concentration

$$\kappa_{\mathrm{D}}^2 = \frac{8\pi e^2 c}{\varepsilon k_{\mathrm{B}} T} \,. \tag{A.3}$$

(Note that for room temperature the screening length (in units of nm) is $\kappa_{\mathrm{D}}^{-1} = 3.08\sqrt{c}^{-1}$ where the salt concentration c is measured in units of mole/l).

If the two membranes exhibit the same sign of the surface charge the electrostatic force is repulsive and counteracts the Van der Waals attraction. An extensive discussion of adhesion mediated by attractive electrostatic forces between oppositely charged receptor-ligand pairs can be found in [14].

There are several contributions to the short range steric repulsion. A contribution which becomes only relevant at distances of $\sim$0.2 nm is the dehydration force associated with the removal of hydration water between the interfaces. The potential decays exponentially: $V_{\mathrm{hyd}} = V_0 \exp\{-h/\lambda\}$ with a screening length of $\lambda \sim 0.3$ nm [20]. A steric short range repulsive force arises for the case of the adhesion of cells or cell models on biofunctionalized solid supports due to the repulsion exerted by the surface protein layer. It dominates the short range repulsion at full hydration and can also be expressed in terms of an exponential law $V_{\mathrm{rep}} = K/h \exp\{-h/\tau\}$, where K is the compression modulus of the protein cover and τ the thickness of the biofunctional film [12].

The third and under practical conditions most important short range force is the repulsion mediated by the glycocalix. It depends critically on the surface concentration c_{R} and the mobility of repeller molecules (besides the Van der Waals radius of the headgroup). An approximate expression for the interaction potential can be given for the model systems where lipids with polyethyleneoxide head groups are used as repellers (*cf.* Bruinsma *et al.* [6]). The size of the head group is determined by the Flory radius of the flexible macromolecule. For the specific (saturation) situation that the concentration of the repellers is adjusted in such a way that the head groups (of radius R_{g}) start to overlap ($c_{\mathrm{R}} R_{\mathrm{g}}^2 \approx 1$) the repulsion potential can be expressed as (*cf.* Bruinsma *et al.* [6])

$$V_{\mathrm{rep}}\left(h\right)=\frac{\pi^{2}}{6}k_{\mathrm{B}}Tc_{\mathrm{R}}\left(\frac{R_{\mathrm{g}}}{h}\right)^{2}\exp\left\{-1.5\left(\frac{h}{R_{\mathrm{g}}}\right)^{2}\right\}. \tag{A.4}$$

For cells the situation is more complicated but it is still possible to define an effective Flory radius of the semiflexible head groups such as chains of IgG-like domains and a similar exponential expression is expected to hold.

Much more important than the detailed structure of the membrane proteins forming the glycocalix is the fact that the repellers are in general mobile and can exchange between the adhesion zone and the bulk of the non-adherent membrane fraction. The repeller molecules in the non-adherent (bulk) membrane fraction provide a reservoir of fixed chemical potential $\mu_{\mathrm{R,bulk}}=k_{\mathrm{B}}T\ln\left\{c_{\mathrm{R}}^{\circ}/R_{\mathrm{g}}^{2}\right\}$ which modifies the repulsive interaction potential. In equilibrium $\mu_{\mathrm{R,bulk}}$ must be equal to the repeller chemical potential, μ_{R}, in the adhesion zone. The effective thermodynamic potential of the repellers in the adhesion zone is determined by the sum of the bare repulsion potential V_{rep} (Eq. (A.4)) and the translational entropy $(k_{\mathrm{B}}T\cdot\ln c_{\mathrm{R}}\left(h\right)/R_{\mathrm{g}}^{2})$ where the concentration of the repellers depends on the interfacial distance h. The effective potential which depends now on c_{R}° is

$$V_{\mathrm{R,eff}}\left(h\right)=k_{\mathrm{B}}T\left\{1-\exp\left\{-\frac{\pi^{2}}{6}\left(\frac{R_{\mathrm{g}}}{h}\right)^{2}\exp\left\{-1.5\left(\frac{h}{R_{\mathrm{g}}}\right)^{2}\right\}\right\}\right\}. \tag{A.5}$$

It is now helpful to consider two limiting situations:

For large interfacial distances $h\gg R_{\mathrm{g}}$ one obtains

$$V_{\mathrm{rep}}\left(h\right)\approx k_{\mathrm{B}}Tc_{\mathrm{R}}^{\circ}\left(\frac{R_{\mathrm{g}}}{h}\right)\exp\left\{-\frac{3}{2}\left(\frac{h}{R_{\mathrm{g}}}\right)^{2}\right\} \tag{A.6}$$

which corresponds to the interaction potential at fixed concentration c_{R}° of the repeller.

For $h\ll R_{\mathrm{g}}$ one obtains

$$V_{\mathrm{rep}}\left(h\right)=k_{\mathrm{B}}Tc_{\mathrm{R}}^{\circ} \tag{A.7}$$

which is the maximum value of $V_{\mathrm{rep}}\left(h\right)$ and corresponds to the two-dimensional osmotic pressure of the repeller.

The interplay between Van der Waals attraction and the thermodynamic repulsive interaction potential of the repellers has two important consequences

- The interaction potential can exhibit two minima corresponding to a state of strong adhesion (the Van der Waals minimum) and a state of weak adhesion. On physically or chemically rough surfaces (exhibiting lateral inhomogeneities of the interaction potential V_{VdW}) this can lead to a decomposition of the adhesion zones into domains of tight adhesion separated by regions of weak coupling;

- With increasing repeller concentration the depth of the Van der Waals minimum is shifted to higher energies and reaches the value of the shallow minimum at $c_{\mathrm{R}}^{\circ} R_{\mathrm{g}}^{2} \approx 0.6$. Since the two minima are separated by an activation barrier this leads to a first order unbinding transition of the adhering shell driven by the chemical potential of the repellers;

- The binding strength of the tight adhesion is strongly reduced by the lateral osmotic pressure exerted by the repeller molecules in the non-adherent membrane fraction, a point discussed in the main text (*cf.* Ref. [21] for an experimental verification).

Consider now the effect of undulation forces. This beautiful concept was introduced by Helfrich to explain the spontaneous swelling of lipid multilayers. It is an entropic force associated with the gradual freezing-in of long wavelength bending excitations if a flickering membrane approaches a surface. The decrease in entropy gives rise to a repulsive disjoining pressure P_{und} which can be estimated as follows [22]. The bending excitations are statistically excited local events and local deflections $U(\vec{r}, t)$ decay with a correlation length ζ_{p} [22, 23]. Because ζ_{p} ($\sim$1 μm) is small compared to the size of vesicles or cells (erythrocytes) we can consider the soft shell to be composed of small segments (cushions) of dimension $\zeta_{\mathrm{p}} \cdot \zeta_{\mathrm{p}}$ which exhibit independent Brownian motions in the normal direction. In the Helfrich image a disjoining pressure arises due to the local collisions of the membrane segments with the wall, similar to the pressure generated by an ideal gas through the momentum transfer onto the wall. Since equipartition theory predicts an energy transfer of $k_{\mathrm{B}}T$ per collision the undulation pressure is

$$P_{\mathrm{und}} = \frac{k_{\mathrm{B}}T}{2\zeta_{\mathrm{p}}^{2} \cdot \langle h \rangle} \tag{A.8}$$

where $\langle h \rangle$ is the average distance between the two interfaces.

The correlation length ζ_{p} is proportional to the average distance $\langle h \rangle$ according to $\zeta_{\mathrm{p}} \approx \sqrt{\kappa / k_{\mathrm{B}}T}\, \langle h \rangle$. This relationship follows from the conditions that the mean square amplitude (the so-called roughness) $\langle u_{\mathrm{L}}^{2} \rangle$ of a membrane segment $L * L$ is $\langle u_{\mathrm{L}}^{2} \rangle = (k_{\mathrm{B}}T/\kappa)\, L^{2}$. This relationship follows by integrating over all bending modes of amplitude $u_{\mathrm{q}}^{2} = k_{\mathrm{B}}T/\kappa q^{4} L^{2}$ from $q = \pi/L$ to $q = \pi/\Delta$, with Δ the membrane thickness. The roughness of a membrane at average distance $\langle h\rangle$ must be $\left\langle u_{\mathrm{L}}^{2}\right\rangle \leq \langle h\rangle^{2}$. It therefore follows

$$P_{\mathrm{und}} = c\frac{k_{\mathrm{B}}T}{\langle h^{3}\rangle}\cdot \tag{A.9}$$

For one-component membranes the constant c has been more rigorously calculated by renormalization group calculations [23] and by Monte Carlo simulations to $c = 0.1$ and is considered as a universal number. The situation may be quite different for mixed membranes due to concentration fluctuations induced by the interaction of soft membranes with surface (S. Marx *et al.*, unpublished data of authors laboratory).

References

[1] H. Lodish, D. Baltimore, A. Berk, S. Zipursky, P. Matsudaia and J. Darnell, *Molecular Cell Biology*, 3rd ed. (W.H. Freeman, New York, 1995).

[2] T.A. Springer, *Cell* **76** (1994) 301-314; M.S. Diamond and T.A. Springer, *Curr. Biol.* **4** (1994) 506-517.

[3] P.Y. Chan, M.B. Lawrence, M.L. Dustin, L.M. Ferguson, D.E. Golan and T.A. Springer, *J. Cell Biol.* **132** (1996) 465-474.

[4] J.G. Garcia and K.L. Schaphorst, *J. Invest. Med.* **43** (1995) 117-126.

[5] E. Sackmann and M. Tanaka, *Trends Biotechn.* **18** (2000) 58; R. Bruinsma, A Behrisch and E. Sackmann, *Phys. Rev. E* **61** (2000) 4253-4266.

[6] Z. Guttenberg, A. Bausch, B. Hu, L. Moroder, R. Bruinsma and E. Sackmann, *Langmuir* **16** (2000) 8984-8993; G. Wiegand, K. Neumaier and E. Sackmann, *Appl. Opt.* **37** (1998) 6892-6905.

[7] Bin Hu *et al.*, *Biochemistry* **39** (2000) 12284-12294.

[8] W. Helfrich, *Z. Naturf*, **33A** (1978) 305-312; U. Seifert, *Phys. Rev. Lett.* **74** (1995) 5060-5063; J. Rädler, T. Feder, H. Strey and E. Sackmann, *Phys. Rev. E* **51** (1995) 4526-4531.

[9] A. Zilker, M. Ziegler and E. Sackmann, *Phys. Rev. A* **46** (1992) 7998-8001.

[10] R. Lipowsky and U. Seifert, *Langmuir* **7** (1991) 1867.

[11] P.S. Swain and D. Andelmann, *Langmuir* **15** (1999) 8902.

[12] G.I. Bell, M. Dembo and P. Bongrand, *Biophys. J.* **45** (1984) 1051; P. Bongrand, *Rep. Progr. Phys.* **62** (1999) 921.

[13] If we start from a spherical shell with contour $U(\vec{r})$ the deformed shape of the adhering shell is $U(\vec{r}) + \zeta(\vec{r})$ (where $\zeta(\vec{r})$ is the local deformation). One can then write $\sigma\Delta O = \frac{1}{2}\int\int\{\mathrm{grad}\,\zeta(\vec{r})\}^{2}\,\mathrm{d}x\mathrm{d}y$ and $H = \Delta\zeta(r)$

[14] J. Nardi, R. Bruinsma and E. Sackmann, *Phys. Rev. E* **58** (1998) 6340-6354.

[15] L.D. Landau and E.M. Lifschitz, *Theory of Elasticity* (Pergamon Press, New York, 1986).

[16] R. Bruinsma and E. Sackmann, *Comptes Rendus de l'Académie des Sciences Ser. IV – Phys.* **2** (2001) 803-815.

[17] From the apex angle ϕ of the triangular shape one can determine the tangential force $f_{\|}$ according to $f_{\|} = (\kappa\cdot\lambda)^{1/2}\,\theta_{\mathrm{c}}^{2}\,(\pi/2-\phi)$ (*cf.* Albersdörfer *et al. Europhys. Lett.* **42** (1998) 227 for details).

[18] R. Simson, E. Walraff, J. Faix, J. Niewöhner, G. Gerish and E. Sackmann, *Biophys. J.* **74** (1998) 514-522; M. Schindl, E. Walraff, B. Deubzer, W. Witke, G. Gerish and E. Sackmann, *Biophys. J.* **68** (1995) 1177-1190.

[19] V.A. Parsegian and D. Gingell, *Biophys. J.* **12** (1972) 1192.

[20] S. Marcelja and R. Radic, *Chem. Phys. Lett.* **42** (1976) 129-133.

[21] Z. Guttenberg, B. Lorz, E. Sackmann and A. Boulbitch, *Europhys. Lett.* **54** (2001) 826-832.

[22] W. Helfrich and R.M. Servuss, *Il Nuovo Cimento D.* **3D** (1984) 137.

[23] R. Lipowsky in "*Handbook of Biological Physics*", Vol. 1B, Ch. 11 (Elsevier, Amsterdam, 1995).

COURSE 8

BIOLOGICAL PHYSICS IN SILICO

R.H. AUSTIN

Department of Physics,
Princeton University,
Princeton NJ 08544, USA

Contents

BIOLOGICAL PHYSICS IN SILICO

R.H. Austin

Abstract

The following set of notes outlines the work that my colleagues and I have been doing for the past 10 years which have exploited the opportunities that that silicon micro to nanomachining technologies provide in biological physics. This work started with my graduate student Wayne Volkmuth and has grown to include the efforts of a great many people, including but being limited to the following faculty: Tom Duke (University of Cambridge Physics), Ted Cox (Princeton Molecular Biology), Harold Craighead (Cornell University Applied Physics), Jim Sturm (Princeton University Electrical Engineering), Steve Chou (Princeton University Electrical Engineering), Paul Kohl (Georgia Institute of Technology), Lois Pollack (Cornell University Applied Physics, Bill Eaton (National Institutes of Health), Klaus Gehardt (Bochum University, Germany); and the following students and post docs: Wayne Volkmuth, Jim Brody, Judith Castellino, Rob Carlson, Olgica Bakajin, Jeff Chou, Jonas Tegenfeldt, Richard Huang, Christelle Prinz and Nick Darnton. To everybody, thanks for all the hard days and nights.

This work has been generously supported by the National Science Foundation, The Office of Naval Research, The National Institutes of Health, and the Defense Research Projects Administration.

1 Why micro/nanofabrication?

The integrated circuit revolution, made possible by microfabrication technology, is just now entering the world of biology. I am old enough to have

This work has been generously supported by the following granting agencies of the United States of America: Office of Naval Research, The National Institutes of Health, The National Science Foundation, the Defense Research Project Administration. Without their support, none of this work could have been done. I would also like to thank my lovely wife Shirley Chan, a formidable biological physicist, for helping me deal with all things. Henrik Flyvbjerg deserves so much credit for expertly and patiently organizing the entire proceedings.

built a Heath Kit tube tuner and amplifier, and to notice that the transfer characteristics of a FET resemble a triode. So, I know the revolution that can happen. When I first realized that that micro and nanofabrication technologies could be used to change the face of biology I launched into a major campaign to realize this goal. I have long since been passed by people smarter and more energetic than I, but I must say it has been a good trip.

The world of biology is inherently on the micron and below scale, and this is where microfabrication lives. We can process, examine, move biological objects at their natural length scale. Further, the world of biology is all about heterogenity: no two objects are alike. Sometimes, the rarest one is the most interesting. Microfabricated devices, which basically are "flatlanders" can find those rare ones.

I WON'T speak much in these lectures about "gene chip" technology, which uses spots of hybridized small DNA molecules (olgionucleotides) to measure patterns of gene expression. This technology is under furious development by many gene chip companies and while very powerful is reaching maturity (and past-IPO status). Instead, I want to talk a bit about efforts to utilize physics to probe biological objects. It will be a very parochial view from my own lab, so if what I say doesn't interest you be aware that there are many other excellent efforts in physics departments across the country.

To give focus and motivation to this introuction, let's discuss how we might use micro/nanofabrication to attack the real killer out there: cancer.

Cancer is basically a break-down in the control logic of the cell: instead of behaving like a well behaved, differentiated, loyal part of some organ, in cancer a cell becomes a rebel and ignores the control rules. Sometimes this can happen due to a single gene mutation, but usually it is caused by a what is called genetic instability: many mistakes in the genome and the attendant breakdown in the highly controlled biological networks that control organisms. Cancer is thus difficult because it is not due to some single organism but rather to the summed effects of many perturbants to a system.

It is often a mistake amongst physicists to think that DNA is all of biology. It is true that DNA contains the code for you. There are about 3 billion basepairs in the human genome, or about 1 meter of DNA. The total length of DNA reaches from the center of the sun to about Pluto. Nearly all the 1014 cells in your body contain the SAME genome. Probably each cell has a slightly different genetic sequence. But, it isn't the genome that makes each cell different. Promoter and repressor proteins, which bind to specific parts of the genome, control expression. The CYTOPLASM of the cells contains the DYNAMIC control information. The DNA is the ROM, the proteins are the OS that makes a liver cell a liver cell. That's why there was all the excitement about Dolly: they took the NUCLEUS from a fully differentiated cell in the udder, put it into the egg cell of another sheep which presumably had the right protein content to reset the clock, and transformed a mature cell into a "fertilized" egg. In some respects, this is what cancer does: goes BACK in time, to embryo level. Since I am among non-biologists, I can let you in on a little secret: biology experiments basically never work MOST of the time, once in awhile things work and you publish that one good result.

Shhhhh....keep it quiet.

In the case of Dolly, it took about 300 tries to get a egg that developed into a normal sheep. And I also note that "Dolly", the world's first cloned mammal, has arthritis, one of her creators said today, heightening fears that cloning causes genetic defects that would make animal clones unsafe for use in human medicine. Professor Ian Wilmut, of the Edinburgh-based Roslin Institute, who led the team that cloned the sheep, said defects were possible and that cloning methods were still "inefficient". (Manchester Guardian, January 4, 2002.)

Fig. 1. Dolly(s).

There is a rather new term being used by biologists that gives a word to the realization that the genome itself while important is by now means the final word in what makes all the cells in your body so different... after all, they all contain the same genome but morphologically and chemically are quite different. The word is "Epigenetics", meaning that there is something over and above genomics per se, the sequence of basepairs in the genome. Epigentics can narrowly mean the chemical modification of the DNA by the addition of typically a methyl group, or more broadly the dynamic occupation of promoter and repressor sites on the genome. This is an extremely difficult area to get at using conventional biological tools. Epigenetics almost by definition involves studying the unique behavior of (ideally) single cells as a function of time, watching the way that the system controls the expression of the genome and the subsequent response of the organism. It is an impossible task, rather like trying to compute the wavefunction of the world if I can draw an analogy. However, in principle the technology I discuss, in the hands of people smarter and with more energy that I can, can make a serious attack on this goal. Unfortunately, much of what I discuss at the start must be of a technological nature because that is where the main raodblock is at present.

My lectures at the les Houches summer school consisted of 6 subjects, although I badly misjudged my time and crushed things at the end, helping nobody in the process. Some of these lectures are new, some are heavily based upon previously published work. Unlike many of the les Houches speakers, I am a simple experimentalist who still tries to keep in the lab, and piles on teaching responsibilities and obligations to my community of fellow physicists and biology the work load can be crushing... but exciting, at times.

Lecture 1a: Why silicon micro/nanomachining? What problems are you going after?

Lecture 1b: Hydrodynamic and electrodynamic transport in a 2.5 D world.

Lecture 2: Fractionation of DNA in a micro/nano world

Lecture 3: Going after epigenetics

Lecture 4: Cell Fractionation in a micro world

Lecture 5: Protein folding and dynamics using a micromachined device

I have made a brief stab at the Introduction, now on to Lecture 1.

Abstract

I'm going to start with a basic discussion of the Navier-Stokes equation in a flat world and show how it is actually a rather subtle subject. Applications will be discussed later. We show that by applying a possibly complicated set of boundary values in the $x - y$ plane we can properly model and control flows in planar microfabricated structures. We also show that the hydrodynamic impedance change in moving from one set of confining structure to another greatly influences the streamlines of the flow pattern in somewhat non-intuitive ways. We consider how understanding of hydrodynamics in 2 1/2 D can help us form a thin stream of one liquid imbedded in another liquid, that we refer to as "injector jet". Finally we extend these ideas to possibilities of precisely controlled complex flow patterns in microfabricated fluidic devices.

This talk is directly driven by our work using micro/nanofabrication in biotechnology and molecular biology. I am NOT interested per se in nanofluidics, I only use it as a tool to do biological physics. I have learned, to my sorrow, that if you DON'T understand micro/nanofluidics you make major mistakes. We transport objects in our chips using primarily two forces, hydrodynamic and electrophoretic. I want to stress right away that the two forces are very closely connected because of the counter-ion shielding of polyelectrolytes in solution. This means that ions moving in the solution pull via hydrodynamic drag on the polyelectrolyte and this shearing action basically cuts off long-range hydrodynamic interactions. This has profound consequences. For the second part of lecture 1 I will discuss the dielectrophoretic response of DNA molecules, because I think that this aspect of the response of a polymer to an AC electric field is a fundamental part of how you can move and separate molecules in silico.

1 Introduction: The need to control flows in 2 1/2 D

The problems of fluid transport that are usually considered describe fluid flow in pipes, a system in which one dimension, the length, is much greater than the two other dimensions. There are, however, many occasions where one wants to control fluid flow in a sheet, a system where two of the dimensions, length and width, are much larger than the third, the height. What we call here "2 1/2-D hydrodynamics" describes such sheet flow. A familiar example of a 2 1/2-D system is the flow of air that gives rise to the weather. A typical pattern of clouds shown in Figure 1 arises from the turbulent flow pattern of air in the layer of atmosphere that is only at the most 10 km thick which is much smaller than Earth's diameter of 12 700 km. The turbulent nature of this sheet flow makes the weather so notoriously difficult to predict. The 2 1/2-D flows in the low Reynolds number world of the microfabricated chip can be predicted and controlled, particularly if some basic aspects of hydrodynamics under the unique conditions of low Reynolds number flow are understood.

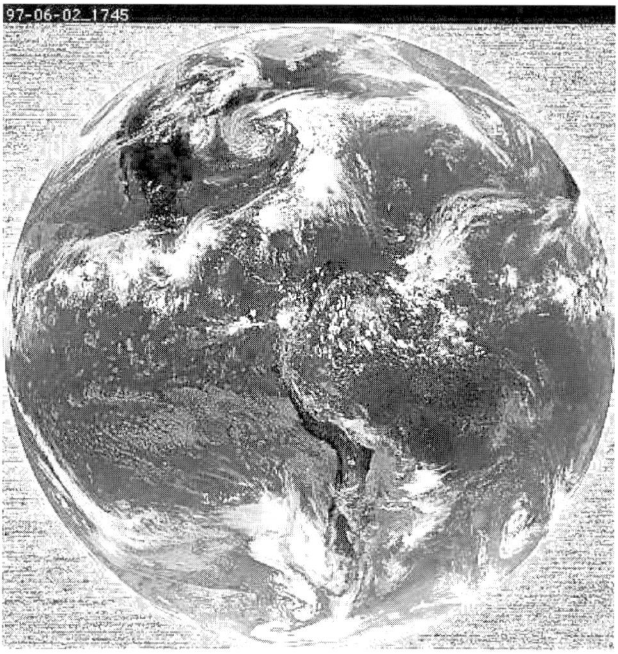

Fig. 1. A snapshot from GEOS-8 of clouds over the surface of the earth. Courtesy of NASA-Goddard Space Flight Center.

Understanding of 2 1/2 D flows is of great importance for the rapidly developing field of bio-device miniaturization. Over the past decade there have been immense efforts directed towards miniaturization of bio-analysis systems through applications of microfabrication. The goal is to integrate different micro components into a single micro total analysis system (μTAS) [1] and run an array of such systems in parallel. So far, many essential components of the μTAS have been demonstrated to perform better than their large scale equivalents. New technologies for sorting and mixing of biomolecules [2] and for fractionation of cells from blood [3] have also arisen.

Highly integrated lab-on-a-chip devices have also been proven feasible [4–6]. For example, capillary electrophoresis (CE) on microchips has been applied to the analysis of nucleic acids, amino acids, and other types of samples [7]. Other microfabricated electrophoretic devices, such as entropic trap arrays [8], could utilize new mechanisms to enable faster fractionation of macromolecules with better resolution than conventional gel electrophoresis. Various types of microfluidic pumps [9] which can potentially be used to transfer samples on microchips have also been demonstrated. Many other applications, such as polymerase chain reaction (PCR) [11, 12] and fluorescence-activated cell sorting (FACS) [13, 14], have been performed on microchips. Microfabrication is the method of choice for integration of all these miniaturized components into a micro-total analysis system. Since microfabrication is basically a set of planar processes applied to planar substrates, the fluid flows in microfabricated devices are often effectively two dimensional. The thickness of the fluid flow sheet, however, cannot be ignored. The flow in these systems is NOT uniform and unbounded in the $x-y$ plane. The z dimension, although small compared to the characteristic sizes of the $x-y$ plane, cannot be set to zero. In this lecture we refer to such geometries (shown in Fig. 2) as "2 1/2" dimensional.

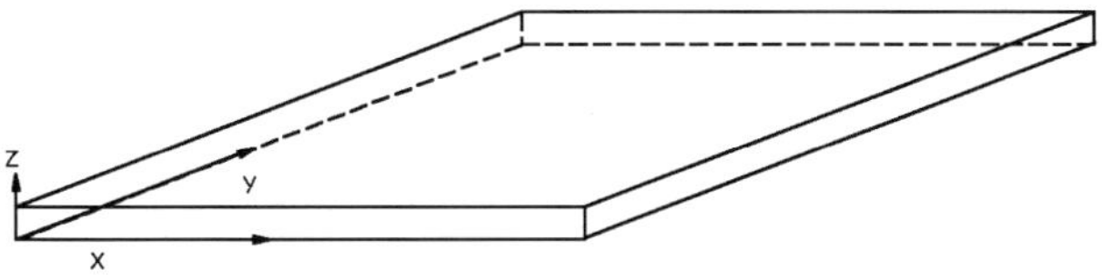

Fig. 2. The xyz coordinate system considered for the rest of this lecture. The etch depth h in the z direction is considerably smaller than any confining boundaries in the $x-y$ plane.

2 Somewhat simple hydrodynamics in 2 1/2 D

In the derivation of the Navier-Stokes equation, Newton's law is of course the starting point [15]. We consider $\vec{F} = m\vec{a}$ for a small element of fluid of mass $\mathrm{d}m = \rho\,\mathrm{d}x\,\mathrm{d}y\,\mathrm{d}z$ under the influence of a force. Since when using Newton's law one always has to follow the motion of the same atoms, the elemental boundaries $\mathrm{d}x$, $\mathrm{d}y$ and $\mathrm{d}z$ enclosing the mass $\mathrm{d}m$ change as the tubing twists and turns out that in order to find the true acceleration one has to worry about the spatial derivatives as well as the simple $\partial\vec{v}/\partial t$. This transforms the simple total derivative into a convective derivative and in the end the acceleration $\vec{a}$ is broken into two terms. Some careful thought, beautifully described in Feynman's Lectures, [16] then yields the famous Navier-Stokes Equation that describes the movement of a viscous fluid of density ρ and viscosity η:

$$-\nabla P + \eta \nabla^2 \vec{v} = \rho \left(\frac{\partial \vec{v}}{\partial t} + (\vec{v} \cdot \nabla)\vec{v} \right). \tag{2.1}$$

The above equation says that the forces per unit volume due to a pressure gradient (∇P) and shear ($\eta \nabla^2 \vec{v}$) are equal to the mass per unit volume (ρ) times the acceleration of the fluid. The right hand side terms are sometimes referred to as the "inertial forces", although they are not forces in the strict physics use of the word. If there are additional forces acting on the fluid (for instance, gravity), these terms are just added onto the left hand side ($\rho \vec{g}$). The Navier-Stokes equation is a non-linear equation in velocity v which makes it analytically unsolvable in a general case. However, when appropriate approximations are made, the Navier-Stokes equation can be reduced and solved.

A useful concept in fluid dynamics is the ratio of the inertial forces (the right hand side of Eq. (2.1)) to the viscous forces for steady flow because it determines what terms of the Navier-Stokes equation can be ignored in a particular situation. In steady flow, the velocity is independent of time, so $\partial \vec{v}/\partial t = 0$. The magnitude of the inertial forces expressed in the term $\rho(\vec{v} \cdot \nabla)\vec{v}$ is $\rho v^2/l$, where l is a typical length scale over which the velocity changes, *i.e.*, it is the effective size of the spatial derivative $\mathrm{d}\vec{r}$ in Navier-Stokes Equation. The viscous force, $\eta \nabla^2 \vec{v}$ goes as $\eta v/l^2$. This ratio, referred to as the Reynolds number is usually written as:

$$R_{\mathrm{e}} = \frac{\rho l v}{\eta} \cdot \tag{2.2}$$

In the microfabricated world that we are interested in, with water as the transporting fluid, the typical dimensions of the system set some limits to R_{e} and allow us to significantly reduce the Navier-Stokes equation. The length over which velocity changes l is at most 0.01 cm, the density ρ is 1 gm/cm^3 and the viscosity η is 0.01 gm/cm/s. Hence, R_{e} is less than 1 for fluid velocities less than 100 cm/s, a huge velocity on a chip. Since R_{e} in the chip is low, we can safely drop the non-linear inertial term $\rho(\vec{v} \cdot \nabla)\vec{v}$ from equation (2.1). In addition, we are interested in the steady state flow with no time dependence of the velocity fields ($\partial \mathbf{v}/\partial t = 0$). The vector equation that describes our system is then:

$$\eta \nabla^2 \vec{v} = \nabla P. \tag{2.3}$$

The second equation that our system obeys follows directly from the incompressibility of water. At the modest pressures and flows found in the world of microfabricated chips the compressibility of water is negligible.

Therefore, all solutions to the equation (2.3) will also be divergence free:

$$\nabla \bullet \vec{v} = 0. \tag{2.4}$$

The type of flow encountered in microfabricated structures has certain unintuitive characteristics. It is known by various names: "Stokes Flow", "Creeping Flow", "Potential Flow", or most simply and clearly "Low Reynolds Number Flow". At low R_e the fluid flow is determined entirely by the pressure distribution and, of course, the boundary conditions ($v = 0$ at the walls). The boundaries of a flow chamber have a profound effect on the flow profile. Such effects determine the relationship between pressure gradients and volumetric flow rates, the uniformity of velocity profiles transverse to the mean flow direction, as well as aspects of advection-enhanced diffusion (that is, enhancement of mixing by shearing flow – Taylor diffusion). Note also that equation (2.3) contains no time derivatives, unlike the general case of the Navier-Stokes equation (2.1). Because of this, under low Reynolds number conditions *all motion is symmetric in time*, meaning that if the pressures or forces exerted on the fluid are reversed the motion in the fluid is completely reversed [17]. While at high Reynolds number, it is virtually impossible to move a fluid and then return it to its original state, at low Reynolds numbers (<1) this is easily done.

We start solving the Navier-Stokes equation by considering the most important physical situation encountered: the flow through a long channel. Surprising predictions for the flow profile in this case are easy to obtain. For both pipe (cylindrical) and channel (rectangular) flow, we consider a conduit of infinite extent in the z-direction, with a (fixed) pressure gradient:

$$\frac{\partial P}{\partial z} = -G, \tag{2.5}$$

with $G > 0$. The velocity field is $\vec{v}(x, y) = u(x, y)\hat{\vec{e}}_z$, and is automatically divergence-free for any scalar function $u(x, y)$. In the case of flow through a cylindrical pipe, the azimuthal symmetry yields

$$\frac{1}{r}\frac{\partial}{\partial r}\left(r\frac{\partial u}{\partial r}\right) = -\frac{G}{\eta}\cdot \tag{2.6}$$

With boundary condition $u(R) = 0$. The solution is:

$$u(r) = \frac{G}{4\eta}\left(R^2 - r^2\right). \tag{2.7}$$

This is the usual parabolic flow profile. It is of great interest to note that the flow is proportion simply to the gradient in the pressure with distance.

Thus, we can find a very simple relationship between the total rate of fluid flow Q from a pipe of length L and the pressure drop δP across the pipe:

$$ZQ = \delta P; \; Z = \frac{8\eta L}{\pi R^4} \cdot \tag{2.8}$$

This is nothing more than Ohm's law. Ohm's law is simply the flow of electrons at low R_e!

Although we derived this simple proportionality between flow and pressure using the special case of an infinitely long cylindrical pipe at low Reynolds number, the relationship holds for any long conduit of any cross section, and even up to moderate Reynolds numbers. This original discovery that $Q \propto \delta P$ dates back to Henry Darcy's experiments with sand traps in the 1850's [18]. Provided the Navier-Stokes equation (2.1) can be reduced to a *linear* form, we will always find that $\delta P \propto v$, and of course $Q \propto v$ as well. The pesky nonlinearity on the right hand side of (2.1) is caused by curving flows. When confined to a long, straight channel in the $\hat{z}$-direction, any fluid will eventually achieve a uniform flow with $\vec{v}(x, y) = u(x, y)\hat{\vec{e}}_z$. The nonlinear term in the Navier-Stokes equation $(\nabla \cdot \vec{v})\vec{v} = (\partial u / \partial z)\vec{v}$ vanishes because u does not depend on z. With the nonlinear term removed, we will always find that $A \propto \delta P$. At intermediate Reynolds numbers ($R_e > 1$, but still laminar flow), the flow entering the mouth of a pipe will take longer to settle into a steady state. This simply means that we must expand the definition of a "long" pipe to allow a greater region at the entrance and exit of the pipe where transients will produce a deviation from the linear law. The hydrodynamic circuits at low R_e can, therefore, be modeled by simple resistive networks. This a huge computational advantage compared to those poor souls trying to predict the weather.

So far we have solved a 1-D flow problem in the pipe and we are now going to discuss flows in systems that are wide and shallow. We found that the flow in a pipe is a linear function of the pressure drop across the pipe. Much of what we want do using microfabricated chips, however, involves large 2-D areas (Fig. 2) rather than narrow pipes. What happens when a pipe injects fluid into a large area? You might be tempted to think that in the case of laminar flow that a "jet" of fluid injected into the area would continue onwards into the area as a well defined "jet stream". That is actually a case in a high R_e situation where the inertial terms dominate. In low R_e situations the viscous damping terms dominate and fight against any high shear environment. Consequently the jet "instantly" slows down to the average rate of flow in the new area. Due to the conservation of mass, it then broadens enormously.

We now make a subtle step by moving from truly three dimensional flow to flow in wide area in the $x - y$ plane which is thin along the z-axis. In this wide area we assume that the thickness of the area h is much smaller than the lateral dimensions h_x and h_y (Fig. 2). We hope that we can separate variables by writing $\vec{v}(x, y, z) = \vec{u}(x, y) \cdot f(z)$. By considering the flow profile in the thin dimension z, we move from a case in 2-D in which the flow would be **infinitely** deep to the 2 1/2 D case in which the flow is shallow. Since the sheet flow is much thinner than it is wide or long, the most important boundary condition is the stick condition at the top and bottom surface. It is the shear forces against these surfaces which restrict how fast the fluid can move. The sides that are a thousand times farther away than the top and bottom control the direction in which the fluid moves. The stick boundary conditions on the top and bottom surface thus will dominate and set up a standard parabolic profile:

$$f(z) = 1 - (2z/h)^2. \tag{2.9}$$

where h is the depth of the chip: 7 microns in the chip used below. This is sometimes known as the Darcy approximation [18]. Substitution of this form for $\vec{v}$ into equation (2.3) gives:

$$\nabla^2 \vec{u}(x, y) \cdot f(z) + \frac{2}{h^2} \vec{u}(x, y) = \nabla P(x, y), \tag{2.10}$$

where ∇^2 now refers *only* to the dimensions x and y. Note that you can get the same equation by making the approximation

$$\frac{\partial^2}{\partial z^2} \gg \frac{\partial^2}{\partial x^2}, \frac{\partial^2}{\partial y^2} \tag{2.11}$$

directly in equation (2.3), separating variables, deducing the parabolic form for the z-dependence, and eliminating the z-coordinate.

Given that $\vec{u}$ has to be divergence free, it is convenient to define the scalar stream function $\Psi(x, y)$ which allows us to simplify the equation we need to solve for a shallow chip even further and end up with a simple Laplace's equation. The stream function is defined so that $\vec{u} \sim \nabla \times (\Psi \hat{\vec{e}}_z)$. By taking the curl of equation (2.10) and substituting this form of $\Psi(x, y)$ after some vector identities we obtain:

$$\left(f(z)\, \nabla^2 + \frac{2}{h^2} \right) \nabla^2 \Psi = 0. \tag{2.12}$$

Considering that the entire z-dependence of the problem has already been absorbed into the statement (2.9), the Laplace operator in equation (2.12) only applies to the x and y dimensions. The statement $\nabla^2 \ll 2/h^2$ follows as long as the boundaries in x and y are far away compared to h. The governing equation is then simply

$$\nabla^2 \Psi = 0. \tag{2.13}$$

This is a pleasure to solve using your favorite computer program, Fourier analysis, or conformal mapping technique.

By assuming that the Laplace's equation is valid we make some major approximations and we need to consider their consequences. If we have "sharp" turns in the jet (near a right-angle turn or an inlet, for instance), then the equivalent assumptions (2.9) and (2.11) will fail. We expect that there will be a region of size about h in which equation (2.13) will not be valid. However, for most places in a large array the solution should be applicable.

Since the end result is a well-behaved two-dimensional stream function $\Psi(x, y)$, it is tempting to posit such a form from the beginning, apply the same definition $\vec{v} \sim \nabla \times (\Psi \hat{\vec{e}}_z)$, substitute this into the Navier-Stokes equation, and perform the standard vector calculus manipulations. This quickly yields the biharmonic equation $\nabla^4 \Psi = 0$. Comparison with equation (2.12) reveals that this corresponds to the limit of *large* h: a very deep chip. This problem occurs if one specializes to a 2-D stream function without imposing parabolic velocity profiles. By dropping the third dimension from the start nothing is allowed to vary with z. In particular, all the pressures and velocities are then set to be invariant in z which is not a case in a shallow device. The case described by biharmonic equation actually describes an *infinitely deep* chip. In fact, one of us (RHA) had a student write an entire Senior Thesis at Princeton using the biharmonic solution, and it was a dead-wrong thing to do because in fact we are in the 2 1/2 D limit, not the 2.0 D limit. This work arose from studying ultra-fast mixing of vertical sheets of moving fluids [20] and was maybe of some utility in the case where the fluid jet is very thick [21] but not for the case of a large flat array.

For the (fourth-order) biharmonic equation (Eq. (2.12)), we must specify four boundary conditions (per dimension) to get a unique solution. The physical constraints are the no-slip conditions $\vec{v} = 0$ at the boundary wall ∂D. Since $\vec{v} = \nabla \times \phi \hat{z}$, we naturally impose the no-slip condition by making $\nabla \phi|_{\partial D} = 0$. This is easier to write as $\nabla \phi \cdot \hat{n} = 0$ and $\nabla \phi \times \hat{n} = 0$, where $\hat{n}$ is the unit normal to the boundary. If we impose this condition everywhere, we get the very simple solution $\phi =$ constant, which correctly predicts that a completely enclosed body of fluid doesn't move. When we add an inflow channel we construct the long-channel flow solution and convert it to a stream function. We then impose such a ϕ on the bit of ∂D where we want fluid to flow in. For a boundary along the x-axis, the relevant stream functions could be $\phi = x - x^3/3, -1 \le x \le 1$ (for parabolic flow in the deep etch) or $\phi = x$ (for plug flow in the shallow etch). Since the long channel has translational invariant flow, it must also have $\nabla \phi \cdot \hat{n} = 0$, where $\hat{n}$ points along the direction of flow. Since we already know that $\nabla \phi \times \hat{n} = 0$ on all the parts of ∂D where there is no channel flowing in or out, we can just integrate this up along ∂D. This gives a ϕ which is constant between the channels, and analytically specified in the channels. We have, thus, constructed the eight boundary conditions for the biharmonic equation that determine ϕ everywhere: we know ϕ on all four walls, and we impose $\nabla\phi \cdot \hat{n} = 0$ on all four walls.

When we move to the second-order Laplace equation (Eq. (2.13)), the system is *overconstrained* and we have to give up some of the boundary conditions resulting with a solution that is valid everywhere but in the narrow region near the walls. We have to give up either $\nabla\phi \times \hat{n} = 0$ or $\nabla\phi \cdot \hat{n} = 0$. The first of these guarantees that $v_\perp = 0$: *i.e.* that no fluid flows in or out through the wall. If we give this up, we will lose conservation of mass! The second guarantees $v_\parallel = 0$: *i.e.* the no-slip condition. If we lose the no-slip condition we allow infinite shear on the fluid at the boundary. In our argument we retain the first condition, and give up the second. This *definitely* and unavoidable violates the no-slip condition but it is just the mathematical consequence of assuming an infinitely shallow chip. For instance, an infinitely shallow, straight channel would have *perfectly* plug-like flow, with $\vec{v}$ = constant right up to the wall, where it would fall discontinuously to zero, giving an infinitesimal region of infinite shear. Again, it is precisely this physically dubious region next to the wall which we know violates the approximations which got us to the Laplace equation in the first place. A full 3-D model would correct the solution next to the wall, forcing $v_\parallel$ smoothly to zero. If we attempt to impose the usual boundary conditions of stick at the walls, so $\vec{v} = 0 \rightarrow \nabla\Psi = 0$ we find that we get a problem which although annoying does not cause major problems. Inside a channel, we assume we have established plug flow: $\vec{v} = c\hat{x} \rightarrow \Psi = cy$. At the very edge of a channel we no longer have plug flow, so $\vec{v}$ is not parabolic in z and the collapse from 3 dimensions to 2 is no longer meaningful. This situation falls under the caveat above: there will be a region of size h where Laplace's equation does not exactly apply. On the plus side, this means we should not get too concerned about the shape of Ψ near the edge of the channel, since we know our differential equation is not really accurate there. In particular, we do not have to worry that the boundary condition $\Psi = cy$ produces an infinite shear at the wall of the channel.

The solution we propose here arises from a well established Darcy's approximation to which we apply complex $x - y$ boundary conditions and explicitly solve $\vec{v}(x, y)$. In the Darcy approximation we assume a vector field $v(x, y)$ and a parabolic profile in z between the plates. The Laplacian potential function equation is really the Saffman-Taylor problem at zero tension (*i.e.*, no surface interface between two different fluids) and in that sense it is a "trivial" example of the much more difficult problems that Bensimon *et al.* tackle in their beautiful article in *Rev. Mod. Phys.* [19].

What IS new here is application of complex $x - y$ boundary conditions that allows us to find the explicit $\vec{v}(x, y)$ solutions using potential functions for two components of the same fluid moving at different initial speeds into a common area.

Making jets

Let us now turn to a practical problem of making fluid run in a narrow stream of constant width across an open area. We will see how our understanding of "2 1/2 D" hydrodynamics can help us design a more uniform jet.

One might think that the structure shown in Figure 3 would work. The philosophy behind this particular chip design was that the two wide side channels would squeeze the already narrow fluid flow from center channel and create a thin jet of fluid moving across the array. Figure 4 shows a wide spreading of the jet upon entering the open area. The problem arises because in a low R_e configuration abrupt changes in fluid velocities are not allowed. The velocity of the fluid near the entrance changes slowly resulting in initial broadening of the jet profile. The jet ultimately narrows as determined by net fluxes but it has to spread first. The resulting profile, thus, is not a jet of uniform width over its entire length.

Simulations of the flow profile using the potential function solutions to equation (2.12) show how well the "2 1/2 D" Darcy approximation agrees with the data. Figure 4 shows a view of a 3 port jet with the predicted streamlines shown as dark lines and the actual flow as outlined by a stream of fluorescent balls. Laplace solution outlined above does an excellent job of modeling these flows. Note that the solution works well even near the boundaries of the area. The simulation confirms that if there are only three channel inputs, there is always a large bulge in the initial flow pattern of the jet upon its entrance to the main open area.

3 The N-port injector idea

There is, however, a way to establish uniform and confined flow patterns by careful consideration of boundary conditions and at a cost of microfabrication complexity.

The key concept here is that the boundary values explicitly determine the flow patterns in the area and that by controlling them we can control the flow. The boundary values in hydrodynamics are determined by the fluxes of fluids at the surface. Normally, as we discussed above, we assume that the boundaries have no fluid flow across them, but this does not have to be the

Fig. 3. A three-port injector. The center channel has a width of 10 microns and the two wide side channels have widths of 100 microns. The entire device is 7 microns deep.

case. Suppose that the sides of the area have a large array of microchannels feeding a large number of streams into the area, the boundary conditions are now changed to a given flux of fluid. In principle the flux itself is an unknown quantity and difficult to posit *a priori*. However, if the fluid flux is injected into a region from a device with high fluid compliance then the flux is constant to a high degree of accuracy and independent of the subsequent restrictions of the area into which the fluid is injected. In this way the fluid pattern becomes "ballistic" as is true of high R_e flow without the problems of turbulence. A high compliance fluid source is nothing more than a current source. It can easily be constructed in a hydrodynamic array in the same way current sources are constructed in electronics. By passing a fluid at an initial high pressure through a long narrow pipe one assures that most of the pressure drop occurs over the pipe and that only a small fraction of it occurs over the subsequent hydrodynamic "circuit". In this way the fluid

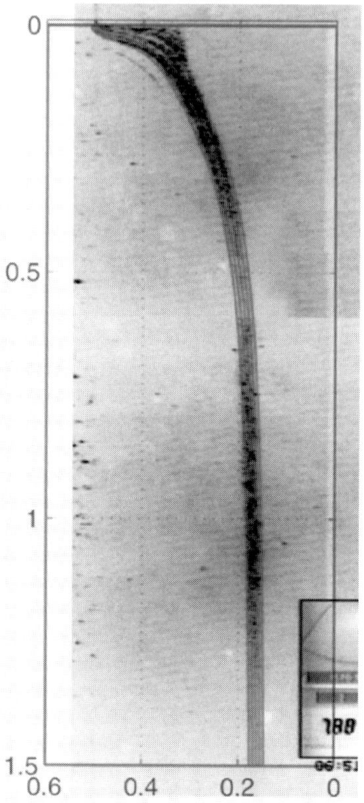

Fig. 4. Comparison of data (bead tracks) and best fit model flowlines (solid dark lines) for the fluid jet observed in a three port jet. This figure shows the center jet formed by higher pressure from the two side inputs. The curvature of the jet arises from the asymmetry of the position of the center jet relative to side jet, and the blooming of the center jet upon entering the open area is due to the change in hydrodynamic impedance.

flux becomes independent to first order of the low impedance part of the circuit and the boundary conditions are fixed.

Figure 5 shows a microphotograph of a device in which the above described principles have been applied. The large open area is an array of posts on a hexagonal lattice used for the fractionation of genomic DNA [22]. The side channels that determine the boundary value flux conditions can be seen as the series of lines feeding into the main are. These long channels act as pressure dropping fluidic resistors and establish the current source nature of the boundary values. One of the channels is special and is connected to a

Fig. 5. A corner of a N-port array showing the array of channels feeding into an open area consisting of a hexagonal array of 2 micron posts. The width of the narrow channels feeding fluid into the array is 4 microns.

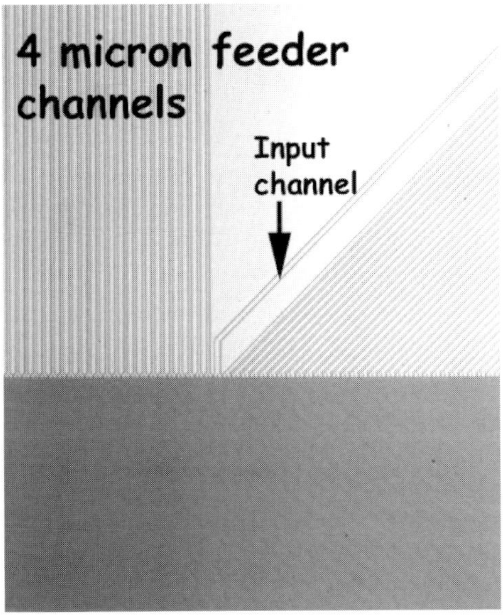

Fig. 6. A view of the input channel which brings a confined stream of liquid into the array. The input channel is 6 microns wide.

unique reservoir which contains a the material to be transported across the array, in our case DNA molecules. Figure 6 taken 1/4 of the way down the side of the array shows the special channel, hydrodynamically isolated from

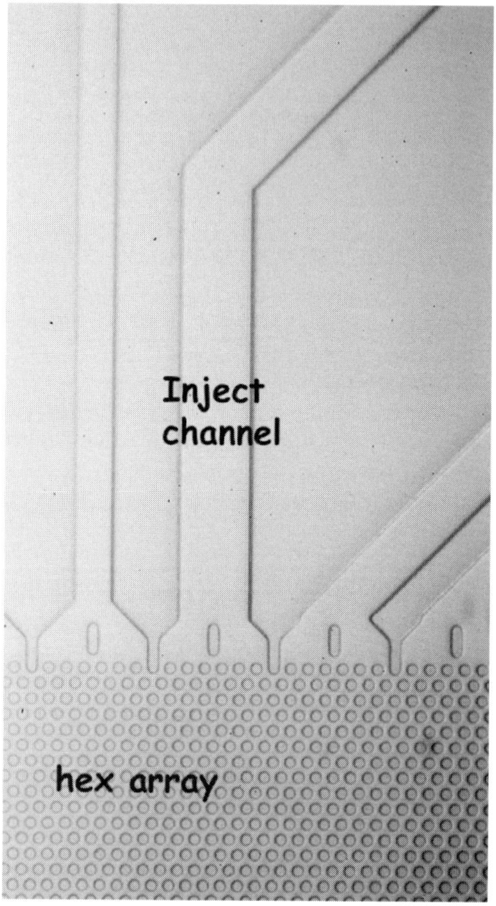

Fig. 7. An enlargement of the injector channel showing the interface of the injector channel with the hexagonal array. The round posts are separated by 2 microns center to center.

the rest of the channels which brings in the biological molecules that will be fractionated in the array. Figure 7 shows a view of this sample injection line. By properly adjusting the pressure heads on the channel reservoirs we now can inject a straight stream of fluid across a large open area without the blooming problem seen in a 3 port injector seen in Figure 4. Figures 8a and b show the propagation of such a flow across the hexagonal array of posts.

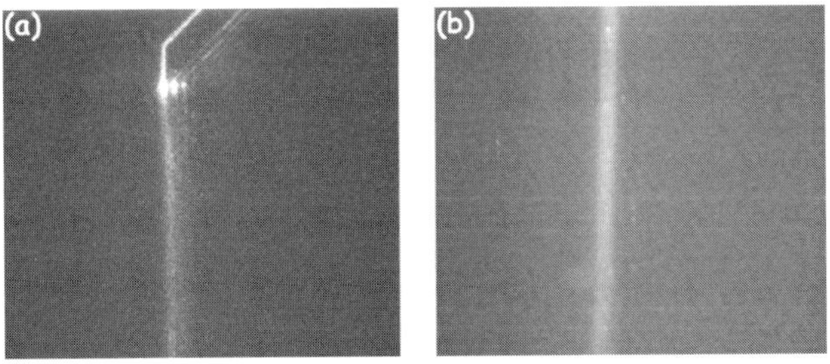

Fig. 8. a) A stream of DNA molecules injected into the post array. Some DNA molecules have been seeded into adjacent channels so that the stream lines of the injected currents can be seen. **b)** The same injected jet now seen about 500 microns across the stream. The broadening seen is due to diffusion but the streamlines of the jet remain intact and focused.

4 Conclusion

The above described technology can be scaled to get increasingly complex flow patterns. By having the channels divided up into a large number of independently adjustable pressure sources the complex boundary value flux patterns can be established. The elegance of the 2 1/2 D hydrodynamics can then be exploited to solve critical problems of transport in miniature bio-devices.

References

[1] A. Manz, N. Graber and H.M. Widmer, *Sensors and Actuators B* **1** (1990) 244-248.

[2] C.-F. Chou, R.H. Austin, O. Bakajin, J.O. Tegenfeld, J.A. Castelino, S.S. Chan, E.C. Cox, H. Craighead, N. Darnton, T. Duke, J. Han and S. Turner, *Electrophoresis* **21**, 81-90.

[3] R.H. Carlson, J.P. Brody, S. Chan, C. Gabel, J. Winkleman and R.H. Austin, *Phys. Rev. Lett.* **79** (1997) 2149-2152.

[4] M.S. Ibrahim, R.S. Lofts, P.B. Jahrling, E.A. Henchal, V.W. Weedn, M.A. Northrup and P. Belgrader, *Anal. Chem.* **70** (1998) 2013-2017.

[5] L.C. Waters, S.C. Jacobson, N. Kroutchinina, J. Khandurina, R.S. Foote and J.M. Ramsey, *Anal. Chem.* **70** (1998) 158-162.

[6] M.A. Burns, B.N. Johnson, S.N. Brahmasandra, K. Handique, J.R. Webster, M. Krishnan, T.S. Sammarco, P.M. Man, D. jones, D. Heldsinger, C.H. Mastrangelo and D.T. Burke, *Science* **282** (1998) 484-487.

[7] D. Schmalzing, N. Tsao, L. Koutny, D. Chisholm, A. Srivastava, A. Adourian, L. Linton, P. McEwan, P. Matsudaira and D. Ehrlich, *Genome Res.* **9** (1999) 853-858.

[8] J. Han and H.G. Craighead, *Science* **288** (2000) 1026-1029.

[9] G. Fuhr, T. Schnelle and B. Wagner, *J. Macromech. Microeng.* **4** (1994) 217-226.

[10] A.V. Lemoff and A.P. Lee, *Sensors and Actuators B* **63(3)** (2000) 178-185.

[11] M.U. Kopp, A.J. de Mello and A. Manz, *Science* **280(5366)** (1998) 1046-1048.

[12] M.A. Northrup, B. Benett, D. Hadley, P. Landre, S. Lehew, J. Richards and P.A. Stratton, *Anal. Chem.* **70** (1998) 918-920.

[13] A.Y. Fu, C. Spence, A. Scherer, F.H. Arnold and S.R. Quake, *Nat. Biotechnol.* **17** (1999) 1109-1111.

[14] A. Wolff, U.D. Larsen, G. Blankenstein, J. Philip and P. Telleman, *Micro Total Analysis Systems Proceedings, Banff Canada* (Kluwer Academic Publishers, 1998).

[15] J.P. Brody, P. Yager, R.E. Goldstein and R.H. Austin, *Biophys. J.* **71** (1996) 3430-3441.

[16] R. Feynman, M. Sands and R. Leighton, The Feynman Lectures on Physics, Vol. II (Addison-Wesley, Redwood City, CA, 1989).

[17] E.M. Purcell, *Am. J. Phys.* **45**, 3-11.

[18] A. Biswas, History of Hydrology (North-Holland Publishing Co., Amsterdam-London, 1970) (p. 288-291, 308-310, 314, 319).

[19] D. Bensimon, L. Kadanoff, S. Liang, B.I. Shraiman and C. Tang, *Rev. Mod. Phys.* **58** (1986) 977-99.

[20] J.B. Knight, A. Vishwanath, J.P. Brody and R.H. Austin, *Phys. Rev. Lett.* **80** (1998) 3863-3866.

[21] L. Pollack, M. Tate, N. Darnton, J.B. Knight, S.M. Gruner, W.A. Eaton and R.H. Austin, *Proc. Natl. Acad. Sci. USA* **96** (1999) 10115-10117.

[22] A.J.T. Duke, R.H. Austin, E.C. Cox and S.S. Chan, *Electrophoresis* **17** (1996) 1075-1079.

Abstract

There are two ways to move biological objects: hydrodynamics (which I have just discussed in Lecture 1a) and with some sort of an external force field. Although usually the force field is a static electric field, a more intriging possibility is AC fields, in which case you probe both the usual electrophoretic response (translation by the force on a charged object) AND dielectrophoretic response (translation by the force on a polar object). Clearly, the dielectrophoretic response is more complex and potentially more powerful in terms of the number of possible parameters, so we will explore that here. This lecture is a mixture of theory and experiment, as are others, so the reader is asked to tolerate the experimental details, in which can be found the Devil himself. They are important.

1 Introduction

Dielectrophoresis (DEP) is the translation of neutral matter caused by polarization effects in a nonuniform electric field [1]. Measuring and understanding the magnitude of the dielectrophoretic force exerted on important biopolymers such as DNA is a difficult fundamental problem that we address in this lecture.

An electrically polarizable object will be trapped in a region of a focused electric field, provided there is sufficient dielectric response to overcome thermal energy and the electrophoretic force. The standard way to make a DEP trap is to create an electric field gradient with an arrangement of planar metallic electrodes either directly connected to a voltage source [2,3] or free-floating [4,5] in the presence of an AC field. In this paper we use a constriction or channel in an insulating material instead of a metallic wire to squeeze the electric field in a conducting solution, such as ionic buffer, thereby creating a high field gradient with a local maximum. The advantages of the electrodeless DEP (EDEP) technology introduced here are: (1) no metal evaporation during the fabrication is needed, (2) the structure is mechanically robust and chemically inert, and (3) a very high electric field may be applied without gas evolution due to electrolysis at metal DEP electrodes. Figure 1 outlines the differences between the metal electrode and the confined field technology of this paper. The simplicity of the device and the lack of metallic objects which cause electrochemical reactions involving gas evolution enable us to probe the response of DNA molecules well below 1 kHz, revealing a huge increase in the dielectric response at low frequencies (below 1 kHz) difficult to observe using metal electrodes as trapping structures.

The subject of this letcture is not only the basic physics of dielectrophoresis, but it also has applications to biotechnology. One of the great

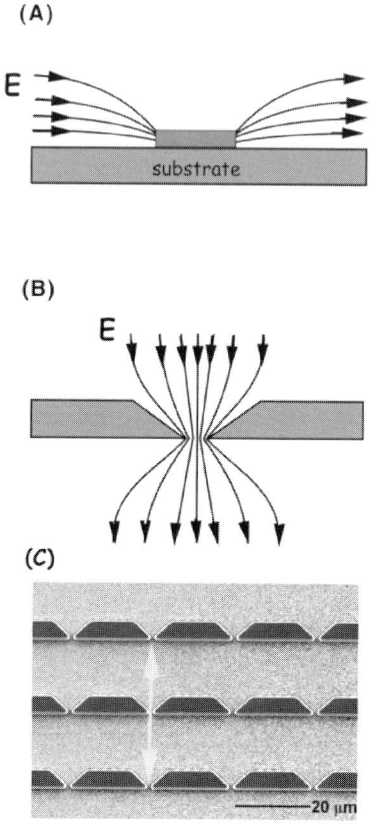

Fig. 1. Schematic of a microfluidic DEP trap **A)**. A metallic DEP trap made of microfabricated wire(s) on a substrate. The wire(s) may be either free-floating or connected to a voltage source **B)**. An electrodeless DEP trap made of dielectric constrictions. The solid lines are electric field lines E **C)**. A scanning electron micrograph of an electrodeless DEP device consisted of a constriction array etched in quartz. The constrictions are 1 microns wide and 1.25 microns deep. The whole chip measures 1×1 cm. The applied electric field direction z is shown by the double headed arrow.

challenges in biotechnology is to move and concentrate molecules in a microfabricated environment. Notable applications of DEP include the separation of colloidal particles [6], DEP ratchets [7,8], the separation of biological objects such as yeast cells [9] viruses [10] and cancer cells [11, 12] and the trapping and manipulation of DNA molecules [4]. EDEP can be used in all of the above listed applications.

2 Methods

The devices were fabricated on quartz wafers using reactive ion etching techniques, and sealed with a glass coverslip coated with an elastomer thin film to act as a sealing gasket. DNA dissolved in electrophoretic buffer was introduced into the sealed space and external gold electrodes attached to a high voltage source provided the external currents. It is important to understand a fundamental aspect of current flow in a basically *insulating* fluid such as water, namely that the current density **J** is proportional to the ion flux since the ions carry the charge. Since it is the electric field that makes the ions move in the solution, in a backward way of saying it that we hope makes some sense the electric field **E** is thus proportional to a hydrodynamic flow of charged ions. Thus, the electric fields are everywhere *parallel* to the surfaces of the constrictions in the insulating quartz and the relative dielectric constants of the quartz and the water are irrelevant. Calculation of the electric fields is thus relatively easy, and no electric fields penetrate the insulating structures.

2.1 Fabrication

The device (see Fig. 1c) was fabricated using UV lithography and reactive ion etching on 3″ (76 mm) crystalline quartz wafers polished on both sides (Hoffman Materials). The gaps in the quartz obstacles are 1 micron wide. Chips were diced out of the wafer and were 1 cm in length.

A 200 nm thick aluminum film was thermally evaporated onto the quartz wafer and treated with hexamethyldisilazane (HMDS) in a Yield Engineering Systems LP-III Vacuum Oven to promote adhesion of the photoresist. Shipley S1813 (Microchem Corp., Newton, MA) photoresist was spun on the aluminum coated quartz wafer at 4000 rpm in 60 s with a 3 s linear ramp. A pre-exposure bake at 115 °C for 60 s was used. The wafers were exposed in a projection aligner (GCA 6300 DSW Projection Mask Aligner, 5× g-line Stepper). After development in MicroPosit CD26 (tetramethylammoniumhydroxide solution in water) (Shipley) for 60 s the aluminum is etched using a modified PK1250 ion etcher from PlasmaTherm. The PlasmaTherm PK1250 was also used to etch the quartz. Etch times of 33 min resulted in an etch depth of 1.25 microns as determined by a Tencor AlphaStep 200 Surface Profilometer. The device was then sealed with a glass coverslip coated with silicone (polydimethylsiloxane) elastomer (RTV 615 A&B, General Electric, NY). Both the coated coverslip and the device were pretreated with oxygen plasma to make the surfaces hydrophilic and wettable when sealed. The top-sealed device is then wetted by capillary action with buffer solutions (pH 8.0, 0.5X Tris-borate-EDTA buffer (TBE), 0.1M dithioDTT, and 0.1% POP-6). POP-6 is a linear polyacrylamide of proprietary formula provided to us by Applied Biosystems and serves to eliminate the transport of fluid in a sealed device due to bound charges on the quartz surfaces, known as electroendoosmosis (EEO). In the absence of POP-6, fluid is transported by ionic currents due to this surface effect greatly complicating the forces acting on objects because of the added hydrodynamic transport.

2.2 *Viscosity*

Tests of the viscosity dependence of the EDEP force were carried out in 0.5X TBE buffer (pH = 8.0 with 0.1% POP-6 and 0.1 M DTT). The viscosity was adjusted by adding sucrose to the buffer without changing the dielectric constant of the buffer [13]. The buffer viscosity of 3.7 cP was prepared by adding 46 g of sucrose to 100 grams of buffer (31% w/w). The viscosity of 5.9 cP was prepared by adding 62.5 grams of sucrose to 100 grams of buffer (38% w/w). Viscosities were checked by viscometry at 20 °C.

2.3 *Electronics and imaging*

A Kepco BOP 1000M amplifier with 1 kHz bandwidth provided the ±1000 V driving voltage. The input to the Kepco BOP was provided by a HP 3325A signal generator which was connected via a GPIB interface to a MacIntosh computer running LabView (National Instruments) software. External gold electrodes driven by the Kepco BOP were immersed in liquid troughs which contacted the liquid wetting the sealed chip. All voltages quoted in the text are the amplitude of the sinusoidal output of the BOP as measured by a HP 34401A digital multimeter. DNA was stained with TOTO-1 (1 dye molecules/5 bp) in 0.5X TBE buffer (pH = 8.0 with 0.1% POP-6 and 0.1 M dithiotreitol (DTT). The images were gathered with a Nikon Microphot-SA microscope using an oil immersion objective lens (60X, N.A.1.4), a cooled CCD camera (Hamamatsu C4880, NJ), and excitation at 488 nm by an Ar-Kr ion laser. Images of the DNA in the chip were taken by epifluorescence. The C4880 camera was run at −20 C.

2.4 *DNA samples*

We used 5 different double stranded DNA with lengths of 368, 1137, 4361 and 39 936 bp, and a single stranded DNA of length 1137 nucleotides. These were prepared as follows:

Double-stranded DNA

The 368 bp DNA was produced from an initial 54 base sequence. Both ends of the monomer duplex had complimentary 4-base overhangs. The monomer was kinased and ligated to create a multimer. Then a ligation step was done at about 15 C to create a multimer ladder by varying the time of ligation from 6 to 24 hours. Then a duplex which had a 20-base primer sequence and a 24-base linker sequence was added to complete the ligation at both ends of the multimers. A quick spin column was used to wash away the monomer duplex, the primer-linker duplex and the short ligated fragments (less than 200 bp). Then the multimers were precipitated by cold ethanol, dried and resuspended in TE buffer. A PCR step was performed on this ladder and analyzed by 1.8% agarose mini gel. When two or three clean bands could be observed in an analyzing gel, a preparatory gel was run and the bands were cut to extract the fragments individually. Then PCR was repeated in preparatory quantity to produce sufficient amounts (10–50 μg) of each fragment for experiments. The resulting product was cloned and sequenced. The GC content was 50% and the sequence is available from the authors. The 1137 bp DNA was prepared by PCR amplification of positions 2457 to 3594 of bacteriophage λ DNA. Before use, the PCR product was purified by standard methods from agarose gels.

The 4361 bp sample was prepared by digesting pBR322 DNA with BstII and purifying the linearized DNA from an agarose gel. The 39 936 bp DNA is bacteriophage T7 DNA, and was purchased from Sigma Chemical and was used without further purification.

1137 nucleotide ssDNA

Single stranded λ DNA was prepared by amplifying the 1137 base pair fragment (above). The primer homologous to the 2457 sequence was labeled at the 5′ terminus with biotin (BiotinTEG phosphoramidite, Glenn Research, Sterling, Virginia). The PCR reaction contained fluorescein-11-dUTP (Amersham Pharmacia, Piscataway, NJ) at a dTTP/dUTP-fluorescein ratio of 1:1. Single stranded product was isolated by adsorbing the reaction mixture to Dynal-streptavidin beads (Dynabeads M-280, Dynal A.S., Oslo, Norway) and isolating single stranded DNA from the beads by incubation at 0^0 C in 100 mM NaOH for a few min. Under these conditions the biotin-labeled strand remains attached to the Dynal beads, which are removed magnetically. The fluorescein-labeled single -stranded product in the supernatant was then concentrated and purified by ethanol precipitation and resuspension in buffer.

3 Results

3.1 Basic results and dielectrophoretic force extraction

We first present a typical image of the basic data. Figure 2 shows the image of trapped DNA density *vs.* applied voltage for 368 bp long fragments at an applied voltage of 1 kV across the cell as a function of frequency. At low frequencies there is basically no trapping, as the frequency is raised the DNA molecules are attracted to the gap between the constrictions and the concentration of the DNA molecules in the gap increases. Clearly, the confinement of the electric field lines within the 1 micron gaps of the structures results in a powerful trapping of the molecules. The apparent force clearly rises with increasing frequency for this 368 bp long sample. However, there are many parameters that must be explored to fully understand and exploit the ability of EDEP to trap and fractionate DNA molecules. Before we can proceed with explaining the way that EDEP can trap DNA molecules as a function of applied electric fields, field frequency, and size (length) of the molecules it is important to have a quantitative way to analyze the trapping force felt by the molecules so that a physical model of the phenomena can be attempted.

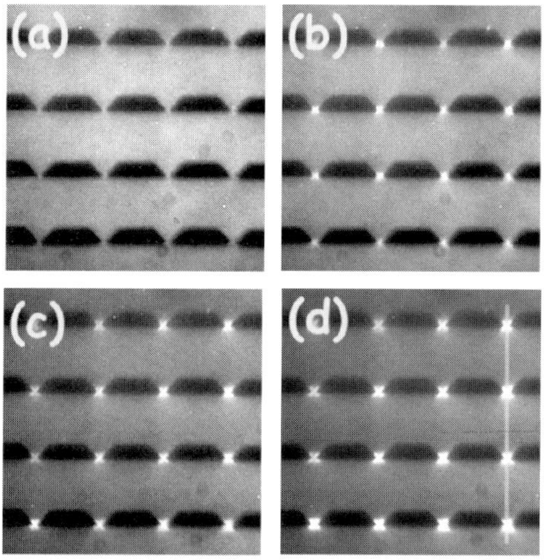

Fig. 2. A–D) Optical micrographs of DEP trapping of 368 bp dsDNA with driving voltage of 1 kV (corresponding to 5 V p-p across each unit cell) and applied frequencies of 200, 400, 800 and 1000 Hz. The frame size is 80 × 80 microns. The images shown here were each averaged over 3 consecutive frames started with the first one taken 1 min after the AC electric field parameters being changed, and 1 min interval for each following images to allow equilibrium densities to be achieved. Equilibration typically occurred in a few seconds at each new field value. Each frame was exposed for 10 s and the light source was shut off when the camera shutter was closed to reduce photobleaching. The line shown in Figure 2d shows the pixel swath used to analyze the density of the molecules in the trap.

In the absence of electrophoretic forces the *molecular* forces acting on single DNA molecules can be extracted from the images shown in Figure 2. Since DNA at neutral pH is charged due to the phosphate groups it also is transported by a DC electric field (electrophoresis), the following analysis is oversimplified and can give rise to misleading effective "forces", but does help to catalog the data. We will attempt to briefly discuss corrections later in this lecture.

The trapping shown in Figure 2 is due to the force a polarizable object feels in a field gradient. Charged polymers such as DNA at pH 7 are electrically neutral in the absence of an external electric field **E** because of the counterion cloud that surrounds the polymer. However, in the presence of an external field two things happen: (1) the movement of ions in the fluid shears away the counterions at the zeta potential surface giving rise to a net charge density σ along the length of the polymer and (2) the counter-ion charge distribution becomes polarized along the length of the molecules, giving rise to a dielectric moment **p**. Since the origin of the dipole moment is due to electrophoretic movement of counterions within the zeta potential surface, the induced dipole moment is a function of the applied electric field, the time over which the field is applied and the size of the polymer. Typically the induced dipole moment **p** is opposite to the the direction of the applied field **E**, but this is not always the case. The Clausius-Mosotti (CM) ratio [14] , which relates the sign of the dielectric force $\mathbf{F}_{\mathrm{d}}$ to the gradient in the electric field energy density, can be either positive or negative, depending on the response of the material to the field [9], although in our case the induced polarization is more complex in origin than the relatively simple displacement of charge within a molecule. Figure 3 shows a cartoon of the way that the counter-ion cloud around a molecule of length L becomes polarized in an external field, leading to an induced dipole moment.

Let the distance z be the distance of a particle between the two external electrodes. The potential energy $U_{\mathrm{p}}(z,\omega)$ of a polar but uncharged molecule in an applied field $\mathbf{E}(\mathbf{z},\omega)$ is:

$$U_{\mathrm{p}}(z,\omega) = -\mathbf{p}\cdot\mathbf{E} = -\alpha(\omega)/2(E)^2 \tag{3.1}$$

where α is the in-phase component of the complex polarizability of the molecule [15] and includes the CM term. DNA trapping only occurs when $U > kT$ and the CM factor is positive (positive EDEP).

The gradient in the potential energy $U(z,\omega)$ gives rise to a dielectrophoretic force F_{d}:

$$\mathbf{F}_{\mathrm{d}} = -\mathbf{grad}(U) = (\alpha/2)\mathbf{grad}(E^2) = \alpha|E|\frac{\mathrm{d}\mathbf{E}}{\mathrm{d}z} \tag{3.2}$$

where $|E|$ is the scalar magnitude of the field. This equation is the low-frequency limit of a generalized theory that can explain both DEP and the

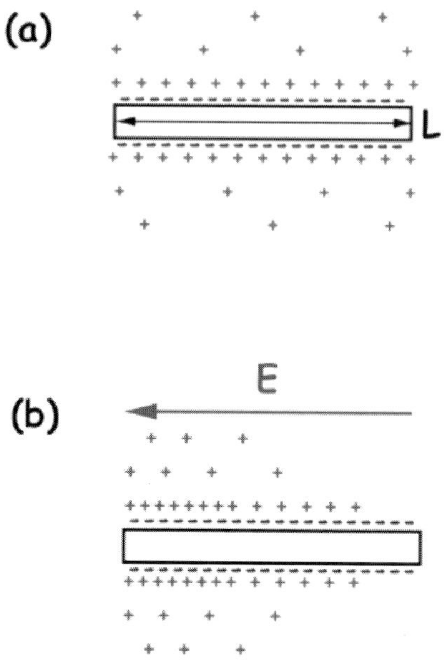

Fig. 3. A) Cartoon of the positive counter-ions surrounding a negatively charged DNA molecule of length L. **B)** Distortion of the counter-ion cloud due to an externally applied electric field **E**.

"laser tweezer" trapping observed when the wavelength of the radiation is smaller than the object [16]. If the electric field has a local maximum then a potential well is formed that traps the molecule since the field gradient changes sign around the maximum. At a finite temperature T the thermal energy kT broadens the distribution of molecules trapped in the potential well. In our system, a DNA molecule is driven by diffusional motion and an average drifting velocity $\mathbf{v}$ due to the external EDEP force $\mathbf{F_d}$ and the external electrophoretic force $\mathbf{F_e}$. The diffusion coefficient D and the average velocity $\mathbf{v}$ of a particle in the presence of an applied force $\mathbf{F}$ are linked through Einstein's relation: $\mathbf{v} = D\mathbf{F}/kT$, where D is the Brownian diffusion coefficient of a DNA molecule and kT the thermal energy. The flux of DNA molecules $J(z,t)$ at point z is governed by the modified Fick's equation:

$$\mathbf{J}(z,t) = \frac{D\mathbf{F}}{kT} \bullet n(z,t) - D\mathbf{grad}n(z,t). \tag{3.3}$$

where $n(z,t)$ is the local concentration of DNA molecules. At equilibrium, $\mathbf{J}(z,t) = 0$, the distribution of DNA molecules $n(z,t)$ obeys a Boltzmann distribution: $n(z,t) = n_o \exp[-U(z,t)/kT]$, where n_o is the density of DNA molecules at the minimum of the potential well.

In the limit of thermodynamic equilibrium the flux $\mathbf{J}(z,t)$ is zero and equation (3.3) allows us to analyze the local density of molecules $n(z,t)$ and extract the DEP force acting on them. The image analysis program NIH Image was used to extract a contour plot of the average density of the DNA across the constrictions (the scan region of the density plot is defined in Fig. 2), and computation of effective force from the density follows from equation (3.3):

$$F(z,t) = kT \frac{\mathbf{grad}[n(z)]}{n(z)} \tag{3.4}$$

where $\mathbf{grad}n(z)$ is the spatial gradient of the density distribution. Determination of the EDEP force is independent of n_o, provided a dilute DNA solution is used in which intermolecular interactions are negligible across the unit cell of the device. Note that the force is determined in absolute units, femtonewtons (fN), since we need only kT to get absolute units.

Many biological molecules are charged as well as polarizable [17] and this complicates our analysis since there is also an electrophoretic force acting on charged molecules during dielectrophoresis. The net force is the sum of the two, and this complicates the analysis because the dielectrophoretic force always points towards the region of high field gradient and thus does not oscillate with the field direction change, while the electrophoretic force points along the direction of $\mathbf{E}$ and thus oscillates with the field direction change. If the electrophoretic force locally is greater than the dielectric force, the net translation $\Delta z \sim v_e \Delta t \sim \mu_e E \frac{2\pi}{\omega}$ can be large compared to the size of the dielectric trap. In that case, the assumption of thermodynamic equilibrium breaks down and the forces are not correctly determined. Thus our analysis correctly describes the high frequency response. At low frequencies the molecule is pulled out of the well by electrophoretic forces, and the apparent force is reduced due to a finite particle flux out of the trap.

4 Data and analysis

Figure 4 shows the dielectrophoretic force exerted on the 368 bp molecule for a given field strength as a function of frequency and distance from the center of the trap. The force was extracted from the density distribution gradient using the pixel swatch shown in Figure 2d. Note how the force reaches its maximum not at the position of the strongest field (the center of the gap) but rather where the product of $E\mathrm{d}E/\mathrm{d}z$ is largest. For the remainder of the analysis we will quote these peak values. It is clear from

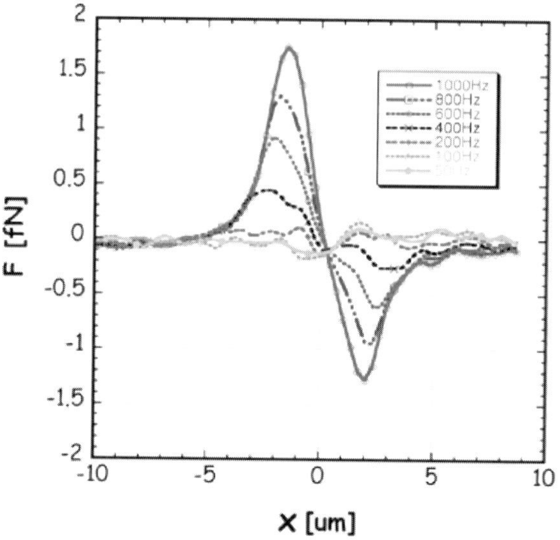

Fig. 4. EDEP force response curve of 368 bp DNA with applied field 1000 V pp/cm (5 Vp-p per unit cell) as a function of frequency. Each curve is an average of all the unit cells in the microscope field of view.

Figure 4 that the dielectrophoretic force is a strong function of frequency. For the 368 bp sample, it rises with frequency to the 1 kHz limit of our amplifier. Figure 4 shows how at 1 kV the maximum force in the trap rises monotonically with frequency. Although for 368 bp molecules we cannot measure the maximum frequency at which the EDEP response peaks with our current apparatus, longer lengths of DNA do show peaks in the trapping response with increasing frequency, as we shall show. We thus we believe the trapping frequency for this sample must also peak at higher values.

We discussed above the basic origin of this force, and in equation (3.2) showed that the trapping force should vary as the square of the electric field. For this to hold experimentally, it is necessary to ensure that the length of the molecule and the frequency of the applied field is such that the density of the trapped molecules has a spatial width great enough for us to easily extract the maximum force without the DNA concentrating into a band narrower than our optical resolution of about 0.5 microns. Figure 5 illustrates this problem. At 1 kV and 200 Hz the 368 bp sample is barely trapped and data analysis is very difficult, while a 39.9 kb long sample at 100 V and 100 Hz is trapped so tightly that analysis again is impossible. Figure 6 shows that within our margin of error the force does scale as E^2 for the 1.1 kB sample using a 200 Hz frequency.

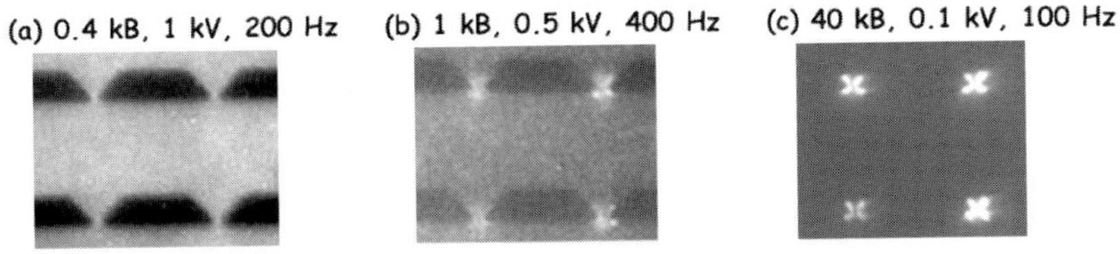

Fig. 5. Images of the trapped DNA density as a function of length, voltage and frequency.

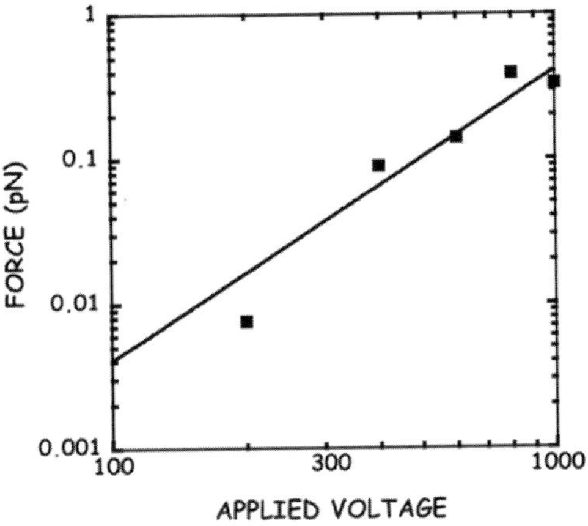

Fig. 6. The measured peak force *vs.* the applied voltage for the 368 bp DNA sample at 200 Hz plotted on a log-log scale. The solid line is a fit to this data assuming that the force varies as E^2. The only variable was the scaling parameter for the force magnitude.

We further show that the dielectrophoretic force greatly depends on the length of the DNA molecules. Figure 7 shows the extracted forces for 368 bp and 1 kB samples as a function of frequency at 1 kV. Next, we show in Figure 8 the peak forces as a function of frequency for 4361 bp and 39.9 kB DNA at 200 V. These measurements were done at a relatively low driving voltage of 200 V as opposed to the 1 kV values used in Figure 7 because at the higher voltages the trapping force for long DNA molecules is so strong that we cannot accurately measure the width of the distribution (see above). Unlike the shorter fragment data, which show a monotonic rise in the trapping force with frequency, there is a hint of a maximum in response for the 4361 bp DNA and a very clear maximum in response for 39.9 kB DNA. Note that there is great dispersion in the force with length, hence by appropriate choice of parameters one can envision selectively trapping one range of DNA molecules while removing others.

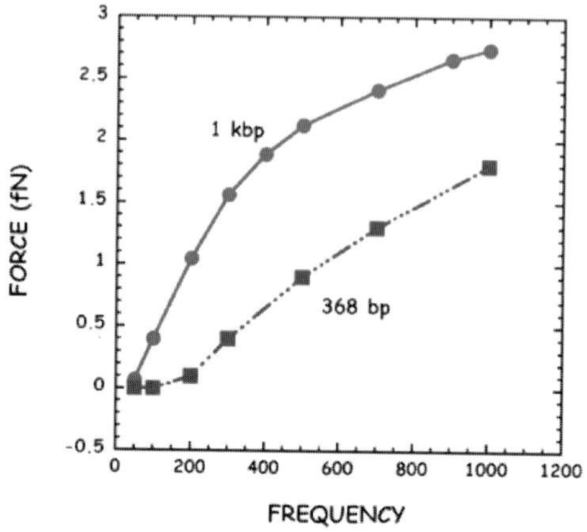

Fig. 7. Force *versus* frequency for 1137 kb and 368 bp long DNA molecules at an applied voltage of 1 kV.

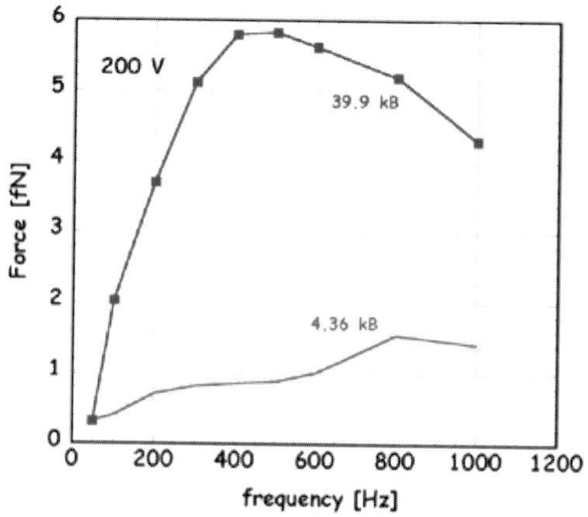

Fig. 8. Force *versus* frequency for 4.36 kB and 39.9 kB long DNA at an applied voltage of 200 volts.

We next examined the effect of solvent viscosity on the frequency dependence of the force. These experiments address the issue of the origin of the observed dielectric response of DNA: internal charge transport down the backbone of the DNA molecule, as would happen if DNA were a conductor, would be expected to result in a very fast response, while counter-ion flow within the Debye sheath of counter ions near the DNA would be dominated by viscous drag. Figure 9 shows the dependence of the EDEP force on our longest DNA molecule over the three viscosities studied: 1 cP, 3.7 cP and 5.9 cP. There is a clear shift of the frequency of maximum force response to lower frequencies with increasing viscosities.

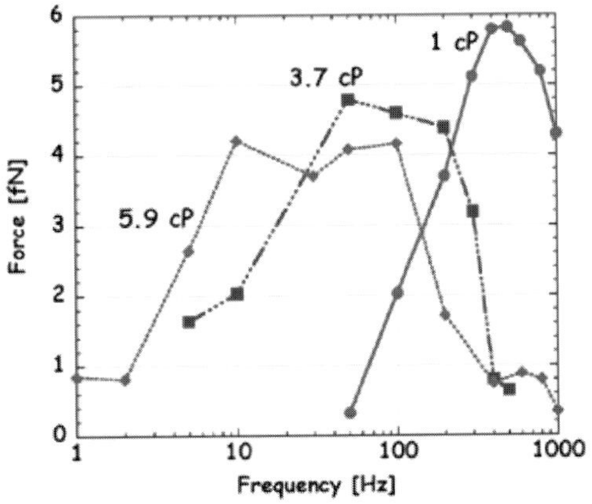

Fig. 9. Force *versus* frequency and viscosity for a 39.9 kB DNA molecule.

Finally, we briefly explore the dependence of the EDEP force on single stranded DNA (ssDNA). One would expect to find differences in the dielectrophoretic forces acting on two DNA molecules of identical molecular length, one ssDNA and the other dsDNA, because (1) ssDNA has half the linear charge density of dsDNA, (2) a different stacking conformation, and most importantly (3) a greatly different persistence length. The persistence length of ssDNA is believed to be much shorter than dsDNA. The persistence length of dsDNA is close to 50 nm, and somewhere between 1 to 6 nm for ssDNA [19–21, 28]. Figure 10 compares EDEP forces on dsDNA and a ssDNA molecule which have the same number of nucleotide units (basepairs for dsDNA, bases for ssDNA). Clearly, ssDNA experiences a substantially smaller force than dsDNA of the same number of nucleotide units.

5 Origin of the low frequency dielectrophoretic force in DNA

We now offer a simplified explanation for the length and low frequency dependence of the EDEP experiments, a complex subject which we only address briefly so that the reader can obtain an intuitive basis for the effects seen in the experimental section. We take as our fundamental starting point that the DNA backbone is an insulator consisting of fixed charges on the backbone with a surrounding layer of counterions. By noting the frequency dependence of the dielectrophoretic force on viscosity, we can further assume that when a DNA molecule is exposed to an externally imposed

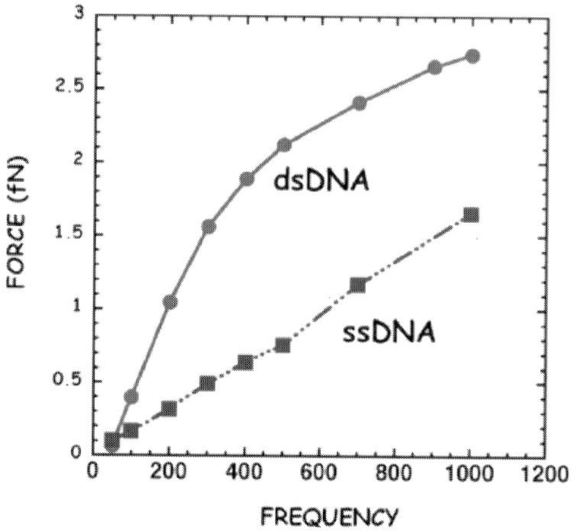

Fig. 10. Force *versus* frequency for ssDNA and dsDNA of 1137 basepairs (dsDNA) and 1137 bases (ssDNA) at an applied voltage of 1000 volts.

electric field **E** the surrounding counterion cloud becomes distorted by the diffusion of the counterions along the backbone [22]. This diffusion results in the formation of an electric dipole moment, but lagged in phase with the applied voltage. The resulting frequency dependence (dispersion) of the phase shift of the dipole moment on the polymer relative to the applied voltage from the external electrodes gives rise to a Debye-like relaxation process which can be used to explain a large part of the frequency dependence of the dielectrophoretic force. In the words of the excellent paper by Foster *et al.* [14], we confine ourselves to a dispersive object (the DNA polymer) in a non-dispersive solvent (water).

When charge moves along the length of a polymer, the polymer behaves like a capacitor C which is "charged" by the movement of the counterions along the backbone and from the surrounding solvent, resulting in an effective charge couple $\pm Q$ separated by some characteristic length d (see Fig. 3). The time for this charge couple to develop we call τ, rather like the charging time RC of a resistor in series with a capacitor, only in this case there is no resistor in series with the capacitor but rather a time-dependent charge build-up on the plate due to diffusion of the counterions, giving rise to the dispersion in the dielectric response.

The solution in the frequency domain for the frequency dependent induced charge $Q(\omega)$ across the polymer is:

$$Q(\omega) = Q_o \left[\frac{1}{1+(\omega\tau)^2} + i\frac{\omega\tau}{1+(\omega\tau)^2} \right] \tag{5.1}$$

where Q_o is the DC induced charge. In the frequency domain the charge response of the molecule thus has an in phase (real) response and an out of phase (imaginary) response due to the lag time of the polarization of the polymer as the ions diffuse along the backbone. The effective dipole moment $\mathbf{p} = Q(\omega)\mathbf{d}$ thus has an in phase (real) and out of phase (imaginary) response to the applied field, where d is some characteristic molecular distance that defines the separation of the charge on the molecule. Both Q_o and d are functions of the length of the polymer and the persistence length γ of the polymer. The in phase component is the component parallel to the applied field and is the component that gives rise to the dielectrophoretic force. In terms of the notation used in equation (3.1), we have:

$$E\alpha(\omega) = Re[Q(\omega)]d. \tag{5.2}$$

Note that the polarizability as given by equations (5.1), (5.2) and (3.2) goes to zero at frequencies large compared to $1/\tau$ and has a finite value at zero frequency.

The relaxation time τ of the system can be viewed as the relaxation time RC of the capacitance of the polymer viewed as a charged object, and the resistance R of the counterion cloud which allows the charge separated on the ends of the molecule to flow together. Thus, the relaxation time τ of the response must be the diffusion time of the counter-ions across a distance x which represents the mean size of the molecule. The fundamental relationship between τ and x is:

$$< x^2 > = 2D\tau \tag{5.3}$$

where D is the diffusion coefficient of the ions. The diffusion coefficient of the ions is related by Einsten's relationship to the ratio of the thermal energy $k_B T$ to the frictional coefficient ζ of the ion:

$$D \sim \frac{\zeta}{k_B T}; \zeta \sim 6\pi\eta a \tag{5.4}$$

where η is the viscosity of the medium and a is some mean hydrodynamic radius. Monovalent ions such as Na^+ have diffusion coefficients on the order of 10^{-5} cm^2/s at room temperature in water of viscosity 1 cP [23]. In addition to the diffusion coefficient of the counter-ions, in order to estimate τ we have to have some idea of the size of the polymer which separates the charge. We should point out here that we consider only the diffusion of the counter ions, not the diffusion of the center of mass of the polymer. The ability of the counter-ions to diffuse freely through the polymer is due to the free-draining nature of the hydrodynamics of a polymer undergoing electrophoresis [24].

We need to point out here that the above analysis is surely oversimplified. We assumed that the dipole moment relaxes due to pure diffusive motion of the counterions, but of course the electric field generated by the dipole moment should enhance this relaxation rate. However, electric fields in a ionic medium are shielded by the counterions and this greatly reduces the actual field due to the dipole across the molecule. An excellent review paper by Hoagland *et al.* [25] gives a clear description of the physics of counterion shielding. The basic length scale for shielding by counter ions is the Debye length $\lambda_{\rm D}$:

$$\lambda_{\rm D} = \left[\frac{\epsilon k_{\rm B} T}{4\pi k e^2 n_{\rm b}}\right]^{1/2} \tag{5.5}$$

where ϵ is the dielectric constant of the fluid, e is the electron charge, k is Coulomb's law constant (9×10^9 Nm^2/C^2) and $n_{\rm b}$ is the number of ions/volume in the bulk solvent. For a 0.1 M salt concentration, $\lambda_{\rm D}$ is about 3 nm, so the screening distance is very short relative to the length of our molecules and perhaps the field enhanced diffusion is not important.

Given then that the purely diffusive relaxation may overestimate relaxation times, we continue with it since it seems to give bascially order of magnitude correct relaxation rates. We can consider easily two extreme cases: (1) long polymers, whose persistence length γ is much less than the extended length L of the polymer, and (2), short polymers whose length L is much less than the persistence length γ.

In the case of a very long polymer, the diffusion distance x can be approximated by the mean separation between the two ends of the polymer $R = (2L\gamma)^{1/2}$ of the polymer. In the case of a short polymer, we can use $x \sim L$ since the polymer is simply extended roughly to its full length. Thus, we have for long polymers the relaxation time $\tau_{\rm long}$:

$$\tau_{\rm long} = \frac{L\gamma}{D} \tag{5.6}$$

while for short polymers $\tau_{\rm short}$:

$$\tau_{\rm short} = \frac{L^2}{2D}\,. \tag{5.7}$$

These two expressions can be roughly used to predict the relaxation times, but should be taken with a grain of salt. For example, consider the T7 dsDNA data shown in Figure 9, which shows the EDEP force as a function of frequency and viscosity. Measurements at the two higher viscosities (3.7 and 5.9 cp) clearly show that the dielectrophoretic force decreases at high frequencies. This occurs at approximately 1 kHz for 1 cP, 300 Hz for 3.7 cP and 150 Hz for 5.9 cP. Since T7 at 39.9 kB is definitely in the $L >> \gamma$

case for long polymers, we can use equation (5.6) to estimate the relaxation time of these polymers. Figure 11 shows the satisfactory agreement between the observed relaxation times and the ones predicted for a long polymer, considering the simplicity of the model used. Short polymers can be expected to have faster relaxation times. In the case of our 368 bp fragment, equation (5.7) predicts a relaxation time in water of about 10^{-5} s, substantially beyond the present 1 kHz bandwidth of our high voltage power supply.

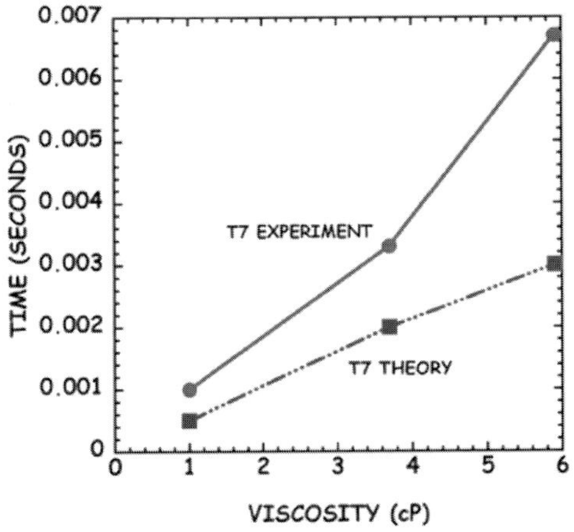

Fig. 11. The measured relaxation time of 39.9 kB (T7 phage) DNA *vs.* viscosity (solid line) *vs.* the predicted relaxation time (dot-dash line).

Rough calculation of the force F felt by the polymer is more difficult as we have mentioned. From equation (3.2) we know that the dielectric force is proportional to the product of the polarizability of the molecule α times $|E|\mathrm{d}\mathbf{E}/\mathrm{d}z$. The polarizability α is equal to Cx^2, where C is the effective capacitance of the molecule and x^2 is the mean squared separation of the two charged ends of the molecule. In the rough approximation that the capacitance C is equal to $\epsilon\epsilon_o A/x$, where A is the area of the charged ends of the molecule, we once again find that the force also depends on the statistical mechanics of the polymer. If (1) $L >> \gamma$, we get that $\alpha_{\text{long}} = \epsilon\epsilon_o[(2L\gamma)]^{3/2}$, while if (2) $L << \gamma$ we find that $\alpha_{\text{short}} = \epsilon\epsilon_o A_o L$, where A_o is an area which characterizes the end area of rigid length of the molecule. These numbers are rather poorly defined. The backwards way to do this is simply to calculate from the measured force at a given $\mathbf{E}$ and $\mathrm{d}\mathbf{E}/\mathrm{d}\mathbf{z}$ the polarizability α. As we have shown, α is a strong function of length and conformation of the molecule so there is no single intensive parameter that characterizes DNA.

There is still a problem with this analysis. Equations (3.2) and (5.1) together imply that the EDEP force is effectively zero at high frequencies (which is not true because of other processes that come into play [18, 26]), rises at a frequency given by $1/\tau$ and then remains constant down to DC. In fact, all our data show the apparent force falling to zero at DC frequencies. The problem is that we have ignored the electrophoretic force. The total force acting on a polyelectrolyte in an external electric field is the sum of the electrophoretic force $\mathbf{F}_\mathrm{e}$ due to the net effective linear charge density β of the polymer, and the dielectrophoretic force $\mathbf{F}_\mathrm{d}$ due to the induced dipole moment $\mathbf{p}$ discussed above. The electrophoretic force $\mathbf{F}_\mathrm{e}$ on a polyelectrolyte in the presence of an electric field is proportional to the local applied electrical field $\mathbf{E}$ and gives rise to a constant velocity $\mathbf{v}_\mathrm{e}$:

$$\mathbf{v}_\mathrm{e} = \mu_\mathrm{e}\mathbf{E}; \quad \mathbf{F_e} = \zeta\mathbf{v}_\mathrm{e} = \zeta\mu_\mathrm{e}\mathbf{E} \tag{5.8}$$

where μ_e is the electrophoretic mobility of the polymer, v_e is the electrophoretic velocity and ζ is drag coefficient between the electrophoretic velocity and the force. The origin of the electrophoretic force F_e in polyelectrolytes has been intensively studied [27] and is characterized by the surprising fact that the electrophoretic mobility of a polyelectrolyte is basically independent of the length of the polymer in free solution, hence we can treat μ_e as a constant independent of length. We then have a final expression for the total force acting on a charged, polarizable polyelectrolyte:

$$\mathbf{F}_\mathrm{tot} = \zeta\mu_\mathrm{e}\mathbf{E} + \alpha|E|\frac{\mathrm{d}\mathbf{E}}{\mathrm{d}z} \cdot \tag{5.9}$$

An interesting aspect of the dielectric force is that it is a nonlinear force as a function of $\mathbf{E}$, and hence at sufficiently high field strengths and sufficiently low ratios of μ_e/α a gradient can trap a molecule even in a static DC field, since the dielectrophoretic force will ultimately be greater than the linear electrophoretic force. By combining the electrophoretic and the dielectrophoretic response, we show in Figure 12 the forces and potential surfaces that charged, polarizable objects experience going through a gap similar to one of our devices. The parameters for the polarizability α and the electrophoretic mobility μ_e were chosen here to roughly correspond to our longest molecules studied, the 40 kB dsDNA. Note that the nonlinear dielectrophoretic component of the trapping force gives rise to a short-range trapping potential. If the field direction is switched, the electrophoretic potential surface will slope in the opposite way while the dielectrophoretic potential is invariant, so that only the dielectrophoretic component of the force serves as a trap.

Since the free flow electrophoretic mobility μ_e is basically independent of length of the DNA molecule, the effect of electrophoresis of the molecule

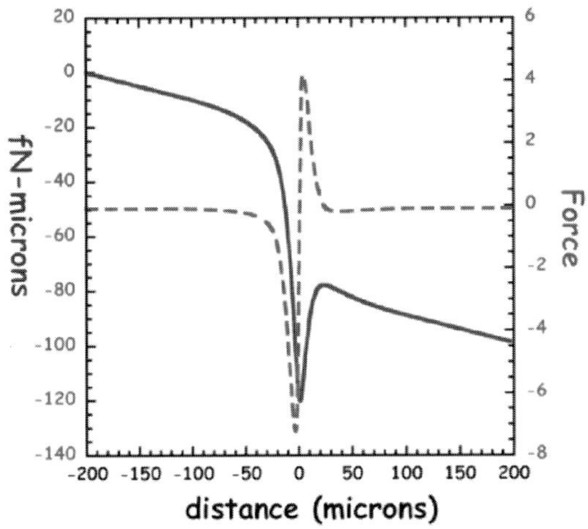

Fig. 12. The total force, electrophoretic and dielectrophoretic, experienced by a particle passing through the gradient trap shown in Figure 1 is presented as the dashed line in the figure. The potential $U(z)$ surface that the particle moves along is shown by solid line.

is an *apparent* decrease in the force at low frequencies if the electrophoretic force is greater than the dielectrophoretic force, which seems to be the case for DNA. In fact, our entire model which we used to analyze the dielectrophoretic force acting on the molecules basically breaks down at low frequencies, since we do not have an equilibrium condition. At present, we have no way of disentangling the true dielectrophoretic force at low frequencies from the electrophoretic force.

6 Conclusion

We have used electrodeless EDEP to trap and concentrate single and double stranded DNA. The analytical simplicity of the field pattern in a electrodeless trap has allowed us to characterize the length and frequency dependence of the EDEP force. We showed the strong dielectrophoretic response of the DNA in the audio frequency range. We also demonstrated that for the given trapping voltage applied, the dielectrophoretic force dramatically increases with the increase of the length of the DNA molecule. There is actually a great dispersion in the force with length, hence by appropriate choice of parameters one can envision selectively trapping one range of DNA molecules while removing others. By measuring dielectrophoretic force under different solvent viscosity conditions, we were able to determine that movements of counterions in the Debye layer are responsible for the dielectrophoretic response of the DNA for 2 reasons: (1) a strong dependence of relaxation times on solvent viscosity indicates that the charge redistribution occurs via movement through the solvent; (2) the expected relaxation times due to diffusion of ions across the radius of gyration of the polymer are in rough agreement with the observed relaxation times.

The dielectrophoretic trapping of the DNA in electrodeless traps has a great potential for use in biotechnology. The EDEP force may be adjusted accordingly by varying the shape and cross section of the constriction. Position of the constriction also can be controlled at will. Since EDEP trapping occurs in high field gradient regions, EDEP allows easy patterning of DNA by appropriate geometrical obstacle design. Other potential applications of EDEP method are selective trapping of specific ranges of DNA, concentration of DNA molecules to very tight bands before launch into a fractionating media, PCR cleanup, concentration of DNA in gene array chips to enhance sensitivity of the detection limit by increasing local S/N, or acceleration of gene hybridization rates by concentration of single stranded DNA, and in general for any reaction for which the rate scales with concentration or any power of the concentration greater than 1.

References

[1] H.A. Pohl, Dielectrophoresis: The Behavior of Neutral Matter in Nonuniform Electric Fields (Cambridge University Press, Cambridge, UK, 1978).

[2] M. Washizu and O. Kurosawa, *IEEE Trans. Ind. Appl.* **26** (1990) 1165-1172.

[3] T. Muller, G. Gradl, S. Howitz, S. Shirley, T. Schnelle and G. Fuhr, *Biosens. Bioelectron.* **14** (1999) 247-256.

[4] M. Washizu, S. Suzuki, O. Kurosawa, T. Nishizaka and T. Shinohara, *IEEE Trans. Ind. Appl.* **30** (1994) 835-843.

[5] C.L. Asbury and G. van den Engh, *Biophys. J.* **74** (1998) 1024-1030.

[6] N.G. Green and H. Morgan, *J. Phys. Chem. B* **103** (1999) 41-50.

[7] J. Rousselet, L. Salome, L. Adjari and J. Prost, *Nature* **370** (1994) 446-448.

[8] L. Gorre-Talini, J.P. Spatz and P. Silberzan, *Chaos* **8** (1998) 650-656.

[9] R. Pethig, *Crit. Rev. Biotechnol.* **16** (1996) 331-348.

[10] H. Morgan, M.P. Hughes and N.G. Green, *Biophys. J.* **77** (1999) 516-525.

[11] F.F. Becker, X.B. Wang, Y. Huang, R. Pethig, J. Vykoukal and P. Gascoyne, *Proc. Nat. Acad. Sci.* **92** (1995) 860-864.

[12] J. Yang, Y. Huang, X.B. Wang, F.F. Becker and P. Gascoyne, *Anal. Chem.* **71** (1999) 911-918.

[13] P. Chinachoti, M.P. Steinberg and D.A. Payne, *J. Food Sci.* **53** (1988) 580-583.

[14] K.R. Foster, F.A. Sauerm and H. Schwan, *Biophys. J.* **63** (1992) 180-190.

[15] J.D. Jackson, *Classical Electrodynamics*, 2nd Ed. (John Wiley & Sons, New York, 1975).

[16] K.F. Ren, G. Grehan and G. Gouesbet, *Appl. Opt.* **35** (1996) 2702-2710.

[17] S. Takashima, *J. Mol. Biol.* **7** (1963) 455-467.

[18] S. Takashima, *Electrical Properties of Biopolymers and Membranes* (IOP, Philadelphia, PA, 1989).

[19] S.B. Smith, L. Finzi and C. Bustamante, *Science* **258** (1992), 1122-6.

[20] S.B. Smith, C. Yujia and C. Bustamante, *Science* **271** (1996) 795-799.

[21] B. Tinland, A. Pluen, J. Sturm and G. Weill, *Macromolecules* **30** (1997) 5763-5765.

[22] D. Porschke, *Biophys. Chem.* **22** (1985) 237-247.

[23] American Institute of Physics Handbook, Third Edition (American Institute of Physics, College Park, MD, 1972).

[24] W.D. Volkmuth, T. Duke, M.C. Wu, R.H. Austin and Attila Szabo, *Phys. Rev. Lett.* **72** (1994) 2117-2120.

[25] D.A. Hoagland, E. Arvanitidou and C. Welch, *Macromolecules* **32** (1999) 6180-6190.

[26] F. Oosawa, *Biopolymers* **9** (1970) 677-688.

[27] Capillary Electrophoresis: Theory and Practice, Paul D. Grossman, Joel C. Colburn (ed.) (Academic Press, 1997).

[28] C. Desruisseaux, D. Long, G. Drouin and G.W. Slater, *Macromolecules* **34** (2001) 44-52.

Abstract

Now, I want to try and apply some of the principles I have outlined in the design of a real working device. Long DNA molecules in agarose gels and other polymer matrices, get hooked on many gel fibers simultaneously, exhibiting complex motion, and confounding theory and experiment alike. I'll describe how a hex array allows better prediction of pulsed field parameters for a given range of molecular sizes, and a relatively simple theory describes the motion. The analytical nature of the motion is a real advantage of the technique since it may allow us to separate unlabeled molecules.

1 Introduction

This story is about 8 years long, and it is so long purely because of the stupidity of the author. Had I been a bit more subtle in my understanding of Nature the story would be shorter and more interesting I suspect.

As I originally mentioned, the shearing boundary at the zeta potential means that DNA molecules undergoing electrophoresis are free-draining, that is, the solvent effectively passes through the random coil of the polymer. In "normal" hydrodynamics, if you move a polymer there is a velocity vector flow pattern $V(r)$ radiating out from a point on the polymer that couples different parts of the object together, modifying the drag coefficient of the polymer from a strictly linear sum of terms (typically a $1/\ln(L)$ correction), $L =$ length of the object.

For a free-draining polymer, that is, a polymer undergoing electrophoresis, the linear sum works, and the drag is proportional to simply L and not $L/\ln(L)$. Since the drag and the force both scale as L, the electrophoretic mobility is independent of the length and you cannot fractionate DNA molecules in bulk solution using electrophoresis. Thus, the ubiquitous presence of some sort of retarding medium (a gel) in most molecular biology labs. The retarding medium adds a length-dependent additional force as the random coil try to squeeze through the medium.

Thus, the basic idea of the microfabricated arrays was simply to simulate a gel by microfabricating obstacles which add another length dependent term to the drag on the polymer. However, there are two things at work here that kill the technology:

(1) The pore sizes ($a = 1$ micron) are much bigger than typical gels, where the pore sizes range from 10 nm to 500 nm or so.

(2) the self hydrodynamic forces acting on a random polymer in a "thin" slit of thickness d are MUCH greater than the self-forces felt by a polymer in bulk solution.

Let's now consider each of these points in turn. However, before we can launch into this subject we need to carefuly at how polymers move through structures.

The effect of large pore sizes is to increase the effective force acting at the contact points between polymer parts and the posts, these larger forces serve to enhance stretching of the polymer. That is, consider if you will that there is an electric field **E** in the solution. The applied force **F** on a stretched fragment spanning the pore of size a is $a\rho E$, ρ = linear charge density of the polymer. If the polymer is randomly coiled, the effective amount of polymer in the pore is increased leading to an even greater force acting on the polymer. There is of course an entropic "spring" constant

The end-end distance R_z of a "hung" polymer in an electric field E is roughly:

$$R_z = \kappa \ln[\sin h(L/\kappa)/(L/\kappa)] \tag{1.1}$$

where L = total length of the polymer, made of N pieces of length $2p$ (p = persistence length of the polymer) and:

$$\kappa = 2\rho p E/k_b T \tag{1.2}$$

where ρ = charge/length of the polyelectrolyte.

If you play with this equation you learn a Big Lesson: in a "large" pore environment, DNA is highly aligned at low fields! A further fact, which can be ascertained from my earlier notes on hydrodynamics in 2.5 dimensions, is that in my thin "slit" of thickness h there is a hydrodynamic coupling to the surface via stick boundary conditions. The stick boundary condition slows down the entropic relaxation time of the polymer. This can be a big effect, and the slow relaxation time enhances elongation. The bottom line is that elongation increases with decreasing etch depth.

Once the polymer is elongated, the mobility becomes length independent because the drag acting on the polymer is now proportional to the length. This is a disaster!

This is why you can't run a gel at high voltages to speed up fractionation times, and why I didn't believe the Human Genome Initiative could succeed as originally planned. Tom Duke in collaboration with our group has suggested two ways to get around this problem. I will first talk about the pulsed field/hex array idea. The idea here is if you can't beat the physics, use the physics: that is, if the polymer physics wants to elongate the polymer, figure out a way to use elongated molecules. Tom's idea was a take on Ed Southern's idea for using PULSED TRANSVERSE fields to fractionate elongated polymers.

Transverse pulsed field electrophoresis in hexagonal arrays uses an array of 2 μm pillars with 2 μm spacings arranged in a hexagonal lattice and takes advantage of the DNA elongation that occurs in microfabricated arrays [5]. Application of a pulsed field along alternating axes of the array, separated by 120°, causes net motion of the DNA molecules along the bisector of the axes, with average migration speeds that depend on their length (Fig. 1).

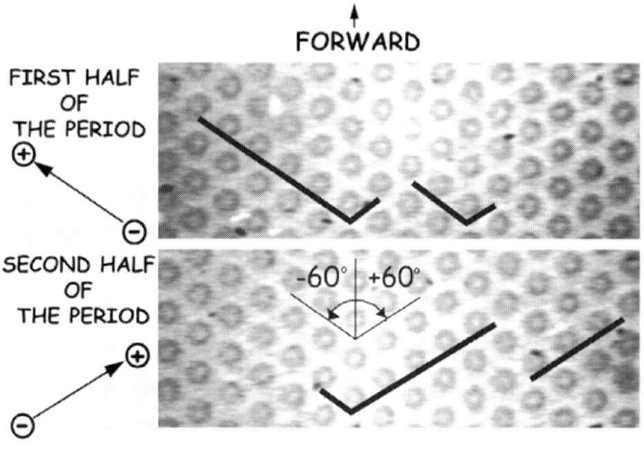

Fig. 1. 2 μm wide pillars were used in these experiments. Cartoon DNA molecules are drawn to illustrate the motion of individual molecules of different lengths. Each period T consists of two pulses aligned along the channels created by the posts in the array, giving a net angle between the two field directions of 120°. Shorter molecules move farther in the array because once they have reoriented along the axis of the field they move in an unhindered straight line for the duration of the pulse. Longer molecules, on the other hand, spend most of the pulse period retracing their paths. In the example shown here, the longer of the two will never advance.

A useful separation device, in addition to using an effective separation mechanism, must also collect and launch molecules in a narrow zone, since initial zone broadening destroys resolving power. In our device the DNA was entropically trapped and released as a band using the principle described by Han *et al.* [6]. We used an entropic barrier placed in the path of the DNA near the entrance to the array (Fig. 2). There is a small gap between the barrier top and the cover slip that seals the array. The gap between the top of the barrier and the cover slip was smaller than the radius of gyration of the DNA molecules to be fractionated (150 nm in these experiments). When very low DC fields are applied to move DNA molecules into the array, the molecules do not have enough applied force applied to them to squeeze through the gap, and hence they get trapped (Fig. 2A). Under higher fields they get stretched and move through the gap with mobilities that are independent of molecular length.

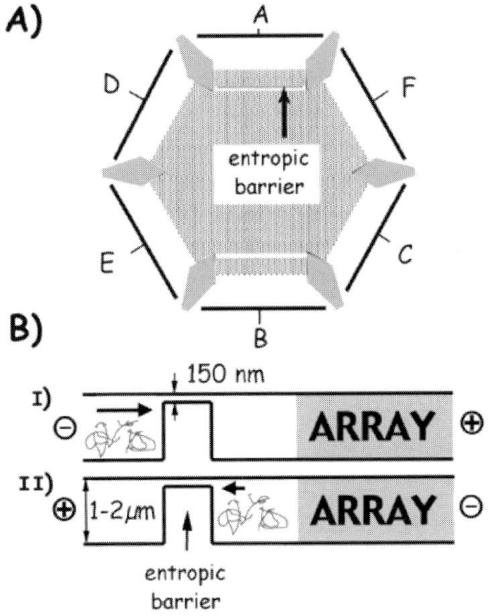

Fig. 2. A) Sketch of the device. The diameter of a typical device was 1 cm to 3 cm. The pulsed field was applied through two pairs of external electrodes (C-D and E-F). The remaining electrode pair (A-B) was used for entropic trapping. The electrodes were insulated from each other by six silicone structures (lozenge shaped in panel **A**)). **B)** Cross section of the device showing entropic trapping. The arrows point in the direction of DNA motion, while their lengths correspond to the strength of the applied field. B-I): the beginning of DNA transport across the barrier using a high electric field. B-II): the concentration and cleanup step where the molecules are forced back against the barrier at low field before they are launched into the array.

In our device molecules were first transported into the array through the gap using a high electric field. They were then concentrated against the barrier by applying a low electric field oriented in the opposite direction. Concentrated molecules were then launched into the array by reversing the field direction. The alternating fields were not applied directly at the entropic barrier but rather were applied only after the DNA band had been moved several millimeters into the array. This was done to avoid the field curvature seen at the corners of the array and ensure that the fields were uniform in the directions necessary for predictable fractionation.

2 Experimental approach

The devices were made of quartz using standard photolithography and reactive ion etching techniques. They contained an array of pillars oriented in a hexagonal lattice whose height was 2 μm. The arrays were sealed using glass coverslips with a spun on thin layer silicone elastomer [7] (RTV615A and RTV615B, GE Silicones, Waterford, NY). The silicone elastomer surface was treated for one minute in an oxygen plasma to make the silicone hydrophobic, necessary for wetting of the sealed device. The devices were shaped as hexagons 3 cm in diameter (Fig. 2B) to allow easy application of electric fields oriented at 120° [8]. They were mounted on a plastic holder that contained the outside electrodes. Two pairs of outside electrodes were used to apply pulsed fields (C-D, E-F Fig. 2) while the remaining pair was used for entropic trapping (A-B).

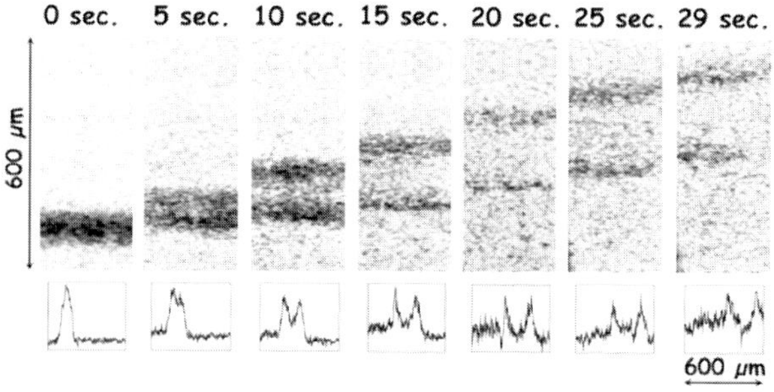

Fig. 3. Video clips of λ and T4 DNA pulsed at 244 V and with period $T = 1$ s after release from the entropic trap.

The DNA fluoresence stain TOTO-1 (Molecular Probes) was used at 1 μg/ml concentrations to stain T4 and λ DNA molecules which were loaded into the arrays in concentrations of 15 μg/ml and 5 μg/ml, respectively, and observed by epifluorescence using the 488 nm line of an Ar/Kr laser. The 0.5 $\times$ TBE electrophoresis buffer (45 mM Tris/borate, 1mM EDTA, pH 8.0) contained 0.1% POP-6 (Perkin Elmer Biosystems) to reduce electroendosmosis and 0.1 M DTT to reduce bleaching.

T4 (168.9 kbp) and λ (48.5 kbp) DNA were separated in a very short time with high resolution. The mixture (see Experimental Protocol) was resolved into two bands in 10 s (Figs. 3 and 4). The position of the peaks is plotted *vs.* time in Figure 4. From $t_1 = -10$ s to $t_2 = 0$ s, the applied voltage was 100 V and the pulse period T was 1 s. Under these conditions the bands separated by less than the bandwidth. At time t_2 the voltage was increased to 244 V, keeping the period the same. The two bands were then

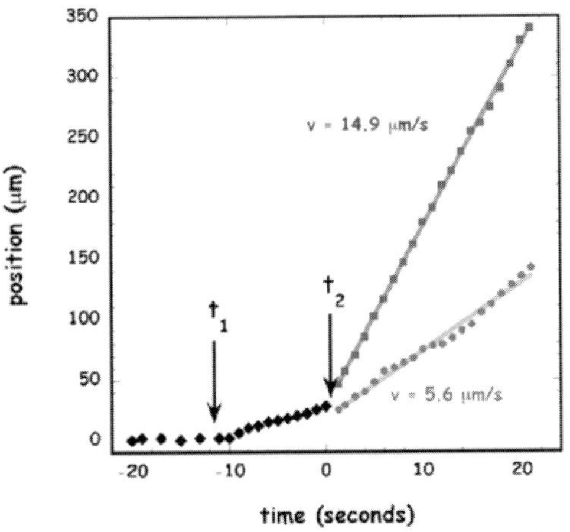

Fig. 4. The position and speeds of the band peaks of λ (squares), T4 (circles) and the mixed band (diamonds) *vs.* time. From t_1 until t_2, the pulsing parameters were 100 V and $T = 1$ s. The peak of the mixed band moved at 2.3 μm/s. At time t_2 the voltage was increased to 244 V. The two bands were cleanly resolved in approximately 10 s. T4 DNA moved at 5.6 μm/s, while λ DNA moved at 14.9 μm/s. Speeds were obtained by fitting the straight lines shown through the band centroid positions.

resolved in approximately 10 s. Figure 5 confirms that under our pulsed field conditions the shorter molecules move faster than the longer molecules.

The widths of the bands after 11 min were approximately the same for both DNAs (*FWHM* was ~200 μm). For a diffusion constant $D \sim 1\ \mu\text{m}^2/\text{s}$ the expected diffusional broadening ($\sim[2Dt]^{1/2}$ due to self diffusion in $t = 11$ min is ~30 μm, while the observed broadening is 100 μm. This dispersive broadening possibly occurs because individual molecules of equal lengths get stretched by different amounts, or some times not at all, when encountering posts. The location of the bands after 11 min was consistent with the band separation derived from the microscopic migration velocities, $\Delta x = (V_\lambda - V_{\text{T4}})t = 6100\ \mu\text{m}$. The band capacity [9] n_{C} in this experiment is estimated to be $\Delta x/[1.5[\sigma(\lambda) + \sigma(T4)] = 20$, *i.e.* 20 bands could be resolved in the 50 to 170 kbp range under the conditions used here. Since the separation is approximately linear in molecular weight (see below), this means that molecules which differ by 6 kbp can in principle be distinguished.

Observation of the microscopic dynamics confirms that the separation is a consequence of "switchback" motion of the DNA molecules for opposing field directions greater than 90°, as reported in preliminary investigations [10, 11] and illustrated in Figure 1. When the field direction is

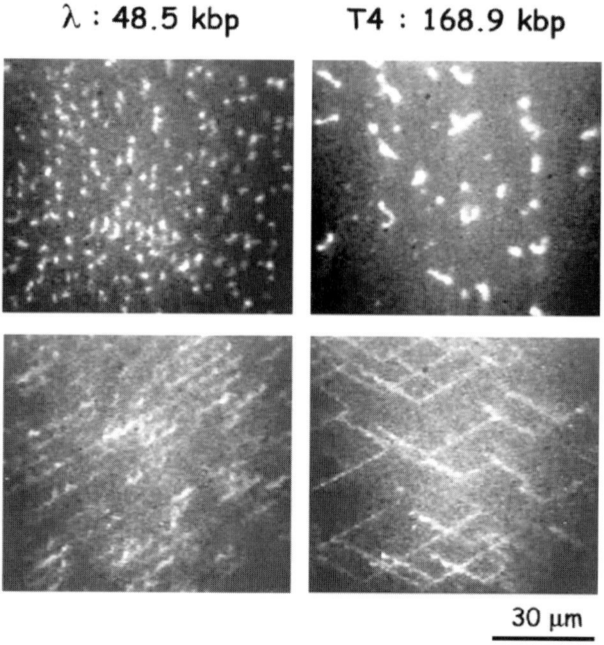

Fig. 5. Images of single molecules taken at 11 minutes after the start of pulsing. These images were gathered 10 mm (left) and 4 mm (right) from the entropic barrier. In the images at the top, the field was turned off and the molecules were at thermal equilibrium, while the bottom panels show the separated molecules elongated under pulsed field conditions. Note that the radius of gyration of the relaxed molecules in the top panels clearly establishes that the faster moving species is λ DNA.

switched, each molecule moves off in the new field direction, led by the end which was previously trailing. As a result, the molecules retrace part of the path they have traveled. Since longer molecules backtrack further than shorter molecules, the rate of advance along the bisector of the field is slower.

The overall speed of migration in a pulsed hexagonal array is length-dependent. Under the simplifying assumption that the molecules remain uniformly stretched during this motion, the net velocity V_{L} of molecules of length L in a pulsed field array can be described by the simple equation [10]:

$$V_{\mathrm{L}} = \mu_{\mathrm{o}} E \cos(\theta/2) \left[1 - \frac{L}{L^*}\right] \tag{2.1}$$

where the angle θ between transverse fields in our case is 120° and μ_{o} is the continuous-field mobility, which is independent of length and L^* is a critical cut-off parameter. The critical cut-off length L^* arises from that fact that molecules which do not have time to reorient completely during a single pulse repeatedly retrace the same path and, since they make no progress, their velocity is effectively zero. The reorientation time t_{or} is to first order simply $L/\mu_0 E$, or more empirically given by:

$$t_{\text{or}} = c_1 L/(\mu_0 E) \tag{2.2}$$

where c_1 is a parameter which takes into account the fact that the DNA molecules are not fully aligned. Roughly, in order to fractionate molecules of length L the pulse period T should be set to t_{or}. Thus, the upper limit of separation, L^*, is proportional to the pulse time T and to the speed $\mu_o E$ of a molecule along a free channel:

$$L^* = \frac{T}{2}\frac{\mu_0 E}{c_1} . \tag{2.3}$$

It is a strength of the analytical nature of the dynamics of DNA molecules in synthetic arrays that L^* and T can be predicted. Note that we can obtain expressions for $\mu_o E$ and L^* from our experimental data at 244 V with a pulse period T of 1 s for T4 and λ DNA. Some algebra gives:

$$\mu_o E = \frac{L_{\text{T4}} V_\lambda - L_\lambda V_{\text{T4}}}{(L_{\text{T4}} - L_\lambda)\cos(\theta/2)} \tag{2.4}$$

$$L^* = \frac{L_{\text{T4}} V_\lambda - L_\lambda V_{\text{T4}}}{(V_{\text{T4}} - V_\lambda)} . \tag{2.5}$$

These expressions can then be used to check the actual parameters used in the experiment. We can assume (somewhat incorrectly) that the lengths of the DNA in the above expression are the fully stretched lengths taking into account that one Kuhn length b contains 300 basepairs and is 130 nm long for TOTO-1 stained DNA [12], giving $L_{\text{T4}} = 168.9$ kbp $= 73$ μm and $L_\lambda = 48.5$ kbp $= 21$ μm. The measured values of the retarded velocities are $V_{\text{T4}} = 5.6$ μm/s and $V_\lambda = 14.9$ μm/s; from these parameters we get that $\mu_o E = 37$ μm/s and $L^* = 104$ μm. The pulse periods T that should be used at L^* are then roughly $t_{\text{or}} = 3$ s. Thus, pulse periods on the order of 1 s in this particular protocol are appropriate.

We can check the consistency of the experimental data at the two field strengths. Since the migration velocity in a continuous field is proportional to the applied voltage, we expect that at 100 V, μ_0 E $= 15$ μm/s. Equation (3) then predicts that $L^* = 43$ μm. Equation (1) predicts that lambda DNA should move at velocity $V_\lambda = 3.9$ μm/s, which is what we observe. Since the T4 DNA is longer than L^*, the theory predicts that it will not move, but in practice it advances at the low speed of 2 μm/s. This migration occurs because the backtracking motion is not ideal. We observe that the molecules do not remain uniformly extended, but usually get stretched by the field and subsequently relax during each pulse. Because of this inherent stochasticity, even the longest molecules do not retrace the same path indefinitely. Our measurements yields $c_1 = 0.18$. Duke *et al.* [10] calculated the value $c_1 = 1.39$, assuming that molecules are extended to their full contour length at all times. If the molecules are stretched to only a fraction of their length, we would expect the value of c_1 to be proportionately reduced, because the reorientation time would be faster. The lower value of c_1 that we measure is therefore consistent with our observations that, at the field strengths used, the molecules are rarely extended to more than 30–40% of their full contour length.

3 Conclusions

Separation in the hexagonal arrays will be even faster when higher electric fields with shorter periods are applied. In our experiments we were limited to relatively low electric fields and long periods because we wished to record single molecule images. Higher fields would cause greater molecular extension, which would enhance the regularity of the "switchback" motion and improve the discrimination between molecules of different size. Reduction of the depth of the device is also expected to increase the extension of the molecules [12].

We can compare these results with those of others. Pulsed field capillary gel electrophoresis [13, 14] achieves fast separation, but this method is severely limited by the tendency of high molecular weight DNA to form supramolecular complexes that interfere with separation. Chou *et al.* [15] have proposed a single molecule sizing device in which molecules in the 2 to 200 kbp range are sized one at a time. This method, however, cannot be used to separate many thousands of molecules simultaneously. Other methods, such as recently developed arrays that separate molecules based on their diffusion coefficients [16], suffer from rapidly deteriorating resolution as the molecules get bigger. Han *et al.* [6] have shown that entropic trapping can be used to separate molecules in reverse order, with the largest molecules moving the fastest. This method, although faster than conventional methods, is still at least an order of magnitude slower than separation in hexagonal arrays using quantitatively understood pulsed field parameters that we report here.

The separation principle that we demonstrate here is not limited to DNA molecules of a particular size. By appropriately adjusting the applied electric field, the pulse time and array parameters such as the pillar size or spacing, the technique could be extended to the separation of polymers of all lengths. Nor is it limited to the separation of fluorescently-stained molecules. Because the motion of the molecules is predictable, it should be possible to operate the device "blind", without the addition of dye stains which can contaminate subsequent processes such as polymerase chain reactions. Alternatively, the dye can be removed by standard techniques using ion exchange resins. In the future, it should be feasible to dispense with fluorescence-based detection methods and detect the DNA molecules electronically using nanosensors constructed on the chip [17].

References

[1] D.C. Schwartz and C.R. Cantor, *Cell* **37** (1984) 67-75.

[2] G.F. Carle, M. Frank and M.V. Olson, *Science* **232** (1986) 65-68.

[3] E.C. Cox, C.D. Vocke, S. Walter, K.Y. Gregg and E.S. Bain, *Proc. Natl. Sci. USA* **87** (1990) 8247-8251.

[4] M.J. Orbach, D. Vollrath, R.W. Davis and C. Yanofsky, *Mol. Cell. Biol.* **8**(4) (1988) 1469-1473.

[5] W.D. Volkmuth and R.H. Austin, *Nature* **358** (1992) 600-602.

[6] J. Han and H.G. Craighead, *Science* **288** (2000) 1026-1029.

[7] D.C. Duffy, J.C. McDonald, O.J.A. Schueller and G.M. Whitesides, *Anal. Chem.* **70** (1998) 4974-4984.

[8] G. Chu, D. Vollrath and R.W. Davis, *Science* **234** (1986) 1582-1585.

[9] J.C. Giddings, Unified Separation Science (John Wiley & Sons, New York, 1991) pp. 101-106.

[10] T.A.J. Duke, R.H. Austin, E.C. Cox and S.S. Chan, *Electrophoresis* **17** (1996) 1075-1079.

[11] O. Bakajin, T.A.J. Duke, C.F. Chou, J. Tegenfeldt, S.S. Chan, R.H. Austin and E.C. Cox, Third International Biophysics Symposium Proceedings, *AIP Conf. Proc.* **487** (1998) 243-248.

[12] O.B. Bakajin, T.A.J. Duke, C.F. Chou, S.S. Chan, R.H. Austin and E.C. Cox, *Phys. Rev. Lett.* **80** (1998) 2737-2740.

[13] Y. Kim and M.D. Morris, *Anal. Chem.* **67** (1995) 784-786.

[14] S. Magnusdottir, H. Isambert, C. Heller and J.L. Viovy, *Biopolymers* **49** (1999) 385-401.

[15] H.P. Chou, C. Spence, A. Scherer and S. Quake, *Proc. Natl. Acad. Sci. USA* **96** (1999) 11-13.

[16] C.F. Chou, O.B. Bakajin, S.W. Turner, T.A.J. Duke, S.S. Chan, E.C. Cox, H.G. Craighead and R.H. Austin, *Proc. Nat. Ac. Sci.* **96**(24) (1999) 13762-13765.

[17] X. Duan, Y. Huang, Y. Cui, J. Wang and C.M. Lieber, *Nature* **409** (2001) 66-69.

[18] A. Manz, N. Graber and H.M. Widmer, *Sensors Actuators B* **1** (1990) 244-248.

Abstract

The previous lecture discussed how using crossed fields greater than 90° can be used to quickly fractionate DNA. However there are problems with the technique we described in the past lecture: it requires running DNA as a single run of prepurified DNA, so that if one is lysing and extracting genomic DNA as part of a continuous process it is very difficult with the technology described. Further, we have not really exploited the ideas of 2.5 D hydrodynamics that we described earlier. Now it is time to correct that problem with what we call a DNA prism, which uses the idea of transverse pulsed field fractionation but with a twist: *asymmetric* pulsed fields. We now tie things together.

1 Introduction

The previous lecture discussed how using crossed fields greater than 90° can be used to quickly fractionate DNA. However there are problems with the technique we described inn the past lecture: it requires running DNA as a single run of prepurified DNA, so that if one is lysing and extracting genomic DNA as part of a continuous process it is very difficult with the technology described. Further, we have not really exploited the ideas of 2.5 D hydrodynamics that we described earlier. Now it is time to correct that problem with what we call a DNA prism, which uses the idea of transverse pulsed field fractionation but with a twist: *asymmetric* pulsed fields. We now tie things together.

The basic principle of prism separation is shown in Figure 1. As we have discussed, when an electric field is applied in one direction, molecules of all sizes migrate between the SiO2 posts with similar mobility (Fig. 1A). When the field is switched 120°, all molecules must backtrack through channels formed by the post array (Fig. 1B). The longer molecules take longer to reorient, and thus at each change in field direction, the separation between small and large molecules increases (Fig. 1C). This is the physical basis for pulsed field gel electrophoresis in microfabricated devices, and probably also for conventional gels. The prism device incorporates these principles and in addition allows continuous fractionation by biasing either the strength of the field from pulse to pulse, or the duration of the pulse at constant field strength (Fig. 1D). Because of the microfluidic design the field should be constant across the device.

2 Design

In our design the microfluidic channels of Figure 2 act as resistors and serve to shape a uniform electric field. Whether or not the device produces

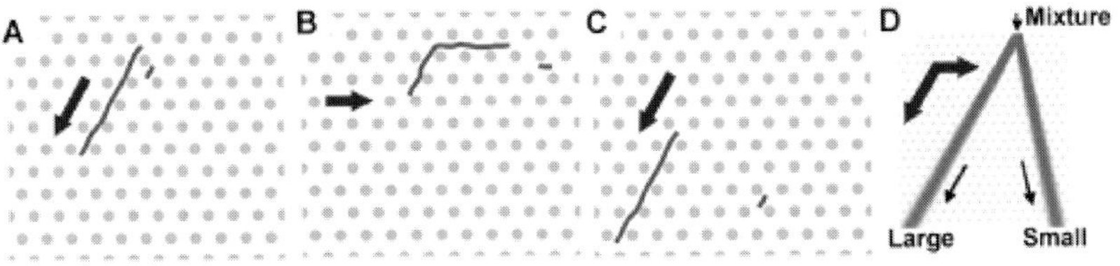

Fig. 1. Schematic diagrams showing the different behavior of small and large DNA molecules in microfabricated arrays with asymmetric alternating-angle electric fields; **A–C)** illustrate the sequential motion of a long and a short molecule through a full cycle of alternating field. **A)** The high field moves both small and large molecules in a channel (arrow shows direction of motion). **B)** A low field rotated 120° causes reversal of the leading and trailing ends, and the low field (or short time) prevents the long molecule from sliding off the posts and reversing direction. **C)** The original field reapplied. The ends again reverse, the large molecule resumes its original track while the small molecule is now in a new track. **D)** Net motion after multiple cycles of a mixture of large and small molecules injected into the array at the same point. The small molecules follow the average field while large molecules follow the stronger field. Here and elsewhere the vectors point in the direction of DNA migration, rather than the direction of the electric field as traditionally defined.

uniform electric fields at the desired angles was tested by tracing the trajectory of fluorescently stained DNA. Predetermined voltages were applied to the reservoirs to create a DC field at −30° with respect to the vertical axis (Fig. 3 inset). A mixture of bacteriophage λ and T2 DNA was loaded into the DNA reservoir and then injected electrophoretically. The injected band should flow along the electric field line. A straight and narrow band ~50 microns wide was formed. If the electric field had not been uniform in direction, the band would have curved. If the electric field had not been uniform in strength, the band would have been tapered or dispersed. We therefore conclude that the electric field is uniform. The voltages were then switched to create a DC field 120° from the previous one, and the band moved at constant speed (Fig. 3).

3 Results

By tracking individual molecules in the bands, we found that the electric field was indeed horizontal. The field is uniform in strength as well, since if it were not, the band would curve or smile as it moves. The injected band is slightly S shaped, but this does not interfere with separation. The slight S-shape near the array boundary results because the non-infinite resistance

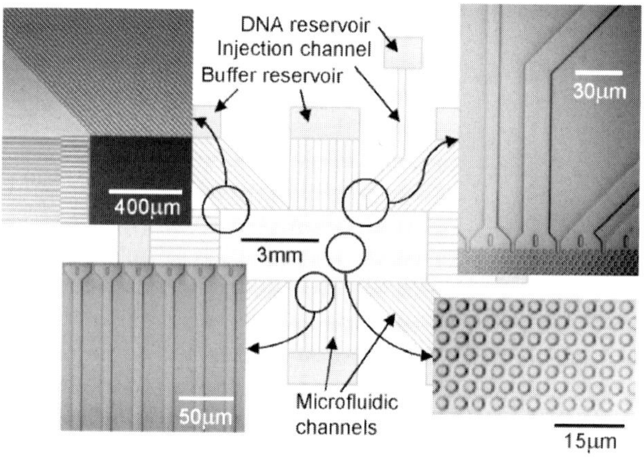

Fig. 2. Structure of the device illustrating the microfabricated sieving matrix integrated with the microfluidic channels. Insets are micrographs of 2 micron microposts with 2 micron spacing, microfluidic resistive channels connecting the post array to electrolyte buffers, and a single special channel connecting the post array to the DNA reservoir.

of the channels makes them imperfect current sources, a well-understood design compromise. In a second experiment designed to test the separation power of the array, restriction enzyme digests of BAC and PAC preparations were injected and electrophoresed with constant pulse times, in a manner similar to conventional pulsed field electrophoresis [1]. A mixture of 61 kbp and 158 kbp BAC inserts was cleanly resolved in 7 s (Figs. 4B,C). The resolution in this experiment, defined as the full width at the half maximum of a band 7, was ~77 kbp at 7 s, and ~36 kbp at 14 s (Fig. 4). Although the resolution is not yet as sharp as can be achieved with standard methods (Fig. 4A), it is clear that the resolution increases with separation distance. Moreover, the current separation time was more than 4 orders of magnitude faster (7 s *vs.* 16 hr). We then continuously loaded and sorted a mixture of four BAC inserts using asymmetric pulsing (the prism mode) (Fig. 5A). The pulsing conditions empirically determine the angles and the widths of the bands. Although the resolution of the four species was achieved in a 1 mm long matrix in 4 s under certain pulsing conditions (Fig. 5A), the two smallest species (61 kbp and 114 kbp) were not resolved under other pulsing conditions tested (Fig. 5B). For molecules larger than ~100 kbp, however, continuous separation does occur with high resolution (Figs. 5B,C).

The resolution achieved in this weight range is about 10 to 15 kbp, comparable to conventional methods. Note, however, that the separation

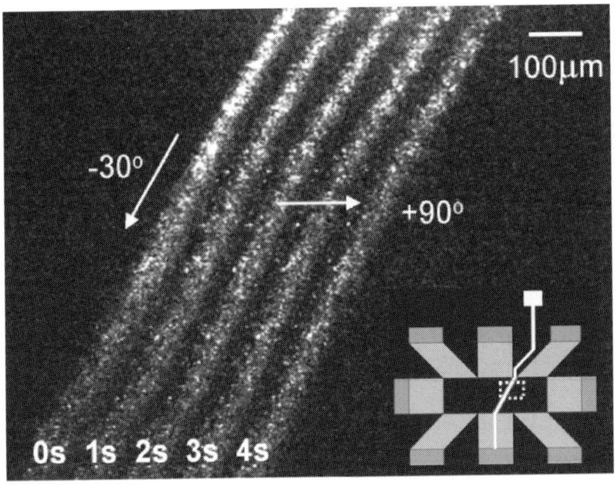

Fig. 3. Overlay of sequential fluorescent images of bacteriophage T2 DNA (164 kbp) mixed with l DNA (48.5 kbp) to show the spatial uniformity of the electric field. The band on the left (0 s) was imaged just after DNA injection using a 52 V/cm electric field -30° with respect to the vertical axis. The other four bands are from images taken at one second intervals later at a field strength of 38 V/cm in the horizontal ($+90^{\circ}$) direction. The inset is a low power schematic of the device to orient the reader. The area of the field in Figure 3 is boxed with a dotted line.

is achieved in ~10 s, using a 2.5 mm long array. These results can be understood with reference to Figure 2. The separation angle is simply the angular difference between the band formed by very small molecules, which follow the average field direction, and the band followed by larger molecules, which tend more toward the direction of the stronger electric field pulse. As random coils, molecules smaller than the spacing between posts follow the average field direction, because they are too small to interact with the posts. For the device used here, with 2 micron post spacing, we observed a cut-off ~100 kb (Fig. 5C). The 61 kb insert in Figure 5C is below this cutoff, while the inserts above the cutoff (~100 kb) interact with the posts and move in a manner depicted by Figure 2. Of course, a major reason for designing microfabricated devices of the kind discussed here is that we can control both the size of the posts in the array and their spacing, tuning both to suit the size range of the DNA molecules.

We note a second valuable feature of this device, its micron scale features. Apart from the advantages mentioned above small volumes, speed, a very shallow chamber depth and hence no heating at high field strengths, and

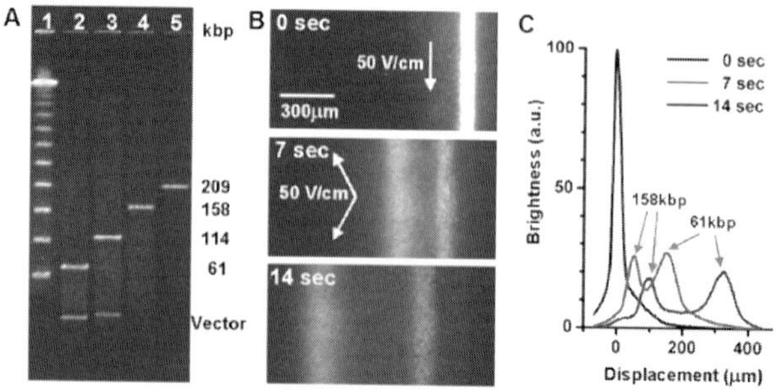

Fig. 4. Pulsed-field electrophoresis of BAC and PAC inserts in a microfabricated array. **A)** Characterization of the BAC and PAC inserts by PFGE. The first lane is a l-multimer marker, and the next lanes are NotI-digested artificial chromosomes of insert sizes 61, 114, 158, and 209 kbp, respectively. Running time, 16 hr at 6 V/cm. **B)** A time series showing the separation of a mixture of the 61 and 158 kb inserts. The initial DNA concentration was 10 micrograms/ml. The sample was injected at 50 V/cm along the vertical axis (0 s). The field was then pulsed symmetrically (60° with respect to the horizontal axis, 167 ms pulse duration, 50 V/cm field strength). **C)** Fluorescence profiles of bands at 0, 7, and 14 s.

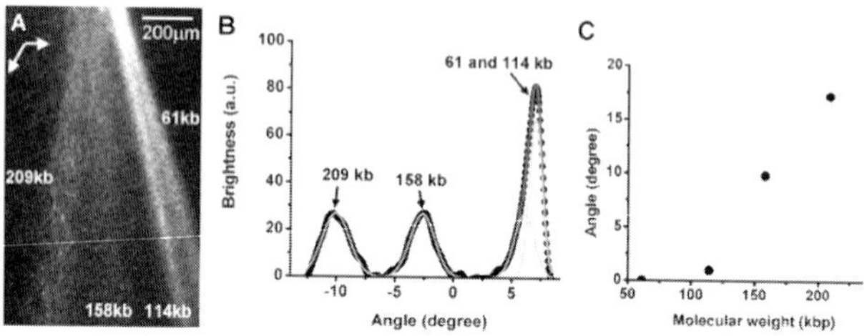

Fig. 5. Prism separation. **A)** Four bands (61, 114, 158, 209 kbp, 10 micrograms/ml total) were resolved using 41 ms pulse times and asymmetric voltages of 231 V/cm and 137 V/cm. **B)** Profile of bands resolved using 100 ms pulse times and asymmetric voltages of 196 V/cm and 108 V/cm, observed 2.5 mm below the injection point fitted with Gaussian distributions. The horizontal axis is in units of degrees with respect to the vertical. **C)** Experimentally observed dependence of the peak deviation angle on molecular mass.

the ability to tune post size and placement there are very low shear forces in micron scale devices. Consequently, very large DNA molecules do not break as they move through them. This is because DNA shearing is a function of turbulence [8] and there is no turbulence at the low Reynolds numbers at which these devices operate [2]. The lack of shearing is evident from the data of Figures 4B and 5B. The bands remain discrete and well separated, with some diffusional broadening, and the optical density trace in Figure 4C shows that the peaks remain symmetrical with time, and the areas under the peaks remain constant. Because there is unavoidable shearing during the preparation of BAC and PAC inserts from bacterial cultures, the baseline in Figure 4C is not zero, as one might expect; the baseline in the conventional separation shown for comparison in Figure 4A is also not zero. In fact an optical density scan of Figure 4A shows that the base-line is higher than Figure 4C. Thus the lack of shearing forces at low Reynolds numbers is an important general advantage.

Although band broadening (dispersion) in the prism mode is not yet fully understood, the various molecular sizes in Figure 5 are broader than can be accounted for by diffusion alone. It seems likely that the angle of sample entry and, more importantly, the degree to which all molecules are fully stretched and therefore back track as required for high resolution (Fig. 2), currently limits resolution. Since an important strength of the microfabrication approach used here is the ability to define and modify parameters such as these, it is likely that resolution in the prism mode can be greatly improved with further work. At the same time, we note that the resolution required for many BAC and PAC sizing experiments, where the size of the insert is the criterion for further use, can now be easily carried out in seconds in the non prism mode (Fig. 4). This represents not only a vast saving in reagents and time sensitive costs, but in space as well.

These results may be compared to other reports in the literature, also designed to separate or analyze high molecule weight DNA by unconventional methods. In previous work from this laboratory, we showed in principle that pulsed field separations in a microfabricated array of posts was possible [3,4] and that diffusion arrays could be used to separate DNA molecules in the many kb range [11]. Evidence for separation in both these cases was gathered by tracking individual molecules at high magnification. These devices are not yet practical tools, however, chiefly because the electric fields were shaped by a few discrete platinum electrodes which resulted in a highly spatially non-uniform electric field distribution. Further, the diffusion arrays are very slow. Other valuable recent reports either operate in a much lower molecular weight regime and on a time scale of many minutes or hours using entropic trapping [5], or use a variant of FACS sorting, where size is judged by fluorescence [6]. These latter approaches are essentially single molecule devices, and although FACS sorting is fast, the total number of molecules that can be harvested per unit time is low, and the macroscopic versions suffer from very high shear forces.

Finally, the success of these experiments has hinged largely on our discovery that microfluidic channels can be used to shape electric fields. In pulsed field gel electrophoresis the electric field must not only be switched on a programmable basis, but the array boundary conditions must be set so that the field lines are straight. This latter principle, first established by Davis and colleagues [1], led directly to the clamped homogeneous electric field device now widely used for the sizing and separation of DNA molecules greater than ~25 kb. The electric field is set by clamping the electric potential at the gel boundary with a series of electrodes surrounding the gel. This principle cannot be carried over to micro and nanofabricated devices because these devices are sealed with a cover slip or, as reported here, a silicone coated cover slip, and even at low field strengths the electrodes evolve hydrogen and oxygen, which very quickly obscures the observational field. This problem is solved here, where microfluidic channels serve to shape the field by acting as large electrical resistors, and thus also act as current sources to inject current uniformly across the array boundaries, rather than set the potential at the array boundaries. There is no gas evolution in the viewing area, and this design principle can now be used wherever current must be carried by buffer in a microfabricated device, for example, if molecules are to be trapped by dieletrophoresis [7].

4 Conclusions

In summary, a new method for the continuous sorting of DNA in a microfabricated device has been realized using the principles outlined in this lecture series. Its distinctive features are its tiny size and consequent fluid volume, control of field shape by a new microfluidics principle, high speed, the complete replacement of standard sieving matrices and electronic control of field shape with structures fabricated on a wafer, and the ability to operate the device at high field strengths without the need for external cooling. We expect that the separation range can be extended both to much larger molecules using the current version, and to much smaller molecules, perhaps as small as 1 kb, in future versions.

The lower bound is set by the sieving power of the post array, where the clear channel has to be smaller than the DNA in its randomly coiled state. Although the current clear channel width is ~1.5 micron, comparable to a ~100 kbp randomly coiled molecule, modern fabrication facilities routinely mass-produce feature sizes smaller than 200 nm by optical lithography. When a DNA molecule of contour length L is randomly coiled, its radius of gyration RG is $R_{\mathrm{G}}^2 = pL/3$, where p is the persistence length [8,9], $p \sim 50$ nm for double stranded DNA. The width of a ~1.5 micron channel is comparable to the size of a 100 kbp DNA molecule ($2R_{\mathrm{G}} \sim 1.5$ micron). Thus in principle, a 200 nm post array could effectively separate 1 kbp DNA molecules. Longer arrays allowing longer separation distances should improve resolution for all molecular weights, since band broadening is a random process. This is clear from Figures 4 and 5. Using the same pulsing conditions used in Figure 5, resolution comparable to conventional pulsed field gel electrophoresis (~10 kbp) should be achieved with a 2 cm-long array.

Theoretical analysis suggests that the resolution can be further improved with higher field strengths and shorter pulse times, currently limited by the power amplifier in our laboratory. Importantly, the methods used here to realize asymmetric pulsed-field fractionation, such as the generation of tunable uniform electric fields over larger arrays by current injection, will be practical tools for the realization of many lab-on-a-chip systems [10].

References

[1] G. Chu, D. Vollrath and R.W. Davis, *Science* **234** (1986) 1582-1585.

[2] J.P. Brody, P. Yager, R.E. Goldstein and R.H. Austin, *Biophys. J.* **71** (1996) 3430-3441.

[3] T.A. Duke, R.H. Austin, E.C. Cox and S.S. Chan, *Electrophoresis* **17** (1996) 1075-1079.

[4] O. Bakajin *et al.*, *Anal. Chem.* **73** (2001) 6053-6056.

[5] J. Han and H.G. Craighead, *Science* **288** (2000) 1026-1029.

[6] H.P. Chou, C. Spence, A. Scherer and S. Quake, *Proc. Natl. Acad. Sci. USA* **96** (1999) 11-13.

[7] C.L. Asbury and G. van den Engh, *Biophys. J.* **74** (1998) 1024-1030.

[8] M. Doi and S.F. Edwards, The Theory of Polymer Dynamics (Oxford University Press, Oxford, 1989).

[9] S.B. Smith, L. Finzi and C. Bustamante, *Science* **258** (1992) 1122-1126.

[10] M.A. Burns, B.N. Johnson and S.N. Brahamsandra, *Science* **282** (1998) 484-487.

Abstract

We close this series of lectures on fractionation with this lecture on how to do brownian ratchets RIGHT, and it involves very careful consideration to the boundary value issues I talked about in Lecture 1.

1 The problems with insulators in rachets

It has been proposed that an array of spatially asymmetric obstacles could operate as a continuous-flow Brownian ratchet [1–3]. As particles drift down through the obstacle array, they are selectively displaced perpendicular to their drift direction based on diffusion in the asymmetric structure. Experiments [4,5] showed that such asymmetric obstacle arrays can fractionate DNA molecules of high molecular weight ($\sim 10^5$ base-pair). However, because there are discrepancies between the theory [1,2] and the experimental results [4,5], it is not clear how the fractionation efficiency scales with molecular weight. While early theories proposed that smaller molecules should fractionate faster [1,2], a recent analysis [6] suggests that arrays of ion-impermeable obstacles could not serve as continuous-flow Brownian ratchets for point-like particles. Thus it remained questionable whether an ion-impervious obstacle array could in principle fractionate small macromolecules (oligonucleotides, proteins, etc.), which are of great biological importance.

The conventional understanding of continuous-flow sorting by Brownian motion using asymmetric obstacle arrays [1,2] is depicted in Figure 1a. Consider particles emerging from gap A driven towards gap B by electric fields. Executing biased random walks towards gap B, the particles spread out over the parabolic shaded region. While particles taking path 1 are blocked and deflected back to gap B, particles taking path 2 are deflected to gap B+. The probability of being deflected depends on how likely a particle diffuses past the corner of an obsticle (point C in Fig. 1a), and thus is a function of the ratio between the width of the parabolic shaded region and the characteristic obstacle dimension [1]. This ratio can be written mathematically as a dimensionless parameter D/va, where D is the particles diffusion coefficient, v is its drift velocity, and a is the gap width between the obstacles. Since the deflection probability depends on the diffusion, particles of different diffusion coefficients should migrate at different directions. Based on theoretical calculation, Duke *et al.* suggested that for the particular array geometry (Fig. 1a), the D/va of the molecules being separated should be between 0.02 and 0.3 for the best resolution, and the largest deflection probability should occur at $D/va \sim 0.7$ [1].

Two major assumptions are made in the above model [1,2]: (i) There is no deflection of the electric field lines by the obstacles. (ii) The particles are assumed to have no physical size (point-like particles).

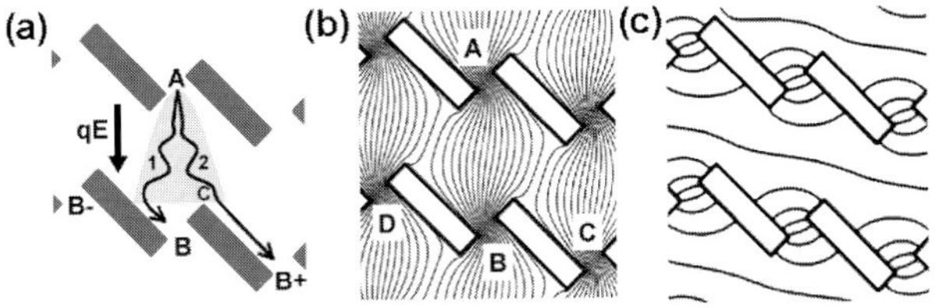

Fig. 1. a) Basic principle of continuous sorting in an asymmetric obstacle array. **b)** Electric field lines (and streamlines) in an ion-impermeable array. The boundary conditions used in this simulation is that all field lines flowing into gap A continue through gap B. **c)** Equipotential lines corresponding to the electric field under same condition.

In the actual implementation of these arrays by microfabrication techniques [4,5], however, the obstacles consist of fused silica or other materials impervious to the ions in the fluid. Because the ions flow around the obstacle and the electric field $\mathbf{E}$ is related to the ion flow $\mathbf{J}$ by Ohms law, $\mathbf{J} = \sigma\mathbf{E}$, where σ is the conductivity of the electrolytic fluid, the electric field lines go around the obstacles (Fig. 1b), violating the first of the above assumptions.

To isolate deflection due to diffusion, it is required that all field lines through an upper gap (A in Fig. 1b) map through a lower gap (B), which is aligned to the upper gap. If the field lines are misaligned so that some field lines through gap A leak to gap C or D, we will not be able to distinguish whether a particle migrating from gap A to C is by diffusion or by following the field. This requirement has to hold over the entire array area. This occurs only for a single choice of the angle of the equipotential lines. The proper equipotential contours (Fig. 1c) in our array were determined by numerically solving Poissons equation using the said field requirement as boundary conditions. Note that although the average current flow is in the vertical direction (from A to B as shown in Fig. 1b), and the equipotential lines are always perpendicular to the local electric field by definition, they are not perpendicular to the average current direction and are not horizontal.

2 An experimental test

Assuming the upper and lower array edges are held at equipotentials, they should be along such an equipotential direction to generate the desired aligned current distribution in the array (in direction of gap A to gap B in Fig. 1b). Therefore we designed our structure with properly slanted top and bottom edges (Fig. 2). The array was etched in fused silica by

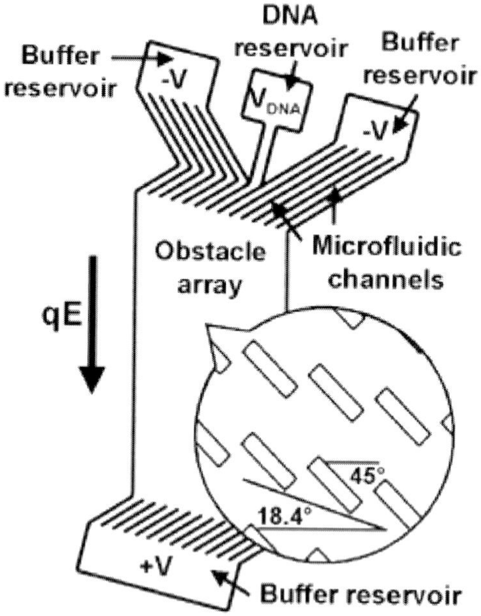

Fig. 2. Schematic diagram of the device. The obstacles are 1.4 microns wide, 5.6 microns long, and 5 microns tall. The etched fused silica substrate was capped with a glass cover slip to form enclosed microfluidic channels. The array is 12 mm high and 6 mm wide.

standard microfabrication methods and sealed with a glass cover slip to form the channels. For fabrication convenience, the structure was designed so that the direction of the calculated equipotential lines was parallel to the rows of obstacles, which we found by modeling not to be the case in general. Further, in practice, the boundary condition is not implemented as a voltage source along the boundary, but rather as an array of current sources which inject (and extract) current along the top (and bottom) boundary. The current sources were implemented as an array of microfluidic channels with a high electrical resistance (compared to that of the array) connected to common fluid reservoirs (Fig. 2) [7]. Holes through the substrate allowed access to the reservoirs. The similar high voltage drop across all channels (compared to the small voltage drop in the array) leads to the same current flowing in each channel at the boundary. The microfluidic channel arrays ensure that in case of imperfect dimension control during microfabrication, the current pattern will still be highly aligned to the obstacles. A single extra channel connected to a special reservoir (with voltage calculated to give the same current as the other channels) was used to inject a 90-micron wide band of the molecules to be separated.

A mixture of Coliphage λ DNA (48.5 kbp, $\sim$5 μg/ml) and Coliphage T2 DNA (167 kbp, $\sim$2 μg/ml) in Tris-borate-EDTA buffer was injected into the array at various speeds using electric fields. At high fields (>5 V/cm), diffusion was negligible ($D/va < 0.05$). In calculation of the diffusion constants, we assume that Coliphage λ DNA and T2 DNA molecules adopt random coil conformations. According to the approach used in Reference 1 for calculating the diffusion coefficients, $D_\lambda = 0.64$ micron^2/s, and $D_{\mathrm{T2}} = 0.35$ micron^2/s. The D/va values of experimental data are calculated using the measured velocities and a = 1.4 microns. As expected no lateral separation occurred (Fig. 3a) for these diffusion constants. The fact that the band did not curve even at the boundary of the array shows that the equipotential boundary conditions were properly imposed, and the current direction was well-aligned to the obstacle array. Lateral separation of the two species was observed at a field strength lower than $\sim$2 V/cm ($D/va > 0.13$ for Coliphage λ DNA), with λ molecules being deflected from vertical more than those of T2. The separation became larger ($\sim$1.3° T) as the electric field was lowered to 0.8 V/cm (Fig. 3b, $D/va \sim 0.32$). The drift velocity at fields less than 0.8 V/cm was so low (<1 micron/s) that the stability of the separated bands became hard to maintain. The two species could be separated into two cleanly resolved bands 11 mm from the injection point, and the density profile of these bands was well fitted by two Gaussian peaks (Fig. 3c). The resolution in the range of $\sim$50 kbp is $\sim$30 kbp, or $\sim$60% [8]. Although we did observe separation of molecules in the D/va range proposed by Duke *et al.* [1], the measured separation was much smaller than the theoretical predictions (Fig. 3d).

To examine the scaling of the deflection to very small molecular sizes, a mixture of 411 bp (PCR product, $\sim$1 μg/ml) and λ DNA ($\sim$20 ng/ml) was injected into the array, using electric fields ranging from 6 V/cm to 120 V/cm. At these field conditions, λ DNA molecules do not deviate from the field direction (Figs. 3a and 3d), and thus are used to label the field direction. The measured D/va for 411 bp molecules using 6 V/cm is $\sim$2.7. Because 411 bp DNA molecules are not random coils, the parameter D/va was obtained by experimentally measuring the band broadening instead of using the method in [1]. Assuming that band broadening is not influenced by the obstacles, we used the following equations: $t = y/v$ and $\Delta x^2 = 2Dt + \Delta x_{\mathrm{o}}^2$, where y is the length of the band, Δx_{o}^2 is the initial width of the band, and Δx is the width of the band after time t. Therefore, $D/va = (\Delta x^2 - \Delta x_{\mathrm{o}}^2)/2ya$. The measured D/va value at low field conditions (6 V/cm) is $\sim$2.7, which implies that the mean diffusion distance during the time when a molecule moves from one row to the next is larger than

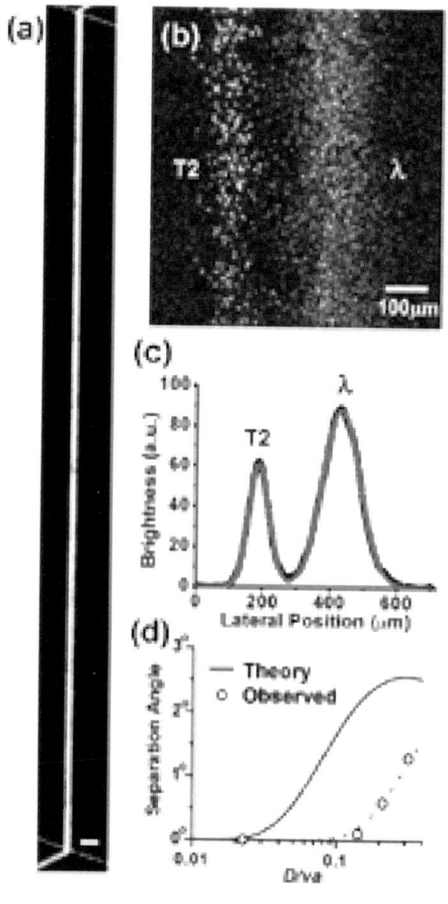

Fig. 3. a) Fluorescence micrograph of Coliphage λ and T2 DNA stained with fluorescent dye TOTO-1 forming a band of ~90 microns wide and 12 mm long at 12 V/cm. The slanted lines on the top and bottom mark the boundary of the obstacle array. Scale bar = 300 microns. **b)** Fluorescence micrograph of the two species separated into two band at ~11 mm from injection using 0.8 V/cm. **c)** Fluorescence profile of **b)**. Experiment data (thick black line) fitted with two Gaussian peaks. **d)** Separation angle between λ and T2 bands as a function of the dimensionless parameter D/va of λ molecules. The solid curve is the theoretical prediction from reference 1. Circles mark experiment data using electric fields of 12 V/cm, 1.8 V/cm, 1.2 V/cm, and 0.8 V/cm. The dash line is the theoretical curve calculated according to reference 1 using an effective gap size 4.1 times larger than the physical gap size (5.7 microns instead of 1.4 microns). Note the log scale of the horizontal axis.

the size of the obstacles. Therefore, the assumption holds. The D/va values for high field conditions were derived assuming constant mobility. so the corresponding D/va range tested is from 0.14 to 2.7. This covers the entire range that the theory suggests the maximum deflection ($D/va \sim 0.7$) [1]. Therefore deflection should be observed for these smaller molecules. However, contrary to all expectations based on the theory [1, 2], absolutely no lateral deflection was observed.

We believe the reason the array failed to deflect small molecules (411 bp DNA) lies in the fact that small particles can precisely follow the electric field lines as they flow through the obstacle geometries (Fig. 4a). Contrary to the basic principles of diffusion array [1, 2], where particles could widen out over the parabolic shaded region in Figure 1a only via diffusion, small molecules will be spread out by the electric field. Particles are now drifted towards the vicinity of boundary L (*via* field line a in Fig. 4a) as well as to boundary R (*via* field line c). Thus small molecules injected from a gap will have a much higher chance to diffuse to the left than what the old theory suggested. In fact, a recent analysis showed that point-like particles are equally likely to diffuse in both directions [6]. We summarize the argument as follows. For small particles that precisely follow electric field lines, their flux density $\mathbf{J}_{\text{particle}}$ can be written as $\mathbf{J}_{\text{particle}} = \rho\mu\mathbf{E} - D\nabla\rho$, where ρ is the particle density, μ is the mobility, and D is the diffusion coefficient. The first term of the flux density is due to the electric field, whereas the second term is from diffusion. According to the continuity equation, we have:

$$\frac{\partial\rho}{\partial t} = \nabla \bullet \mathbf{J}_{\text{particle}} = \mu\mathbf{E} \bullet \nabla\rho - D\nabla^2\rho. \tag{2.1}$$

Note we have used $\nabla \bullet \mathbf{E} = 0$ because the electrolytic solution is neutral. If there is a high field so that the second term in equation (1) becomes relatively small, we find:

$$\mathbf{E} \bullet \nabla\rho = 0 \tag{2.2}$$

at steady-state. This says the particle density is approximately constant along any field line. Thus if one has a uniform concentration of particles arriving across all field lines entering a given gap (originating from a reservoir of uniform concentration), as the field lines (particle streamlines) widen out after the gap, the particle density will remain unchanged. This is illustrated in the fluorescence image of 411 bp DNA molecules in the array (Fig. 4a), which shows that DNA under high fields uniformly fills the entire space between rows of obstacles. Now, consider uniform injection of particles into all gaps at the top of the array using high fields, leading to uniform particle distribution in the array, and then we lower the field strength so that diffusion becomes important. Since all spatial derivatives of ρ in

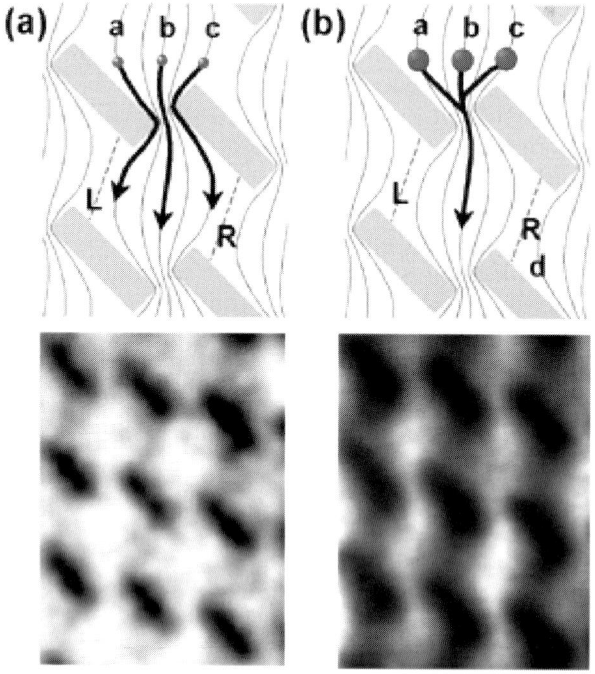

Fig. 4. Schematic flow diagram and fluorescence image of particles in center of band for particles of size **a)** 411 bp at average field of 120 V/cm and **b)** 48.5 kbp (1.2 V/cm). The exposure times were 2 s so that the image brightness shows the particle density.

equation (1) are zero in our case of uniform density distribution, the particle density stays uniform according to equation (1), and thus the diffusion flux of particles across any field line must be equal to the inverse flux. Combined with translational symmetry, this implies that the probability of a particle diffusing across boundary L in Figure 4a equals that across boundary R, a result which must hold for any distribution, not just for the assumed uniform distribution of particles. Given that there is no preferred direction of diffusion, there is no physical basis for ratcheting.

When a much larger λ DNA molecule approaches a gap, it is physically deflected by the obstacle and centered to the gap, because of its finite size (a random coil of $\sim$1 micron) compared to the gap width ($\sim$1.4 microns). Thus molecules initially following field lines a, b, and c in Figure 4b will all tend to leave the gap region on line b. The fluorescence image in Figure 4b clearly shows this shadowing in contrast to Figure 4a for the case of small molecules. Unlike 411 bp molecules, which are spread out in the space between rows of obstacles by the field, λ DNA molecules can only reach boundaries L and R in Figure 4b by diffusion. Because boundary R is farther than boundary L from the gap where molecules emerge, molecules are more likely to diffuse across boundary R. Once a molecule reaches field line d, it will be driven to the right. Therefore the obstacle array acts as a Brownian ratchet. The reason that the observed deflection is smaller than what the theory predicted (Fig. 3d) is because the electric field funnels back most molecules that the theory counts as being deflected, such as particle C in Figure 4b. In fact, our experimental data (Fig. 3d) suggest that the old theory can take into account the funneling effect of the field by using an effective gap size that is 4.1 times larger than the physical gap size for the calculation of D/va (Fig. 3d).

3 Conclusions

In summary, this lecture clearly shows (at least I hope I clearly showed) that in a continuous-flow Brownian ratchet array, there exists a critical particle size threshold, which is related to the size of the narrowest feature through which the particles must pass in the array. Particles below this threshold maintain their flow along electric field lines through the gaps and are thus incapable of being ratcheted. Particles above this threshold size will be deflected from their original field lines by the obstacles, and can thus be ratcheted. Once above this size, larger particles are ratcheted less because of their lower diffusion coefficient. This points to the importance of very narrow gaps in the obstacle array if the separation of small particles is desired.

References

[1] T.A. Duke and R.H. Austin, *Phys. Rev. Lett.* **80** (1998) 1552.

[2] D. Ertas, *Phys. Rev. Lett.* **80** (1998) 1548.

[3] A. Oudenaarden and S.G. Boxer, *Science* **285** (1999) 1046.

[4] C.F. Chou, O. Bakajin, S. Turner, T. Duke, S.S. Chan, E.C. Cox, H.G. Craighead and R.H. Austin, *Proc. Natl. Acad. Sci. USA* **96**(24) (1999) 13762.

[5] M. Cabodi *et al.*, *Electrophoresis*, in press.

[6] R.H. Austin, N. Darnton, R.L. Huang, O. Bakajin, T. Duke and J. Sturm, *App. Phys. A*, in press.

[7] L.R. Huang *et al.*, *IEDM Technical Digest* (2001) 363-366.

[8] J.C. Giddings, Unified Separation Science (Wiley, New York, 1991) 101-106.

Abstract

I discussed in the introduction to this lecture series that there are two aspects to modern genomics: deciphering the actual sequence of basepairs of a particular organism and then deciphering out of all the possible genes that are present the actual genes that are expressed and the interactions between them... a daunting task. In this lecture I will discuss a possible technique, taken from nanotechnology, that claims to be able to determine the actual occupation of various control elements in the genome.

1 Introduction

The goal of this project is to develop a robust experimental approach to reveal the logic of the transcription code of the DNA of a living cell, because this is the process which controls the function and life of biological organisms. The most immediate challenge for the scientific community, to identify the individual genes in each genome, is well underway, with both experimental and bioinformatic approaches reasonably in hand for the purpose. What is not in hand are intellectual approaches and technologies to understand how the genetic program is interpreted, which is at the heart of the question of how do biological organisms really work?

In broad terms, the genetic program encoded in the DNA is interpreted by the "expression of genes" – *i.e.* the genes encoded in the DNA each lead to the creation of a specific protein, and the proteins in turn regulate the activity of the cell. But an understanding in precise terms the code or logic by which the cell selects which genes to express, and exactly how this happens in a temporally and spatially appropriate manner is not known. That fundamental challenge is the subject of this proposal.

It is well established that expression of genes requires the interaction of proteins with DNA in the vicinity of the gene, the most significant of which is known as a "transcription factor". While all cells in an organism have the same DNA, it is different transcription factors bound in different locations on the DNA, which cause one cell to evolve and function different that the next. Furthermore for a relatively small number of genes that have been studied in detail, it is clear that the efficient expression of a gene requires the simultaneous binding of multiple proteins. However how those proteins are chosen, and in what combinations they may productively interact with one another, is unknown.

There are four main directions to our approach for understanding the expression of genes.

I. Preparation of DNA with different tagged transcription factors.

II. Identification of the location of transcription factors on DNA.

III. Regulation and Functional Analysis.

IV. Proteomic Analysis.
Our general approach will be to combine the expertise of three groups in nanofabrication, molecular biology, and informatics to analyze this problem in the following way.

First, we will continue to develop and adapt a near field device invented at Princeton which is capable of detecting single fluorescent molecules with a spatial resolution of 50 nm.

Second, we will construct strains of E. coli carrying regulatory proteins fused to the Green Fluorescent protein (GFP) that will bind with high affinity to a specific site on the genome. A red-shifted variant of GFP will be designed to bind to a second well-defined site, and the near field detector will be used to measure the distance between them as the DNA flows through the detection channels. These are test of principle strains and apparatus.

Third, a large number of strains for high throughput analysis of the combinatorial code will be constructed by a random insertion technique that will fuse GFP to essentially all of the E. coli transcription regulatory proteins one at a time. These strains will then be analyzed in a massively parallel fashion in the nanofabricated device. The localization of each protein for a variety of cell states will then give us a large and important data base with which to understand the combinatorial code.

Fourth, we will use our strengths in biological computing to begin an analysis of the data base, using clustering algorithms, hierarchical ordering schemes, and vector support machines.

I will examine in this lecture just the near field scanner which is central to the entire scheme, because it is the one subject we have been attacking most vigorously at present.

2 The nearfield scanner

There is a great need in in our approach to obtain the highest possible spatial resolution images of biological structures in the shortest amount of time. The information obtained from such images ranges from mapping of the genome on chromosomal length DNA molecules [1, 2] to obtaining images of surface components on cells. In the case of the work of Schwartz *et al.* [1] and Bensimon *et al.* [2] molecular combing techniques are used to map DNA molecules by observing the fragment lengths of restriction cut DNA molecules. Although there are robust computer algorithms used in reconstructing the DNA map, ultimately the information is limited by the spatial resolution R of the microscope objective used, given by $R = 0.6\lambda/N.A.$, where λ is the wavelength of the light used and $N.A.$ is the numerical aperture of the objective. Although R can approach 0.3 microns for a very well designed high $N.A.$ objective, as R decreases the field of view of the objective also decreases and scanning times of molecules become prohibitively long and data rates prohibitively low. Similarly, although high resolution point scanning techniques such as the near-field optical microscope (NSOM) [3] or the atomic force microscope (AFM) [4] can obtain very high spatial resolution images, it is at the cost of relatively slow scanning speeds and the need to find the object to be scanned in a very small field of view compared to the area where the object may be located.

3 The chip

In this lecture we discuss a technology which provides high optical resolution as well as very high data rates for *moving molecules*. The basic idea is simple: we bring molecules past a near-field slit, the spatial resolution is not determined by the wavelength of light but by the width of the slit. Long slits were nanofabricated in an evaporated aluminum film using electron beam lithography and reactive ion etching. Laser light from a collimated Ar:Kr ion laser was focused onto the back side of the film, creating a non-radiating near field on the other side of the film. When macromolecules pass over the slit in the vicinity of the near-field, fluorophores in the object are excited and can radiate in the far field. A microscope objective captures this far-field radiation which is recorded in a time-dependent manner by a photomultiplier. This time-dependent signal then gives a record of the passage of the object over the slit with a spatial resolution roughly equal to the width of the slit, independent of the resolution R of the objective used. Figure 1a shows a scanning electron microscope image of a completed device. The aluminum floor was 200 nm thick and protected by 100 nm of SiO_2, which ultimately gives the limiting resolution for this device. Future devices will have thinner insulating floors. The channel which is oriented perpendicular to the slits is 5000 nm wide and etched to a depth of 1000 nm, much too deep to truly confine the DNA molecules to under their persistence length of 50 nm [5] but easy to fabricate in this first device. An array of posts was used at the entrance to the slits in anticipation of the need to stretch DNA molecules before they entered the slit region. Three slits were used so that the velocity of the molecules as they passed over the slits could be measured by arrival times of molecules at slits of a known spacing. The multiple slits also act as a good rejection of uncorrelated events since we could demand that the signal from a molecule show that 3 slits were traversed with equal time intervals in order to be counted as a valid event.

An important aspect of this device is the dependence of both the near field radiation pattern and the far-field transmission of a thin slit on the polarization of the incident light, which has been discussed theoretically in the

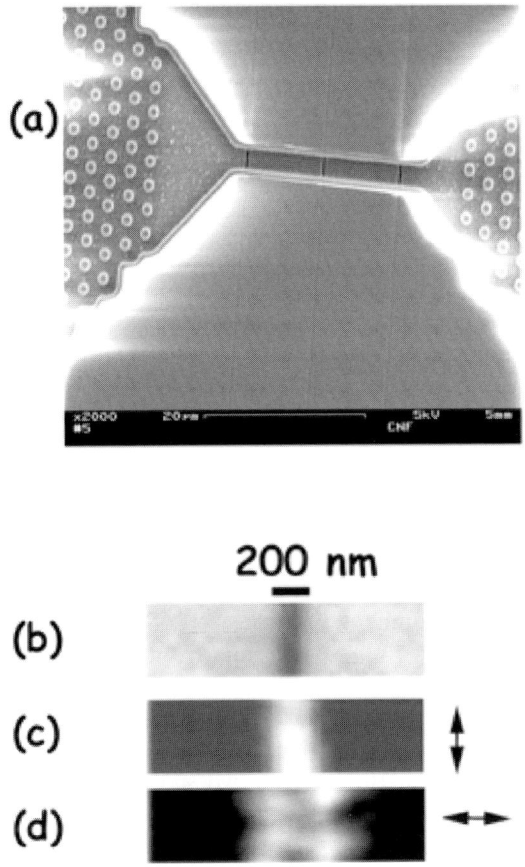

Fig. 1. a) Scanning electron microscope image of three 100 nm wide slits and hydrodynamic channels nanofabricated into a etched quartz structure. **b)** Shear force feedback signal image from the NSOM of a 100 nm slit. **c)** NSOM imaging of the evanescent light from a 100 nm slit. Electric field polarized parallel (p-polarization) to the slit. A 532 nm diode pumped solid state laser illuminated the back (glass side) of the chip. An Al coated tapered fiber probe of approximately 50 nm diameter positioned $\sim$50 Å from the chip was used to collect the evanescent wave from the slit. The polarization of the incident light was rotated by a $\lambda/2$ plate. The light collected by the tapered fiber probe was sent into a photon counting TE-cooled PMT. **d)** NSOM signal for light polarized perpendicular to the slit (s-polarization).

literature but never before measured. There is a substantial literature on near field optics, starting with Bethe's classic calculation of the transmission of light through a small hole [6]. Unfortunately, most work has concerned itself with optimizing the transmission of the far-field light through a structure, in our case we are concerned with *minimizing* the amount of far-field radiation, since our use of fluorescent probes means that any transmitted light will result in excitation and loss of resolution. Thus, although Betzig *et al.* [7] have done an exhaustive calculation of the electric fields for a slit with **E** polarized perpendicular to a slit in fact this is exactly the wrong polarization for our work since this polarization in fact maximizes far field radiation. In the case of a slit, it is convenient to use Babinet's principle [8] and view the slit problem as the re-radiation by a metallic line of width w. Intuitively, we expect that when **B** is parallel to the slit (TE excitation) that the transmission will be largest and we would expect that when the **B** of the light is perpendicular to the slit (TM excitation) that the transmission is minimized. We would also expect that the near field pattern would be confined to the region between the slit for polarization parallel to the slit and would radiate as two lobes from the edges of the slit for polarization perpendicular to the slit, resulting in poor spatial resolution.

Calculation of the amount of the intensity of the near field radiation and the far field (radiating) intensity is important. Unfortunately, a finite thickness slit geometry is difficult to do analytically [9]. Those calculations that have been done [7] assumed that the thickness of the metal film T was much greater than the wavelength of the radiation λ and so only considered TE wave propagation. However, as we show below this results in a high background of far field transmission since a TE wave in a rectangular waveguide with infinitely long width has no cutoff frequency [8]. Our situation is apparently new to the literature: the film thickness T (200 nm) is less than the wavelength of the radiation (488 nm) and hence the film is not thick, and we are concerned with TM waves. No closed form analytical expressions seem to exist for the far-field transmission [10] of a slit or the near field pattern. Measurements of the near-field radiation patterns for a 100 nm slit using a NSOM constructed by Fann *et al.* [11], are shown in Figures 1b–d. In agreement with the near field calculations of [7], the NSOM image confirms that for electric field polarization perpendicular to the slit the near-field pattern has two lobes radiating from the edges of the slit, while for polarization parallel to the slit the near-field pattern is confined to the opening between the slits, indicating that the techniques developed in [7] can in principle be used successfully to calculate the near-field patterns for our particular geometry.

However, although it is crude we can recognize two basic regimes at least for the far-field radiation whose background we must minimize. When the slit width w is much greater than the film thickness T the far field TE radiation is from an electric dipole whose magnitude P is proportional to the slit width w. We would then expect that the far-field transmission in this range should vary as P^2 and hence as w^2, a power law. However, as the slit width w becomes less than the film thickness T from analogy to the attenuation of radiation in a waveguide we would expect that the far field will now be further decreased by the transmission coefficient of a waveguide beyond cut-off, where imaginary solutions to the wave equation lead us to expect the far-field transmission will fall off exponentially with (λ/w).

$$I(w) \sim I_o \frac{1}{1 + (T/w)^2} \times \exp\left(-\frac{\Gamma\lambda}{w}\right) \tag{3.1}$$

where T is the film thickness, w is the width of the slit, λ is the wavelength and Γ is the polarization dependent transmission coefficient of the slit. Thus, as a function of $1/w$ we expect first a slow power law decrease in transmission with exponent 2 followed by an exponential decay.

To test these assumptions we carried out far-field transmission measurements using the 488 nm line of a Kr:Ar laser, $\lambda/2$ plate polarization rotation and a cooled CCD camera (C4880, Hamamatsu Corp., Bridgewater, NJ) operated in integrating mode. The data shown in Figure 2a however indicate that there is in fact a cutoff for both TE and TM waves as the slitwidth is decreased. The lines in Figure 2b show fits to equation (3.1), with a value for Γ of 0.8 for TE waves and 1.3 for TM waves, showing that although the TE wave does have higher transmission it ultimately falls exponentially as the slit width gets much less than T. At slit widths of 100 and 50 nm the primary excitation is near-field, in agreement with our near-field NSOM measurements which are shown in Figure 2b. In this data the intensity of TE polarized light was measured as a function of the distance z of the NSOM tip from a 100 nm slit. The exponential decay fit has a value of 100 nm $\pm$ 30 nm, the constant offset intensity we ascribe to farfield radiation for TE polarization. The TM polarization used for the remainder of this paper will have far-field intensity 1/30 of the 10% farfield observed for TE.

4 Experiments with molecules

Since the actual near field (and far field) TM radiation patterns for a finite thickness slit is ultimately not well understood, the spatial resolution of the device was determined by running fluorescent polystyrene balls of a known diameter across slits of decreasing widths from 1000 nm to 50 nm and measuring the time-resolved emission of the balls as they passed across the slit. Figure 3 shows the response of the photomultiplier for 100 nm balls passing 1000 nm, 300 nm and 50 nm wide slits under back illumination with

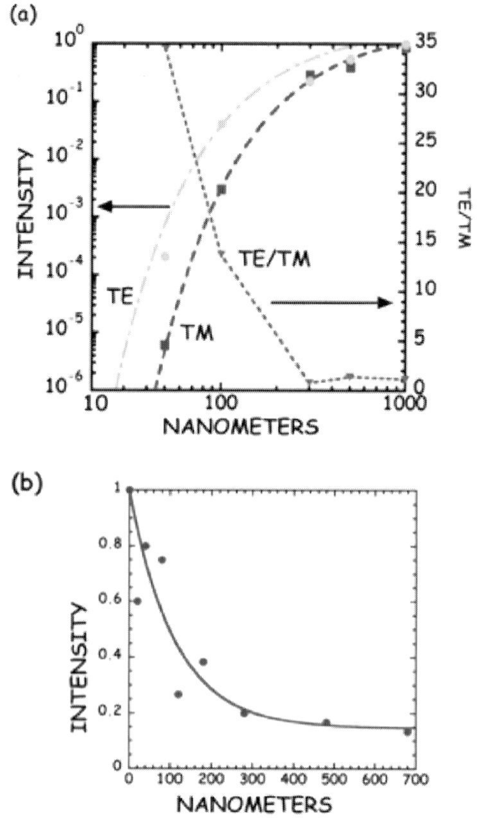

Fig. 2. a) Far-field transmission of the 100 nm wide slit taken using a 5 nm band-pass interference filter, $\lambda_o = 650$ nm. The left side log scale is the transmission for TE (E field perpendicular to the slit, circles) and TM (E field parallel, squares) to the slit. Intensity is normalized to the perpendicular transmission for a 1000 nm wide slit. The right side axis is the ratio of the TE/TM (triangles) for far field transmission. Solid lines are curve fits to the data using equation (3.1). **b)** Normalized NSOM intensity from a 100 nm wide slit as a function of distance z from the slit surface for p polarization. The solid curve fit is to an exponential decay plus constant offset.

488 nm wavelength laser excitation, with a $N.A. = 0.9$ objective as the light gathering optic. Our present slits have 100 nm of SiO_2 over the slits to protect the aluminum, the result is that at present our resolution is limited to 200 nm spatial resolution. The fact that the pulse width of the light from a bead passing over a slit decreased with the width of the slit and exceeded

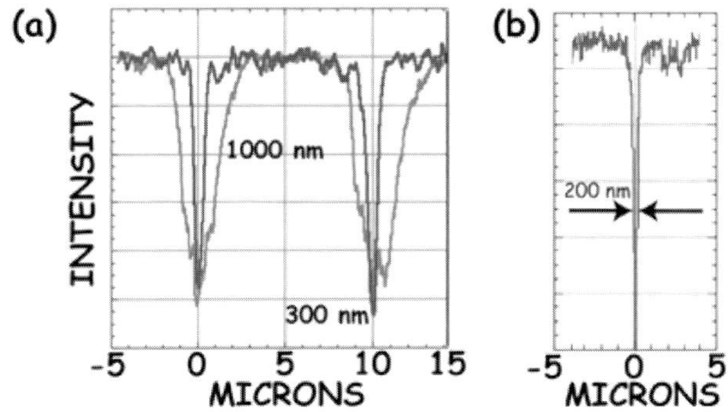

Fig. 3. a) Fluorescence *vs.* distance as determined by time of flight for a 100 nm diameter fluorescent ball passing a 1000 nm (red line) slit and 300 nm slit (green line). Typically, 50 mW of excitation light was used on the back side, focused to a 10 micron spot size using a 50 mm focal length lens (Gradium lens, Newport Optics). **b)** Fluorescence *vs.* distance as determined by time of flight for a 100 nm ball passing by a 50 nm wide slit. The *FWHM* of the signal is 200 nm.

the resolution of the objective used for this measurements ($R = 350$ nm) indicates that the beads were excited by the near-field pattern. Note that resolution observed here is *independent* of the *N.A.* of the light gathering microscope objective.

Linear polymers such as double stranded DNA (dsDNA) consist of long sequences of basepairs whose sequential ordering comprise the genomic content of the organism from which the DNA came from. We examine here the passage of T4 phage dsDNA molecules, T4 has a length of 169 kilobases, or a stretched length when loaded with our intercalating dye TOTO-1 (Molecular Probes, Eugene, OR) of 7.4×10^4 nm [12]. In the absence of a shearing force field DNA undergoing electrophoretic transport T4 would be expected to form a Gaussian coil of radius $R_{\mathrm{g}} = \sqrt{\frac{Lp}{6}}$, where p is the persistence length of DNA, about 50 nm [5]. This yields an expected $R_{\mathrm{g}} = 700$ nm for T4 DNA.

The forces acting on charged polymers in microfabricated environments have been discussed in reference [12]. In our case the channels are 5000 nm wide and 1000 nm deep, both dimensions are large compared to the persistence length of dsDNA (approximately 50 nm). The forces acting on the polymer create a tension $T(z)$ in the polymer where z is the distance from the tethered end. The forces acting on a polymer which is momentarily hooked on an obstacle create a tension in the chain which declines from a maximum value at the hooked end to zero at the free end; as a result, the

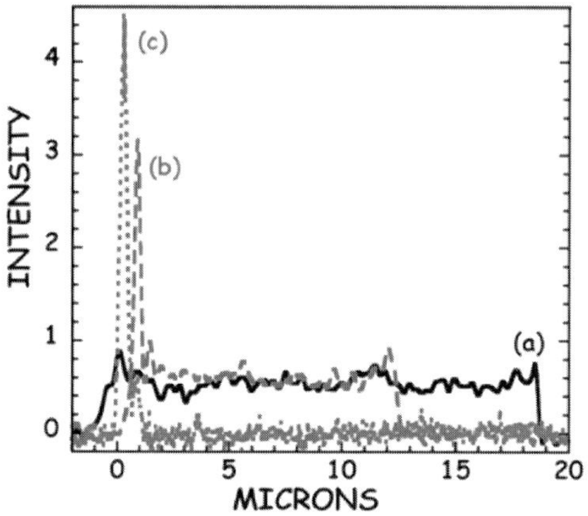

Fig. 4. Fluorescence signal from T4 dsDNA molecules passing across a slit. T4 molecules were pre-incubated with the fluorescent intercalating dye TOTO (Molecular Probes, Eugene, OR) before entering the array using electrophoresis. 3 different molecules are shown under different states of elongation: **a)** a relatively extended molecule, **b)** molecule with entropically coiled leading head, **c)** compact molecule.

polymer appears as a highly stretched "stem", surmounted by a "flower" of increasingly disordered random coils. The terminal blob of the DNA free end should have a diameter $D \sim (\frac{kT}{\eta v})^{1/2}$ where η is the viscosity of the medium and v is the velocity of the polymer in the medium [12].

Typically in these experiments the molecules had electrophoretic velocities of ~100 μm/s, with a predicted disordered head size of 200 nm.

Figure 4 shows selected time traces of T4 molecules passing by a 100 nm slit. Data selection was needed here because with the present channel width the DNA molecules can come off the posts in a variety of entropic configurations, and the deep etch of the channel gives rise to a background of low intensity poorly resolved molecules. The molecules chosen gave an example of the range of signals of a molecule which has been extended by the applied force field and then released. The brief light pulse at the heads of the molecules is the disordered head of the extended molecule, the diameter of the head is 200 nm, within the resolution limits of our apparatus at present, and also approximately the expected size from the parameters of the stretching force parameters we are using. In addition to the disordered head of the molecule there is also visible small "lumps" of disordered sections which are generated when the tension in the molecule is suddenly released. The faster the DNA molecules are run through this structure the more extended the molecules are and the less time there is for the taut end to form a disordered region. Thus, for DNA mapping applications using the posts one can use electrophoretic high velocities and simply work backwards from the taut end and still maintain high spatial resolution.

5 Conclusions

The real use of this technology will be realized when the channels which the molecule traverses are made on the same scale as or smaller than the persistence length of the polymer, so that the molecule must enter the channel not as a coil but as an extended strand. Under those conditions truly high spatial resolution of the length of the molecule will occur allowing this technique to do high resolution dynamic mapping of single dsDNA molecules. Given the high degree of discrimination between polarized light oriented parallel and perpendicular to the slit (Fig. 2), it also seems likely that this device can be used to measure fluorescence polarization [13] of single molecules as they pass successive slits. This can be achieved by placing the polarized light source on the solution side of the device, and the detector below the slit. In this configuration the relaxation time of a polarized signal [14] could also be followed as a function of slit number.

References

[1] J. Lin, R. Qi, C. Aston, J. Jing, T.S. Anantharaman, B. Mishra, O. White, J.C. Venter and D.C. Schwartz, *Science* **285** (1999) 1558-1562.

[2] X. Michalet, R. Ekong, F. Fougerousse, S. Rousseaux, C. Schurra, N. Hornigold, M. van Slegtenhorst, J. Wolfe, S. Povey, J.S. Beckmann and A. Bensimon, *Science* **277** (1997) 1518-1523.

[3] Optics at the Nanometer Scale, E. Nieto-Vesperinas and N. Garcia (eds.) (Kluwer Acad. Publ., 1996).

[4] M. Miles, *Science* **277** (1997) 1845-1847.

[5] M. Hogan, R.H. Austin and J. LeGrange, *Nature* **304** (1983) 752-754.

[6] H.A. Bethe, *Phys. Rev.* **66** (1944) 163-182.

[7] E. Betzig, A. Harootunian, A. Lewis and M. Isaacson, *Appl. Opt.* **25** (1986) 1890-1900.

[8] Classical Electrodynamics, J.D. Jackson, 3rd edition (John Wiley & Sons, N.Y., 1998).

[9] T.W. Ebbessen, H.J. Lezec, H.F. Ghaemi, T. Thio and P.A. Wolff, *Nature* **391** (1998) 667-669.

[10] M. Born and E. Wolf, Principles of Optics, 6th edition (Pergamon Press, Oxford, 1984).

[11] P.K. Wei, R.L. Chang, J.H. Hsu, S.H. Lin, W.S. Fann and B.R. Hsieh, *Opt. Lett.* **21** (1996) 1876.

[12] O.B. Bakajin, T.A.J. Duke, C.F. Chou, S.S. Chan, R.H. Austin and E.C. Cox, *Phys. Rev. Lett.* **80** (1998) 2737-2740.

[13] F. Perrin, *J. Phys. Radium* **7** (1926) 390-401

[14] X. Chen, L. Levine and P.-Y. Kwok, *Genome Res.* **9** (1999) 492-498.

Abstract

I discuss in this lecture a magnetic separation idea which utilizes several ideas from microfabrication and nanomagnetics. The basic idea comes from our earlier work using asymmetry in obstacles and brownian motion to effect separation of objects [10] by moving them in streams whose angle to the hydrodynamic average velocity is a function of the diffusion coefficient of the object.

1 Introduction

Cellular and molecular biologists have developed a variety of ways to sort living cells according to certain characteristics. It is frequently desirable to sort cells according to their chemical content, enzyme activity, surface antigens, or size. Such steps allow physicians to purify a patient's blood sample for further analysis and allow scientists to isolate rare cell types to study biological processes. The possibility of isolating rare cells, such as hematopoietic stem cells and metastatic cancer cells, carries important medical applications as well. Density centrifugation, fluorescence-activated cell separation (FACS), and magnetic-activated cell separation (MACS) have proven to be very effective cell sorting methods to this point. However, the great importance of such methods requires constant innovation and refinement of the technique.

The device I propose here is not technically a brownian ratchet device but uses the idea of force which acts at angle to the hydrodynamic flow. In our case, the force is generated by a magnetic field gradient which comes from an array of magnetized wires which lie at an angle θ to a hydrodynamic field flow. The sum of the hydrodynamic force and the magnetic force create a new vector which as in the case of the brownian ratchet moves the cell out of the main stream direction. Figure 1 shows how the two ideas are correlated.

2 Blood specifics

Blood carries nourishment and oxygen to, and waste products away from, all parts of the body through the arteries, veins, and capillaries. Blood also mediates the immune system, recognizing foreign macromolecules and mounting an attack against them. Humans contain approximately five liters of blood, which accounts for 7% of our body weight. There are three main types of blood cells: erythrocytes (red blood cells), leukocytes (white blood cells), and platelets. However, all three arise from precursor cells in the bone marrow, called hematopoietic stem cells.

Erythrocytes are the most common type of blood cell, existing at concentrations around 5×10^{12} cells per liter [2]. They constitute approximately

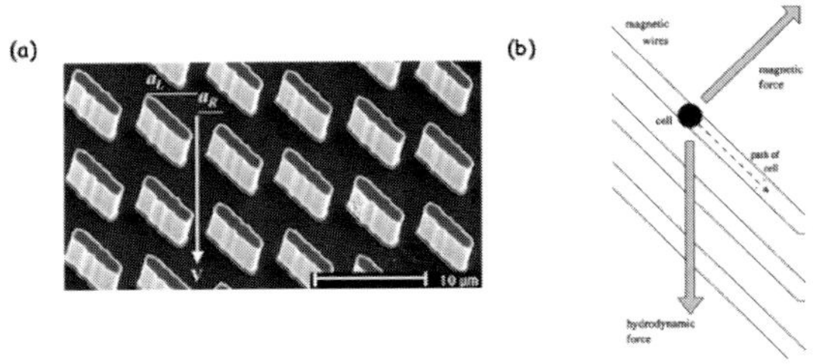

Fig. 1. a) The original thermal ratchet concept. As molecules are moved down in a flow field, the odds of moving to the left or right are not equal. **b)** The magnetic force separation idea. High magnetic field gradients provide forces at an angle to the flow of cells.

45% of the total volume of blood. Erythrocytes are very small (7.8 μm) and are normally shaped as biconcave disks. Aside from rare exceptions, mature erythrocytes have no nuclei or internal membranes. Their purpose is the first listed above: they are primary components of the circulatory system. Leukocytes exist at nearly one-thousandth the concentration of erythrocytes, and they serve an entirely different purpose. Leukocytes protect the body from infection, in cooperation with the organs of the immune system. They are typically classified into three types: granulocytes, monocytes, and lymphocytes. Granulocytes, so named because of granules in their cytoplasm, make up the majority of leukocytes at 5×10^9 cells per liter. They range from twelve to fifteen microns in diameter. Monocytes exist at only 4×10^8 cells per liter, but they range from fifteen to eighteen microns across. Lymphocytes, at 3×10^9 cells per liter, are only slightly larger than erythrocytes, though each has a very large nucleus which occupies most of the cell.

All adaptive immune responses are mediated by B-lymphocytes (B cells) and T-lymphocytes (T cells). All lymphocytes bear variable cell-surface receptors to detect antigens, or foreign macromolecules and cells. Of special interest to cellular and molecular biologists are the blood-borne proteins called antibodies, for they can be used as highly-specific probes to identify and distinguish between different cell populations. Antibodies are Y-shaped proteins of the immunoglobulin (Ig) family. The body produces antibodies as a defense against extracellular materials. For instance, an antibody may bind to a virus or toxin to prevent it from infecting a cell, or it may coat a foreign bacterial cell and mark it for destruction. However, a particular antibody can only bind to select molecules that fit into its antigen-binding site. Although the amino acid sequence among all antibodies is mostly constant, the end of each "arm" of the Y-shaped molecule sports a variable region. These arms form the binding site of an antibody, and it is their variability that accounts for the specificity in what they can bind to. Today, antibodies can be obtained which distinguish between two polypeptides that differ by only a single amino acid [3]. Once families of antibodies have been produced, they can be conjugated to small fluorescent molecules, such as fluorescein or rhodamine, by which they can be detected under a fluorescent microscope.

Cell biologists have isolated families of antibodies that selectively recognize different subpopulations of leukocytes by specific proteins contained in the cells' outer membranes. Though they are natural components of the leukocytes, these cell-surface molecules are called "antigens" because antibodies can be raised against them. A more fitting name for them is "markers" because they are characteristic of specific cell populations. Markers can be grouped into multiple categories; some are specific for cells of a particular lineage, while the presence of others may vary according to the stage of differentiation of cells of the same type [4]. Any cell surface marker that identifies a particular lineage or differentiation stage and is recognized by a group of monoclonal antibodies is part of a "cluster of differentiation". All leukocyte surface markers whose clusters are defined are designated with a CD, followed by a number.

The CD system has allowed immunologists to identify cells participating in various immune responses. For instance, most helper T cells are $CD3^+$ $CD4^+$ $CD8^-$, while cytotoxic T cells are $CD3^+$ $CD4^-$ $CD8^+$ [4]. Fluorescent molecules can be conjugated to the antibody clusters that recognize their specific markers, and then incubated with the cells in a sample. However, the CD system also enables the fractionation of blood cells according to their specific surface antigens. Clusters of antibodies can be used to selectively bind fluorescent molecules or magnetic beads to certain leukocytes, which are then isolated by flow cytometry techniques. This system is crucial to our method of separating cells as well as to the existing technologies. Figure 2 gives a picture of how antibodies can be labeled and attached to cells.

Fluorescence-activated cell separation provides scientists with one way to isolate cells of a uniform type from a tissue or cell suspension. Specific cells in a sample are labeled with antibodies of an appropriate cluster, coupled to a fluorescent dye. The cells are sent single file in a fine stream through the path of a laser beam. As each cell passes the beam, it is monitored for fluorescence. A nozzle then forms droplets containing single cells and gives each a positive or negative charge, depending on whether the cell it contains is fluorescent. Finally each droplet is deflected by a strong electric field into a collection chamber [2]. This process results in two collection tubes, one containing labeled cells and one containing unlabeled cells. Consequently, the specific cells in the sample that were labeled are isolated from the rest

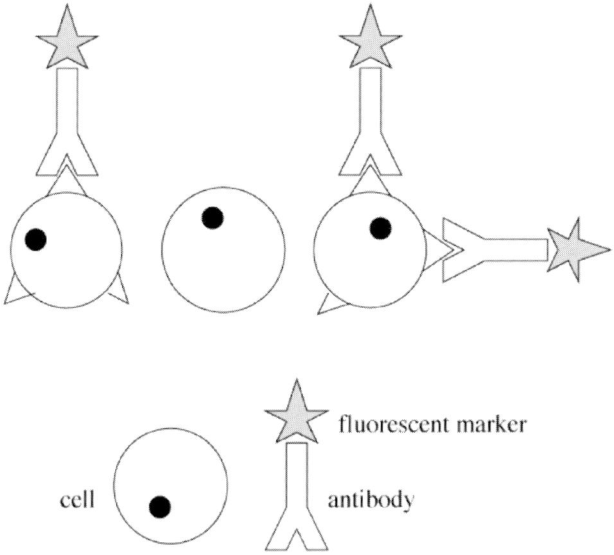

Fig. 2. Labeled antibodies and their attachment to a cell membrane.

of the sample. A schematic view of a FACS machine and a magnetic sorting device is shown in Figure 3.

The beads themselves are composed of iron (II) oxide nanocrystals approximately 50 nm in diameter, coated with a polysaccharide which provides functional groups for the attachment of antibodies [6]. Thus, each bead is smaller than the average leukocyte by a volume factor of more than 10^6. Their extremely small size makes them very gentle on the cells to which they attach, and they bear apparently effect on the cells' function or viability. However, although only a few dozen beads are needed to separate a cell, several thousand may be bound to a particular cell after incubation [7].

An interesting physics result of the nanosize of the beads is that they are superparamagnetic, which means that they are single domain but too small to form a stable magnetic moment (ferromagnetic materials have stable magnetic moments in the absence of an external magnetic field $\mathbf{H}$). Ferromagnetism comes from the Fermi exchange interaction between two atoms, each of which has a net electronic unpaired spin $\mathbf{S}$, and the net odd symmetry that the total wavefunction must have. The sign of the exchange interaction is such that the spins have a parallel alignment energy on the order of 400 K, and if the system is big enough can be ferromagnetic at room temperature. However, for small volumes thermal fluctuations will be sufficient to overcome the anisotropy energy and cause the spontaneous loss

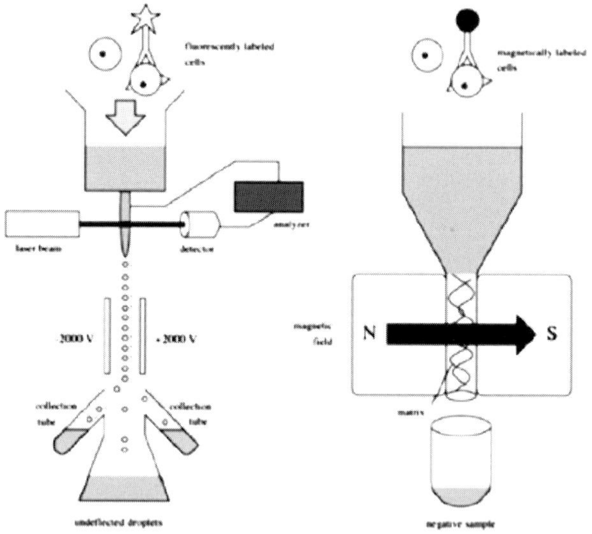

Fig. 3. Two conventional ways to sort labeled cells from whole blood.

of a permanent magnetic moment. Under such conditions the material is classified as superparamagnetic. In zero external field, the net magnetic moment of the superparamagnetic beads is zero. In the presence of an external magnetic field the beads can be highly magnetized.

The high specificity and efficiency of magnetic methods have made them quite useful in obtaining rare cell types. Hematopoietic stem cells, residual tumor cells, and antigen-specific B and T cells can be isolated and used in a variety of functional assays. Hematopoietic stem cells can be isolated by their expression of the CD34 antigen [8]. Stem cell purification techniques are of great value for both science and medicine. Pure populations of stem cells will make possible scientific studies of blood cell formation and differentiation. Additionally, they are necessary for successful transplantation procedures. Stem cells from the bone marrow and peripheral blood are transplanted in combination with chemotherapy for the treatment of certain malignant and genetic disorders. The success of a transplant procedure depends on the effectiveness of techniques that are used to isolate the cells for the transplant. Any additional cells lingering in the preparation may pose a risk to the recipient.

Cancer patients benefit from cell separation techniques by their capacity to remove residual tumor cells from the bone marrow. Since conventional cancer therapy is toxic to bone marrow stem cells, a fraction of bone marrow must be removed from a patient before high dose therapy can be given. The bone marrow can subsequently be reinfused into the patient. But before this occurs, it is desirable to eliminate all tumor cells from it. Tumor cells can be selected by specific antibodies to surface markers and then removed from the sample. Scientists have achieved removal of 99.9% of the malignant cell population using magnetic separation techniques [9].

We describe an alternative way to isolate biological cells from a larger sample. Our method is similar to MACS in that specific cells are attached to antibody-coated magnetic beads in a high magnetic field gradient. However, our device is entirely confined to one single microchip. Additionally, it confers the advantages of continuous input and two-dimensional separation. It is believed that our device could potentially yield a higher efficiency and a greater degree of purification than existing cell separation techniques.

3 Magnetic separation

In our device, cell fractionation is made possible by an array of very thin magnetized "wires" which are aligned at an angle to a net hydrodynamic flow direction. There are two advantages to these microfabricated wires: the extremely thin ferromagnetic layer forces the spin system to be single domain, or at least "few domain" with resultant very high magnetic fields. Secondly, because of the small length scale of the wires such small structures have large magnetic field gradients at their edges. Since magnetic force depends on field gradients, the path of a paramagnetic object exposed to this array of wires will be altered. Thus, our device can separate paramagnetic objects from diamagnetic ones. Paramagnetic beads attached to cells are attracted to the wires and are deflected away from unlabeled cells.

Since there are no magnetic monopoles, forces result from field gradients acting on magnetic dipole moments. If an object is ferromagnetic with a permanent magnetic moment, $\vec{\mu}$, it will feel a force in the presence of a magnetic field gradient given by:

$$\vec{F_{\mathrm{m}}} = (\vec{\mu} \cdot \nabla)\vec{B}. \tag{3.1}$$

The magnetic moment of a paramagnetic object is induced by an external field. A paramagnetic object with magnetic susceptibility χ will feel a force given by:

$$\vec{F_{\mathrm{m}}} = (\chi\vec{B} \cdot \nabla)\vec{B}. \tag{3.2}$$

In our apparatus, beads attached to the cells have an induced magnetic moment $(\chi\vec{B})$ aligned parallel to the field produced by the wires. The preferred magnetization direction of the magnetic stripes is to place $\vec{B}$ magnetized perpendicular to the plane of the wafer, so the field points in the

$+y$-direction. Since beads travel through our device at a fixed height y_o above the wires, the force that they encounter is:

$$F = \mu \partial B_y / \partial x. \tag{3.3}$$

Our beads are single domain, 50 nm particles of ferrous iron oxide (FeO). They are superparamagnetic, which means that they are too small to sustain a stable dipole moment but that they will exhibit a net moment in an external applied field.

4 Microfabrication

Microfabricated devices are capable of accessing the small length scales of biological cells and providing magnetic field gradients high enough for cell fractionation. Microfabrication was performed at the Princeton Center for Photonics and Optoelectronic Materials (POEM). The facility offers a full range of sophisticated processes and equipment for scientific research, which enables the fabrication of devices that could not have been made several years ago.

The fabrication of our device required a series of steps to make it suitable for magnetic cell separation. First, the outer design was exposed onto a silicon wafer by photolithography. This included the channels, inlets, outlets, and central chamber, constituting the framework of the apparatus. Then, this design was etched 16 μm into the wafer and the remnant photoresist was stripped off. A separate pattern for the diagonal magnetic wires was exposed onto the central chamber, again by photolithography. Grooves for these wires were then etched an additional 0.2 μm so that the wires would be countersunk and not impede the flow of cells. A ferromagnetic metal alloy was deposited onto the wafer in a uniform coat. The unexposed photoresist was then removed from the wafer, lifting off the overlapping metal with it. This left behind only the thin wires on the wafer. Finally, a protective layer of SiO_2 was deposited on top of all structures.

Masks for positive imaging were made commercially by Adtek Photomask of Montreal, Quebec. Our device required two masks: one for the outer channels and outlets, and one for the magnetic wires. Each mask was a square quartz plate, $5'' \times 5'' \times 0.09''$. The masks were designed on L-Edit, a computer graphics program, and they were fabricated by electron beam lithography. A pair of alignment marks at opposite ends of each mask ensured that the patterns overlapped each other precisely.

To form magnetic wires, a cobalt-chrome-tantalum (Co-Cr-Ta) alloy was sputtered onto silicon wafers that had been exposed to the diagonal wire pattern. Co-Cr-Ta was used because of its high remanent magnetization. Co-Cr-Ta was chosen as a suitable alloy for our magnetic structures because it has an exceptionally high remanent magnetization of 6000 Gauss [16]. This should be compared to the *saturation* magnetization of its main component, cobalt (1400 Gauss) [15]. As an alloy, Co-Cr-Ta has the ability to acquire and retain a greater magnetic moment than each of its three components. Co-Cr-Ta occupies a hexagonal close packed (hcp) lattice structure, and, like most sputtered films, it is isotropic [17]. Additionally, as the wires in our device are very thin films, it is expected that they are single domain. Evidently the wires are capable of acquiring a large, uniform magnetization in an applied field and retaining it when removed from the field. As they are isotropic, they may be magnetized either in the plane of the wafer or perpendicular to it.

Wires were magnetized perpendicular to the plane of the wafer, in a uniform external field. The magnetic field was provided by a large electromagnet and measured at 5 kG. It is believed that wires achieved their saturation magnetization in this field and retained a nearly uniform internal field of 6000 Gauss as the electromagnet was disabled.

The strength of the magnetic remanence of the wires was tested using paramagnetic and latex beads. First, fluorescent paramagnetic beads were suspended above the wires. Beads were attracted by the wires in a striking fashion, accumulating at the edges where the field gradients were the highest. As a control, nonmagnetic polymer beads were then suspended. They disseminated uniformly, showing no preference for the magnetic wires.

5 Magnetic field gradients

Magnetic fields are produced by Co-Cr-Ta wires positioned at a 45° angle to the input stream of cells. Each wire is 10 μm wide and 0.2 μm thick, and neighboring wires are separated by 25 μm. The following calculations presume uniform magnetization of the wires, with a remanent field of 6000 Gauss.

Wires may be magnetized in either of two directions: in the plane of the wafer or perpendicular to the wafer. The two orientations are shown in Figure 4, with a consistent coordinate system. For neither method is there a force directed along the length of the wire; $\partial B/\partial z = 0$ in each case. However, for both orientations, there are very large gradients in the magnetic fields near their edges. These gradients attract paramagnetic objects to the wires and redirect their flow through the chamber. The orientation that can provide higher field gradients should be chosen for optimal results.

Field gradients were also calculated analytically. Since cells travel at a fixed height y_{o} above the wires, only $\partial B/\partial x$ is significant. Although gradients for both in-plane and perpendicularly magnetized wires are comparable at large y_{o}, gradients at lower heights are significantly greater for the case

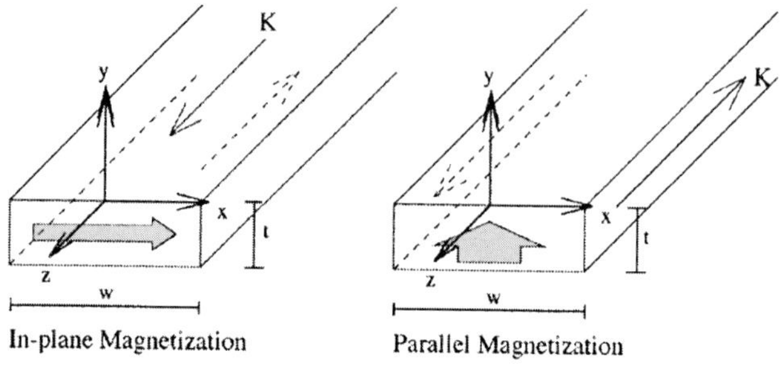

Fig. 4. Two possible directions for magnetization.

of perpendicular magnetization. Thus, in order to maximize the magnetic force for separation, wires were magnetized *perpendicular* to the plane of the wafer.

A bead that is not attached to a cell will roll along the floor of the chamber. The bead is separated from the wires by only a 0.2 μm layer of SiO_2. Since its magnetic moment aligns with the field and $\vec{F}_{\rm m} = (\vec{\mu} \cdot \nabla)\vec{B}$, the bead will experience horizontal forces of up to 3×10^{-11} Newtons, or 3×10^{-6} dynes. A bead attached to a cell will feel a smaller force because it is elevated above the floor of the chamber. Lymphocytes are approximately 8 μm in diameter. Beads will therefore be, on average, 4 μm above the wires in the chamber. At this height, fields are 0.0045 Tesla and gradients are 1300 T/m. Consequently, beads will experience forces up to 1.5×10^{-13} N, or 1.5×10^{-8} dynes. For comparison, a constant force of 1.5×10^{-8} dynes will cause a leukocyte to reach a terminal velocity in water of 1 μm/s, a modest speed for a cell in a microchip. In reality, though, the forces on the cells will be considerably *higher*. Because the beads are so small, several thousand may bind to each cell. Also, many of these beads will be closer to the wires than 4 μm, where forces are significantly greater. Cells will experience the net effect of the magnetic forces on all beads bound to them.

It is important to note that magnetic forces point in the x-direction, *perpendicular* to the wires. The magnetic force is greatest at the edges, and it is always attractive. Since the cells flow at a 45° angle to the wires, there is an x-component to the hydrodynamic force which should exactly cancel the magnetic force. All that is left is the z-component to the hydrodynamic force (and viscous drag). As a result, the cells flow in the z-direction, along the lengths of the wires. Although they continue to be propelled by the input jet stream, they are constrained to follow the wires by the high field gradients at their edges.

6 Device interface

In order for the deflected cells to be resolved, the input jet stream must be confined to a narrow width through the entire chamber. If the jet stream broadens out over time, magnetic separation becomes impossible. However, for a single jet that stands alone, this is exactly what happens. As soon as the jet enters the chamber, it experiences an abrupt change in environment. It slows down almost at once to the velocity of the surrounding fluid and, by conservation of mass, broadens immediately. Designed with this in mind, our device contains an alternative to the single input jet: the N-port injector [23]. The foundation for this idea can be derived from elementary fluid dynamics, as I discussed in the first lecture.

The apparatus consists of a series of channels and chambers etched into a silicon wafer (Fig. 5). Channels are sealed by a glass coverslip. Large hydrodynamic forces produced by a pressure gradient drive cells through the device, where they are collected in nine outlet portals. All structures in the apparatus are etched 16 μm deep, allowing ample room for cells to flow.

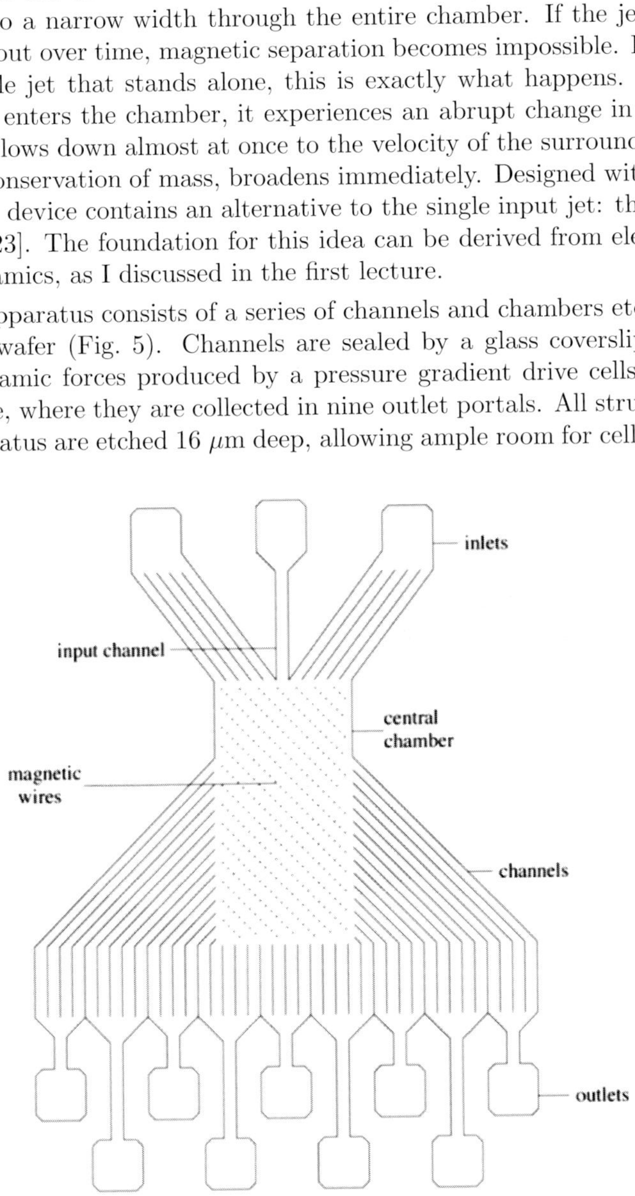

Fig. 5. The basic schematic of the microfabricated cell fractionation chip.

Once cells are labeled, the sample is fed into the main input channel at a high hydrodynamic pressure. The input channel is 40 μm wide and feeds into a central rectangular chamber. Buffer flows through multiple channels on both sides of the input at an even higher pressure. This confines the sample to a narrow stream as it flows through the chamber. The rationale behind this "N-port injector" has been discussed in earlier lectures, *ad nauseaum* no doubt.

The central chamber (5 mm $\times$ 18 mm) contains the magnetic structures and is where magnetic separation occurs. Wires of a ferromagnetic material are countersunk into the floor of the chamber at a 45° angle to the flow of cells. The wires are 10 μm wide and spaced 25 μm apart. The stripes were magnetized by placement in a 10 kG external magnetic field from an electromagnet. The high magnetic field gradient at the edge of each wire imposes a force on the superparamagnetic beads at an angle to the hydrodynamic force. The component of the magnetic force perpendicular to the flow causes the lateral deflection of all labeled cells. Cells are constrained to move along the magnetic wires by the large field gradients that exist at their edges. Figure [2] shows fluorescent images of magnetic beads.

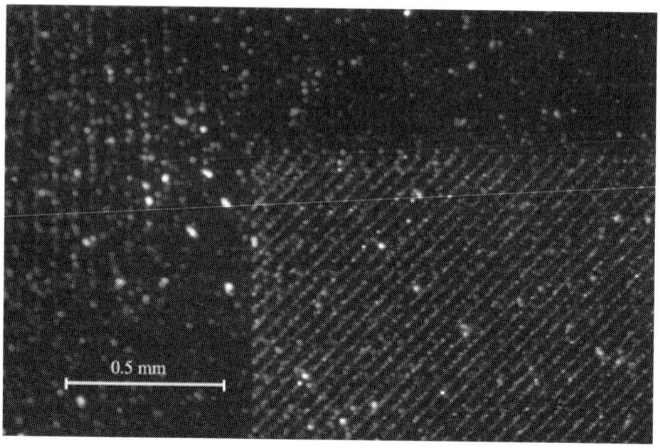

Fig. 6. Magnetic wire array with paramagnetic beads bound to the edges of the strips where the magnetic field gradient is highest.

However, the undeflected cells must stay confined to a narrow stream across the length of the chamber in order for perpendicular deflections to be resolved. Channels are positioned at the sides and end of the chamber to capture cells as they exit. More than 350 channels, 24 μm wide, feed into nine outlets where cells are collected. All unlabeled cells will flow into the central outlet. In the outlets to the right, the fraction of labeled cells will be enriched. No cells should migrate to the left.

Fig. 7. Microscopic views of etched structures, before sputtering.

If cells can indeed be sorted by the high gradients in our device, rare cell types expressing known surface antigens can be isolated. Though its implications are the same as for FACS and MACS, our method comes with several unique advantages. First, it allows for the continuous input of cells. It therefore holds the promise of tolerating a large volume of cells with high efficiency. Second, its two-dimensional design allows simultaneous isolation of different cell types. By labeling different populations of cells with particles of differing magnetic susceptibilities, one could expect different degrees of lateral separation. Third, its small size would make it a cheap and convenient mechanism for separating cells.

The central chamber (5 mm $\times$ 18 mm) contains the magnetic structures and is where magnetic separation occurs. Wires of a ferromagnetic material are countersunk into the floor of the chamber at a 45° angle to the flow of cells. The wires are 10 μm wide and spaced 25 μm apart. The high magnetic field gradient at the edge of each wire imposes a force on the superparamagnetic beads at an angle to the hydrodynamic force. The component of the magnetic force perpendicular to the flow causes the lateral deflection of all labeled cells. Cells are constrained to move along the magnetic wires by the large field gradients that exist at their edges. However, the undeflected cells must stay confined to a narrow stream across the length of the chamber in order for perpendicular deflections to be resolved.

Channels are positioned at the sides and end of the chamber to capture cells as they exit. More than 350 channels, 24 μm wide, feed into nine outlets where cells are collected. All unlabeled cells will flow into the central outlet. In the outlets to the right, the fraction of labeled cells will be enriched. No cells should migrate to the left. Figures 7–9 show images of various features of the fabricated wafer.

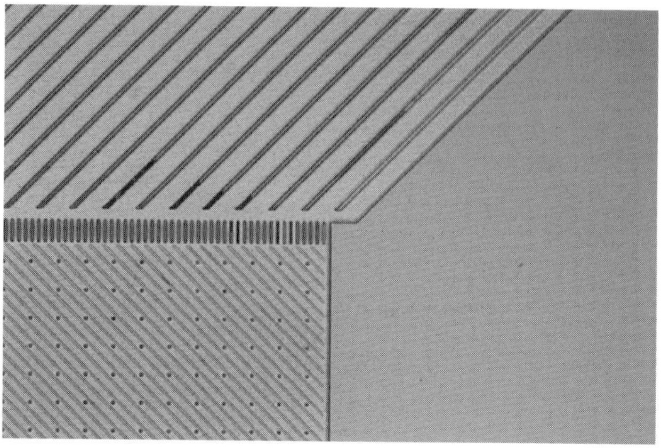

Fig. 8. Entrance to the central chamber. Magnetic wires are at a 45° angle to the input jet stream. Round dots are posts to prevent the coverslip from collapsing.

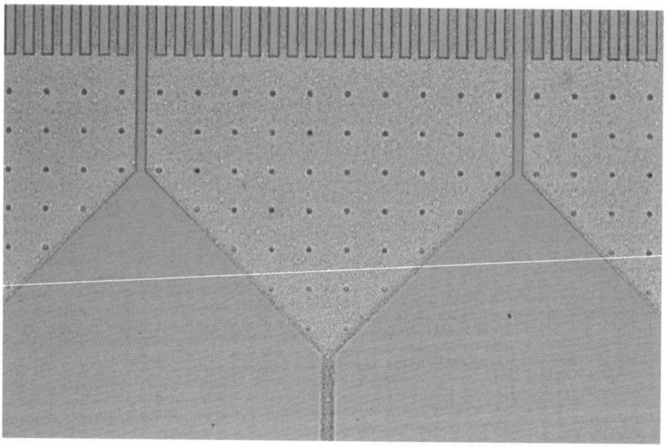

Fig. 9. Channels and outlets at the end of the central chamber.

Once a chip IS prepared with a coverslip, it is loaded onto a lucite chuck which mediated the flow of fluid (Fig. 10). The chuck was machined in the Princeton Physics Department machine shop. It is designed so that a high pressure could be applied to the fluid in all three inlets to drive it through the chip. Twelve tubes are drilled in the chuck so as to exactly line up with the holes in the chip. Fluid can enter the chip from the bottom, through the holes drilled at the inlets. It can then exit through one of the nine outlet holes in the chip, eventually filling up the tubes in the chuck that were exposed to atmospheric pressure. From there, the cells can be

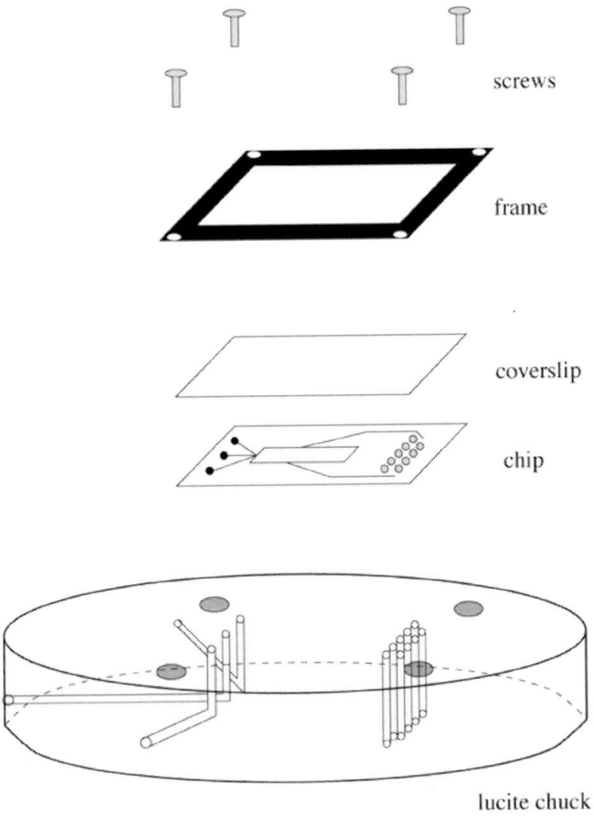

Fig. 10. Chuck assembly.

collected. The sealed chip is placed on top of the o-rings and held down by a rectangular steel frame. It is recommended that spring washers be used with the screws that hold down the frame so that the chip does not crack under stresses. However, controlling the flow through the input channels is not easy. Bubbles are a big obstacle to establishing hydrodynamic flow. Small bubbles that became trapped in the apparatus blocked off channels to flow. To work around this problem, the empty chuck was submerged in buffer before it was even assembled. The chip was then loaded onto the chuck, completely underwater. Additional problems came from loading the inlets with the syringe. The right angles in the input tubes turned out to be an awkward design for filling them. When bubbles are encountered, vacuum pumping of the chip under buffer can often liberate them. Otherwise, the chuck can be soaked overnight to let the bubbles gradually disappear.

7 A preliminary blood cell run

Since this is really a progress report on the development of a technology as we develop this idea it is unfortunate that we cannot present at this time pictures of cells labeled with magnetic beads separating in the device. However, we have been able to show at this point that whole blood samples can be run in the device and that ultra-narrow jets of cells can be run across the entire 2 cm length of the chip with precision, and that white blood cell adhesion to the device walls can be controlled.

White blood cell adhesion is a major problem because of the high concentration of proteins presented at the cell surface, particularly white blood cells. Our chip was fabricated out of silicon and has approximately a 0.2 micron thick silicon dioxide overcoat. However, silicon dioxide surfaces are highly charged and strongly bind cells to the surface. We have been exploring ways to avoid cell adhesion, and one of the most promising materials is the tri-block copolymer polyethylene oxide-polypropylene oxide-polyethylene oxide (PEO-PPO-PEO), also known as Pluronics under the manufacture of BASF. A good reference on the use of Pluronic surfactants to prevent the adhesion of cells to surfaces can be found in the papers of Karin Caldwell [27]. The great advantage of the Pluronic system is that the hydrophobic center bloc (PPO) can bind to hydrophobic surfaces of silicone elastomers (polydimethylsiloxane, PDMS) while the end group polyethylene oxide (also commonly known as polyethylene glycol, PEG) is a very unusual polymer that is neutral but very hydrophilic and strongly hydrogen bonding. If the PPO end groups and the center PPO group are chosen to be of the appropriate molecular weights, a very robust surface on a virgin PDMS surface can be created which does not bind proteins due to a combination of the shielding of the hydrophobic surface of PDMS by the Pluronics and an entropic repulsion of the surface PPO groups of the protein.

In our case, we used a 0.1% solution of Pluronic F108 from BASF (BASF Corporation, Mount Olive, NJ 07828) in a saline buffer to wet our magnetic array wafer that was sealed with a cover slip upon which a thin layer of PDMS had been spun and polymerized. Although PDMS is very hydrophobic and normally water will not penetrate a sealed PDMS structure, a saline buffer solution which contains 0.5% F108 will over a period of 24 hours penetrate and wet a sealed PDMS structure. This then provides a very interesting surface which is hydrophilic, uncharged and not "biofouling".

Once the array was successfully wet and the surface passivated a 20 microliter sample of blood from a finger prick was put into Becton Dickinson Microtainer tube treated with lithium heparin to prevent clotting of the blood (Becton Dickinson, Franklin Lakes, NJ 07417-1885). The white blood cells in the blood sample were then stained with the vital

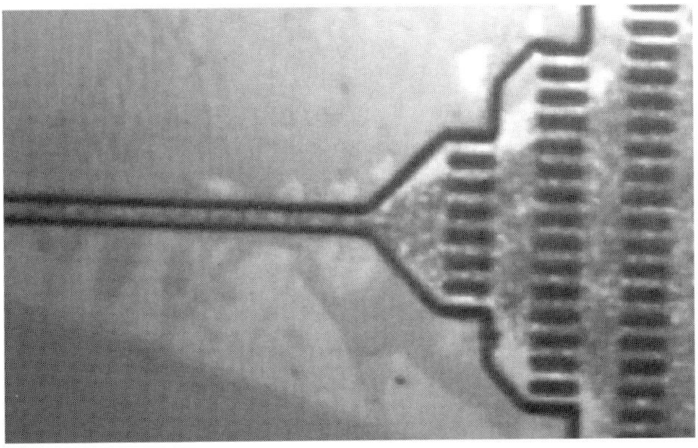

Fig. 11. Brightfield image of blood at entrance to the array.

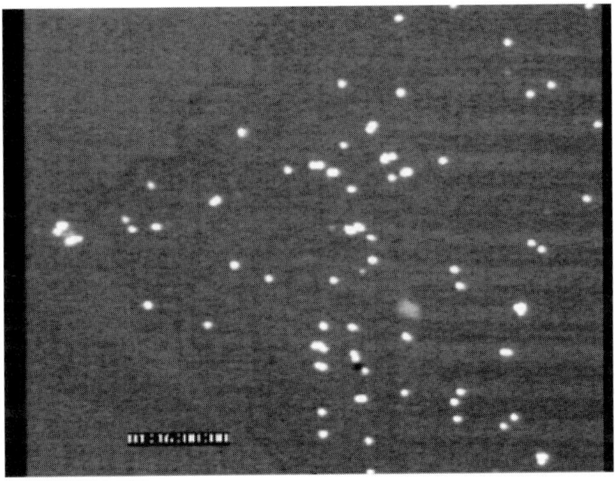

Fig. 12. Epifluorescence image of labeled white blood cells at entrance to the array.

nuclear stain Hoechst 33342 (Molecular Probes, Eugene, OR 97402-9165) by incubation at 37 °C for 30 min. The sample of stained blood was then loaded in the center chamber and positive air pressure of approximately 0.2 psi where applied to the center jet and the side fluid curtain flow.

Figure 11 shows in epi bright field illumination the blood sample at roughly ×40 magnification leading up to the injection jet. Figure 12 shows in epifluorescence the same blood sample only now the labeled white blood cells are visible. Figure 13 shows the stream of blood cells flowing from right to

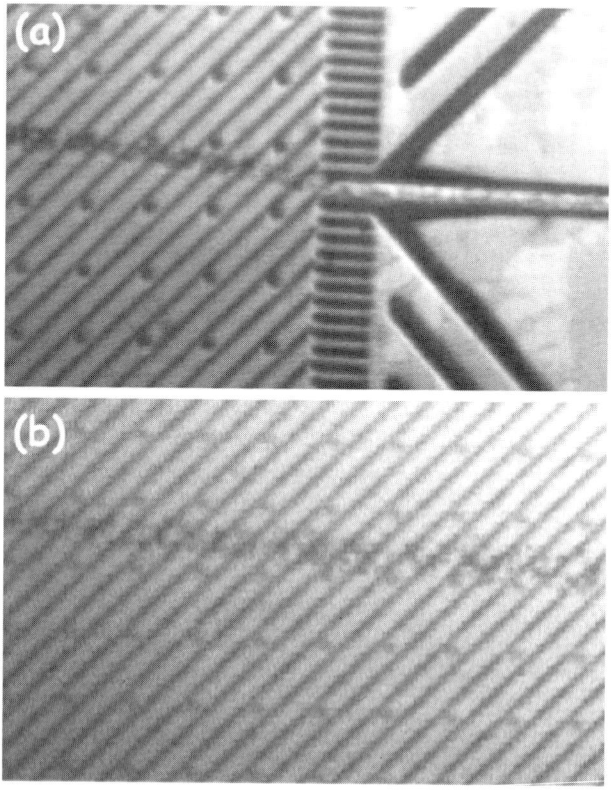

Fig. 13. Brightfield image of blood cells flowing **a)** from injection port and **b)** in the center of the array about 1 cm in from the injection port.

left into the magnetic line array, illustrating how the hydrodynamics ensures a smooth non-expanding flow into a large open area, both at the entrance (a) and well within the array (b). No evidence of red blood cell adhesion to the device was seen. Figure 14 shows an image of two labeled white blood cells constrained within the red blood cell stream seen in Figure 13b, showing how the white cells move smoothly with the stream of red cells and show very little adhesion to the surface. Clearly, the next step now is to label these cells with paramagnetic antibody beads and deflect them from the main stream, but alas due to the deadline constraints of this paper that project remains to be done in the next several months.

Fig. 14. Epifluorescence image of two white blood cells in (b) of Figure 13.

8 Conclusions

Progress has been made in constructing a magnetic cell separation device. However, its potential to fractionate cells remains unfulfilled. Several procedures still need to be optimized, and several experiments still need to be run. In this section, a few main areas are highlighted.

It is important to determine how to optimize the hydrodynamic forces in scale with the magnetic forces produced by the wires. Magnetic separation of beads must be demonstrated. If the hydrodynamic forces are so strong that they overshadow magnetic deflections, the pressure gradient must be diminished in a way that does not compromise the jet width. If instead the beads adhere to the wires, pressures may need to be increased. The electrostatic adhesion of beads to wires is a concern, in which case the SiO_2 layer on top of the wires should be made thicker. At all points along the way, careful control experiments need to be conducted. Paramagnetic beads can be run alongside latex beads, preferably of two different colors.

Also, it will be important to have more confidence in handling cells. Labeling cells is crucial for achieving magnetic fractionation. Experimenting with different stains and different markers will likely determine a good test sample for the apparatus. Further, multiple markers and fluorescent stains will be necessary to conduct control experiments. Surface passivation measures must also be optimized in order to prevent the adhesion of cells to silicon structures.

References

[1] R. Wind *et al.*, *Appl. Phys. Lett.* **78** (2001) 2205–2207.

[2] B. Alberts, J. Lewis, K. Roberts, D. Bray, M. Raff and J. Watson, *Molecular Biology of the Cell*, Third edition (Garland Publishing, Inc., 1994).

[3] G. Karp, *Cell and Molecular Biology Concepts and Experiments*, Second edition (John Wiley & Sons, Inc., 1999).

[4] A.K. Abbas, A.H. Lichtman and J.S. Pober, *Cellular and Molecular Immunology*, Fourth edition (W.B. Saunders Company, 2000).

[5] C. Janeway, *Immunobiology: the immune system in health and disease* (Garland Publishing, Inc., 1999).

[6] S. Miltenyi, Methods and materials for improved high gradient magnetic separation of biological materials, U.S. Patent Number 5411863, 1995.

[7] Magnetic Cell Sorting and Separation of Biomolecules, Miltenyi Biotec Catalog, 2000.

[8] P. Lansdorp, Device for separating magnetically labelled cells, U.S. Patent Number 5514340, 1996.

[9] B. Haukanes and C. Kvam, *Bio/Technology* **11** (1993) 60–63.

[10] T.A.J. Duke and R.H. Austin, *Phys. Rev. Lett.* **80** (1998) 1552–1555.

[11] C. Chou, O. Bakajin, S. Turner, T. Duke, S.S. Chan, E.C. Cox, H.G. Craighead and R.H. Austin, *Proc. Natl. Acad. Sci. USA* **96** (1999) 13762–13765.

[12] M. Madou, *Fundamentals of Microfabrication* (CRC Press, 1997).

[13] C. Kittel, *Introduction to Solid State Physics*, Seventh edition (John Wiley & Sons, Inc., 1996).

[14] M. Blois, *J. Appl. Phys.* **26** (1955) 975.

[15] J. Smit, *Magnetic Properties of Materials* (McGraw Hill Book Company, 1971).

[16] J. Castelino, *Sorting White Blood Cells in Microfabricated Arrays*, Doctoral Dissertation (Princeton University, 2000).

[17] C. Denis Mee, *Magnetic Recording Technology*, Second edition (McGraw Hill, 1996).

[18] P. Chaikin and T. Lubensky, *Principles of Condensed Matter Physics* (Cambridge University Press, 1995).

[19] H. Hoch, L. Jelinski and H. Craighead, *Nanofabrication and Biosystems* (Cambridge University Press, 1996).

[20] K. Healy, *Curr. Opinion Solid State Mater. Sci.* **4** (1999) 381–398.

[21] A. Braem, Manipulating Adsorbed Polymer Layers Via Polymer-Surfactant Complexation, Poster Session (Carnegie Mellon University, 2000).

[22] D. Lide, *CRC Handbook of Chemistry and Physics*, 72nd edition (CRC Press, 1991).

[23] N. Darnton *et al.*, Hydrodynamics in 2 1/2 Dimensions: Making Jets in a Plane (1999).

[24] R. Feynman, M. Sands and R. Leighton, *The Feynman Lectures on Physics*, Volume II (Addison-Wesley Publishing Company, 1989).

[25] G. Binnig and C. Quate, *Phys. Rev. Lett.* **56** (1986) 930–933.

[26] L. Kong and S. Chou, *J. Appl. Phys.* **81** (1997) 5026–5028.

[27] K. Caldwell, in Polyethylene glycol: Chemistry and Biological Applications, J. Harris and S. Zalipsky (eds.), ACS Symposium Series 680 (American Chemical Society, Washington, DC).

Abstract

The continuous-flow mixing chips presented in this lecture open up new dimensions for spectroscopy of protein reactions. The time resolution of 400 microseconds is about 1000 times faster than recent IR stopped-flow setups [20]. Additionally, the miniaturisation reduces the sample consumption by an even higher magnitude. For a reactant with a higher diffusion constant than TFE, the time resolution can be further improved by using a smaller IR focus spot and a higher flow velocity. With the current design and the 10 mm spot of a synchrotron IR source [29], a time resolution on the order of 50 microseconds is feasible.

1 Introduction

The rough energy landscape of a protein includes not only the steep free energy funnel that guides the unfolded protein into its compact native state [1, 2], but also traps of misfolded intermediates. In some cases this rough surface can lead to distinct conformations of the same protein that consist of either α-helix or β-sheet. An erroneous transition from α-helix to β-sheet structures has fatal consequences in prion and other amylogenic diseases. Since these distinct conformations are both highly compact, an important question is the size of the energy barriers and the interconversion times for α-helix to β-sheet structural changes. An excellent model system to study helix/sheet transitions is b-lactoglobulin (BLG), a major component of mammalian milk. BLG-A consists of nine antiparallel b-strands (51%) and one α-helix (7%) in its native state (Fig. 1) [3]. The high β-sheet content sharply contrasts with secondary structure predictions for the amino acid sequence that indicate a distribution of 48% helix and 13% sheet [4]. In fact, BLG has the remarkable property of refolding to 80% helical content under the influence of TFE [5, 6]. The transition occurs cooperatively between 15% and 20% TFE.

So far, protein folding has been mostly probed by UV/Vis, fluorescence and circular dichroism (CD) spectroscopy as well as NMR spectroscopy. Each one of these techniques probes different structural changes. UV-Vis and fluorescence spectroscopy monitor the interactions of a prostetic group or a fluorophore like tryptophan with its microenvironment. CD spectroscopy is mainly sensitive to helical secondary structure. NMR spectroscopy can provide residue resolved information [7, 8] but spectra of denatured states are difficult to interpret and the time resolution in real time measurements is low. Fourier-transform infrared spectroscopy on the other hand is able to clearly distinguish between all types of secondary structure [9], it is sensitive to tertiary structure [10] and capable of detecting single residue reactions on a nanosecond time scale [11, 12]. The most

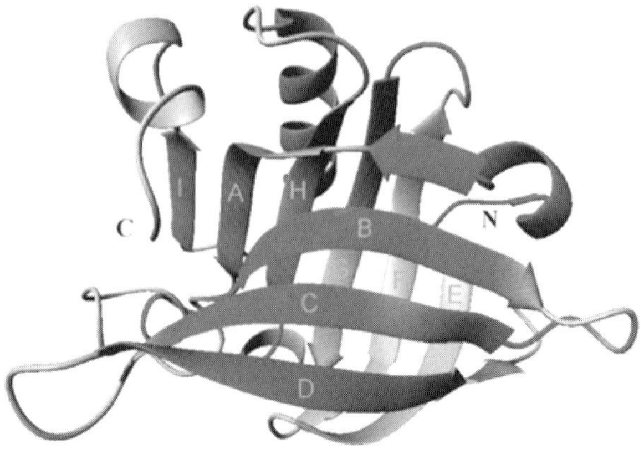

Fig. 1. Backbone representation of β-lactoglobulin A. Regions with high helical preference are marked.

frequent approaches to initiate folding reactions in kinetic experiments have been rapid mixing techniques in either stopped-flow or continuous-flow operation mode [13]. FTIR-spectroscopy, despite its sensitivity, has only been rarely used for kinetic mixing experiments. The main obstacle has been the necessity to mix solutions in a few micron thick samples to keep the background absorption of water and guanidine low. These thin layers make it impossible to generate turbulent flow, the phenomenon utilised by macroscopic mixers to facilitate mixing [13]. Here, we overcome this problem by silicon microfabrication of a diffusional IR mixer which allows microsecond mixing in thin films.

2 Technology

Silicon micromachining has great potential to establish FTIR spectroscopy as a new method for microsecond mixing experiments since silicon is transparent in the mid-infrared. When the dimensions of liquid channels are reduced to a few microns, the flow will be laminar at virtually all velocities [14]. Therefore, there are no turbulent vortices to enhance mixing. In micron dimension channels, diffusion is the only way reactants can be brought together and the flow pattern has to be designed such that diffusion length scales are kept small. It has been shown that micro scale mixing devices can decrease the characteristic mixing time from milliseconds down to 10 microseconds [14–16]. Based on the idea of Brody *et al.* we have designed a continuous-flow mixing chip for FTIR microscopy shown in

Figure 2. The protein solution in the centre and two streams of mixing buffer enter through 80 mm deep inlet channels (Fig. 2a, green) which intersect with the 8 mm deep observation channel (Fig. 2a, red). Because the three inlet channels are a factor of ten deeper than the observation channel, at the merger there is almost equal pressure over the whole width of the observation channel. Due to the viscosity determined laminar flow, no turbulence is induced when the second and third channel merge in, but a layer of the centre (protein) solution between two buffer layers over the whole width of the observation channel results. A fluid dynamics simulation was performed to verify the desired flow pattern.

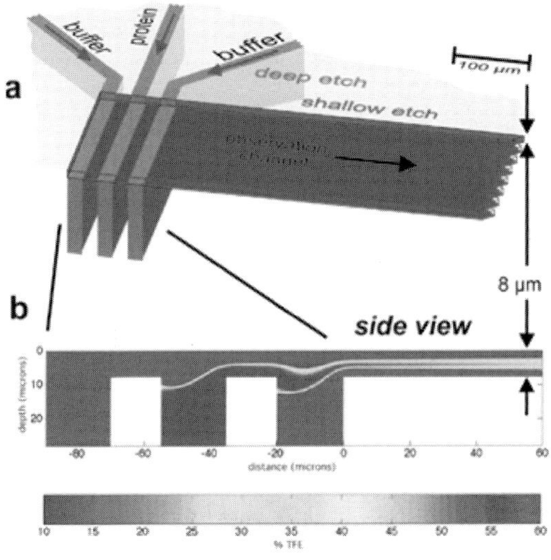

Fig. 2. Design of the mixing chip. **a)** Top view on the chip. The 80 micron deep inlet channels are shown in green and the 8 micron deep outlet channel is shown in red. **b)** Two dimensional fluid dynamics simulation with false color representation of the TFE concentration. A jet of the center (protein) solution between two layers of buffer solution is formed. The TFE concentration is computed by adding the advective term to the diffusion equation.

Figure 2b shows a side view of the intersection of the channels. By approximating the observation channel as "infinitely wide", we can make a two-dimensional model of the fluid flow. At low Reynolds number, the stream function for the flow must obey the biharmonic equation. The resulting flow pattern (Fig. 2b) verifies the desired formation of a protein layer between two layers of buffer solution. Just because the protein layer is so thin, diffusion of the reactant molecules from the buffer into the protein jet and thus mixing is fast. The time resolution is achieved by scanning along

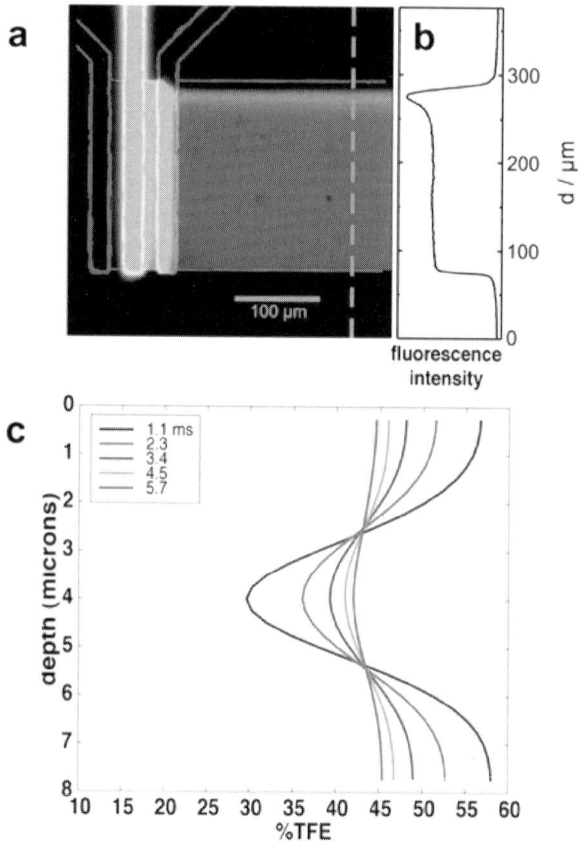

Fig. 3. Epi-fluorescence image of the flow pattern with superimposed contour. Fluorescein solution is flowed in the center channel and non-stained buffer in the side inlet channels. The fluorescence cross section **b)** shows a homogeneous jet over 180 microns width. The peak in intensity in the upper part results from the small flow velocity component in the direction of the in let channels. **c)** TFE concentration profiles through the observation channel as extracted from the Figure 2b.

the observation channel with the focussed beam of an FTIR microscope [17]. Since the protein only moves in about the centre third of the observation channel only marginal blurring of the time axis due to the parabolic flow profile occurs. The flow pattern is visualised by a fluorescein solution flowing in the centre inlet channel and buffer in the two side channels (Fig. 3a). The fluorescence cross section of the observation channel (Fig. 3b) shows the formation of a homogeneous jet over a width of about 180 microns. In the top 20% the formation of the jet is distorted which can be explained by the velocity component in the direction of the inlet channels.

3 Experiments

To induce the refolding of BLG, a protein solution containing 10% TFE (below the refolding transition) entered the center channel and mixed with 60% TFE in the side channels. The mixing time was estimated using a numerical solution of the flow and diffusion equations. The advective term is added to the diffusion equation and solved for the concentration of TFE everywhere in the jet. The TFE concentration is depicted in false-colour in Figure 2b. Once the three inlet streams have combined, the TFE profile gradually relaxes to a constant level as the jet travels down the observation channel. The evolution of the TFE concentration, as a function of channel depth, is pictured in Figure 3c. At the first data point (taken at 100 microns or 1.1 ms), the TFE concentration in the middle of the channel exceeds 30% which is well above the cooperative transition threshold of 15–20%. The characteristic mixing time, extracted from the simulation is 0.4 microseconds. When the IR focus is further confined, this is the maximum time resolution.

Figure 4a shows the time-resolved FTIR spectra of BLG after mixing with TFE between 1600 cm^{-1} and 1700 cm^{-1} where the amide I band (the protein backbone carbonyl groups) absorb. The amide I spectrum is sensitive to the secondary structure of the protein. In the time course of the reaction the maximum of the amide I band shifts from 1632 cm^{-1} to 1652 cm^{-1}. In order to resolve the overlapping bands of secondary structure elements the second derivative spectra are calculated (Fig. 4b). Absorbance maxima correspond to negative peaks in the second derivative spectra. The initial state spectrum (black) shows bands at 1632 cm^{-1} (strong) and 1691 cm^{-1} (weak). These are characteristic of β-sheet structures [9], as expected. In the course of the transition, the signal at 1632 cm^{-1} decreases in favour of the more intense peak at 1652 cm^{-1} which is assigned to α-helices [9]. Additionally, the weak band at 1692 cm^{-1} first shifts to 1686 cm^{-1} and finally disappears. The spectrum of the final state indicates an almost complete loss of β-sheet structure, consistent with CD and NMR results [4, 18, 19]. There is no isosbestic point between the two bands at 1632 cm^{-1} and 1652 cm^{-1} in Figure 4b, revealing the population of at least one intermediate state along the conformational pathway.

The dataset was kinetically analysed by two methods, singular value decomposition (SVD) and least squares curve fitting. SVD of the spectra between 1620 cm^{-1} and 1700 cm^{-1} resulted in three linear independent

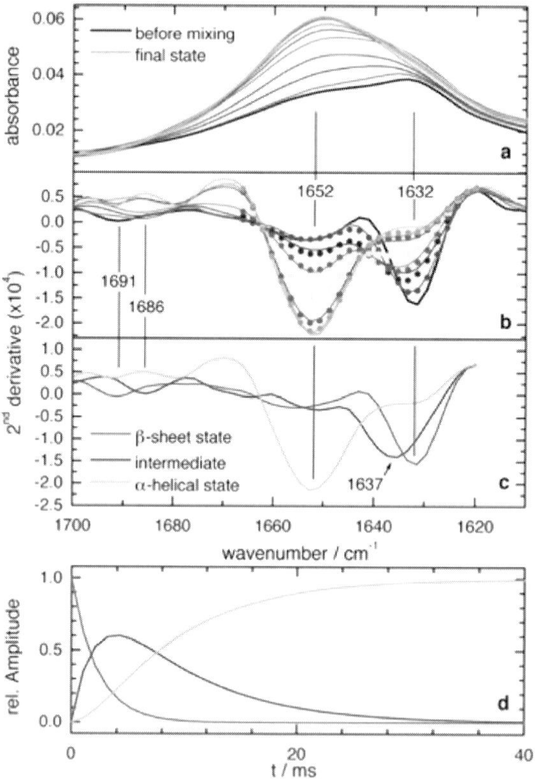

Fig. 4. Time-resolved FTIR spectra and kinetic analysis. **a)** Time-resolved absorbance spectra taken along the observation channel. Time-resolved spectra at 1.1 ms, 3.4 ms, 5.7 ms, 10.2 ms, 21.6 ms, and 103 ms. Spectrum before mixing (black line) and final state spectrum (green line). **b)** Second derivative spectra of **a)** (solid lines) and results of a three-state exponential fit (dots, plotted only up to 1670 cm^{-1}). Line colouring same as in **a)**. **c)** The three basis spectra, resulting form the fit. **d)** Time course of the three states as deduced by the fit.

basis spectra. Similarly, least squares fits yielded at least three spectral states necessary to model the data adequately. The kinetic analysis provided no clear preference for either an exponential, stretched exponential or exponential model which includes back reactions for the time dependence of the states. The result of the three-state fit with two simple exponential functions is shown in Figures 4b to 4d. Figure 4c depicts the three basis spectra: the β-sheet state spectrum (maximum at 1632 cm^{-1}), the helical state spectrum (maximum at 1652 cm^{-1}) , and the intermediate state spectrum. The amide-I peak of the intermediate is centred at 1637 cm^{-1} (Fig. 4c) and the high wavenumber β-sheet peak has shifted to 1686 cm^{-1}. These wavenumbers are within the typical range of antiparallel β-sheet structures [9]. The intermediate clearly is not unfolded which would correspond to a amide I band centred at 1645 cm^{-1} [20]. However, an upshift from 1632 cm^{-1} to 1637 cm^{-1} points in the direction of a more loosely packed β-sheet structure. A similar amide-I frequency is observed for the molten globule of pressure denatured RNAse A, another predominantly β-sheet protein [21].

This suggests a molten β-sheet structure of the intermediate. The frequency upshift can be explained by intruding of TFE into the hydrophobic core of the protein and weakening backbone hydrogen bonding. The structure of the kinetic intermediate differs from the structure of the equilibrium intermediate detected by CD spectroscopy [18] which was proposed to have 53% helical and only 2% sheet content. However, in the equilibrium intermediate detected by heteronuclear NMR about half of the β-sheet structure persists [19]. The kinetic intermediate state structure deduced from our FTIR measurements resembles more the NMR intermediate.

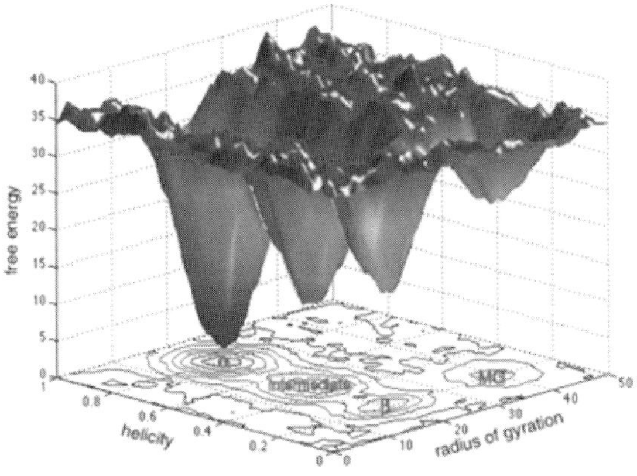

Fig. 5. A schematic of the free energy surface of β-lactoglobulin A as deduced by this work and extrapolated from other work on the local conformational distribution of barrier heights and radii of gyration of protein structures of similar molecular weight.

The intermediate state is formed with a time constant of (2.2 ± 0.1) ms and disappears with a time constant of (7.7 ± 0.1) ms. The β to α transition of BLG is is more than three orders of magnitude faster than the unfolding in 4 M guanidine hydrochloride or the refolding to the native state [22,23]. This is surprising because the change in free energy from the native (β-sheet) state to the (helical) TFE-state almost equals the free energy change from the guanidine denatured state [24]. Hence, similar rates would be expected [25]. However, since the β to α transition proceeds *via* a compact state, in contrast to the unfolding, the transition might be performed mach faster from a phase space argument [26]. The question thus arises, how does TFE reshape the energy landscape to facilitate an interconversion between compact structures without unfolding the protein? The well known equilibrium effect of TFE is the weakening of hydrophobic interaction in the core of the protein and the enhancement helical structures [27]. Additionally, TFE forms hydrophobic clusters in water, providing a hydrophobic micro environment [28]. This facilitates a movement of hydrophobic groups from the hydrophobic inside to the outside of the protein. By providing a hydrophobic microenvironment and by favouring helical structures, the protein seems to be able to rearrange within a rather compact state with low activation barriers.

4 Conclusions

Our results with this new technology indicate that specific changes in the (micro) environment can make the transition between compact structures much more likely because the protein does not have to unfold. From this finding it may be postulated that a hydrophobic microenvironment is generally a crucial factor in helix/sheet transitions. It is known that chaperones like GroEL expose hydrophobic side chains in their cavity to allow for rearrangement of misfolded structures [30]. One may speculate further that a yet unknown microenvironment might be the missing link to explain the formation of prion β-sheet structure.

References

[1] J.N. Onuchic, Z. Luthey-Schulten and P.G. Wolynes, *Ann. Rev. Phys. Chem.* **48** (1997) 545-600.

[2] A.G. Ladurner, L.S. Itzhaki, V. Daggett and A.R. Fersht, *Proc. Natl. Acad. Sci. USA* **95** (1998) 8473-8478.

[3] S. Brownlow *et al.*, *Structure* **5** (1997) 481-495.

[4] K. Shiraki, K. Nishikawa and Y. Goto, *J. Mol. Biol.* **245** (1995) 180-94.

[5] D. Hamada, Y. Kuroda, T. Tanaka and Y. Goto, *J. Mol. Biol.* **254** (1995) 737-46.

[6] D. Hamada, S. Segawa and Y. Goto, *Nature Struct. Biol.* **3** (1996) 868-73.

[7] A.K. Bhuyan and J.B. Udgaonkar, *Curr. Sci.* **77** (1999) 942-950.

[8] H.J. Dyson and P.E. Wright, *Ann. Rev. Phys. Chem.* **47** (1996) 369-395.

[9] J.L. Arrondo, A. Muga, J. Castresana and F.M. Goni, *Progr. Biophys. Mol. Biol.* **59** (1993) 23-56.

[10] D. Reinstadler, H. Fabian, J. Backmann and D. Naumann, *Biochemistry* **35** (1996) 15822-15830.

[11] K. Gerwert, G. Souvignier and B. Hess, *Proc. Natl. Acad. Sci. USA* **87** (1990) 9774-9778.

[12] K. Gerwert, *Curr. Opin. Struct. Biol.* **3** (1993) 769-773.

[13] H. Roder and M.C.R. Shastry, *Curr. Opin. Struct. Biol.* **9** (1999) 620-626.

[14] J.P. Brody, P. Yager, R.E. Goldstein and R.H. Austin, *Biophys. J.* **71** (1996) 3430-3441.

[15] J.B. Knight, A. Vishwanath, J.P. Brody and R.H. Austin, *Phys. Rev. Lett.* **80** (1998) 3863-3866.

[16] L. Pollack *et al.*, *Proc. Natl. Acad. Sci. USA* **96** (1999) 10115-10117.

[17] R. Rammelsberg, S. Boulas, H. Chorongiewski and K. Gerwert, *Vibr. Spectr.* **19** (1999) 143-149.

[18] J. Mendieta, H. Folque and R. Tauler, *Biophys. J.* **76** (1999) 451-457.

[19] K. Kuwata, M. Hoshino, S. Era, C.A. Batt and Y. Goto, *NMR. J. Mol. Biol.* **283** (1998) 731-9.

[20] A. Troullier, D. Reinstadler, Y. Dupont, D. Naumann and V. Forge, *Nat. Struct. Biol.* **7** (2000) 78-86.

[21] N. Takeda, M. Kato and Y. Taniguchi, *Biochemistry* **34** (1995) 5980-7.

[22] M. Arai *et al.*, *J. Mol. Biol.* **275** (1998) 149-162.

[23] K. Kuwajima, H. Yamaya and S. Sugai, *J. Mol. Biol.* **264** (1996) 806-822.

[24] M. Buck, *Quart. Rev. Biophys.* **31** (1998) 297-355.

[25] K.W. Plaxco, K.T. Simons, I. Ruczinski and D. Baker, *Biochemistry* **39** (2000) 11177-11183.

[26] H. Lu, M. Buck, S.E. Radford and C.M. Dobson, *J. Mol. Biol.* **265** (1997) 112-7.

[27] D. Hamada *et al.*, *Nature Struct. Biol.* **7** (2000) 58-61.

[28] D.P. Hong, M. Hoshino, R. Kuboi and Y. Goto, *J. Am. Chem. Soc.* **121** (1999) 8427-8433.

[29] L.M. Miller, C.S. Carlson, G.L. Carr and M.R. Chance, *Cell. Mol. Biol.* **44** (1998) 117-127.

[30] A. Horovitz, *Curr. Opin. Struct. Biol.* **8** (1998) 93-100.

[31] K.R. Harris, P.J. Newitt and Z.J. Derlacki, *J. Chem. Soc. Faraday Trans.* **94** (1998) 1963-1970.

COURSE 9

SOME PHYSICAL PROBLEMS IN BIOINFORMATICS

E.D. SIGGIA

Center for Studies in Physics and Biology, Rockefeller University, 1230 York Avenue, New York, NY 10021, U.S.A.

Contents

SOME PHYSICAL PROBLEMS IN BIOINFORMATICS

E.D. Siggia

Abstract

Bioinformatics is a data driven field, in which a significant number of problems require statistical modeling. The flood of data emerging from genome centers uses sequence comparison to delimit and assign function to genes, and in very limited ways infers gene control from approximately repeated sequence motifs near to the genes themselves. Traditional topics in computer science such as coding theory, natural language processing, and old fashioned cryptography all impinge on the problem of deducing regulatory information from the genome, but are not probabalistic enough to cope with the fuzziness of biological patterns. The means by which living things encode information is a problem common to both neural biology and the regulation of gene expression by the genome. Physical analogies are employed to highlight some of the problems and opportunities in this area.

1 Introduction

The intent of this very condensed summary is to stimulate the curiosity of students in the physical sciences for a nascent field where a medley of techniques are required for success. Bioinformatics is most fruitfully situated as a branch of natural science: merely publishing a clever algorithm is not enough, it has to be used on real data to solve a real problem. The most significant problems will probably emerge by looking at genome-wide data rather than reading biology texts, though they are essential. Their authors are in most cases not quantitatively trained and do not know what can be done computationally. The flood of quantitative information in the form of genomic sequences, gene expression, and protein interactions, provides for the first time in molecular biology a realm where the primary discoveries could emerge from analysis of public data.

The author's work has been supported for many years by the Division of Materials Research of the National Science Foundation.

Biological sequence data is growing at a faster rate than Moore's law. Sequencing is now an industrial enterprise carried out by robots and venture capitalists with not a graduate students in sight. Biologists flock to lectures with titles such as "Drowing in data, thirsty for knowledge" (S. Brenner & Rockefeller 2001) hoping to learn what the genome teaches us about the large scale organization of life. While there is no question that an organism's genome is of immediate utility to experiments targeting individual genes, and the comparison of genomes provides glimpses into the evolution of homologous genes, there is nothing immediately evident in the genome about how all the genes are coordinated. The various celebratory articles announcing a new genome give little more than lists of genes in this or that category by way of exegesis. It is as if those searching for extraterrestrial life obtained a telephone book (or, to be charitable, the yellow pages) of some remote civilization and tried to reconstruct the social system.

One route into the problem of how the genome defines the organism is through development and specifically how the genome dictates the expression of genes listed therein (E. Davidson). (All steps in the process by which a segment of a eukaryotic genome is transcribed into nuclear RNA; the introns spliced out; the mRNA is capped; exported to the cytoplasm; translated; and the nascent peptide chain chemically modified; are subject to regulation. The details fill several chapters of the major molecular biology texts and the student needs to become familiar with them). Among the commentaries surrounding the publication of the human genome (which will not be "complete" in the usual sense of this word for many years), was how few genes we have (30–35 k), roughly twice the number of a model plant, nematode, and fly (16 k). (Single celled organisms such as yeast have 6000 genes and bacteria have typically between 1000–4000 genes.) Man is not the center of the genomic universe, anymore than he is the center of the celestial one. The realization of the commonness of man's genomic endowment recalls an earlier "paradox" that a number of seemingly simpler organisms (salamander, tulips, and water lilies) have larger genome sizes than we do by a factor of ~ 8. The resolution of this paradox was the category of "junk DNA", that with no obvious function; *they* may have more total DNA, but we have more genes. (Because of their genetic over-endowment, the number of tulip genes may never be known directly, since they are expensive to sequence.)

The next line in defense of man's uniqueness is gene control, as revealed most clearly in development. Here at least the numbers mark a big jump in the fraction of the genome available for regulatory purposes (80% *vs.* 20%) when comparing model multicellular organisms (plant, nematode, and fly) with genome sizes in the range of 150 megabases (Mb) compared with 12 Mb for yeast and 1–4 Mb for bacteria. (For humans, genes, narrowly defined as protein coding regions, make up less than 3% of the 3000 Mb genome, and manifestly repetitive and perhaps parasitic DNA another 50%.) Thus multicellularity, at least, calls for a big increase in regulatory depth. (How this regulation is achieved is also the subject of many texts and actively studied, but suffice it to say there is no Cartesian system in the genome giving coordinates for this or that body part, but rather a seemingly haphazard medley of space and time dependent signals that define one part relative to others.)

Everyone realizes but sometimes forgets to say, that cells make cells, genomes do not. There are no genes coding for lipids *per se*, but hundreds of different lipids, specific to particular locations in the cell, are built by a variety of enzymes. The cell is highly compartmentalized, traffic between compartments is regulated, and proteins with correlative activities are clustered. Particular subsystems can be reconstituted *in vitro* from purified components, but even the biochemists would not call this life.

About references: in addition to the books cited at the end of this chapter, a number of other authors are mentioned in the text. Their past and future contributions along with abstracts can be found by searching in the *medline* data base [1] which any student must be familiar with. The references are restricted to a few common texts.

2 New technologies

Bioinformatics deals with such issues as efficient archival, retrieval, and dissemination of information (*e.g.*, gene ontologies); how to effectively compare sequence; automatically assign function to stretches of the genome (annotation); how to organize sequencing projects and assemble the ~600 base pair (bp) fragments that are the immediate output of the sequencing machines into whole genomes. A snapshot of the field can be found on the web sites and proceedings of the major meetings (*e.g.*, ISMB, and RECOMB), however the tone of these contributions is closer to a technology essential to biology rather than theoretical biology.

A number of bioinformatic problems such as locating genes in raw sequence, have a heavy statistical component [2]. Regulatory sequences pose different problems, since they occur in 100–500 bp clumps of 2–20 sites each of 5–15 bp. In the fly these so called "modules" can be up to 100 kb from the genes they control, though under 10 kb is typical. Their discovery is akin to deciphering a language without knowing either the words or the grammar and in the presence of much variability. The algorithms are again statistical and involve difficult search problems to which physicists have much to contribute.

Biology just as physics, progresses through the application of new technologies. Sequencing is one such technology whose costs have decreased (to about \$0.10 per finished base) so much that a recent cover of *Science* pictured a Noah's ark of organisms at a cocktail party discussing whose genome would be sequenced next. (Cornell is coordinating a canine project, so fido will be pleased.) There is an interesting technical history to be written about the advances in instrumentation (capillary electrophoresis), chemistry (the end-labeling dyes, and preparation of clones), computer science (the assembly algorithms), and process control (there were over 10^7 clones amplified and sequenced for the human project) that made all this possible. Success depended on engineering in the best sense of the term, since the costs and accuracies of all the technologies that intervened between organism and finished sequence had to be balanced against each other.

Another technology essential to my lectures is mRNA gene expression. Currently there are both artesanal small-lab approaches and high tech industrial ones competing for acceptance and commercial success. The technology depends on an enzyme that copies RNA to DNA, used by certain viruses such as HIV; now productively harnessed to copy in one reaction all the mRNA produced by a population of cells to chemically labeled DNA.

The problem is then how to assay the level of all 6000 potential transcripts in yeast say, which are mixed in one tube. The key is, of course, to exploit the base-complementarity of DNA. In the laboratory-scale spotted array technology [3] as applied to yeast, pairs of gene specific primers are used in 6000 separate reactions to amplify genomic DNA (with the aid of another product of biotechnology, PCR, itself made possible by another hijacked enzyme, this time from hot spring bacteria). Then $\sim$50–100 μm spots of $\sim$500 bp double stranded (ds) DNA for each gene are arrayed on a specially surfaced glass slide by a robot, and anchored down. The fluorescently labeled single-stranded (ss) cDNA (complementary to the mRNA) is then allowed to hybridize with the slide and the fluorescent level of each spot is a measure of that gene's expression. However the hybridization kinetics of ssDNA with surface bound dsDNA is unknowable, so the fluorescence is calibrated by processing a reference sample of mRNA identically to the real sample, but labeling it with another color. The color ratio is then the mRNA expression ratio. Clearly genes with similar sequences will cross hybridize and can not be distinguished.

The alternative high-tech approach to measuring mRNA levels synthesises about 20 tags of length 26 bp for each gene directly on the chip, by methods inspired by lithography in the semiconductor industry [4]. The redundancy is necessary for controls and the company supplies black-box software (that the mathematically literate would want to modify (M. Magnasco, submitted)) to reduce the multiple oligo-readings to a single number. The kinetics of hybridization again has to be calibrated by color ratios, which now go on separate chips (each costing hundreds of dollars, so this is a technology aimed at medical applications). There are still many other technologies vying for attention (*e.g.*, `http://www.rii.com/home.htm`); let a hundred flowers bloom! Because there is money to be made, lawsuits are common (*e.g.*, the saga of Ed Southern *vs.* Affymetrix).

DNA chips are a versatile genome-wide readout device. They have been utilized (P. Brown and M. Snyder) to measure where certain proteins bind on regulatory DNA, by crosslinking all proteins attached to DNA, fragmenting the DNA, extracting the protein (plus DNA) of interest with an antibody; undoing the crosslinks; and assaying the liberated DNA on an array spotted with all the intergenic regions of yeast.

3 Sequence comparison

The comparison between two sequences was probably the first "killer application" that drew many computer scientists into molecular biology, and as a measure of their success, it would be impossible to imagine modern biology without it. This subject is well described in textbooks [5, 6], so I will merely state the general ideas which recur in other problems and emphasize the shortcomings. General expositions have such a preemptory tone that the student might infer that it is a closed subject, whereas many obvious questions are not resolved and provide problems for the statistically inclined. The experts are well aware of these questions, but seldom write about them.

The first ingredient of sequence comparison for proteins is a scoring "matrix" which quantifies, for pairs of amino acids, the differing penalities to be assigned to the replacement of one amino acid by another, (slight for similar residues and large when a hydrophobic residue is substituted for a charged one). Values are derived from collections of aligned homologous protein domains where there are no gaps or deletions. Treat positions in the alignment as independent and compute $P_{a,b} = \langle \rho_a \rho_b \rangle$ where the a and b run over the 20 amino acids and the normalizations are such that $\sum_b P_{a,b} = \langle \rho_a \rangle$, the fraction of a-residues in the data set. There is an implicit time parameter τ induced by grouping with weight one, all sequences with percent identity over some value (this also prevents biases in the protein data base from overly influencing the scoring matrix). That τ indeed acts as a time within the correlation function $P_{a,b}$, can be seen from two limits; $P_{a,b}(\tau \to 0) = \delta_{a,b}\langle \rho_a \rangle$ and $P_{a,b}(\tau \to \infty) = \langle \rho_a \rangle \langle \rho_b \rangle$. One then defines the transition probability $T(a \to b) = P_{a,b}/\rho_a$ and the scoring or substitution matrix $s_{a,b} = \ln(P_{a,b}/\rho_a \rho_b)$. (Thus for short times there are no transitions, while for long times the probability for obtaining a given residue is independent of where it came from.) I have added these few details to make evident there is nothing very subtle in the construction of the scoring functions (*e.g.*, BlosumXX) that everyone uses.

Two sequences (not necessarily of the same length) are brought into correspondence and thus scored, either by making point mutations or creating gaps (or intervals of deletions) which are penalized by one parameter to create and another to extend. Global alignment finds the optimal score accounting for the entire sequence. It is constructed in a time of $\mathcal{O}$ (product of the two sequence lengths) by a recursive calculation. Place the two sequences along the x- and y-axis, and map each alignment between them into a path through the rectangle thus defined. A diagonal bond means bases (or residues) i, j are paired, a horizontal bond means i (on the x-axis) is paired with a gap, and a vertical bond means the reverse. Starting from $(0,0)$, find the best scoring path up to the perimeter of the sub rectangle defined by (i, j), and then fill in the next row and column from these values and proceed to the end of either sequence. Local alignment finds the highest-scoring subsequences in a pair of sequences in comparable time. Various shortcuts to complete pairwise comparison are essential to practical applications because there are over 10^{10} bases deposited in GENBANK, and go under names such as BLAST [7] and FASTA [8].

The first thing a biologist does with a new sequence, is compare it with the huge database of known sequences. Thus it is important to know the probability of obtaining a certain score by chance from uncorrelated sequences, which is best done by first determining the functional form of $P(s \geq s_0)$, the probabality of a score larger than s_0. This is done by a stationary phase argument (Yu and Hwa) very analogous to the passage from a microcanonical ensemble to a canonical one. One finds for large s_0, $P \sim N_1 N_2 \, \mathrm{e}^{-\lambda s_0}$, where $N_{1,2}$ are the lengths of the two sequences (or sequence times data base) being compared and for ungapped alignment $\lambda = 1$ because of the definition of the scoring function, while for gapped alignment λ has to be computed numerically as function of the gap penalties. Rapid ways of doing this akin to importance sampling in Monte Carlo have been developed by T. Hwa and coworkers. Some theory is necessary here, since a probabality has to be placed on events that are rare, but become possible when looking through a sample size of 10^{10}.

Now for the problems. The scoring function is designed for ease of computation, the iterative algorithm ignores history (prior resides on the optimal path) other than whether a gap is being created or extended. There are no block rearrangement moves for instance. The most widely used algorithms return the best local alignment (*i.e.*, entropy is ignored) rather than the probability of transforming one sequence into another in all possible ways, which is more relevant biologically even within the impoverished move-set of current algorithms. The scoring optimizes the contiguous interval with the highest total score without regard to length, but this does not mean that a shorter region of greater similarity might not give a more significant probability score under some other scheme.

The scoring parameters are not contingent on the species being compared and, more importantly, not optimized for maximum discrimination. To make this clear by analogy, imagine a substrate with patches of material (the sequences) with different affinities for water. If the regions are to be distinguished based on their ability to adsorb water, what is the optimal point in the phase diagram at which to work? Clearly the condition where water wets one substrate and not the other will provide optimal discrimination, *i.e.*, near a phase transition point small inhomogeneities can have large effects. In reality there are, of course, a continuum of substrate affinities and a cost to be paid for small domains. Nevertheless, working at a random point in the phase diagram is not a recipe for optimal discrimination.

Two other issues are addressed in part by an extension of the BLAST algorithm, PSI-BLAST. The typical scoring function is position independent, yet certain regions of proteins are more constrained than others, and they should be weighted differently (*i.e.*, the catalytic region is more constrained than the loops which tether domains together). BLAST also looses information by using only pair scores in matching against a data base. A marginal match to several *unrelated* database entries may be significant even if any pair is not. However, separate database entries for a human, mouse, and rat protein do not add much to comparisons against an unknown fly protein. So it is not trivial to put a significance measure on the comparison of several species at once.

Given a genome, the first question asked is what are the genes. Most attention has been focused on protein-coding genes; those encoding functional RNA's (*i.e.*, not messages) are very interesting, but require different algorithms (S. Eddy). The primary modeling tool is Hidden Markov Models (HMM's) [2]. To illustrate just a simple Markov model, imagine one is presented with a long string σ_i of 0's and 1's which is not obviously periodic. One might model it by letting the ith bit occur with a probability that depends on several previous ones. So in the simplest case, where only memory of the previous bit matters, the model is entirely specified by a 2×2 matrix of transition rates $T(\sigma_1 \rightarrow \sigma_2)$ where $\sum_{\sigma_2} T_{1,2} = 1$, *i.e.*, the sum of all probabilities for leaving a state, must be 1. The probability of observing 0,1 satisfies $\sum_1 p(\sigma_1) T_{1,2} = p(\sigma_2)$. Thus we have given the right and left eigenvectors of the matrix T with eigenvalue 1, which is in fact the largest eigenvalue, because all the entries of T are positive. (Under these definitions, the usual nearest-neighbor Ising model in one dimension would not be Markov since the correlations in spin are not strictly limited to a finite number of lattice sites).

Hidden Markov models originated in speech-recognition where the computer was presented with sounds and needed to infer the phonemes that the speaker was uttering. So in our context, assume there were two hidden states, coding (C) and non-coding (N), with transitions between them as defined for our Markov model. For each hidden state there are separate probabilities for "emitting" 0, 1, and it is only 0, 1 that one observes. The inversion problem has two levels; first of inferring the model parameters from data, and then partitioning the data into domains corresponding to hidden states. In the case of gene-finding, one has a large training-set where the hidden state is known, and one can fit the emission-probabilities, and also the transitions between hidden states. Then real data can be scored and probabilities assigned to where the coding and non-coding regions lie. A HMM is well suited to gene-finding since the biological structure can be built in. Promoters are followed by exons, exons by introns, successive exons must maintain a common codon-phasing, and various splice signals must fall in the correct place etc. See C. Burge and S. Karlin for the current state of the art.

The task of determining parameters directly from data is done by iteration. The basic idea is to note that a transfer-matrix -like calculation (by summing over all paths through the hidden states starting from either end) will supply the total probability for the data, given the model. Work from both the right and left ends, and compute the probability for observing a certain base and hidden state (or transition between them) at a given point in the data. A suitable spacial average of this "profile" value gives the next iterate for the parameters. (When parameters have converged or are directly fit, a profile calculation will reveal where the hidden states are.)

4 Clustering

Many problems in bioinformatics call for grouping similar things together–the task of clustering. These may be genes whose behavior is monitored in a series of chip experiments, or a series of samples of cancer tissues for which the expression of a palette of genes is observed and one wants to group the cancers into types. Clustering can be effected along one dimension as in these examples, or in two when for instance one wants to find blocks in the array of *genes*×*samples* which isolates subsets of genes that are most indicative of particular samples.

Algorithms can be categorized by a series of Levi-Straussian binaries; hard *vs.* soft (is cluster membership binary or probabilistic?); one-pass or annealed; agglomerative *vs.* divisive (do clusters grow from the primary elements by fusion, or do clusters derive by fission from larger sets?). Phylogenetic (family tree) clustering is hard, one-pass, and agglomerative. The k-means algorithm is a descent-scheme in which each point is assigned to the nearest center, and the centers repositioned to be the geometric centers of the points assigned to them. It becomes a divisive algorithm if new centers are added to eccentric clusters. Another scheme assigns a Potts variable to each element, and the coupling constant between two elements is a monotone function of their degree of similarity. As the temperature is lowered, groups within which the Potts degree of freedom is more correlated than some value, define clusters (E. Domany).

There is no clustering algorithm optimal for all problems. Often essential to success is the choice of metric. For gene expression, frequently only a small percentage of the genome has a meaningful response. If one measures the correlation between genes by summing over all experiments (assuming many are available) the real signal from a few experiments is washed out by the noise from the others. Thus the metric should weight experimental values by their significance determined from the noise level.

Given a metric, cluster membership can be based on the average pair-score of the new element with other cluster members, the best score with any single cluster member, or some other cluster-wide score which is not a sum of pairs. Clustering is bedeviled, as are many other optimization problems, by multiple local optima, and it is frequently unclear when, if ever, one has hit upon the best one. Another shortcoming of most schemes is the absence of a statistical model from which to assign significance to a particular clustering. Most algorithms will cluster random variables.

Some of these issues are illustrated by a clustering scheme for sequences developed at Rockefeller (van Nimwegen) which has an obvious bearing on motif finding and illustrates aspects of Bayesian statistics (named after an 18th-century English cleric now the object of cultic veneration [9]). Assume an alphabet of size A and letter probabilities p_a. Then the probability of a particular string of letters (n_a of each, $\sum_1^A n_a = N$) is $P(\text{data}|\text{model}) = P(n_a|p_a) = \prod_1^A p_a^{n_a}$. This is properly normalized since the sum over all possible strings of data just reduces to $(\sum_1^A p_a)^N = 1$. To compute $P(p_a|n_a) = P(n_a|p_a)P(p_a)/P(n_a)$ (the definition of conditional probability), we have to make an assumption about $P(p_a)$, namely that it be uniform, *i.e.*, $P(p_a) \propto \delta(1 - \sum_1^A p_a) \prod_1^A \mathrm{d}p_a$. To compute $P(n_a)$ we have to evaluate the integral $I(x) = \int_0^\infty \delta(x - \sum_1^A p_a) P(n_\alpha|p_\alpha) \prod_1^A \mathrm{d}p_a$ for $x = 1$ (in which case the upper limit can be replaced by 1). By homogeneity, $I(x) = x^{N+A-1} I(1)$; multiply both sides by e^{-x} and integrate from zero to infinity on x, to find, $I(1) = \prod_1^A n_a!/(N + A - 1)!$. From this we can derive $P(p_a|n_a)$ and for instance show $\langle p_a \rangle = (n_a + 1)/(N + A)$, *i.e.*, the average value of the model parameter p_a, given a finite sample drawn from the distribution, is not the most probable value n_a/N, which for instance can be 0.

To apply this to clustering, consider a large number, S, of sequences, each of length ℓ obtained by sampling M unknown frequency matrices, w_a^i, where $i = 1, 2..\ell$, $\sum_1^A w_a^i = 1$ (*i.e.*, the entries in the matrix give for each column i the frequencies of the letters). The problem is then to group together the sequences from a common weight-matrix and recover, within the errors imposed by the finite sample-size, the set of w_a^i. Consider a subset of N sequences, then the probability $P(C)$ that they were drawn from single weight matrix is the product of $I(1)$ over all the ℓ columns. A probability distribution can be defined over the entire set of S sequences by allowing all possible partitions into clusters, each with a weight $\prod_i P(C_i)$. Thus there is a competition between all ways of partitioning S things into subsets and the "energy" which favors putting sequences from the same frequency matrix together, one can show. This weighting scheme can be used, either in a one-pass phylogenetic clustering, or more correctly with Monte Carlo sampling which will generate soft clusters and allow an assignment of significance.

Intuitively, for given S, there is a limit to how many frequency matrices can be resolved (which depends also on their degree of polarization). Discrimination obviously improves if more samples from the same matrix are supplied. Finally, there is a very interesting regime where it is possible to classify most sequences if the set of M frequency matrices is known, yet it is impossible to cluster these sequences if one knows nothing about the matrices. The former problem is the one faced by the cell, since it "knows" the proteins which do the site-recognition, whereas sequence-clustering is only a problem for the bioinformatician.

5 Gene regulation

The extraction of the sites active in transcription control from the genome is a more daunting task than gene identification, since the individual protein-binding sites are much smaller than typical exons, and their arrangement is not so choreographed as the promoter-exon-intron pattern of genes. Three types of data can be brought to bear on the problem, and all appear necessary. For a single genome, one can search for repetition between the regulatory regions of different genes. The repeats can be at the level of specific strings (perhaps with a few spelling errors) or groups of similar strings that occur in clusters. In all cases it is assumed that improbability under some model implies function, and for the calculations to be tractable there needs to be some vestige of the signal on scales short enough to be searched exhaustively. (The hard cases are those where the motif is long and mutated and where there is no statistically significant signal in just a few copies.)

The issue raised above, that the cell can function by merely classifyng sites while there may not be enough copies to allow clustering, is clearly relevant here. The application of one genome-wide algorithm to yeast was discussed by H. Bussemaker.

The second source of data is comparative genomics, namely we exploit the fact that what is functional is more constrained and evolves less rapidly than what is not. The protein coding regions serve as landmarks for the regulatory regions to compare since they are much larger and evolve more slowly than the regulatory sites. In reality there are merely degrees of constraint and the scale in bp on which compensatory mutations (preserving fitness) can occur is also unknown. The ideal case is individual protein binding sites immersed in a sea of random sequence. In bacteria where the total regulatory region of a gene is a few hundred bases, the conserved domains are typically larger than a single binding site (N. Rajewsky). The current state of the art (McCue and C. Lawrence) in this area is to examine the regulatory regions for one gene from several species. One is then faced with the task of clustering sites for individual genes into families recognized, one hopes, by a single protein.

Finally there remains mRNA expression data. If the question being asked is how expression follows from sequence, there is little reason to first cluster genes based on similarity of expression, and then look for common sequence motifs. The clustering should follow from the sequence. Following the idea that the polymerase which makes mRNA is recruited to the promoter by equilibrium binding to certain sites (or other proteins attached to these sites), we have fit the log of the expression ratio, R_g for gene g, to the sum of contributions F_m for motif m by minimizing $\sum_g (R_g - C - \sum_m (F_m N_{g,m}))^2$ with respect to F_m and C, where the integer $N_{g,m}$ is the number of copies of motif m upstream of gene g. (H. Bussemaker). This scheme is sensitive to combinatorial control. Genes which do not respond, but carry a functional site, are informative about potential compensatory factors. All genes are fit and when the residuals are Gaussian it is easy to assign significance to the sequence motifs that correlate with expression.

References

[1] `http://www4.ncbi.nlm.nih.gov/PubMed/`

[2] R. Durbin, S. Eddy, A. Kroch and G. Mitchison, *Biological Sequence Analysis* (Cambridge Univ. Press, 1998).

[3] `http://cmgm.stanford.edu/pbrown/`

[4] `http://www.affymetrix.com/index.shtml`

[5] M.S. Waterman, *Introduction to Computational Biology* (Chapman & Hall N.Y., 1995).

[6] D. Gusfield, *Algorithms on Strings Trees, and Sequences* (Cambridge Univ Press, N.Y., 1997).

[7] http://www.ncbi.nlm.nih.gov:80/BLAST/

[8] http://www.people.Virginia.EDU/wrp/pearson.html

[9] http://www-groups.dcs.st-andrews.ac.uk/history/Mathematicians/Bayes.html

COURSE 10

THREE LECTURES ON BIOLOGICAL NETWORKS

M.O. MAGNASCO

Center for Studies in Physics and Biology, The Rockefeller University, 1230 York Avenue, New York, NY 10021, USA

Contents

THREE LECTURES ON BIOLOGICAL NETWORKS

M.O. Magnasco

Abstract

This course comprised three lectures whose uniting thread was the study of some biological networks of interest to the physicist: enzymatic, gene transcription and neural.

Lecture 1: Enzymatic networks. Proofreading knots: A conjecture on how DNA topoisomerases disentangle DNA. It is vitally important to living cells to be able to manage the topology of their DNA. Topoisomerases are the enzymes in charge of handling knotting and supercoiling of DNA. It was believed for a long time that they did so by permitting random strand passage, rendering DNA effectively a ghost-like polymer. But it has been shown experimentally that this is not so: topoisomerases do quite substantially better than random strand passage. This then begs the question of how an enzyme may survey the topology of a DNA strand thousands of times larger than itself. We discuss some possible mechanisms for this.

Lecture 2: Gene Expression Networks. Methods for analysis of DNA chip experiments. We outline and discuss the most important features behind "gene chips", hybridization arrays in widespread use for gene expression. We concentrate on one of the two most popular technologies, the GeneChip arrays. We discuss various methods for reconstructing RNA concentrations from the measured fluorescence in the arrays.

Lecture 3: Neural and gene expression networks. Song-induced gene expression in the canary brain. We outline the basic features of immediate early gene expression following stimulation. We show how it has been used, via large-scale histological mapping, to dissect the rules of representation of song elements in canary brain.

These lectures are about pieces of research with which I have been personally involved over the past few years. Their subject matter goes from enzymology, through bioinformatics, to neuroscience; and spans purely theoretical work, through data analysis, to experimental work. Yet though apparently dissimilar they do have an overarching theme: the study of biological networks, in their various incarnations and hugely diverse timescales and lengthscales. Networks are interesting, in that their function does not reside in any given part, but in the way the whole is assembled together–too many times do we think that biological function is carried out by a specific and single-purpose piece of hardware which does "just that".

I opened with a theme with which physics and mathematical audiences would feel probably more at ease. Topoisomerases are enzymes which manage the topology of DNA, a vital function to living beings. Experimental evidence shows that topoisomerases are somehow able to unknot DNA better than random strand crossing–*i.e.*, they obtain information about the topology of DNA. How an enzyme may survey the topology of an object several orders of magnitude larger is a clearly defined problem for a physicist. That it happens to be of some biological importance does not detract from the clarity of the definition. One could conceive a machine of measuring topology. But one could also conceive that the measurement is accomplished by the enzymatic network of chemical reactions defined by the dynamics of the enzyme, and this is the possibility explored here.

The next subject was meant to exemplify the dire need for analytically-minded people in the gene expression network area. Gene expression arrays, or gene chips, have become hugely popular tools to try to infer gene regulation interactions. I describe a bit the general ideas and then plunge into a problem we studied in detail, that of obtaining a measurement of concentration from the GeneChip arrays manufactured by Affymetrix.

I close with what I consider to the be frontier, in every sense. Neuroscience is, to my mind, the most deeply fascinating branch of science; also the most deeply disturbing. For a physicist, the extent to which the discussion is ill-defined is simply unsettling; yet the mystery is so deep that one cannot but feel excited and awed. I hope I have been able to illustrate this by means of our studies of representations of song fragments in the brain of canaries, a study which we carried out by looking at gene expression in the auditory nuclei.

1 Enzymatic networks. Proofreading knots: How DNA topoisomerases disentangle DNA

In many instances, biochemistry shows us little specific machines which undertake a particular job: they cut *one* specific bond in one specific configuration, or they take such an arrangement of atoms and rearrange it exactly *thus*. So it is usually thought that a specific job is carried out by the enzyme, just like a little clockwork thingie; and so the doctrine of *one enzyme, one function* evolved. On the other hand, we don't think of *thought* as something done on a single neuron level: it is a task collectively carried out by a network of neurons. In this lecture I will tell the story of something in between: a very specific job, the untangling of DNA, which may be a job carried out, not by a specific machine doing one specific rearrangement, but by a network. There is, of course, an underlying machine, the topoisomerase, which is the one carrying out the strand rearrangements necessary to actually change the topology; but untangling does not entail strand crossings alone, but also the ability to discriminate the topology of the system being untangled. Such discrimination between the knotted and unknotted states could not have been done by one particular machine; we propose that it is, rather, a property of the *network* of chemical reactions the topoisomerase carries out. The network of chemical reactions is structurally similar, and quantitatively behaves similarly, to the kinetic proofreading reaction networks proposed by Hopfield and Ninio to understand the specificity of biological polymer replication–hence the title of this lecture.

1.1 Length scales and energy scales

DNA is a long and thin polymer, and living beings carry a whole lot of it. In bacteria, as a rough guide, DNA is about 1000 times longer and 1000 times thinner than a typical bacterium. So if *E. coli* was the size of a small classroom, about 5 m, its DNA payload would be about 5 kilometers' worth of 5-mm-thick wire; or about as much ethernet wire as there is in a whole small building. A polymer in a fluid freely fluctuates under the action of thermal agitation, bending and writhing to the extent compatible with thermal energetics. A natural comparison between its bending stiffness and thermal energy scale may be introduced by a lengthscale, called the persistence length, which intuitively is the "typical" radius of curvature of the polymer strand when agitated by thermal motion. It is defined, equivalently, as the correlation length of the tangent vector to the polymer, or as the length of polymer that can be bent in a circular arc of a radian angle change with a cost of 1 kT worth of bending energy. The persistence length of DNA is less than a tenth of the size of a typical bacterium. Thus DNA is quite flexible within the scale of the bacterium, and it can be easily fit within one, as far as elastic energy is concerned. On the other hand, if left to its own devices, the bacterial genome would form a Gaussian loop of string about 10 times the size of the bacterium (10 000 persistence lengths' worth of DNA would like to random walk around a heap of approx 100 persistence lengths in diameter, or 10 times the size of the bacterium).

Stuffing a lot of DNA within the small confines of a bacterium is a problem, as can be seen in electron micrographs showing punctured bacteria: DNA literally geysers out of them by the loopful. But as we have argued above, confining DNA to the inside of a bacterium does not poise an elastic energy problem, but an *entropic* problem. This unusual characteristic will stay with us through this lecture: pretty much everything which we shall discuss below carries the strange flavor that bending is not so much the issue as confining entropically to a small region.

1.2 DNA topology

So there's a lot of DNA and it is bending and writhing under the kicking of thermal energy, which is a recipe for a topological nightmare. In addition to this, it should be remembered that DNA is a double helix. The individual strands are twisted around one another, about 1 turn per 10 base pairs, or half a million turns for *E. coli*. This means one strand may not be easily separated from the other. They could, in principle, if the ends were free–but free ends of DNA are chemically fragile and hard to replicate, so bacteria either carry their DNA in a loop, or they anchor DNA free ends to the wall. (We have a sophisticated structure in place at the free ends of our DNA, the *telomere*, deserving much lengthier description). In either case, the ends are rigidly held, so we can speak about the *topology* of the DNA: while a knot may be smoothly removed from a piece of string with free ends (which is the reason one should first locate the free ends in order to untangle a knot) a loop of string, or a piece of string tethered to a wall, has no free ends, and so a knot may not be smoothly removed from it. If in addition, the string has an internal structure such as being made of two twisted cables, then the number of *links* between the cables is a topological invariant.

In order to reproduce, a bacterium needs to duplicate its DNA. To do so, it shall make a copy of each of the two strands. But in order for each cell to go its separate way, the two strands need to be separated from one another–unlinked, in topological terms. Thus it is necessary, just to be able to reproduce, to perform topological operations; at least a few hundred thousand per reproductive cycle, since there are about half a million links between *E. coli* strands. These operations must be performed with utmost care, as is anything affecting the integrity of the DNA backbone.

The enzymes charged with managing the topology in DNA in living cells are called topoisomerases. The name itself requires an explanation. In biochemistry talk, a widdigly-ase is an enzyme charged with catalyzing a reaction of widdigly into something else; *i.e.*, widdigly is a substrate of the widdiglase. You don't choose to name the enzyme after whatever is in common between the before and after the catalysis states, but after the most prominent difference, that which has something to do with whatever is it that the enzyme has changed, with the reaction that took place. Now, sometimes you have stuff that is identical except in some particular way: such stuff are called isomers. In particular, topoisomers are two or more structures which are identical except in topological terms. A 10 572 base pair loop of DNA in an untangled configuration, and the identical strand tangled in a knot cannot be considered the same chemical substance, since there is no transformation short of cutting the covalent bonds of the backbone, untangling and then soldering the covalent bonds again that will change one into the other. They are thus called *topoisomers*, since they are identical except for their topology. An enzyme whose job is the catalytic transformation of one topoisomer into another is, thus, called a *topoisomerase*. Then another deep-seated tradition steps in, and topoisomerases become “topos”.

1.3 Topoisomerases

Topoisomerases are classified according to whether they do or do not use energy to perform their jobs, a very basic distinction between enzymes in the biological world. Enzymes which do not use energy are *catalysts*, passively accelerating the reaction rates of reactions that otherwise would take place naturally. They can change the timescale of the reaction (say by binding the reactants in close-together spots in their surface that favor the reaction to take place; however, they cannot change the equilibrium probabilities of finding the reactants in this or that configuration, for this is given by the Boltzmann distribution. Once the Boltzmann equilibrium has been achieved, and in the absence of energy consumption, the discussion is that of one isothermal bath and the Second law forbids moving away from equilibrium. Powered enzymes, on the other hand, can perform chemical tasks which are thermodynamically “uphill” [8]. They do so by coupling the uphill reaction to a “downhill” reaction–the consumption of some form of fuel or energy currency–which makes the *overall* reaction be downhill [9]. An example is a motor protein, which consumes energy by hydrolyzing ATP (the universal energy currency in cells) and may exert mechanical work by moving against an external force. This has been a major subject during the previous Lectures, and hence I'll refer back to them. Here I shall point out that it is not always evident what the energy being spent is being used for, or in which way; I think this lecture will provide an example.

Energetically passive topoisomerases are called “class I”, while topos which couple to energy consumption are “class II”. Class I topoisomerases catalyze the following reaction: they bind to DNA, they temporarily nick or cut the covalent backbone in one strand of DNA; they allow the free ends to rotate for a while around the un-nicked covalent bond, then solder them again. Thus they change the number of links between the two DNA chains, and they do so in the strictly energetically downhill direction: if the two strands had been overwound, type I topo relaxes this back to the equilibrium torsion. Strictly speaking, this is all that would be needed to allow a circular

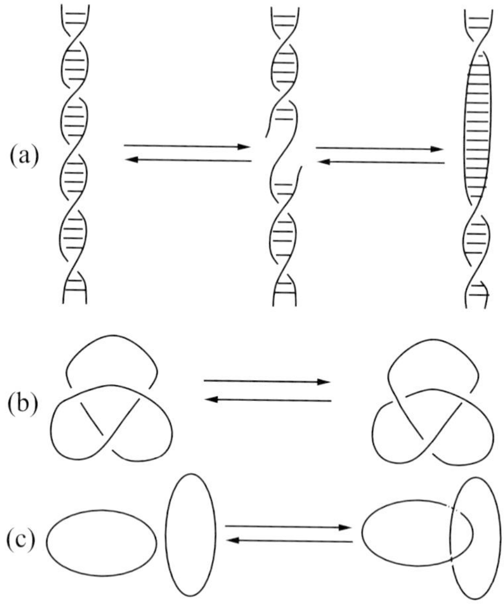

Fig. 1. Some of the topological operations performed by topoisomerases. **a)** Class I topos can nick a strand, rotate around the unnicked one, and religate, changing the linking number of the two strands. Class II topos permit double strand passage and so can **b)** remove knots and **c)** link and unlink (catenate and decatenate, in topospeak) circular strands. From [15].

piece of DNA to be replicated and the two daughter strands to be separated from one another, but there are other problems and situations to solve.

Type II topoisomerases perform more sophisticated jobs. All known type II topos allow segments of two-stranded DNA to cross through another segment of two-stranded DNA. To do so, they bind to the DNA, cut through the two strands (while keeping a hold on the ends), and let the other segment through. One specific kind, *gyrases*, perform this task in a particular way: they bind to DNA, *twist around* the DNA segment, and then they let the top segment though the bottom segment; in this way, they change the linkage of the DNA strands *always in the same direction*, rather than in the energetically downhill direction. This direction is that of *unwinding*, making DNA less twisted; this is important, for all of the machinery that needs to access the letters of DNA needs to open the strands apart, thus pushing the helical turns closer together outside of the open bubble; unwinding the DNA makes it easier to open the strands. Other type-II topos simply permit the passage of far-away segments, allowing, for instance, the disentanglement of knots.

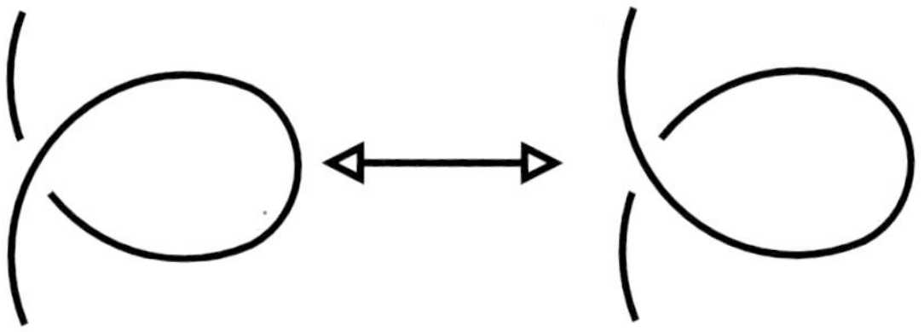

Fig. 2. This reaction is "futile" from the viewpoint of a single-stranded polymer; however, it changes the linking number between the two strands of a double-stranded polymer by ± 2. This is the reaction catalyzed by gyrases.

Now, it should be clear to the reader that topoisomerases are a matter of life and death to the organism. This may sound like an abstract idea, but it is not abstract at all. At the turn of the eighties, patient care for the treatment of gram-negative infections was completely revolutionized. Prior to that point, gram-negative bacteria, being hard to target with conventional antibiotics, were being treated with large dose intravenous antibiotics; this required a month's stay in the hospital, where the patient was bound to get more intrahospital infections. The worst, toughest, most resistant strains of any given infection are most easily found at a hospital. A gram negative infection was thus akin to some ancient curse, for patient and caregivers alike. But then, *gyrase inhibitors* were introduced into clinical treatments–fluoroquinolones, for instance, can treat the same infection in seven days, by taking one pill a day in the privacy of home, with no risk of intrahospital strains and no IV line, at a hundredth of the cost for patient care. A gyrase inhibitor operates by targeting bacterial gyrases, which are evolutionarily distinct enough from vertebrate gyrases that they can be targeted specifically. The fluoroquinolone inserts itself into the gyrase during the time the enzyme is bound to DNA and has cut one of the two strands; it immobilizes the enzyme in that state, not allowing it to move forwards or backwards along its chemical cycle. Thus, the enzyme does not perform its job, and the signal that caused the enzyme to act in the first place stays on, and more and more enzymes are sent to the job and immobilized while attempting. Pretty soon there's a large number of double stranded breaks in the DNA, held on to by gyrases with little monkey wrenches in their works. As they fail and fall apart, the bacterial genome is blasted into little pieces. This has caused literally a revolution in patient care. Needless to say, quinolones and fluoroquinolones have meant huge incomes to the pharmaceutical companies that marketed them. *Raxar*, the star antibiotic of Glaxo-Welcome, was a huge best seller for years for its ability to target rare respiratory infections. *Cipro* has been one of the best selling items from Bayer–originally for its ability to treat otherwise-resistant urinary tract infections. Recently it appeared in the cover of the New York Times every day for weeks: Cipro is the *only* antibiotic cleared by the american F.D.A. to treat *anthrax*. It became a best-selling item in the aftermath of bioterrorist threats in Oct. 2001 in the U.S., when the anthrax-laced-letter scares caused the population of New York City to deplete the city's stock of Cipro in a matter of days and caused the stock price of Bayer to soar. Thus we see a clinical application of topology: how knot theory is really a matter of life-and-death, for bacteria and patient alike.

1.4 Knots and supercoils

Knowing now of their importance, let's review briefly what knots and supercoils are. Knottedness is a topological property of the embedding of circles in 3-space. There are no knots in 2 or 4 dimensions: there is no way to tangle a loop in 2 dimensions without self intersections, while in 4-D any 1-D structure, loops in particular, may be smoothly untangled without ever going through a self-intersection. We know from basic differential topology the "dimension" formula from intersection theory, which tells us that the dimension of the intersection between two submanifolds is the sum of their dimensions minus the dimension of the ambient manifold; so surfaces and lines in an ambient 3D space typically intersect at a zero-dimensional set: a discrete number of points. So in the case of knots, if we visualize a line trying to move "though" another line that is left static, the first line traces out an object of one more dimension, a surface; and hence they typically will intersect at a discrete number of points in 3D (thus knotting) or not at all in 4D (thus no knotting). Notice that the topological dual of this situation is a point moving through a surface, which is the case of ions moving through an ion channel; thus, topologically, topoisomerases are in a sense "topological dual" ion channels, since a DNA segment bars the passage of another DNA segment in, topologically, the exactly dual way to a membrane barring passage to an ion, and a molecule must be there to permit or deny passage.

The typical way to depict knots is by their projections onto the 2D page, where the projection causes strands at different depths to appear to self-intersect; in this case the further-away strand is drawn with a break through it, to give the illusion of an "occlusion" by the nearer strand.

The simplest knot is the unknot, a knot that is not knotted. A circle is the simplest projection of the unknot, but there are infinitely many, arbitrarily complicated projections of the unknot.

I do not think there's any need to stress here that knot theory is hard–I may just refer the curious reader to the many excellent textbooks [1, 2]. I will just point out a couple of points. There is currently no known "running" algorithm that can recognize the unknot. One may compute the Jones

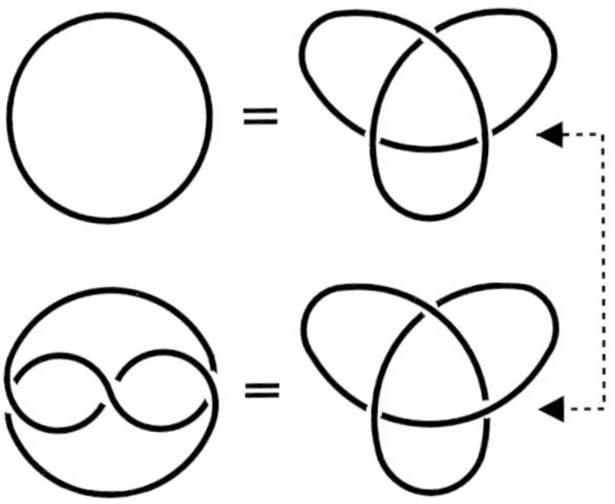

Fig. 3. Knot projections. Top line: two projections of the unknot. The left one is the "canonical" projection, while the right one can be transformed into the left one by lifting the "flap" up and untwisting. Bottom: two projections of the trefoil. Please notice that the right projections of the unknot and the trefoil can be changed into one another by switching the sign of the bottom right crossing.

polynomial, an NP-complete task [3], and check whether it's trivial; but it has not yet been proven whether the Jones polynomial (or its extensions like Homfly) do classify knots; thus there is still the possibility that a nontrivial knot may exist whose Jones polynomial is trivial, though none is known. On the other hand, there is an algorithm, due to Haken & Hermion [4], that can classify all knots; but there is no running implementation of this algorithm that I know of, and it is unclear whether it stands in the complexity hierarchy–it seems to be a *lot* worse than NP-complete! Thus, the seemingly innocuous task of deciding whether a projection of a knot is or is not actually knotted is still an unsettled business, and in the best possible current scenario (that the Jones polynomial or a relative are shown to classify) is NP-complete: exponential in the number of crossings of the projection.

An ideal, infinitely-thin object may be endowed with elastic properties. An object with a finite thickness has, additionally, torsional elasticity: the resistance of a rod to have opposite ends twisted in opposing directions.

Supercoiling results from the competition between torsional elasticity and bending elasticity–since both of them are quadratic the total energy is minimized by appointing a fraction to both rather than most to any one of them. Thus if a torsion is embedded in the polymer, it will spontaneously attempt to relieve some by *writhing, i.e.*, twisting its core into space; see Figure 4.

Supercoils are usually seen in desktop telephones, in the cables joining the handset to the body. Taking a message is a sure way to imbed torsion into such a cable: a right handed person will typically take the handset with

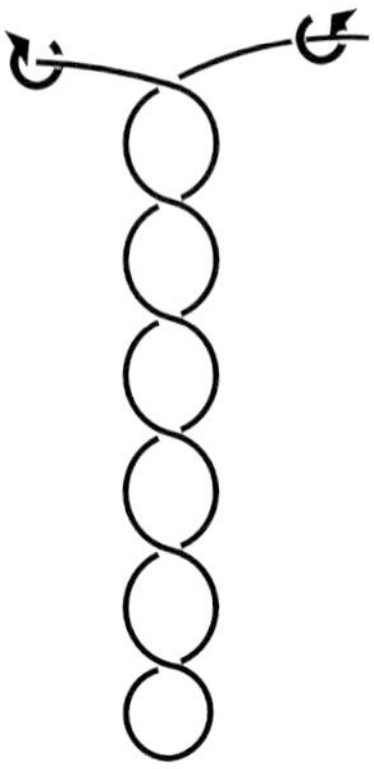

Fig. 4. Supercoiling is produced when an object with torsional elastic degrees of freedom suffers overall torsion; in this case part of the elastic energy from twisting around the axis is relieved by the axis itself *writhing* around in space. Try this with any cable with appreciable resistance to being twisted, like an ethernet cable.

the right hand, and lift it straight up to his ear. If a conversation ensues in which no use is otherwise made of his hands, he shall hang up by the reverse of the original path and nothing will happen topologically. But if he has to take a message, then he will pass the handset over to his left ear, so that the right hand is free for action, and in doing so turn the handset by half a turn clockwise (seen from the cable). After the conversation finishes, he will hang up using the left hand, since the phone is on the left ear, embedding another half turn clockwise. Thus one can generally learn the handedness of a person by looking at the handedness of the supercoils off her phone.

1.5 Topological equilibrium

A *ghost polymer* is a theoretical model of a polymer that has all the normal local properties of a polymer, but can freely pass through itself–*i.e.*, it has no interactions which are long-range along the polymer strand and hence can not "feel" self-intersections. This is the easy, "lazy" thing to do if one tries to implement a computer model of a polymer–the nice description of a polymer is *along its arclength*, and thus a self intersection, which is local in space but nonlocal in arclength, would require, in the most naive implementations, everyone-against-everyone checking. This can be avoided by more sophisticated techniques, *e.g.* 3D Delaunay tesselations, but they are a pain to implement.

One can then implement the simplest polymer–a ghost polymer with bending elasticity–in thermal agitation. If we made it a closed loop, one could then check to see how often the loop was knotted and how often it wasn't. In fact, one may do a histogram of how often the loop is found on *any* given knot. This distribution over the knots is called the *topological equilibrium* distribution [10, 11]: it is nothing other than the Boltzmann distribution integrated over topological classes, and so it is the most basic object of discussion in the statistical mechanics of DNA loops.

For sufficiently short loops the knotting probability is very small, and vanishes as $e^{-\frac{L_0}{L}}$, where L is the length and L_0 is the length at which it first becomes probable to twist so much knots can be made, of the order of 100 persistence lengths. While a circle is obliged to accumulate a total of 2π worth of curvature, and thus the integral of absolute value of curvature needs to be *at least* 2π, it has been shown that any knot has to accumulate at least 6π in total curvature. (A result derived by John Milnor at the age of 17!). This represents an energy barrier to knotting which becomes steeper and steeper the *shorter* the polymer loop is, since the elastic energy is the integral of the square of curvature differential of length, which is homogeneous order -1 in the length.

Experiments can be made to check this theory. They have been done by using equilibrium religation [12, 13] – the nice thing about an equilibrium distribution is that it does not matter how one reaches it, and so the experiments were done by letting DNA loops with "sticky ends" flicker freely between open (*i.e.*, linear) and closed (circular) configurations. At some point an enzyme (DNA ligase) is added which solders the sticky ends, thus freezing the mixture at an equilibrium snapshot. If the DNA pieces are run through a gel, they migrate through it at different speeds depending upon their topology, and so it can be quantified with exactness how much of a given topology there is in the mixture. For instance, for 10 kb loops, about 3% are trefoils, 0.1% are figure eight knots, and negligible quantities of higher-order knots; about 97% of the mixture is unknotted loops. For 7kb loops about 1.8% are knots. These experimental results were in complete agreement with the theoretical calculations, and everyone was happy.

1.6 Can topoisomerases recognize topology?

So experiment and theory were in agreement. But then, someone had the idea of checking whether this topological equilibrium was respected by topoisomerases. And thus the trouble started.

The idea people had about class-II topoisomerases was that, by allowing double stranded DNA to pass through itself, they effectively rendered real DNA into a ghost polymer. There is no topology problem in a ghost polymer. It was hard to think that they could do anything else, since topology *is a global property* and the topoisomerases are thousands of times smaller than the DNA they untangle. But becoming a ghost polymer cannot alter the topological equilibrium distribution, since it is just given by the Boltzmann distribution of the elastic polymers. So, adding type-II topoisomerases to a topological equilibrium mixture should not change the distribution, according to these thoughts.

But of course it did. Type-II topos *strongly* suppressed knots, as well as supercoil density fluctuations. Let's look at knotting first.

	Equilibrium	Experimental [5]
10 kb P4	0.031	0.00062
7 kb PAB4	0.017	0.00019
nm P4 Links	0.064	0.04

So from this experiment the suppression of knotting and linking can be as large as a hundredfold. Pretty impressive for an enzyme many thousands of time smaller than the DNA it is unknotting. Furthermore, titration of enzyme concentration showed that full activity was being reached with an average concentration of only one enzyme per DNA plasmid; thus this effect is not the outcome of collective interactions.

Now, the fact that a deviation from the Boltzmann distribution is effected is no problem, since class-II topos consume energy in the form of ATP to perform their job. The Second Law is not the one at risk here. The problem is that topology is a global object, while the enzyme acts locally. It is even more insulting when we think that we do not have a good solution to the unknotting problem.

1.7 Proposal: Kinetic proofreading

Now, the problem rapidly acquires twists. The suggestion in [5] is that somehow topos recognize a few specific configurations and only effect strand passage on them. A suggestion then developed further by Vologodskii is that topos may bend DNA locally into a hairpin, and then strand passage *into* a hairpin would be more likely from a knotted than from an unknotted state. This proposal has its own problems which we will comment upon later. For the time being let us just look at one feature. If the topo binds to DNA in one place, and allows a different segment to cross through it, the overall rate at which this happens will be computable as an integral over all possible locations of the first and second segment–the path integral decomposes into a double integral. Now, it is known that no double integral can discriminate a knot from an unknot; the linking number can be computed as a double integral, but it diverges when evaluated in a single loop rather than on two disjoint loops. No equivalent for knottedness, no matter how crude, has been developed. Higher order binding rapidly becomes complex, and disagreeably chiral. Chirality is of vital importance, since there are *two* trefoils, the right handed and left handed ones, and *both* seem to be strongly suppressed in the experiment, so we have to find a chirally-insensitive mechanism.

It looks like, simply stated, there is a "ground" state (unknotted) and "excited" states (knots), and a way has been found to focus the system onto the ground state. There is one biochemical mechanism that does just that, and it is called *kinetic proofreading*. It was discovered/invented by Hopfield [6], and independently by Ninio [7], in an attempt to explain how the accuracy of DNA replication comes about. I will detail the mechanism below, but the key point is that the signature of kinetic proofreading is the *squaring* of the error rates: if the native mechanism had an error rate of 1%, then by repeating it twice independently in a kinetic proofreading scheme it becomes 0.01%.

Now, examination of the previous table becomes suggestive the moment we add a new column with the *square root* of the experimental data:

	Equilibrium	Experimental [5]	$\sqrt{[expe]}$
10 kb P4	**0.031**	0.00062	**0.025**
7 kb PAB4	**0.017**	0.00019	**0.014**
nm P4 Links	**0.064**	0.004	**0.063**

i.e., the square root of the experimental data is almost *unreasonably* close to coinciding with the topological equilibrium distribution. Even if our detailed model below is wrong, which it might very well be, it sounds like too much of a numerical coincidence to have close to a square of the Boltzmann number for there not to be some form of two-collision process at work.

1.8 How to do it twice

The kinetic proofreading proposal works as follows. Naively, one always expects that if there is a test to weed out unwanted stuff, if the test fails to detect an undesirable with probability p, then repeating the test twice will fail with probability p^2. This is, of course, subject to all sorts of caveats, including the tests being fully random and statistically independent of one another; *i.e.*, for *any* undesirable, it must be the case that it is detected with probability p.

The problem is how to implement this in chemical reactions. In equilibrium statistical mechanics there's no such thing as doing it exactly twice, since everything is a random walk: any attempt at doing it twice will result in doing it once with some probability, and doing it thirteen times with some other probability. This is evident in the specific setting that Hopfield discussed: assembling a biopolymer such as DNA so that the right letters are copied faithfully. Let's say the global strokes are such: there is a letter soup containing the right and wrong letters. Both must be there because next letter will be different and so we need a soupwith all letters. We have a substrate, and then we want to incorporate the right letter to the substrate at some rate:

$$\begin{aligned} S+R &\leftrightarrow SR \rightarrow^{\vartheta} \text{right} \\ S+W &\leftrightarrow SW \rightarrow^{\vartheta} \text{wrong} \end{aligned}$$

the discrimination here is carried out at the level of the double arrow, which establishes different equilibria for the SR. Now, in the case of DNA replication, it can be argued that the left arrow in the reaction has the *same* rate at the top and bottom reaction: the letters in the soup don't have a clue as to what letter is being copied, and thus arrive at the same rates, without regards for whether it's the right or wrong letter. What does change between the right and wrong reaction is the *back* arrow: the right letter sticks longer than the wrong letter. It sticks exactly $\mathrm{e}^{\Delta E/kT}$ longer than the wrong letter, to be precise, where ΔE is the discrimination energy. Then if ϑ is sufficiently slow, then enough time is given to the back reaction to equilibrate and the overall rate at which wrong product is being incorporated is $p = \mathrm{e}^{-\Delta E/kT}$.

In order to proofread these reactions, there must be a way in which the discrimination can be done *twice*. But the discriminating step is the unbinding part, the back arrow; how can this be done twice? Only by adding a new state, SR^*, and allowing this new state to dissociate–otherwise there's no way to add another dissociation

$$\begin{array}{ccc} S+R \leftrightarrow SR \leftrightarrow & SR^* & \rightarrow^{\vartheta} \text{right} \\ & \updownarrow & \\ & S+R & \end{array}$$

and a similar set of reactions for the wrong substrate. But the problem is that the moment we put a down arrow to allow SR^* to decay to $S+R$ again, we *necessarily* must put an up arrow. Therefore, allowing the exit path that permits the reaction to be carried out twice allows the entry into the second stage of the reaction *directly*, without ever doing the first check. When all is said and done, when all reactions satisfy all necessary energetic constraints, no major improvement is achieved from this mechanism: instead of getting a *geometric* improvement like p^2 we get an *algebraic* improvement like $p/2$. This is all good and sound: there should be no way to cheat Boltzmann in an equilibrium situation.

Hopfield and Ninio's observation is that energy expenditure breaks us away from these limitations. Notice that the $\updownarrow$ is not the only arrow we added to the reaction diagram: we necessarily added a $\leftrightarrow$between the SR and SR^* states. In an equilibrium situation this is the no-win place: all forward arrows starting at the left $S+R$ and ending at the bottom $S+R$ must multiply to one to satisfy detailed balance, and simile for the back arrows. Otherwise we would be gaining energy going through a loop, which is forbidden by the first law. Making SR^* higher in energy than SR diminishes the chances of entering through the vertical pathway, but in so doing slows down the horizontal pathway by exactly the same amount because of this balancing requirement. But if we couple the transition from SR to SR^* to degradation of ATP, for example, we shall be exempt of this restriction: we can make the $SR \to SR^*$ reaction effectively unidirectional.

The effect of this energy expenditure is deep: we can then make the SR^* state a higher-energy state than SR, and so make entry through the second pathway arbitrarily difficult.

$$\begin{array}{cccc} S+R \leftrightarrow & SR^* & \to^{\vartheta} \text{right} \\ & \uparrow & \\ S+R \leftrightarrow & SR & \end{array}$$

while the quotient of left and right rates may be the same for the top and bottom arrows, their individual values now will be scaled by a factor of the exponential of the energy difference between the SR and SR^* states.

1.9 The care and proofreading of knots

This is all fine and dandy, but the question remains of how we could possibly proofread a knot. In order to know how to do it twice, we need to know how to do it once. Let's first imagine a topological transition in the ghost polymer setting. The transition between the unknotted U and knotted state K goes through a self-intersecting or singular state S:

$$K \underset{\lambda}{\overset{\kappa}{\rightleftharpoons}} S \underset{\nu}{\overset{\lambda}{\rightleftharpoons}} U.$$

But we have theorems telling us that there is no way to know, when we are in the state S, which way lies the knot and which way the unknot. Thus λ must necessarily be the same in both decays; this feature is going to give our diagrams a weird symmetry. In physical terms, this symmetry comes about because our problem is purely entropic. If elasticity was important, then the strands would be pressed against one another trying to untangle the knot and it would be possible to know which way the unknot lied. This is the most important difference between this model and the standard proofreading models of Hopfield and Ninio: in these models, there was a difference in *energy* of discrimination; in our case the energy of the intermediate state is the same, we must proofread on the basis of *entropy*. But of course this distinction is illusory: while a forward arrow is an energy and a backward arrow an entropy, looking at the diagram in reverse simply exchanges them.

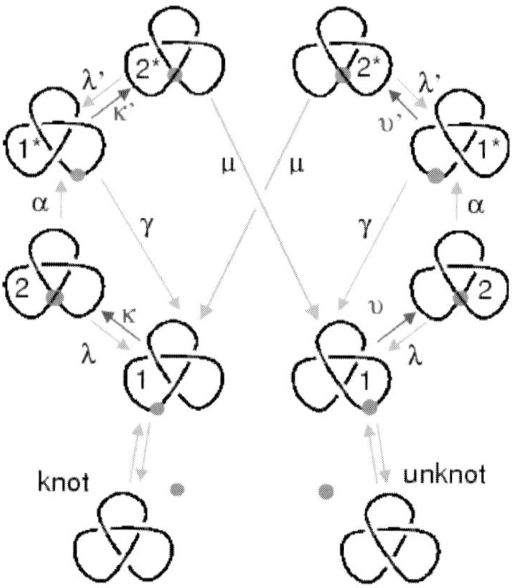

Fig. 5. The kinetic proofreading model for topoisomerases. Notice all rates are mirror-symmetric *except* for the crossing attempts, which are given by κ, κ', ν and ν'. Notice also that the path from right to left crosses both κ and κ'. From [14].

This diagram must be complicated a little bit because the state S going from K to U is distinct from the state S' going from U to K; this distinction comes about because the topo grabs one DNA segment first, and allows the second segment to cross, but cannot chemically allow it backwards: its clamp has an entrance and an exit which are distinct.

In order to proofread we must do something twice that we were doing only once before. What we were doing once is that a DNA segment bangs against a second segment in an attempt to cross it; there happens to be a topo sitting there, and it lets it through. So we need to allow *two* bangs, two crossing attempts; but we couldn't possibly *if the topo lets the segment through on the first try.* So the proofreading model simply says: the topo will insist on getting *two* attempts at crossing (within a small time period) before allowing it.

Thus the enzyme must be able to count to two, and so needs a physical substrate for a one-bit memory. There's of course a large number of different ways that bistability can be built into such a system: enzymes can have more than one conformational state (like ion channels) or can be subject to reversible posttranslational modifications, like phosphorylation. This "bit" has to be strongly coupled to other mechanical properties of the enzyme, since it is on the basis of this bit that the second strand is allowed passage or isn't. In this particular case, the "bit" need only exist during the duration of the enzyme-DNA complex, since the proofreading scheme does not require memory "across" instances of the complex. Furthermore, the chemistry of the topo itself already has a "bit" of information, though unused: whether the covalent backbone of the segment has been cut or not. Thus one possible physical embodiment of this model proposes itself: the "bit" is whether the segment has been cut or hasn't, and so the proofreading translates to stating that segment cutting is triggered by a crossing attempt. A further attractive feature is that if the "high energy state" of proofreading is the DNA segment having been cut, then evidently there's no strand crossing through the low energy state–the segment is still uncut; and there is a need for a γ built in, since we do not want to leave cut DNA lying around too long. Thus this particular implementation of proofreading is attractive because of the simplicity with which everything falls together; but Occam's razor is dangerously blunt in biology so we should not make that much of it.

There is a regime in which the choice of all the rates becomes immaterial; the analytic solution to this diagram is

$$\frac{P_{\rm knot}}{P_{\rm unknot}} = \frac{\nu\nu'}{\kappa\kappa'} \frac{\gamma(\lambda' + \mu) + \kappa'\mu}{\gamma(\lambda' + \mu) + \nu'\mu}$$

so when γ and λ' are much larger than the other quantities the resulting ratio becomes $\kappa\kappa'/\nu\nu'$ which is $P_{\rm eq}^2$, the square of the topological equilibrium probability. All the rates have fallen off the equation! Figure 6 shows then the agreement between this model and the data.

In other words: the model, while still local in space, by insisting on receiving two independent crossing attempts has become nonlocal in *time*.

1.10 Suppression of supercoils

An interesting aspect kinetic proofreading is that the proofreading property is a function of the *network* of reactions, and not of any individual reaction. Furthermore, this network works out to have a neat property: the squaring of probabilities looks exactly like squaring the Boltzmann distribution, which can be done by doubling the energies or halving the temperature. In any case, even though the *internal* arrows of the diagrams are out-of-equilibrium, the diagram as a whole works out to be a pseudoequilibrium, detailed-balance-respecting gizmo.

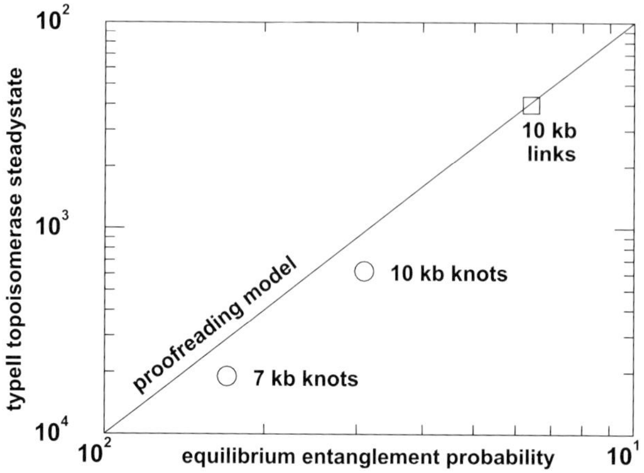

Fig. 6. The proofreading model predicts a quadratic improvement. This plot shows the knot data from [5] graphed against the square line. Please note that this *is not a fit* since the solution of our model *has no free parameters*. From [14].

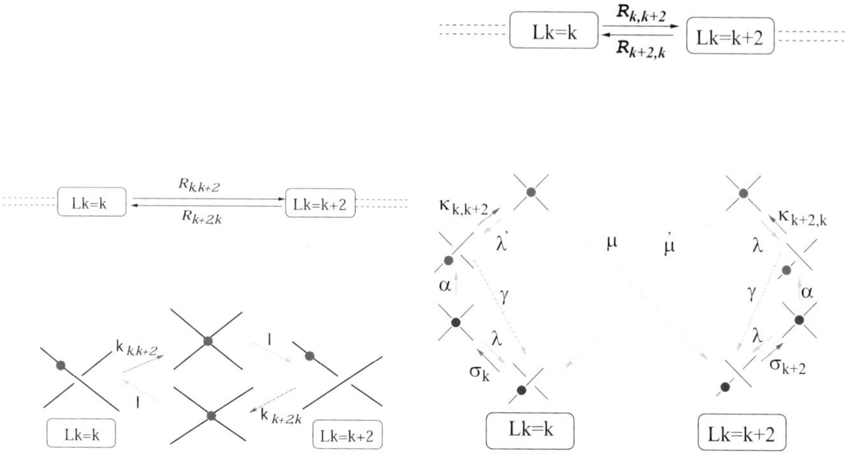

Fig. 7. An infinite stack of linking number change reactions, and how to proofread them. From [15].

Therefore, if all of the relevant reactions are proofread, we can then *kinetically proofread an infinite stack of reactions* and have it work out as if it was still in detailed balance. This is the easiest out-of-equilibrium kind of system one can conceivably get. Notice that because of the temperature-halving analogy, the moment one reaction is not proofread we have a system with two equivalent temperatures, and we're stuck with a full, hideous, out-of-equilibrium, probably-intractable model.

But if all the reactions in an infinite stack are proofread, we just need to square the equilibrium probabilities and we're done. This is the case for linking numbers and supercoiling. In the range of the experiments, the supercoiling energy looks just like a parabolic potential deviation from some minimum: $E = E_{\mathrm{D}}(Lk - Lk_0)^2/2$ where E_{D} is the supercoil discrimination energy and Lk the linking number. Thus the probabilities for the equilibrium model would be Gaussian, and their squares would also be Gaussian, with exactly the same energy form except that the width of the Gaussian distribution will be reduced by a factor of $\sqrt{2}$, or, equivalently, the supercoil discrimination energy would be doubled. Interestingly, the supercoiling data from Rybenkov *et al.* [5] shows quite a good Gaussian behaviour, which can be fit to a supercoiling discrimination energy which is about 1.9 times the actual one.

Thus we can say that all the quantitative data currently in existence is compatible with a kinetic proofreading model; furthermore, we cannot emphasize enough that in this model there is no freedom to fit anything, so we cannot dismiss the agreement between the model and the data on the basis of the scant number of datapoints.

1.11 Problems and outlook

The only competing model to ours which we are aware of has been championed by Alex Vologodskii. As mentioned before, the idea is that topos bend the segment to be cut into a hairpin shape; since they only allow passage of the second segment from the inside to the outside of the hairpin, this is an out-of-equilibrium reaction capable of recognizing knottedness. This model has currently two problems: first, there is no analytical treatment showing the model to be capable of the *amount* of knottedness suppression shown by both experiment and our proofreading model. So even though it sounds plausible that such a mechanism might distinguish knots from unknots, it is not yet known if it can agree with the data. Second, it is unclear why it should suppress supercoiling fluctuations, and Volovodskii's team has not studied this issue. On the other hand, in support of the model, it has been shown that EM pictures of topos attached to DNA bend the DNA locally; which as evidence is somewhat slim, since almost anything that binds DNA will bend it, especially after freezing to produce EM pictures.

Vologodskii has raised in turn a serious objection to our model. Noticing that the model depends crucially on the assumption that the first and second crossing attempts are of the same topological type, he claims that

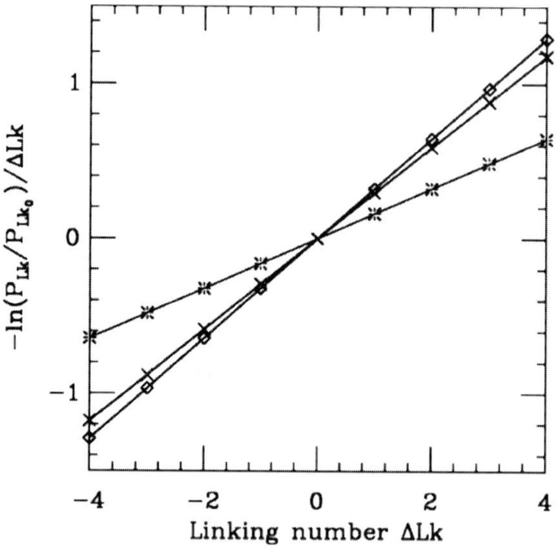

Fig. 8. Comparison between the theory and the data. These plots should be straight lines if and only if the distributions are Gaussian. The slope of the line reflects the width of the Gaussian distribution, which in turn is the supercoiling discrimination energy. The three datasets are the equilibrium distribution, the proofreading model (in which we just multiplied the equilibrium distribution by two, and the experimental dataset, which is in quite close agreement with the model). Once again, this is not a fit, for the model gives a parameterless function with no freedom for fitting. From [15].

this implies that the vast majority of crossing attempts must be topology changing for our model to hold.

His computation of crossing attempt rates shows that the vast majority of such attempts do not change the knottedness state. This part is of course true. Given a circular polymer of only a few hundred persistence lengths, the primary crossing attempt is when the polymer acquires a figure-eight shape; or perhaps I should say "hourglass shape" to avoid confusion with the figure eight knot. Crossing in that state does not lead to knotting–but it is the primary means of changing supercoil number as in Figure 2, and is taken into account in the supercoiling analysis of the previous section, which agrees in detail with the experimental data. Second, the regime in which our knotting analysis is correct is the limit in which the two crossing attempts are close to one another because the de-excitation rate γ is large. Of course, the time required for the polymer to change conformation from a trefoil-like conformation as shown in the figures to an hourglass should be quite bigger than the time between two successive crossing attempts in either state.

1.12 *Disquisition*

I like this model particularly because of two reasons. The first one, is that we have an implementation of a rather precise function in molecular biology which is not being carried out by an enzyme which does exactly this or that–it is the outcome of the dynamics of a network of chemical reactions, and not a result of any individual reaction.

Second, because the model presented here could very well be *wrong*. It may sound strange, but the thing that I personally miss the most from my life as a physicist is the ability to be wrong, which stems from the ability to make a model possessing unambiguous predictions which can be checked against experimental reality without arguing room. Our problem is such: topoisomerases either wait for two bangs or they don't, and this can be, in due time, checked. Rarely this happens in the interaction between physics and biology, because rare is the time when theory can take a leap in front of experimentation–most usually one is left fitting experimental data with models which have seventeen parameters too many.

2 Gene expression networks. Methods for analysis of DNA chip experiments

I will give a fast and loose description of the regulation of gene expression, gene chip technology etc. This introduction is meant to whet the appetite of the physicist considering studying this fascinating branch of technology; it is deliberately fast and loose, so much that biologists may feel annoyed by the lack of precision. The reader interested in continuing the study of this subject are well advised to get into a real textbook on genes and reading the many reviews on the subject of gene chips, like [18–20].

2.1 *The regulation of gene expression*

Cells encode in their genes proteins which carry out the various tasks required to stay alive, be it to digest foods, detoxify dangerous chemicals, or sense and process information. They adapt themselves to circumstances by changing the amounts or even the kinds of proteins which they deploy; a first line of intervention to change these levels is to change the amount of RNA transcription for a given gene.

$$\text{DNA} \quad \overset{\text{transcription}}{\longrightarrow} \quad \text{RNA} \quad \overset{\text{translation}}{\longrightarrow} \quad \text{protein.}$$

A gene, in the genetics sense, is a unit of heredity. There are various inequivalent ways to define such things and the details gets messy if one tries to be rigorous, so we won't try to here. In common parlance, a gene is a "functionally meaningful" region of the genome, and as such defined by its sequence and its position within the chromosome; it has a "coding" region, which is the one in which the code for the gene product (a protein) is spelled out; around the coding region there are "control" regions in which various little sequence snippets act as landing pads for the elements of the transcription machinery and its regulatory entourage. The latter consists of various proteins which either enhance or diminish the chances of getting the transcriptional machinery to transcribe the gene; these are the activators and repressors of transcription. Coding and control refer, in this context, exclusively to the transcriptional process, for the copy of the coding region which we call the RNA transcript contains various control elements for everything that happens later.

So, through transcription, various substrings of the DNA sequence get copied each into individual RNA pieces; the little pieces of RNA for various different genes then float around, in various abundances, and get shuttled around, processed by splicing and other alterations before being used for translation; any particular gene may have from zero or one RNA transcripts to thousands of identical transcripts in any given cell. The transcript abundance for a given gene is established by the competition between two processes: transcription generates more transcripts, while RNA degradation destroys them. Degradation is less specific than transcription, but not unspecific: the RNA transcripts contain sequences which target them for degradation at various rates. Some transcripts are very rapidly transcribed *and* degraded, establishing thus a non-equilibrium steady state that can be controlled on very fast timescales. This is the case with various information-processing enzymes like kinases, whose transcripts have half-lives in the 0.2 to 2 hour range. There are various transcripts which are very slowly degraded, like those of various ion pumps and the like, whose half-life may be days.

The process is similarly complex for the rest of the diagram, *i.e.*, protein synthesis. Zooming in on any portion of this diagram would reveal many complexities we have glossed over. The one piece we shall need to concentrate on is the region above mRNA: the control of mRNA abundance through the control of the arrows around it.

This picture suggests a dynamical network of control for any particular gene. One should imagine a diagram like the above for *every single gene*; these diagrams are then strung together by various interactions, because all arrows in all such diagrams are effected by proteins which are themselves controlled and may be in various states of activation etc.

		DNA	
		$\downarrow$	transcriptional
		transcription	control
		complex	
		bound to DNA	
		$\Downarrow$	*transcription*
		RNA transcript	
		$\downarrow$	splicing etc.
RNA	$\leftarrow$	mRNA	
turnover		$\downarrow$	initiation of
		mRNA in	translation
		ribosome	
degradation		$\Downarrow$	*translation*
of misfolds	$\leftarrow$	peptide chain	
		$\downarrow$	processing
protein	$\leftarrow$	protein	and folding
turnover			

The function of a given protein may be to cut one specific bond in a particular sugar, in which case we call it an *enzyme*; or it may be to bind to the little snippets of DNA sequence and help or prevent the assembly of the transcriptional machinery, thereby inducing or repressing expression of the gene where the snippet lies. In the latter case we call it a *transcriptional regulator* or transcription factor. Gene regulation networks are the networks of interactions caused by all proteins which are transcriptional regulators of other proteins (perhaps including themselves), of which there is a fair amount. Since a transcriptional regulator can regulate many genes and any gene may be (and usually is) regulated by many factors in a combinatorial way, the transcriptional network is capable of sophisticated behaviour; since regulators may be activated or inactivated through processes like phosphoryllation by elements outside the transcriptional network proper (like protein kinases) the network reacts to outside inputs. Regulation of genes not in the network is then the output of this network.

All long-term changes in living beings are mediated through transcriptional regulation programs. The differentiation of genetically identical cells into distinct cell types, like liver cells or muscle cells, is mediated through flip-flop-like switches of transcriptional regulation; the long-term changes to synapses in brain cells that mediate our memory are supported through gene regulation circuits, as will be described in the third part of this course.

2.2 Gene expression arrays

Traditionally it has been extremely laborious to figure out pieces of transcriptional regulation circuits; even now, when in possession of the whole genome sequence, data mining techniques have failed to pop things out brightly, but rather provide long lists of possible suspects to be confirmed by the slow traditional methods [21]. We do not know the whole complement of transcriptional regulator binding sites, we do not know any precedence rules stating, if an activator and a repressor are active at the same time, which one prevails, or, obviously, any of the higher-order combinatorics. The picture is complex because even history effects have to be taken into account. The set of binding sites for the transcription factors for a given gene is *more* than a "logic gate" reacting instantaneously to the inputs, since the chemistry of binding permits, for instance, history effects: overlapping binding sites exclude simultaneous binding by the respective proteins, in which case the factor which was turned on first will bind, and prevent the one turned last from binding. Finally, it bears mentioning that fluctuations are an essential part of this picture since we are not dealing with a mass-action system here: these are single-molecule systems virtually by definition.

It would be interesting to attempt to start reverse-engineering the circuits from measurements of the behaviour, on the assumption that genes whose expression is temporally correlated have a large chance of being co-regulated. This assumption is naive, but the best shot we currently have at a problem whose complexity is otherwise overwhelming, and whose importance overshadows most of Biology.

Gene expression arrays are solid surfaces onto which pieces of DNA have been attached in spatial patterns. This pattern is arranged as an array of regions or spots; within a spot the DNA is chemically homogeneous. Spots may be 1–100 μm and contain many millions of identical copies of DNA. When a drop of fluorescently labeled RNA is placed on such a surface, the individual RNA molecules will bind to the DNA complements attached to the surface. The binding will be sequence-specific: the RNA is expected to bind extremely well to its exact complement, while little or not at all to completely different sequences. The drop is then washed away together with any unbound pieces of RNA. When the array is viewed under a fluorescence microscope, the spots will glow in direct proportion to how many pieces of fluorescently-labeled RNA are bound to its DNA. A measurement of spot fluorescence is then a proxy for a species-specific measurement of RNA concentration.

There are two main kinds of array in existence, following a divide between the *do-it-yourself* approach versus the *ready-made* approach, that somehow mimics similar divides in other areas like operating systems

(Windows *vs.* Linux for example). For a clear introduction (aimed mostly at the biological public) see [18].

	spot arrays	GeneChip (R)
who	do it yourself	buy if from Affymetrix
for	your favorite animal, your favorite tissue	commercially important animals (human, rat...)
what	full-length cDNAs on a glass surface	short (~25) bp DNA oligos on a glass surface
how	deposit a drop from a test tube and let dry	photolithographic chemical synthesis
how many	as many as you have patience for: 100 → 10 000 cDNAs, hopefully distinct	as many as will fit in the wafer at given feature size: 500 000 features at (30–40) features per gene (now)
source	you make the library	sequence database
cost	with a library, 2 $ each + your copious time	about 2000 $ each (Fedex'ed) minus university discounts

Spotted arrays are home-made. Their popularity took a great boost when Pat Brown's group published (open-source style, [24]) the plans and specs for a robot device costing about 13 000 $ that would make batches of hundreds of arrays from cDNA libraries in standard 384-well plates. The robot operates simply by dipping a small array of metal pins into the little tubes, and then impacting the pins upon a glass slide (the same kind used for microscopes). The robot repeats the operation through an array of glass slides, then changes the tubes, until the collection of tubes is exhausted. As the fluid droplets dry, the cDNA from the libraries dries on the glass and somehow bonds to the surface. At any given research institution, it's likely that one of these robots will have already been built at some central facility (or at a nearby institution), so the main expense is the creation, normalization and curation of the cDNA library. This is an arcane branch of black magic so we shall not dwell upon it here–just remark two important points. A cDNA library contains cloned pieces from the expressed mRNA in the cells of the tissue/animal the RNA was purified from. As such, the sequence is unknown, so the data is labelled by spot number and eventually points to a test tube. If something interesting is inferred–well, a bit will be drawn from the tube and get sequenced. The other point is that one may end up with multiple copies of the cDNA clone in many different tubes; which is not known in advance. A process called "normalization" attempts to factor out relative mRNA abundance, but it introduces noise and fragmentation into the collection. Another process called "substraction" attempts to make a library enriched for the mRNAs present in one tissue but not in another; it also generates a lot of noise.

A company called Affymetrix has a broad patent covering a range of techniques having to do with putting lots of different DNAs on solid surfaces. The patents are so broad no one else has been able to sell DNA on a solid surface, so Affymetrix enjoys a dominant position in the marketplace. Affymetrix manufactures RNA hybridization arrays under the trade name *GeneChip*; they are the other popular array, perhaps the most popular by "arrays used" count. The technology is a photolithography adaptation of solid-phase synthesis in broad use for sequence-directed oligonucleotide synthesis. Because photolithography works in parallel, the number of spots (called *features* or *probes* in this context) is not a concern. But the synthesis is directed by sequence, and so the sequence of every spot must be known in advance so that the photolithographic masks can be laid out. Also, since about 4 masks will be needed per position, the sequence can't get too long: a couple dozen letters is a practical limit. This poses a conundrum: short DNA segments are not expected to have the sharp complementary-sequence specificity of longer DNA fragments, and cross-hybridization is expected to happen. To solve this, Affymetrix uses a two-fold approach in GeneChip arrays. First, a differential signal is generated, by taking the difference between a "perfect match" sequence (PM) and a "single mismatch" (MM) obtained by replacing the middle letter in the PM sequence by its opposite letter. Both probes together are called a *probe pair.* The rationale behind this construct is that MMs will bind less well the target sequence, but get full-strength all of the cross-hybridization and other physical noise sources, so the difference between the PM and MM should eliminate cross-hyb. Second, redundancy is introduced, by tiling the target gene sequence with several (sometimes overlapping) PMs. Current chip versions use 16 to 20 PMs of length 25 base pairs; letter number 13 is then changed to its complement to generate an equal number of MMs. The whole set of probe pairs tiling a target is called a *probeset.* The sequences are considered by Affymetrix to be proprietary information and are *not disclosed* to the public like us.

Thus, in GeneChip arrays, we get between 32 and 40 numbers (the brightnesses at each probe in the probeset), out of which we need to reconstruct a single number, the mRNA concentration. There's obviously infinitely many different functions of 40 variables returning identical values for "ideal" measurements, but having inequivalent noise rejection properties on imperfect or noisy data. One standard algorithm is provided in the Affymetrix software suite, and many researchers are completely unaware of its shortcomings or even that it may be bypassed and your own favourite algorithm used instead. I will describe the problems we encountered when studying this issue in the last section.

2.3 Analysis of array data

So, how is this data then used? There are two prototypical experimental designs: time series and condition clustering. Time series works as follows. A culture of cells (say, fibroblasts, or yeast cells, or...) is "synchronized", *i.e.*, all cells are brought to an appropriately similar state. In the case of fibroblasts, they may be starved for a particular serum growth factor; or yeast cells may be arrested at a given stage in their cell cycle. The cultures are then given the appropriate "start" signal, be it the growth factor or nutrient. Samples of the culture are taken periodically, their RNA extracted, amplified and fluorescently labeled, and then hybridized on the chips. The output of such an experiment is an $N \times M$ table of numbers, where N is the number of spots on the array or genes being probed, and M is the number of time points; similar gene expression patterns are then grouped by similarity using clustering techniques (along the N gene directions), and sorted by the relative time order in which activation or repression happens. The hope here is to observe a cascade of transcriptional events unfold.

An apparently simpler design is one in which a number of dissimilar samples are thrown together. For instance, one may collect a number of clinical tissue biopsies may be collected, say polyps from colon biopsies and nearby healthy tissue. In this case, not only are genes clustered together by similar expression profiles, but also the *experiments* get clustered together. In this case one is trying to get a transcriptional signature for a classification of colon tumors: hopefully the top level of the clustering will divide healthy from cancerous tissue, and subsequent branches of the clustering will reveal distinct tumor types. Clustering analysis was fairly well established already before gene chips, but they have provided a strong impetus and so a flurry of new methods has appeared [42].

Analysis of array data differs strongly from established time-series analysis methods, because the data has the wrong aspect ratio for proper time-series analysis. For example, a flurry of methods for dynamical system identification were created in the "chaos" boom of the seventies and eighties. Most of them require a number of time slices which increases *exponentially* with the dimension of the attractor to be reconstructed [23]. This is because high-dimensional spaces are exponentially large: they require many many points to be "filled" so that their volume is sampled throughout. (Consider how many "corners" a hypercube has). But in array analysis, the aspect ratio of the data is all wrong: best cases we know of involve about 10^4 genes in 10^2 experiments. Notice that the number of experiments is not only smaller than the number of genes, but is smaller than the square root of the number of genes. Because of this. it has become apparent recently that the full $N \times M$ data set may be too large an object to analyze *together*. Clearly, even if we had uncorrelated, Gaussian white noise as our only noise source, as we cluster along the "genes" direction, the residual noise goes like the square root of the vector dimension; so we get a disadvantageous signal to noise ratio, $S/N \sim 1$ simply because of the geometry of the data! So methods have evolved to find and cluster *submatrices* of the full thing. This is also in keeping with the notion that gene expression is there for *many* purposes in addition to the one we're looking at in the experiment–transcriptional regulation of colon tissue may respond to the kind of diet the patient had before the biopsy, for instance. In addition, gene expression is so labile that any small change imprints itself on the data: there may be the experimental artifacts one would like to avoid. (Tumor data sets have been known to result in clustering of the *surgeon* performing the biopsy, for instance). Thus, a proper way of selecting a smaller subset of the genes and experiments for analysis is extremely important [43–45].

I hope I've been able to convey the impression that the analysis of array data is by now a thriving subject, and that the interested reader should immerse herself into the growing literature. I will now first describe some of the unwritten concerns that a physicist may want to carry into the subject, and then describe a much more basic problem: that of actually getting the numbers to do clustering on. Most researchers are happy to use whatever numbers the available software spew out without giving them much thought, be it software like *Scanalize* for spot arrays, or the Affymetrix software suite for GeneChip arrays. However, close inspection shows that there are so many unsolved issues at the level of the measurement process that substantial improvements to the quality of the data could be made just by treating the raw measurement data more carefully.

2.4 Some simplifying assumptions

Here we shall detail a few of the common simplifying assumptions that lurk about in this subject ready to catch the unwary analyst. Once again, I would like to stress that, in many cases, it is known that the assumption is wrong, it's just the best shot one has at a problem, and it is otherwise unknown whether its a "safe" wrong assumption or a deadly one.

One way in which simplifying assumptions ruin an otherwise good piece of science is by creeping into the *null hypothesis*. Any quantitative analysis in this subject must be validated by an estimate of its statistical validity, since these experiments generate copious noise together with the signal (whose "copiousness" is unknown *a priori*). No test of statistical validity operates against a vacuum, but as a way of distinguishing the observed experimental data from a null hypothesis. If the null hypothesis is highly artificial, then it is worthless to assure us that the observed experiments are, with high statistical confidence, different from the artificial null hypothesis, because we already knew they were. There is a widespread tendency to accept otherwise unacceptably dumb null hypothesis, because the tests to establish significance against any more refined model are extremely difficult to carry out, and researchers do not agree on a standardized null hypothesis. The result is an escalation of the "significance scores" that are expected. For instance, sequence alignment algorithms are supposed to tell us what the optimal alignment of their input sequence is, against the sequences in a database; and then tell us what is the probability that this alignment arose by pure chance alone. What do we mean by pure chance alone exactly? The usual test is to test against a scrambled sequence of similar length; but of course, a scrambled sequence is spectrally white, while all biological sequences have prominent correlations. A test against a random sequence of similar composition *and* correlation structure would require people to agree on which feature of correlation structure is the important one, and would be much more difficult to carry out. As a result, the simpler null hypothesis gets ingrained, and researchers just expect astronomically small significance scores. But it should always be borne in mind, that significance scores against a more refined null hypothesis are not necessarily monotonous respect to the scores of a simpler hypothesis.

It is not known what a proper null hypothesis would be for expression array data, and it is a matter of current debate. Researchers have by and large used the arrays as sieves, trying to catch low-hanging fruit, to be verified by more conventional methods. Since biology labs have in the past struck gold by discovering and then studying the right molecule for a given process, a list of the 20 top candidates to be "the right molecule" is worth a lot to a biology lab–even though it hardly makes a piece of finished science. In this regard, the current climate favours sensitivity over accuracy: the people running the sieves are worried about catching *some* fruit, and so prefer to get a bigger list with many false positives and all of the right candidates over a smaller list with no false positives but important puzzle pieces missing. So it has been hard to get anyone worked up about the appropriate null hypothesis to quantify things like clustering analysis.

One simplifying assumption is that RNA concentration is a proxy for RNA transcription (just 90 degrees out of phase). The importance of RNA degradation rates cannot be overstressed, since it is roughly homogeneous within a given family of proteins but varies by orders of magnitude across families. Thus, any clustering of the observed raw signals likely will cluster together genes with similar RNA degradation rates even if unrelated–just on the basis of a similar spectral profile inducing spurious correlations against the "spectrally white" null hypothesis.

Another is that averaging over cells preserves the appropriate correlations one is trying to use. Instances of boolean processes which have a graded *population* distribution are known, and any population average is incapable of distinguishing such a thing from a graded response at the cellular level. See [28], where kinase cascades which had been shown to be graded *in vitro* were actually boolean at the single-cell level. Even for graded responses, instances are known in which the exact population distribution carries an enormous amount of information (see the third part of this course). This is a matter of much current concern for many people, who are trying to push array technology to single-cell detection limit.

Finally, gene expression is extremely labile, and it responds to *everything* a living being is in contact with. As discussed above artifacts are easy to come by. The assumption that the gene expression patterns observed only have information related to the parameters the experimentalist is attempting to control has already been shown to be quite dangerous.

2.5 Probeset analysis

> *In theory, theory and practice are the same, but in practice they aren't.* Attributed to Yogi Berra.

If one takes an experimental sample of RNA extracts from tissue, divides them into two identical vials, and then carries out all of the amplification, fluorescence labeling, hybridization to the microarray, washing, fluorescence laser scanning and image analysis of the scanned image required to get the brightnesses at each probe, one observes a curious feature. The probe brightnesses are, by an large, repeatable with a high degree of certainty–for bright probes, within a few percent. There are about 20 probes per probeset, so averaging over them should make things better by a factor of $\sqrt{20} \sim 4$. So the technology holds the promise that, somehow within it, there is the possibility of making measurements precise to two decimal digits. Yet, in actual practice, the final numbers coming out of analysis can scarcely be trusted for changes smaller than a factor of two–"times/over two" is the standard error line for chip data. ne may argue that the differential design is of course sensitive to noise, but still there is evidently something wrong, and we shall now explore what.

The rationale behind using PM and MM sequences is that sequence-specific hybridization will definitely notice a change of one letter, while cross-hybridization, or any other nonspecific effect won't. A statement of this thinking, which I will call the *standard hybridization model,* is

$$PM = I_{\mathrm{S}} + I_{\mathrm{NS}} + B$$
$$MM = \alpha I_{\mathrm{S}} + I_{\mathrm{NS}} + B$$

where I_S is the brightness due to the binding of the specific target, I_{NS} the nonspecific binding, B a background brightness of "physical" origin (photodetector dark current or reflections in the glass), and $0 < \alpha < 1$ is the loss of binding strength due to the single letter change.

The Affymetrix software suite then analyzes these numbers as follows. $\Delta = PM - MM = (1-\alpha)I_S$, so the nonspecific and background contributions should have been obliterated from the probe pair differentials. There are 16–20 such Δ, one per probe pair, per gene being probed. In order to discard outliars, the top and bottom scores are discarded; the rest are then algebraically averaged, as is sometimes practiced in the scoring of some olympic sports. This would eliminate the influence of defective probes.

The implicit assumptions that would allow such an averaging procedure to work are:

1. That $I_S = k[\text{RNA}]$, *i.e.*, that the relationship between *brightness* and RNA concentration be linear. This implies a conversion constant k that translates from concentrations to brightness. Similarly, there should be a constant p relating the nonspecific portion of the brightness with the RNA concentration which causes it;

2. That the conversion constant k be the same for PM and MM, and similarly that p should be the same for PM and MM. Otherwise substraction does not cancel them out;

3. That $0 < \alpha < 1$;

4. That the α and k be relatively homogeneous in magnitude throughout the probeset.

Grabbing a dataset and doing some statistics quickly belies these assumptions. In fact, such an exercise shows that all the assumptions are violated. The most visible violation (and the one that was first noted by researchers in the subject) is that (3) implies that $\Delta > 0$, or $PM > MM$, an assertion that's easy to check. Turns out that probe pairs for which $MM > PM$ were noticed quite early on, since they lead to negative concentrations in the Affy software suite.

Let me repeat that $MM > PM$ is a very *evident* violation of the hybridization assumptions; evident since no more numerical analysis than a substraction and checking for a $-$ sign is necessary. Felix Naef and I have been working on a number of large datasets, including an 86 sample set from human blood from rheumathoid arthritis patients, from Nila Patil at Perlegen Inc., formerly the human genetics division of Affymetrix; 36 Drosophila chips from M. Young's lab at Rockefeller and 24 mouse chips from mouse brain tissue by Dan Lim *et al.* We have found that across different chips, different kinds of tissue, different people carrying out the reactions, etc. all chips with the single exception of yeast chips (arguably a different beast altogether) show pretty much the same statistics: about 30% of all probe pairs show $MM > PM$. Thirty percent is a figure hard to dismiss as negligible or small. We have checked whether these probe pairs are clustered in any way we can figure out. The first naive idea would be that at low intensity levels, noise becomes percentwise higher, and so it might make some probe pairs cross the line. It isn't so: 27% of all probe pairs *in the top quartile of intensity* are $MM > PM$. The "bad" probe pairs are not concentrated into bad or problematic probesets either: 97% of all probesets have at least one such bad guy, and 60% have in excess of 5. (See the table in [27].)

This begs the question of whether there's any interesting feature in the *joint* probability distribution of PM and MM. Figure 1 shows a gray-coded two-dimensional histogram with quite an interesting structure.

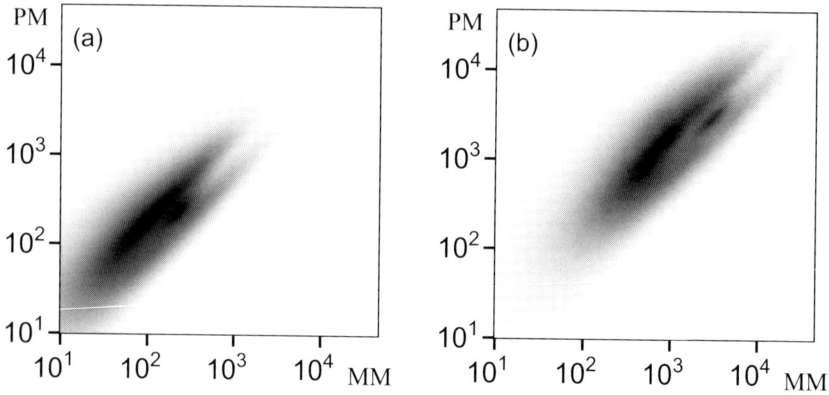

Fig. 9. Histograms of log PM *vs.* log MM for two different datasets: **a)** 86 human chips (HG-U95A), human blood, and **b)** 20 mouse chips (Mu11K/A), mouse brain tissue. The difference in overall brightness scale reflects a change in the standard scanning protocol; data set **a)** is more recent, and was scanned at lower laser power. From [27].

Notice that the joint probability distribution forks out into two branches, leaving a little "button"-like structure at the center of the branching structure. The lower branch and half of the button are completely below the $PM = MM$ diagonal. This plot not only belies the standard model above by showing the deviations to be meaningful–it also indicates that the deviations are likely interesting, since they appear as an elaborate structure. Unfortunately, it is impossible to check the obvious assumptions about sequence-specificity, since the sequences are considered proprietary information. Clearly the single mismatch binding is a much more complex process than naively thought, and a great deal of care should be exercised with exactly how to construct a differential discriminator from the match/mismatch game. Obvious culprits are secondary structure both in the probes and the targets, sequence-specific stacking interactions, and fabrication efficiencies, which are strongly letter-specific and so evidently accumulate exponentially through the 100-odd mask processes that the chips are subject to during fab.

So the MM probes are not doing what they were expected to be doing. A simple method to deal with this problem has been presented in [27]: separate factors are fitted to PM and MM by a singular-value decomposition process.

Assumption (4) in the list above is easily belied too. Brightnesses within a probeset vary by orders of magnitude. The histogram of log(max PM/ min PM) has its mode around 300, so the typical probeset spans two orders of magnitude and a half. The distribution of intensities within probesets can be assayed by normalizing all intensities to the median of the probeset (which is always well defined); since there is an abundance of data, we can build separate histograms for bright, medium and dim probesets.

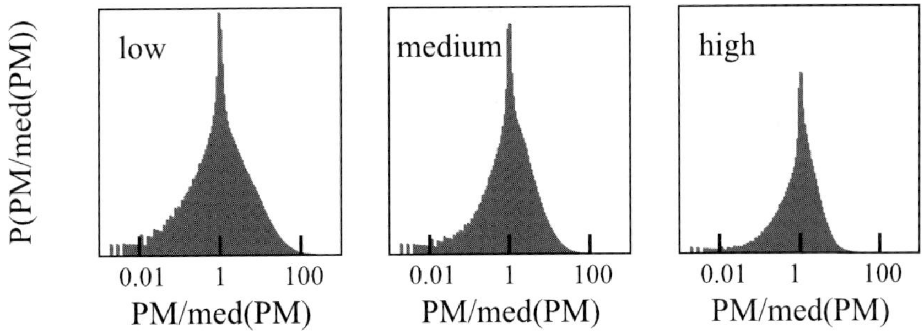

Fig. 10. Histograms of log(PM/median(PM) for the dataset of Figure 1a. Three distinct intensity ranges have been histogrammed separately, this allows to verify that it is not the low-end of the data that contaminates the histograms.

Because of the exponentially-distributed nature of the data, it is clear that *algebraic averages do not converge.* An average over quantities that vary on an exponential scale is dominated by the largest value, which is with high probability an outliar. Simply replacing the algebraic average with a geometric mean does wonders for the reliability of the data, as we showed in [26]. Yet, it is not sufficient for a high-quality method, for it has no built-in method to reject cross-hybridization. A simple way of doing cross-hyb rejection without resorting to the MM was also shown in [26] for the *ratio* comparison between two chips: a sum of two exponentially varying quantities looks mostly like the maximum of the quantities, except in the narrow diagonal range when they are of comparable size. Thus one may assume that a given probe is diplaying *either* mostly specific signal or mostly cross-hyb. So, if we compare all PM probes from one experiment to the corresponding probes in the second experiment, their ratios are likely to be showing either the real ratio between the two RNA concentrations, or nonsense. A histogram (if one had enough data to build one) would show the superposition of two distinct distributions: a sharp "specific" peak on a broad "nonsense" background. Robust estimators to fish out the signal out can be built, on a maximum likelyhood basis or any of many known statistical methods.

Finally, we should like to observe that we have clear indications that assumption (1) is false as well. The change in protocol alluded to in the caption to Figure 1 was introduced because of widespread complaints by GeneChip users that their data was showing saturation and that highly expressed genes which were known from blot assays to vary quite a bit were showing up as unchanging in the Affy data. That individual probe pairs *were* showing optical saturation is clear from Figure 1b–just notice the top and right borders. But a much more interesting problem is that many probes show (to the careful observer) evidence of *chemical* that some probes get chemically saturated, even when they are not optically saturated.

2.6 Discussion

In order to build better methods for extracting the RNA concentrations from this data, clearly a close look at the data is necessary. We've now seen the data and some of its problems, and I hope to have succeeded in making the case that, in all likelyhood, not one single method will be able to mine all relevant information from the data. This is because complex methods are very hard to validate, while simple methods fail to capture all of the complexities. We believe this should be so, and that analysis of this kind of data shall benefit from *many* methods in existence, rather than few–just like the existence of several different clustering techniques enriches the arsenal of the analyst rather than complicating life. Only the purveyors of proprietary software, and the harried scientist who can't be bothered to use anything other than *pret-a-porter* solutions can think otherwise–yet it has been quite problematic for people to, for instance, publish in this area.

3 Neural and gene expression networks: Song-induced gene expression in the canary brain

This lecture is more of a story. Partly because the underlying material is largely unfinished and ill understood; just the beginning of a long tale to be unraveled during many more years. It's a story that happens at the busy intersection between two large avenues of exploration: the corner of "gene expression networks" and "neural networks". This is the place where perception becomes memory. There's too much happening here, and it's a lot of effort to tease apart the pieces of the picture, but it's an interesting and exciting place nonetheless.

As we discussed in the previous lecture, gene expression is regulated; in fact, the point of it is to be regulated. As a response to changes in the environment, transcriptional programs are put in motion that effect long-lasting adaptations to cope with those changes. The nervous system is no exception: in fact, it is the tissue in which these changes are most prominent, varied and clear. All of the consolidations of long-term memory, for instance long-lasting synaptic change, involve transcriptional regulation. The marvel is the swiftness of the response and the ease with which it is put in motion.

So the outline of our story is thus. Imagine a canary sitting in a cage. You place a tape recorder next to it, and you press PLAY. The tape contains a recording of another canary's song–one our particular canary hadn't heard before. Hearing the song causes a blush of gene expression in some auditory nuclei of the canary brain: a transient, yet vigorous response, easily excited. Studying the blush reveals it to be topographically organized, so that different song elements cause geometrically different blushes. In fact, within a small family of stimuli, we were able to invert the map: we could say what was on the tape based on the shape and "color" of the blush. And this is the response to doing nothing more than playing sounds to the canary–we can scarcely say we have "done" anything to the bird, and yet there is a discernible response to just one or two playbacks of the song; the response is visible within 5 min, and lasts for hours.

This level of resolution cannot be achieved with any other technique currently in existence short of large-scale electrophysiological recording. We were able to dissect extremely important differences between similar-sounding natural and artificial sounds. Yet the story has so many open threads that it's hard to foresee how it shall go on.

The story has three main characters: a bird, a song, and a gene; they act on a stage, the brain. I need to introduce these characters now.

Fig. 11. Zebra finches and canaries. Courtesy of A. Alvarez-Buylla.

3.1 The study of songbirds

We intuit as one of the basic features of human beings that we are talking animals. We communicate vocally (the means through which this lecture was originally delivered) and so much of our culture goes through this channel that we hardly think about it as special, or as any more special than an opposable thumb. Curiously, it actually is, for within the animal kingdom there's precious few species which *learn* to vocalize. Lions don't learn to roar from their daddys, nor do dogs learn to bark from mom. Within the great apes there's no other talker than us, and even within mammals, there's hardly more than some marmosets and some species of whales and dolphins. These are not exactly "lab" animals, I dare say. Needless to say, vocal learning is even rarer outside of mammals, with one noticeable exception: three orders of songbirds. There's several dozen species, from canaries, finches, silver sparrows to hummingbirds (which vocalize in the near ultrasound range) which learn vocalizations which get to be extremely elaborate.

So we are left with songbirds as the only viable experimental animal for the study of vocal learning. As is well known, songbirds like canaries are easily bred in captivity and also easily kept and cared for–much easier than whales. As a laboratory animal they are so much nicer than rats that one cannot but feel privileged: they smell much nicer than rats, they sing rather than bite, and are a colorful and cheerful presence in the otherwise drab surroundings of a lab. I shall henceforth talk mostly about canaries, though some of the pioneering studies in gene expression were carried out in several other species, such as zebra finches.

3.2 Canary song

The first thing that has to be understood about songbirds is that their song is as vital to them as the spoken word is to us. They live and die by song: they warn others of danger, recognize one another, call their children. If they don't sing well, they don't get any sex, so they spend most of their spare time practicing. Courtship may be more exhausting than singing the Nibelungen: in some species a male may need to sing for several hours straight to the female before she will accept him.

The song of a songbird satisfies several of the most desirable requirements in a stimulus to be used in neurophysiological studies: it is behaviourally and ecologically relevant, easy to record, and easy to play back to an awake and attentive animal. You may compare with visual stimuli, which while simple to play back to an anesthetized animal, they are hard to play back to an awake and unrestrained animal, for it may easily turn away its gaze. And then what would be an ecologically and behaviourally relevant sound or visual stimulus to play to a rat? On the other hand, we do not yet understand the *space* of song well enough to generate songs ourselves that may pass as "natural" to a canary–not yet at any rate, though the subject is an active area of study, from the basic features of the vocal production [33] to the statistics of the song ensembles. The difference between natural and artificial stimuli is quite relevant, as you'll see shortly.

Canary song is composed of repetitions of the same *syllable,* strung together in phrases. A common way to depict the song is via a *sonogram,* a moving-window Fourier transform that displays energy content as a function of frequency, across time. This two-dimensional plot is a coordinate system not unlike a musical score, in which time is the horizontal axis and frequency the vertical [32]–the main difference being the homogeneity of the sonogram's coordinates. See Figure 12. Canaries have several dozen syllables in their repertoire, though particular strains or social groups may use a fraction only. A most prominent syllable in the strain in the lab of our collaborators, is a *whistle,* a flute-like sound in which the second and higher harmonics have been deliberately muted. (It's known that it is "deliberate" since letting the canaries breath a mixture of air and helium changes the frequency of their whistles–and makes the second harmonic prominent, so it had been filtered in the first place, see [31].)

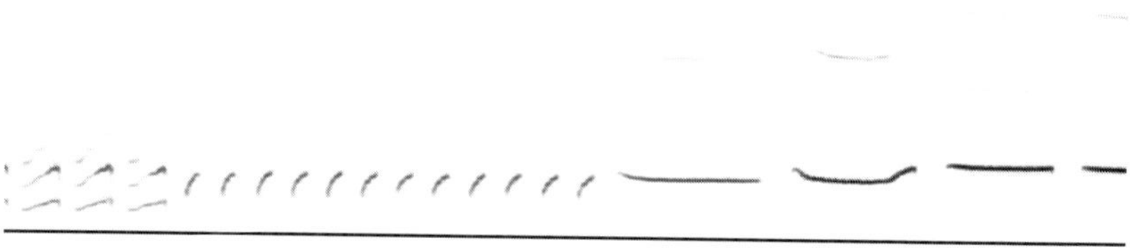

Fig. 12. A sonogram. Thanks to Tim Gardner.

3.3 ZENK

The third character in our story is a gene: a transcriptional regulator of the immediate-early gene (IEG) family. It is variously known by incomprehensibly different names in different species from mice to human: zif268, EGR-1, NGFI-A and krox24. The avian version was named by Claudio Mello by the acronym of the previous names: ZENK. I will refer to it as ZENK henceforth, though everyone calls it by the proper name in their own favorite animals.

ZENK is a transcriptional regulator of the zinc finger family: this means its job is to bind to DNA, using a zinc atom in the joint, and once bound, cause or impede assembly of the transcriptional machinery as described in the previous lecture. (Yes, the Z in ZENK comes from the z in zif268, which is, as you may imagine, "zinc finger 268"... so ultimately from z as in zinc).

In every cell type and tissue where it has been studied, ZENK plays a role mediating plastic changes. The general pattern is that ZENK is not transcribed while conditions are constant, and then upon some sudden change bearing directly on the cell's function, ZENK is rapidly induced for some period, and then turned off. Osteocytes, (bone cells), which strengthen the bone along stress lines, transcribe ZENK when the bone is subject to perpendicular stresses. Endothelial cells, the inner lining of capillary arteries, transcribe ZENK after an injury–when the vascularization has to be remapped around the wound. Lymphocytes (white blood cells) transcribe it upon their *first* encounter with an antigen–when the acquisition of immunity is made–but not during subsequent immune responses. ZENK figures prominently among the response factors activated after lesions to the liver and kidney, during muscle formation, etc. And in every tissue type, ZENK is what is called a "passive tumor suppressant": since ZENK activity signals differentiation, and tumor formation involves de-evolving into less differentiated stages, ZENK has to be inactivated in order for tumors to be able to grow.

But nowhere is the role of ZENK as visible as in neurons, for neurons are in continuous and subtle states of differentiation, and ZENK is one gateway to visualizing their physiology. When rats are placed in "complex environment" (meaning a really ugly mess of a maze, where they may have to walk a tightrope over a chasm filled with burning coals to get their food–you get the picture), they have to develop complex topographical maps of their environment to avoid electric shocks and the like. These maps are believed to be stored in the hypocampus. Some hypocampal neurons exhibit profuse arborization and sprouting of new synapses and connections during the exploration of the "complex environment"–well, the same neurons express ZENK at the same time. But, most meaningfully, ZENK is the only immediate-early gene which is activated by the long-term potentiation (LTP) protocol. LTP is a long-lasting increase in synaptic strength which follows vigorous stimulation of both the pre- and post-synaptic neuron simultaneously. Since LTP is the closest thing anyone has found in neuroscience to the famous Hebb rule of neural nets, it is widely believed (mostly on ideological grounds, I should admit) to be *the* process underlying memory [30]. Briefly after the discovery that the electrical stimulation that induces LTP also induced ZENK (and ZENK alone of all IEGs), Claudio Mello decided that if he could clone the avian homolog of ZENK, play a song to a bird and then stain its brain for the ZENK protein, he might have a standing chance at finding where on earth do canary brains store the memories of song. This story we shall develop in some more detail in the rest of the lecture. Most noteworthy after this work was the tour de force by Sabrina Davis' group that ZENK induction is *required* for long-term memory: a controlled ZENK knockout is long-term-memory impaired [41].

ZENK is in the midst of a high-connectivity area of the genetic and enzymatic circuits of the cell, and thus teasing apart its local net of interactions has proved quite difficult. ZENK is an immediate-early response gene: this means it is at the "input layer" for the gene circuit. The pieces that are known are thus: sustained membrane depolarization in the neuron provokes calcium influx, which causes a number of enzymatic pathways to turn on; in particular, a well-studied pathway is via the cre/creb system (calcium response element/CRE binding protein), protein kinase C, and then ZENK induction. There are in excess of 100 putative binding sites for ZENK in the vertebrate genome, so the potential connectivity is huge. An extremely meaningful confirmed link is that ZENK directly promotes the induction of synapsins–the proteins promoting synaptic sprouting and proliferation.

It may seem strange that we know so little and in such vagueness about this gene–even though a MEDLINE search hits thousands of papers mentioning the names ZENK (or zif268 etc.) in their abstract. The problem is that the effects are so many that they are extremely hard to unravel.

3.4 The blush

Claudio Mello succeeded in cloning the ZENK homolog in birds, coined the ZENK acronym/name (thus contributing to general confusion and to the ever-growing mess that is the biological naming of genes), and then did the following experiment. He kept two canaries in sound-proof cages. At $T = 0$ he played several times a song to one of them. He sacrificed them both at $T = 1$ hour, cut the brains and reacted the tissue with a probe to look for the ZENK gene, comparing the slices to find it differentially in the animal that had heard the song. He found a very robust induction in a hitherto little studied nucleus of the canary brain, the *caudiomedial neostriatum* (NCM).

Some features that are worth noticing. The induction is extremely rapid: statistically significant levels of ZENK transcript are visible within 5 min. The induction is strong: the contrast between the animal kept in silence and that exposed to song is a factor of 12 (in number of visibly labelled cells). In fact, this induction has such high contrast that it can be observed in the wild: Mello, Ribeiro and friends took a tape-recorder to the field station, and played in it a new song, following which they captured a bird, which showed significant amounts of ZENK induction in its NCM. The induction does not require many repetitions: D. Clayton has shown that as little as one exposure to a 2 s song is enough to elicit measurable activation in zebra finches, and full strength is achieved at 3 repetitions [40]. I wish we had known this before our own study.

After induction, the timecourse of ZENK is a typical activation-habituation dynamics; *i.e.*, the levels of gene expression as a function of time show a pulse of gene induction followed by a slower return to zero. The time course for the ZENK *protein* lags behind the ZENK mRNA, of course: the peak of mRNA induction appears to be 30 min after stimulation, while the peak of ZENK protein seems to lie at about 90 min [38].

After habituation, gene induction is *not* reinstated by playing the song all over again. No gene induction at all is elicited. But this habituation is *song specific:* playing a different song induces the original timecourse all over again. This stimulus specificity in the habituation dynamics of ZENK was grounds to speculate that ZENK induction might be part of the circuit in charge of consolidating long-term memories and has now been partly validated by Davis' study.

3.5 Histological analysis

At this stage, Sidarta Ribeiro (in Mello's lab) and Guillermo Cecchi (in mine) decided that they wanted to use this induction as a means to probe the structuring of representations of complex objects. The idea was that the robustness of the ZENK signaling made it a better marker than any other histological marker in use (cFOS, jun, etc.), and the specificity to memory consolidation would give us the "support" of a representation, in the sense that one would observe the set of cells involved in the representation of an object. The direct relevance of canary song to canary life and the evolutionary forces that shaped its brain during millennia made the study of canary song as a "complex object" not an idle exercise made for the sake of getting the word "complex" into the abstract, but one with direct implications for physiology.

But standing in the way of all this highly-charged ideology was a minor technical drawback–one that became the centerpiece of our work, of course, as is always the case with biology. The problem was analyzing the gene expression patterns. The way biologists had been doing it usually entailed a *camera lucida*–a contraption so medieval I shall skip its description. The way we did it entailed writing a lot (about 30 000 lines) of C code, which, while modern, is not that less medieval.

The analysis flow is as follows: after stimulation the bird is quickly (and painlessly) sacrificed, its brain extracted and rapidly frozen in liquid nitrogen. Some time later it is sliced with an ultrasound-driven knife within a cryostat, at -20 °C. The 20 μm slices are then dipped in a strong detergent to solubilize all membranes, and reacted with an antibody to the ZENK protein. This reaction is then amplified to generate a macroscopically observable stain by a method known as ICC; this amplification step is not unlike the development reactions used in photography, in that it is monotonic but nonlinear. The other issue is that the reactants are specific but not infinitely so, and then one is left with a tissue sample that is full of histological debris of all sorts.

Since ZENK protein is a transcriptional regulator, it is immediately shuttled to the nucleus of the cell after production, and so the histological stain is confined to the cellular nucleus. *Please beware: nucleus is being used here in two distinct senses, the nucleus of a cell is the organelle where the chromosomes are contained, while a neuronal nucleus like NCM is a collection of neurons which performs some distinct job in a geographically distinct area–one is a part of a cell or neuron, while the other is made of millions of neurons.* Because the neuronal nucleus has a distinctive shape, it is possible to write image-recognition algorithms which can pick it up from a high-quality photograph of the tissue–provided this photograph is high-resolution enough, which in our case was about 3 pixels per micron. At this resolution, a 2 mm nucleus like the NCM becomes a 36 megapixel photo–we've worked with photos as large as 50 thousand pixels on a side, for marmoset visual cortex. These photographs are constructed by connecting a camera to the microscope and the computer, and then joining to the computer also an XYZ computer-controlled microscope stage; the photos are then stitched together in memory, a job that's very easy to do to a low level of quality, but very hard to do at the quality levels we need. Please see in Figure 13 a screenshot of the program we wrote to deal with the analysis issue.

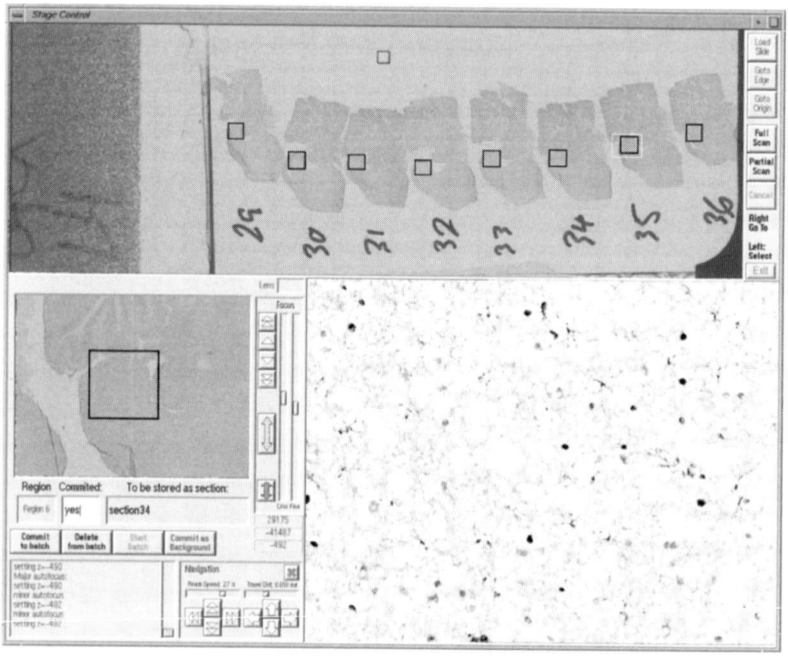

Fig. 13. Snapshot of our program's main window. See [36].

After the photo is reconstructed, we apply image-recognition algorithms to recognize ZENK-labelled nuclei, extract the coordinates and staining levels, and then align the NCM boundary outlines to a prototypical NCM using affine transforms. Maps of labelled cell density are constructed for distinct labeling intensities separately. In the search for discriminants that extracted the maximal amounts of information from our datasets, we found that the histogram of labeling intensities was distinctive enough to allow us to discriminate stimulus family on the basis of single slices alone. So an important component of our analysis was discriminating between a case where a few cells were expressing a lot of ZENK as opposed to many cells expressing a little bit of ZENK, and hence we created a representation where we displayed areal densities of cells in different ranges of gene expression levels.

3.6 Natural vs. artificial

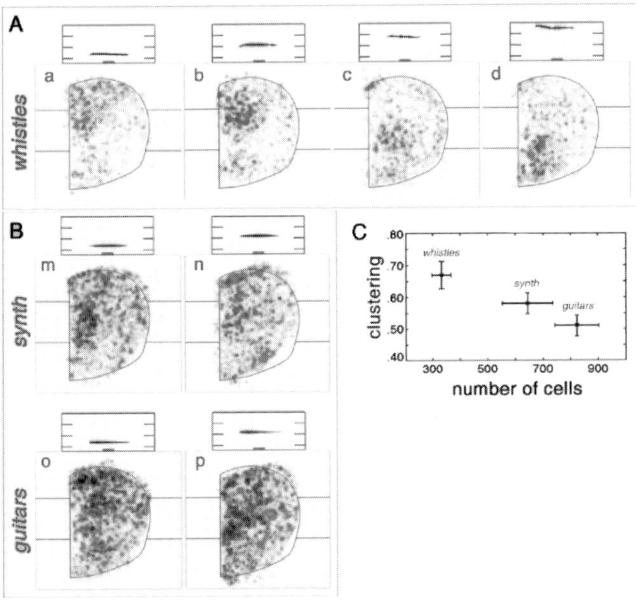

Fig. 14. Activation for natural *vs.* artificial stimuli. From [34].

We found that for natural stimuli such as repetitions of recorded canary whistles, gene induction was confined to a somewhat stereotyped band of intensities, almost a boolean phenomenon: a few percent of neurons showed gene induction in this band, while most neurons did not show any measurable expression level. On the other hand, artificial stimuli elicited a broad band of activation: many cells with *extremely high* level of gene induction, and a whole lot of cells with very faint (but measurable) labeling. Thus, rather than a single well-defined peak, the histogram of gene expression levels becomes broader band. Please notice, in connection with the previous lecture, that this effect would be completely invisible to a gene-chip analysis as currently performed on solubilized tissue: by averaging over all cells the ability to see the different ways in which different cells respond to the stimulus a great deal of violence has been done to the underlying biology. However, at 2000 \$ a shot one can not expect to run a gene chip per neuron either!

There is a similar "compactness" feature occurring, not in the space of gene expression level, but in real geographical layout. Looking at the whistles it becomes clear that a relatively tight clump of cells is activated, and that this clump moves up and down the dorsoventral axis of NCM according to the frequency of the whistle. One would normally label this as a *tonotopic organization*, meaning that there is a geographical organization to the neuronal system according to tone, or physical frequency. However, we then did the following experiment: given that the whistle is almost a pure tone (the second harmonic is more than 20 dB below the main harmonic) we reasoned that, if we were to eliminate all the little irregularities inherent in a natural emission, like the little fluctuations in amplitude envelope or the fluctuations in frequency, then we might get a *tighter* clump of cells being activated. We then constructed a "synthetic" whistle, made by digitally enveloping a pure sine wave with a parabolic envelope. Instead of getting less cells activated, we got *more*. We then proceeded to generate a sound that the bird could not conceivably imitate, yet was similarly narrow-band: we filtered a guitar sample narrowly around its main frequency to eliminate higher harmonics. It's a percussive sound, one that canaries would find most difficult to imitate. The result was even *more* activation than before: both over a larger geographical range, with a larger amount of cells, as well as a larger spread over gene expression intensities. Composite grayscale panels are shown in Figure 14. Please refer to `http://asterion.rockefeller.edu/marcelo/Canario/` for a full-color, Java version with all stimuli and sound.

The conclusions are strange. First, clearly the cells in the NCM are not working, as is naively thought in visual neurophysiology, as "feature detectors". If this were so, we would be hard at ease explaining how a stimulus with a lot less features could elicit so much more activation, in geographic range, gene expression range and sheer numbers. Second, if only in a poetic sense, it seems to be that canaries are not "expanding sound in a Fourier basis". Whatever it is that they are doing, they are doing it in a "canary basis". It seems to be the case that the representations are not tonotopically organized: if they were so, there would be an organization according to *tone*, which is narrowly defined in all three stimuli; the representations seem to be organized according to pitch *only for the natural stimuli*, with the artificial stimuli largely unorganized.

3.7 The Blush II: gAP

We tend to think of electrical activity as the primary function of neurons. Yet neurons have a complex physiology which goes well in excess of electrical activity alone. One of these physiological aspects is that some kinds of electrical activity, which by means currently ill-understood are judged as "novel", excite into motion transcriptional programs which ultimately result in long-term changes underlying memory. Let me belabor the point that as we effortlessly visually and auditorily parse scenes around us, we perceive distinct objects which we immediately recognize–and there is no re-cognition without a prior cognition and the memory of it. Even in the shortest of timeframes, the fractions of seconds required for a cognitive event, the path travelled by electrical activity in the brain does not propagate over virgin ground, but rather over a landscape that was labored by means of complex transcriptional patterns.

David Clayton has nicknamed the transcriptional programs triggered by various kinds of electrical activity the *genomic action potential* (gAP), in analogy with the more familiar electric action potential [35]; he describes the latter as integrating over the dendritic arbor inputs the various forms of electrical activity conveyed by synapses, while the former integrates *in time*, over a much longer timeframe, the changes and adaptations necessary to become memory. Clearly it is no less a neuronal function to adapt than it is to integrate and transmit information–even the dumbest being with neurons has a memory.

3.8 Meditation

Neuroscience is a highly charged, ideologic field. Where else would a choice of preposition denounce an ideological stance? When we were preparing our first manuscript on this subject, my collaborators and I got into a heated argument over the phrase: "the sensory environment is processed **by** the brain". The issue was whether to use "by the brain" as opposed to "in the brain"; in the first case, the brain is the active element which takes the initiative, goes out and finds the world and analyzes it; while in the second case, we have the brain as information processor, brain as computer, brain as Shannon communication channel idea.

The topoisomerase collaboration started when John Marko dropped down for a visit and related the problems in identifying a mechanism raised by the Rybenkov study; I have learned a lot from John in the 12 years we have done research together, and hope to continue. Most of the gene chip data analyis and algorithm development described here was carried out by Felix Naef, working on data of our excellent collaborators: Dan Lim and Arturo Alvarez-Buylla, whose work on neurogenesis in the adult mammalian brain started us trying to refine the analysis techniques available, and Nila Patil and Colleen Hacker at Perlegen (formerly the human genetics division of Affymetrix), first through the collaboration on neurogenesis and afterwards on the rheumathoid arthritis data; the Drosophila circadian rythm dataset by the Young lab, etc. The canary work was done in collaboration with Sidarta Ribeiro and Claudio Mello, then at the lab of Fernando Nottebohm, who has been extremely supportive of all of our nonsense; and on our side, most of the work was done by Guillermo Cecchi; I'd also like to acknowledge the work of Pabel Delgado, in charge of the "prosciutto machine". That work received invaluable ideological support from Roy Crist and the ineffable support of Jim Hudspeth. Finally, I would like to warmly thank the organizers of the Les Houches meeting, the colleagues who taught the other courses and lectures, and, particularly, the students at the Les Houches summer school, for an extremely stimulating time.

References

[1] L. Kauffman, *On Knots, Ann. Math. Stud.* **115** (Princeton University Press, 1987).

[2] K. Murasugi, *Knot theory and its applications* (Birkhauser, 1996).

[3] D.J.A. Welsh, *Complexity: knots, colourings and counting, Lond. Math. Soc. Lecture Note* **186** (Cambridge University Press, 1993).

[4] G. Hemion, *The classification of knots and 3-dimensional spaces* (Oxford University Press, 1992).

[5] V.V. Rybenkov, C. Ullsperger, A.V. Vologodskii and N.R. Cozzarelli, *Science* **277** (1997) 690-693.

[6] J.J. Hopfield, *Proc. Natl. Acad. Sci. USA* **71** (1974) 4135-4139.

[7] J. Ninio, *Biochimie* **57** (1975) 587-595.

[8] B. Alberts *et al.*, *Molecular Biology of the Cell* (Garland Press, 1994).

[9] M. Magnasco, *Phys. Rev. Lett.* **72** (1994) 2656.

[10] M.D. Frank-Kamenetskii, A.V. Lukashin and A.V. Vologodskii, *Nature* **258** (1975) 398-402.

[11] K.V. Klenin, A.V. Vologodskii, Anshelevich V.V., A.M. Dykhne and M.D. Frank-Kamenetskii, *J. Biomol. Struct. Dyn.* **5** (1988) 1173-1185.

[12] Y.S. Shaw and J.C. Wang, *Science* **260** 533-538.

[13] A.V. Rybenkov, N.R. Cozzarelli and A.V. Vologodskii, *Proc. Ntl. Acad. Sci. USA* **90** (1993) 5307-5311.

[14] J. Yan, M.O. Magnasco and J.F. Marko, *Nature* **401** (1999) 932-934.

[15] J. Yan, M.O. Magnasco and J.F. Marko, *Phys. Rev. RE.* **63** (2001) 031909.

[16] H.J. Zhou, Y. Zhang and Z.-C. Ou-Yang, *Phys. Rev.* **E 62** (2000) 1045-1058.

[17] A.V. Vologodski *et al.*, *Proc. Natl. Acad. Sci. USA* **98** (2001) 3045-3049. A.V. Vologodski *et al.*, Microarrays.

[18] J. Lockhart and E.A. Winzeler, *Nature* **405** (2000) 827-836.

[19] M. Chee *et al.*, *Science* **274** (1996) 610-14.

[20] R.J. Lipshutz, S.P. Fodor, *Bold* **21** (1999) 20-24.

[21] H.J. Bussemaker, H. Li and E.D. Siggia, *Proc. Natl. Acad. Sci. USA* **97** (2000) 10096-10100.

[22] Affymetrix Microarray Suite 4.0 User Guide (2000), Affymetrix, Inc., Santa Clara, CA.

[23] reference for "the capacity dimension calculation requires exponentially many datapoints" etc.

[24] `http://cmgm.stanford.edu/pbrown/`

[25] C. Li and W.H. Wong, *Proc. Natl. Acad. Sci. USA* **98** (2001) 31-36.

[26] F. Naef, D.A. Lim, N. Patil and M. Magnasco, Proceedings of the DIMACS Workshop on Analysis of Gene Expression Data 2001, in press; also `physics/0102010`.

[27] F. Naef, D.A. Lim, N. Patil and M. Magnasco, to appear in *Phys. Rev. E*, also `physics/0111199`.

[28] J.E. Ferrell and E.M. Machleder, *Science* **280** (1998) 895-898; C.Y.F. Huang and J.E. Ferrell, *Proc. Natl. Acad. Sci. USA* **93** (1996) 10078-10083.

[29] F. Naef, C. Hacker, N. Patil and M. Magnasco, to appear in *Genome Biol.* (2002) F. Naef, C. Hacker, N. Patil and M. Magnasco, Genomic Action Potential.

[30] E. Kandel, J.H. Schwartz and T.M. Jessell, eds., *Principles of Neural Science* (MacGraw Hill, 2000).

[31] Book on birdsong–try to find at home.

[32] T. Gardner and M. Magnasco, in preparation.

[33] Gardner *et al.*, *Phys. Rev. Lett.* **87** (2001) 8101.

[34] S. Ribeiro, G. Cecchi, M. Magnasco and C. Mello, *Neuron* **21** (1998) 359.

[35] D. Clayton, *Neurobiol. Learn. Mem.* **74** (2000) 185-216.

[36] G. Cecchi, S. Ribeiro, C. Mello and M. Magnasco, *J. Neurosci. Meth.* **87** (1999) 147-158.

[37] C.V. Mello, D.S. Vicario, D.F. Clayton, *Proc. Natl. Acad. Sci. USA* **89** (1992) 6818-6822.

[38] C.V. Mello and D.F. Clayton, *J. Neurosci.* **14** (1994) 6652-6666.

[39] E.D. Jarvis *et al.*, *Nature* **406** (2000) 628-632.

[40] A.A. Kruse, R. Stripling and D.F. Clayton, *Neurobiol. Learn. Mem.* **74** (2000) 179-184.

[41] M.W. Jones *et al.*, *Nature Neurosci.* **4** (2001) 289-296.

[42] G. Getz, E. Levine and E. Domany, *Proc. Natl. Acad. Sci. U.S.A.* **97** (2000) 12079.

[43] A. Floratos *et al.*, in *Proceedings 3rd annual ACM International Conference on Computational Molecular Biology (RECOMB99)* (1999).

[44] N. Barkai, private communication (2001).

[45] E.P. Xing and R.M. Karp, *ISMB Conference* (2001).

AAA
GGC
AGC

COURSE 11

THINKING ABOUT THE BRAIN

W. BIALEK

NEC Research Institute,
4 Independence Way,
Princeton, New Jersey 08540, USA
Present address:
Department of Physics,
Princeton University,
Princeton, New Jersey 08544, USA

Contents

THINKING ABOUT THE BRAIN

W. Bialek

Abstract

We all are fascinated by the phenomena of intelligent behavior, as generated both by our own brains and by the brains of other animals. As physicists we would like to understand if there are some general principles that govern the structure and dynamics of the neural circuits that underlie these phenomena. At the molecular level there is an extraordinary universality, but these mechanisms are surprisingly complex. This raises the question of how the brain selects from these diverse mechanisms and adapts to compute "the right thing" in each context. One approach is to ask what problems the brain really solves. There are several examples–from the ability of the visual system to count photons on a dark night to our gestalt recognition of statistical tendencies toward symmetry in random patterns–where the performance of the system in fact approaches some fundamental physical or statistical limits. This suggests that some sort of optimization principles may be at work, and there are examples where these principles have been formulated clearly and generated predictions which are confirmed in new experiments; a central theme in this work is the matching of the coding and computational strategies of the brain to the statistical structure of the world around us. Extension of these principles to the problem of learning leads us into interesting theoretical questions about how to measure the complexity of the data from which we learn and the complexity of the models that we use in learning, as well as opening some new opportunities for experiment. This combination of theoretical and experimental work gives us some new (if still speculative) perspectives on classical problems and controversies in cognition.

1 Introduction

Here in Les Houches we are surrounded by many beautiful and dramatic phenomena of nature. In the last century we have come to understand the powerful physical forces that shaped the landscape, creating the peaks that reach thousands of meters into the sky. As we stand and appreciate the view, other powerful forces also are at work: we are conscious of our surroundings, we parse a rich scene into natural and manmade objects that have meaningful relationships to one another and to us, and we learn about our environment so that we can navigate even in the dark after long hours of discussion in the bar. These aspects of intelligent behavior–awareness, perception, learning–surely are among the most dramatic natural phenomena that we experience directly. As physicists our efforts to provide a predictive, mathematical description of nature are animated by the belief that qualitatively striking phenomena should have deep theoretical explanations. The challenge, then, is to tame the evident complexities of intelligent behavior and to uncover these deep principles.

Words such as "intelligent" perhaps are best viewed as colloquial rather than technical: intelligent behavior refers to a class of phenomena exhibited by humans and by many other organisms, and membership in this class is by agreement among the participants in the conversation. There also is a technical meaning of "intelligence", determined by the people who construct intelligence tests. This is an area fraught with political and sociological difficulties, and there also is some force to Barlow's criticism that intelligence tends to be defined as what the tests measure [1]. For now let us leave the general term "intelligence" as an informal one, and try to be precise about some particular aspects of intelligent behavior.

Our first task, then, is to choose some subset of intelligent behaviors which we can describe in quantitative terms. I shall have nothing to say about consciousness, but for learning and perception we can go some way toward constructing a theoretical framework within which quantitative experiments can be designed and analyzed. Indeed, because perception constitutes our personal experience of the physical world, there is a tradition of physicists being interested in perceptual phenomena that reaches back (at least) to Helmholtz, Rayleigh, Maxwell and Ohm, and a correspondingly rich body of quantitative experimental work. If we can give a quantitative description of the phenomena it is natural to hope that some regularities may emerge, and that these could form the basis of a real theory.

I will argue that there is indeed one very striking regularity that emerges when we look quantitatively at the phenomena of perception, and this is a notion of optimal performance. There are well defined limits to the reliability of our perceptions set by noise at the sensory input, and this noise in turn often has fundamental physical origins. In several cases the brain approaches these limits to reliability, suggesting that the circuitry inside the brain is doing something like an optimal processing of the inputs or an optimal extraction of the information relevant for its tasks. It would be very attractive if this notion of optimization–which grows out of the data!–could be elevated to a principle, and I will go through one example in detail where we have tried to carry out this program.

The difficulty with collecting evidence for optimization is that we might be left only with a list of unrelated examples: there is a set of tasks for which performance is near optimal, and for each task we have a theory of how the brain does the task based on optimization principles. But precisely because the brain is not a general purpose computer, some tasks are done better than others. What we would like is not a list, but some principled view of what the brain does well. Almost since Shannon's original papers there has been some hope that information theory could provide such organizing principles, although much of the history is meandering rather than conclusive. I believe that in the past several years there has been substantial progress toward realizing the old dreams. On the one hand we now have direct experimental demonstrations that the nervous system can adapt to the statistical structure of the sensory world in ways that serve to optimize the efficiency of information transmission or representation. On the other hand, we have a new appreciation of how information theory can be used to assess the relevance of sensory information and the complexity of data streams. These theoretical developments unify ideas that have arisen in fields as diverse as coding theory, statistics and dynamical systems... and hold out some hope for a unified view of many different tasks in neural computation. I am very excited by all of this, and I hope to communicate the reasons for my excitement.

A very different direction is to ask about the microscopic basis for the essentially macroscopic phenomena of perception and learning. In the last decade we have seen an explosion in the experimental tools for identifying molecular components of biological systems, and as these tools have been applied to the brain this has created a whole new field of molecular neurobiology. Indeed, the volume of data on the molecular "parts list" of the brain is so vast that we have to ask carefully what it is we would like to know, or more generally why we are asking for a microscopic description. One possibility is that there is no viable theory at a macroscopic level: if we want to know why we perceive the world as we do, the answer might be found only in a detailed and exhaustive investigation of what all the molecules and cells are doing in the relevant regions of the brain. This is too horrible to discuss.

One very good reason for looking at the microscopic basis of neural computation is that molecular events in the cells of the brain (neurons) provide prototypes for thinking about molecular events in all cells, but with the advantage that important parts of the function of neurons involve electrical signals which are wonderfully accessible to quantitative measurements. Fifty years of work has brought us a nearly complete list of molecular components involved in the dynamics of neural signalling and computation, quantitative experiments on the properties of these individual molecules, and accurate mathematical models of how these individual molecular properties combine to determine the dynamics of the cell as a whole. The result is that the best characterized networks of molecular interactions in cells are the dynamics of ion channels in neurons. This firm foundation puts us in a position to ask questions about the emergent properties of these networks, their stability and robustness, the role of noise, ... all in experimentally accessible systems where we really know the relevant equations of motion and even most of the relevant parameters.

A very different reason for being interested in the molecular basis of perception and learning is because, as it turns out, the brain is a very peculiar computational device. As in all of biology, there is no obvious blueprint or wiring diagram; everything organizes itself. More profoundly, perhaps, all the components are impermanent. Particularly when we think about storing what we have learned or hope to remember, the whole point seems to be a search for permanence, yet almost every component of the relevant hardware in the brain will be replaced on a time scale of weeks, roughly the duration of this lecture series. Nonetheless we expect you to remember the events here in Les Houches for a time much longer than the duration of the school. Not only is there a problem of understanding how one stores information in such a dynamic system, there is the problem of understanding how such a system maintains stable function over long periods of time. Thus if the computations carried out by a neuron are determined by the particular combination of ion channel proteins that the cell expresses and inserts in the membrane, how does the cell "know" and maintain the correct expression levels as proteins are constantly replaced? Typical neurons express of order ten different kinds of ion channels at once, and it is not clear what functions are made possible by this molecular complexity–what can we do with ten types of channel that we can't do with nine? Finally, as in many aspects of life, crucial aspects of neural computation are carried out by surprisingly small numbers of molecules, and we shall have to ask how the system achieves reliability in the presence of the noise associated with these small numbers.

The plan is to start by examining the evidence for optimal performance in several systems, and then to explore information theoretic ideas that might provide some more unified view. Roughly speaking we will proceed from things that are very much grounded in data–which is important, because we have to convince ourselves that working on brains can involve experiments with the "look and feel" of good physics–toward more abstract problems. I would hope that some of the abstract ideas will link back to experiment, but I am still unclear about how to do this. Near the end of the course I will circle back to outline some of the issues in understanding the microscopic basis of neural computation, and conclude with some speculative thoughts on the "hard problems" of understanding cognition.

Some general references on the optimality of sensory and neural systems are reviews which Barlow [2] and I [3, 4] have written, as well as sections of the book *Spikes* [5], which may provide a useful reference for a variety of issues in the lectures. Let me warn the reader that the level of detail, both in the text and in the references, is a bit uneven (as were the lectures, I suspect). I have, however, taken the liberty of scattering some problems throughout the text. One last caveat: I am almost pathologically un–visual in my thinking, and so I wrote this text without figures. I think it can work, in that the essential ideas are summarizable in words and equations, but you really should look at original papers to see the data that support the theoretical claims and (more importantly) to get a feeling for how the experiments really were done.

2 Photon counting

Sitting quietly in a dark room, we can detect the arrival of individual photons at our retina. This observation has a beautiful history, with its roots in a suggestion by Lorentz in 1911[1]. Tracing through the steps from photon arrival to perception we see a sampling of the physics problems posed by biological systems, ranging from the dynamics of single molecules through amplification and adaptation in biochemical reaction networks, coding and computation in neural networks, all the way "up" to learning and cognition. For photon counting some of these problems are solved, but even in this well studied case many problems are open and ripe for new theoretical and experimental work. I will try to use the photon counting example as a way of motivating some more general questions.

Prior to Lorentz' suggestion, there was a long history of measuring the energy of a light flash that just barely can be seen. There is, perhaps surprisingly, a serious problem in relating this minimum energy of a visible flash to the number of photons at the retina, largely because of uncertainties about scattering and absorption in the eye itself. The compelling alternative is a statistical argument, as first exploited by Hecht *et al.* (in New York) and van der Velden (in The Netherlands) in the early 1940s [8, 9]:

- The mean number of photons $\langle n \rangle$ at the retina is proportional to the intensity I of the flash;

[1]Reviews on photon counting and closely related issues include Refs. [2, 3, 7] and Chapter 4 of Ref. [5].

- With conventional light sources the actual number n of photons that arrive from any single flash will obey Poisson statistics,

$$P(n|\langle n\rangle) = \exp(-\langle n\rangle)\frac{\langle n\rangle^n}{n!}\cdot \qquad (2.1)$$

- Suppose that we can see when at least K photons arrive. Then the probability of seeing is

$$P_{\rm see} = \sum_{n\geq K} P(n|\langle n\rangle). \qquad (2.2)$$

- We can ask an observer whether he or she sees the flash, and the first nontrivial observation is that seeing really is probabilistic for dim flashes, although this could just be fluctuations in attention;

- The key point is that however we measure the intensity I, we have $\langle n\rangle = \alpha I$, with α some unknown proportionality constant, so that

$$P_{\rm see}(I) = \sum_{n\geq K} P(n|\langle n\rangle = \alpha I). \qquad (2.3)$$

  If we plot $P_{\rm see}$ *vs.* $\log I$, then one can see that the *shape* of the curve depends crucially on the threshold photon count K, but changing the unknown constant α just translates the curve along the x-axis. So we have a chance to measure the threshold K without knowing α (which is hard to measure).

Hecht *et al.* did exactly this and found a beautiful fit to $K = 5$ or $K = 7$; subjects with different age had very different values for α but similar values of K. This sounds good: maybe the probabilistic nature of our perceptions just reflects the physics of random photon arrivals.

Problem 1: Poisson processes[2]**.** To understand what is going on here it would be a good idea if you review some facts about Poisson processes. By a "process" we mean in this case the time series of discrete events corresponding to photon arrivals or absorption. If the typical time between events is long compared to any intrinsic correlation times in the light source, it is plausible that each photon arrival will be independent of the others, and this is the definition of a Poisson process[3]. Thus, if we look at a very small time interval dt, the probability of

[2]Many of you have seen this before, so this is just a refresher. For the rest, you might look at Appendices 4 and 5 in *Spikes* which give a fairly detailed step–by–step discussion of Poisson processes [5].

[3]There is also the interesting fact that certain light sources will generate Poisson photon counting distributions no matter how frequently the photons arrive: recall that for a harmonic oscillator in a coherent state (as for the field oscillations in an ideal single mode laser), measurements of the number of quanta yield a Poisson distribution, exactly.

counting one event will be rdt, where r is the mean counting rate. If we count in a time window of size T, the mean count clearly will be $\langle n \rangle = rT$.

a. Derive the probability density for events at times $t_1, t_2, \ldots, t_n$; remember to include terms for the probability of *not* observing events at other times. Also, the events are indistinguishable, so you need to include a combinatorial factor. The usual derivation starts by taking discrete time bins of size dt, and then at the end of the calculation you let $dt \to 0$. You should find that

$$P(t_1, t_2, \ldots, t_n) = \frac{1}{n!} r^n \exp(-rT). \tag{2.4}$$

Note that this corresponds to an "ideal gas" of indistinguishable events.

b. Integrate over all the times $t_1, t_2, \ldots, t_n$ in the window $t \in [0, T]$ to find the probability of observing n counts. This should agree with equation (2.1), and you should verify the normalization. What is the relation between the mean and variance of this distribution?

An important point is that the 5 to 7 photons are distributed across a broad area on the retina, so that the probability of one receptor (rod) cell getting more than one photon is very small. Thus the experiments on human behavior suggest that individual photoreceptor cells generate reliable responses to single photons. This is a lovely example of using macroscopic experiments to draw conclusions about single cells.

It took many years before anyone could measure directly the responses of photoreceptors to single photons. It was done first in the (invertebrate) horseshoe crab, and eventually by Baylor and coworkers in toads [10] and then in monkeys [11]. The complication in the lower vertebrate systems is that the cells are coupled together, so that the retina can do something like adjusting the size of pixels as a function of light intensity. This means that the nice big current generated by one cell is spread as a small voltage in many cells, so the usual method of measuring the voltage across the membrane of one cell won't work; you have to suck the cell into a pipette and collect the current, which is what Baylor *et al.* managed to do. Single photon responses observed in this way are about a picoamp in amplitude *vs.* a continuous background noise of 0.1 pA rms, so these are easily detected.

A slight problem is that van der Velden found $K = 2$, far from the $K = 5$–7 found by Hecht *et al.*. Barlow explained this discrepancy by noting that even when counting single photons we may have to discriminate (as in photomultipliers) against a background of dark noise [12]. Hecht *et al.* inserted blanks in their experiments to be sure that you never say "I saw it" when nothing is there [that is, $P_{\rm see}(I = 0) = 0$], which means you have to set a high threshold to discriminate against the noise. On the other hand, van der Velden was willing to allow for some false positive responses, so his subjects could afford to set a lower threshold. Qualitatively this makes sense, but to be a quantitative explanation the noise has to be at the right level. Barlow reasoned that one source of noise was if the pigment molecule rhodopsin spontaneously (as a result of thermal fluctuations) makes the transitions that normally are triggered by a photon; of course these random events would be indistinguishable from photon arrivals. He found that everything works if this spontaneous event rate is equivalent to roughly 1 event per 1000 years per molecule: there are a billion molecules in one rod cell, which gets us to one event per minute per cell (roughly) and when we integrate over hundreds of cells for hundreds of milliseconds we find a mean event count of $\sim$10, which means that to be sure we see something we will have to count many more than $\sqrt{10}$ extra events, corresponding to what Hecht, Shlaer and Pirenne found in their highly reliable observers.

One of the key points here is that Barlow's explanation only works if people actually can adjust the "threshold" K in response to different situations. The realization that this is possible was part of the more general recognition that detecting a sensory signal does not involve a true threshold between (for example) seeing and not seeing [13]. Instead we should imagine that—as when we try to measure something in a physics experiment—all sensory tasks involve a discrimination between signal and noise, and hence there are different strategies which provide different ways of trading off among the different kinds of errors.

Suppose, for example, that you get to observe x which could be drawn either from the probability distribution $P_+(x)$ or from the distribution $P_-(x)$; your job is to tell me whether it was $+$ or $-$. Note that the distribution could be controlled completely by the experimenter (if you play loud but random noise sounds, for example) or the distribution could be a model of noise generated in the receptor elements or even deep in the brain. At least for simple forms of the distributions $P_\pm(x)$, we can make a decision about how to assign a particular value of x by simply setting a threshold θ; if $x > \theta$ we say that x came from the $+$ distribution, and conversely. How should we set θ? Let's try to maximize the probability that we get the right answer. If x is chosen from $+$ with probability $P(+)$, and similarly for $-$, then the probability that our threshold rule gets the correct answer is

$$P_{\mathrm{correct}}(\theta) = P(+)\int_\theta^\infty \mathrm{d}x P_+(x) + P(-)\int_{-\infty}^\theta \mathrm{d}x P_-(x). \tag{2.5}$$

To maximize $P_{\mathrm{correct}}(\theta)$ we differentiate with respect to θ and set this equal to zero:

$$\frac{\mathrm{d}P_{\mathrm{correct}}(\theta)}{\mathrm{d}\theta} = 0 \tag{2.6}$$

$$\Rightarrow P(+)P_+(\theta) = P(-)P_-(\theta). \tag{2.7}$$

In particular if $P(+) = P(-) = 1/2$, we set the threshold at the point where $P_+(\theta) = P_-(\theta)$; another way of saying this is that we assign each x

to the probability distribution that has the larger density at x–"maximum likelihood". Notice that if the probabilities of the different signals + and − change, then the optimal setting of the threshold changes.

Problem 2: More careful discrimination. Assume as before that x is chosen either from $P_+(x)$ or from $P_-(x)$. Rather than just setting a threshold, consider the possibility that when you see x you assign it to the + distribution with a *probability* $f(x)$.

a. Express the probability of a correct answer in terms of $f(x)$, generalizing equation (2.5); this is a functional $P_{\rm correct}[f(x)]$.

b. Solve the optimization problem for the function $f(x)$; that is, solve the equation

$$\frac{\delta P_{\rm correct}[f(x)]}{\delta f(x)} = 0. \tag{2.8}$$

Show that the solution is deterministic [$f(x) = 1$ or $f(x) = 0$], so that if the goal is to be correct as often as possible you shouldn't hesitate to make a crisp assignment even at values of x where you aren't sure (!).

c. Consider the case where $P_\pm(x)$ are Gaussian distributions with the same variance but different means. Evaluate the minimum error probability (formally) and give asymptotic results for large and small differences in mean. How large do we need to make this "signal" to be guaranteed only 1% errors?

d. Generalize these results to multidimensional Gaussian distributions, and give a geometrical picture of the assignment rule. This problem is easiest if the different Gaussian variables are independent and have equal variances. What happens in the more general case of arbitrary covariance matrices?

There are classic experiments to show that people will adjust their thresholds automatically when we change the a priori probabilities, as expected for optimal performance. This can be done without any explicit instructions–you don't have to tell someone that you are changing the value of $P(+)$. At least implicitly, then, people learn something about probabilities and adjust their assignment criteria appropriately. As we will discuss later in the course, there are other ways of showing that people (and other animals) can learn something about probabilities and use this knowledge to change their behavior in sensible ways. Threshold adjustments also can be driven by changing the rewards for correct answers or the penalties for wrong answers. In this view, it is likely that Hecht *et al.* drove their observers to high thresholds by having a large effective penalty for false positive detections. Although it's not a huge literature, people have since manipulated these penalties and rewards in HSP–style experiments, with the expected results. Perhaps more dramatically, modern quantum optics techniques have been used to manipulate the statistics of photon arrivals at the retina, so that the tradeoffs among the different kinds of errors are changed... again with the expected results [14].

Not only did Baylor and coworkers detect the single photon responses from toad photoreceptor cells, they also found that single receptor cells in the dark show spontaneous photon–like events at just the right rate to be the source of dark noise identified by Barlow [15]! Just to be clear, Barlow identified a *maximum* dark noise level; anything higher and the observed reliable detection is impossible. The fact that the real rod cells have essentially this level of dark noise means that the visual system is operating near the limits of reliability set by thermal noise in the input. It would be nice, however, to make a more direct test of this idea.

In the lab we often lower the noise level of photodetectors by cooling them. This isn't so easy in humans, but it does work with cold blooded animals like frogs and toads. So, Aho *et al.* [16] convinced toads to strike with their tongues at small worm–like objects illuminated by dim flashes of light, and measured how the threshold for reliable striking varied with temperature. It's important that the prediction is for more reliable behavior as you cool down–all the way down to the temperature where behavior stops–and this is what Aho *et al.* observed. Happily, Baylor *et al.* also measured the temperature dependence of the noise in the detector cells. The match of behavioral and cellular noise levels *vs.* temperature is perfect, strong evidence that visual processing in dim lights really is limited by input noise and not by any inefficiencies of the brain.

Problem 3: Should you absorb all the photons? Consider a rod photoreceptor cell of length ℓ, with concentration C of rhodopsin; let the absorption cross section of rhodopsin be σ. The probability that a single photon incident on the rod will be counted is then $p = 1 - \exp(-C\sigma\ell)$, suggesting that we should make C or ℓ larger in order to capture more of the photons. On the other hand, as we increase the number of Rhodopsin molecules ($CA\ell$, with A the area of the cell) we also increase the rate of dark noise events. Show that the signal–to–noise ratio for detecting a small number of incident photons is maximized at a nontrivial value of C or ℓ, and calculate the capture probability p at this optimum. Do you find it strange that the best thing to do is to let some of the photons go by without counting them? Can you see any way to design an eye which gets around this argument? Hint: think about what you see looking into a cat's eyes at night.

These observations on the ability of the visual system to count single photons–down to the limit set by thermal noise in rhodopsin itself–raise questions at several different levels:

- At the level of single molecules, there are many interesting physics problems in the dynamics of rhodopsin itself;

- At the level of single cells, there are challenges in understanding how a network of biochemical reactions converts individual molecular events into macroscopic electrical currents across the rod cell membrane;
- At the level of the retina as a whole, we would like to understand the rules whereby these signals are encoded into the stereotyped pulses which are the universal language of the brain;
- At the level of the whole organism, there are issues about how the brain learns to make the discriminations which are required for optimal performance.

Let's look at these questions in order. The goal here is more to provide pointers to interesting and exemplary issues than to provide answers.

At the level of single molecule dynamics, our ability to see in the dark ultimately is limited by the properties of rhodopsin (because everything else works so well!). Rhodopsin consists of a medium–sized organic pigment, retinal, enveloped by a large protein, opsin; the photo–induced reaction is isomerization of the retinal, which ultimately couples to structural changes in the protein. One obvious function of the protein is to tune the absorption spectrum of retinal so that the same organic pigment can work at the core of the molecules in rods and in all three different cones, providing the basis for color vision. Retinal has a spontaneous isomerization rate of ~1/yr, 1000 times that of rhodopsin, so clearly the protein acts to lower the dark noise level. This is not so difficult to understand, since one can imagine how a big protein literally could get in the way of the smaller molecule's motion and raise the barrier for thermal isomerization. Although this sounds plausible, it's probably wrong: the activation *energies* for thermal isomerization in retinal and in rhodopsin are almost the same. Thus one either must believe that the difference is in an entropic contribution to the barrier height or in dynamical terms which determine the prefactor of the transition rate. I don't think the correct answer is known.

On the other hand, the photo–induced isomerization rate of retinal is only $\sim 10^9\,\mathrm{s}^{-1}$, which is slow enough that fluorescence competes with the structural change[4]. Now fluorescence is a disaster for visual pigment–not only don't you get to count the photon where it was absorbed, but it might get counted somewhere else, blurring the image. In fact rhodopsin does not fluoresce: the quantum yield or branching ratio for fluorescence is $\sim 10^{-5}$, which means that the molecule is changing its structure and escaping the

[4]Recall from quantum mechanics that the spontaneous emission rates from electronic excited states are constrained by sum rules if they are dipole–allowed. This means that emission lifetimes for visible photons are order 1 nanosecond for almost all of the simple cases...

immediate excited state in tens of femtoseconds [17]. Indeed, for years every time people built faster pulsed lasers, they went back to rhodopsin to look at the initial events, culminating in the direct demonstration of femtosecond isomerization [18], making this one of the fastest molecular events ever observed.

The combination of faster photon induced isomerization and slower thermal isomerization means that the protein opsin acts as an electronic state selective catalyst: ground state reactions are inhibited, excited state reactions accelerated, each by orders of magnitude. It is fair to say that if these state dependent changes in reaction rate did not occur (that is, if the properties of rhodopsin were those of retinal) we simply could not see in the dark.

Our usual picture of molecules and their transitions comes from chemical kinetics: there are reaction rates, which represent the probability per unit time for the molecule to make transitions among states which are distinguishable by some large scale rearrangement; these transitions are cleanly separated from the time scales for molecules to come to equilibrium in each state, so we describe chemical reactions (especially in condensed phases) as depending on temperature not on energy. The initial isomerization event in rhodopsin is so fast that this approximation certainly breaks down. More profoundly, the time scale of the isomerization is so fast that it competes with the processes that destroy quantum mechanical coherence among the relevant electronic and vibrational states [19]. The whole notion of an irreversible transition from one state to another necessitates the loss of coherence between these states (recall Schrödinger's cat), and so in this sense the isomerization is proceeding as rapidly as possible. I don't think we really understand, even qualitatively, the physics here[5]. If rhodopsin were the only example of this "almost coherent chemistry" that would be good enough, but in fact the other large class of photon induced events

[5]That's not to say people aren't trying; the theoretical literature also is huge, with much of it (understandably) focused on how the protein influences the absorption spectra of the chromophore. The dynamical problems are less well studied, although again there is a fairly large pile of relevant papers in the quantum chemistry literature (which I personally find very difficult to read). In the late 1970 and early 1980s, physicists got interested in the electronic properties of conjugated polymers because of the work by Heeger and others showing that these quasi–1D materials could be doped to make good conductors. Many people must have realized that the dynamical models being used by condensed matter physicists for (ideally) infinite chains might also have something to say about finite chains, but again this was largely the domain of chemists who had a rather different point of view. Kivelson and I tried to see if we could make the bridge from the physicists' models to the dynamics of rhodopsin, which was very ambitious and never quite finished; there remains a rather inaccessible conference proceeding outlining some of the ideas [20]. Our point of view was rediscovered and developed by Aalberts and coworkers a decade later [21, 22].

in biological systems–photosynthesis–also proceed so rapidly as to compete with loss of coherence, and the crucial events again seem to happen (if you pardon the partisanship) while everything is still in the domain of physics and not conventional chemistry [23, 24]. Why biology pushes to these extremes is a good question. How it manages to do all this with big floppy molecules in water at roughly room temperature also is a great question.

At the level of single cells, the biochemical circuitry of the rod takes one molecule of activated rhodopsin and turns this into a macroscopic response. Briefly, the activated rhodopsin is a catalyst that activates many other molecules, which in turn act as catalysts and so on. Finally there is a catalyst (enzyme) that eats cyclic GMP, but cGMP binds to and opens ionic channels in the cell membrane. So when the cGMP concentration falls, channels close, and the electrical current flowing into the cell is reduced[6]. The gain of this system must be large–many molecules of cGMP are broken down for each single activated rhodopsin–but gain isn't the whole story. First, most models for such a chemical cascade would predict large fluctuations in the number of molecules at the output since the lifetime of the active state of the single active rhodopsin fluctuates wildly (again, in the simplest models) [7, 26]. Second, as the lights gradually turn on one has to regulate the gain, or else the cell will be overwhelmed by the accumulation of a large constant signal; in fact, eventually all the channels close and the cell can't respond at all. Third, since various intermediate chemicals are present in finite concentration, there is a problem of making sure that signals rise above the fluctuations in these concentrations–presumably while not expending to much energy too make vast excesses of anything. To achieve the required combination of gain, reliability, and adaptation almost certainly requires a network with feedback. The quantitative and even qualitative properties of such networks depend on the concentration of various protein components, yet the cell probably cannot rely on precise settings for these concentrations, so this robustness creates yet another problem[7].

Again if photon counting were the only example all of this it might be interesting enough, but in fact there are many cells which build single molecule detectors of this sort, facing all the same problems. The different systems use molecular components that are sufficiently similar that one can recognize the homologies at the level of the DNA sequences which code for the relevant proteins–so much so, in fact, that one can go searching for unknown molecules by seeking out these homologies. This rough universality of tasks and components cries out for a more principled view of how such

[6] Actually we can go back to the level of single molecules and ask questions about the "design" of these rather special channels...

[7] If the cell does regulate molecule counts very accurately, one problem could be solved, but then you have to explain the mechanism of regulation.

networks work (see, for example, Ref. [27]); photon counting is such an attractive example because there is an easily measurable electrical output and because there are many tricks for manipulating the network components (see, for example, Ref. [28]).

At the level of the retina as a whole, the output which gets transmitted to the brain is not a continuous voltage or current indicating (for example) the light intensity or photon counting rate; instead signals are encoded as sequences of discrete identical pulses called action potentials or spikes. Signals from the photodetector cells are sent from the eye to the brain (after some interesting processing within the retina itself...) along $\sim 10^6$ cables–nerve cell "axons"–that form the optic nerve. Roughly the same number of cables carry signals from the sensors in our skin, for example, while each ear sends $\sim 40\,000$ axons into the central nervous system. It is very likely that our vision in the photon counting regime is triggered by the occurrence of just a few extra spikes along at most a few optic nerve axons [5, 29]. For the touch receptors in our fingertips there is direct evidence that our perceptions can be correlated with the presence or absence of a single action potential along one out of the million input cables [30]. To go beyond simple detection we have to understand how the complex, dynamic signals of the sensory world can be represented in these seemingly sparse pulse trains [5].

Finally, the problem of photon counting–or any simple detection task–hides a deeper question: how does the brain "know" what it needs to do in any given task? Even in our simple example of setting a threshold to maximize the probability of a correct answer, the optimal observer must at least implicitly acquire knowledge of the relevant probability distributions. Along these lines, there is more to the "toad cooling" experiment than a test of photon counting and dark noise. The retina has adaptive mechanisms that allow the response to speed up at higher levels of background light, in effect integrating for shorter times when we can be sure that the signal to noise ratio will be high. The flip side of this mechanism is that the retinal response slows down dramatically in the dark. In moderate darkness (dusk or bright moonlight) Aho *et al.* found that the slowing of the retinal response is reflected directly in a slowing of the animal's behavior [25]: it is as if the toad experiences an illusion because images of its target are delayed, and it strikes at the delayed image[8]. But if this continued down

[8]We see this illusion too. Watch a pendulum swinging while wearing glasses that have a neutral density filter over one eye, so the mean light intensity in the two eyes is different. The dimmer light results in a slower retina, so the signals from the two eyes are not synchronous. As we try to interpret these signals in terms of motion, we find that even if the pendulum is swinging in a plane parallel to the line between our eyes, what we see is motion in 3D. The magnitude of the apparent depth of oscillation is related to the neutral density and hence to the slowing of signals in the "darker" retina.

to the light levels in the darkest night, it would be a disaster, since the delay would mean that the toad inevitably strikes behind the target! In fact, the toad does not strike at all in the first few trials of the experiment in dim light, and then strikes well within the target. It is hard to escape the conclusion that the animal is learning about the typical velocity of the target and then using this knowledge to extrapolate and thereby correct for retinal delays[9]. Thus, performance in the limit where we count photons involves not only efficient processing of these small signals but also learning as much as possible about the world so that these small signals become interpretable.

We take for granted that life operates within boundaries set by physics–there are no vital forces[10]. What is striking about the example of photon counting is that in this case life operates *at* the limit: you can't count half a photon, your counting can't be any more reliable than allowed by thermal noise, chemical reactions can't happen faster than loss of quantum coherence, and so on. Could this be a general principle? Is it possible that, at least for crucial functions which have been subjected to eons of evolutionary pressure, all biological systems have found solutions which are optimal or extremal in this physicist's sense? If so, we have the start of a real program to describe these systems using variational principles to pick out optimal functions, and then sharp questions about how realistic dynamical models can implement this optimization. Even if the real systems aren't optimal, the exercise of understanding what the optimal system might look like will help guide us in searching for new phenomena and maybe in understanding some puzzling old phenomena. We'll start on this project in the next lecture.

3 Optimal performance at more complex tasks

Photon counting is pretty simple, so it might be a good idea to look at more complex tasks and see if any notion of optimal performance still makes sense. The most dramatic example is from bat echolocation, in a series

---

[9] As far as I know there are no further experiments that probe this learning more directly, *e.g.* by having the target move at variable velocities.

[10] Casual acceptance of this statement of course reflects a hard fought battle that stretched from debates about conservation of energy in ∼1850 to the discovery of the DNA structure in ∼1950. If you listen carefully, some people who talk about the mysteries of the brain and mind still come dangerously close to a vitalist position, and the fact that we can't really explain how such complex structures evolved leaves room for wriggling, some of which makes it into the popular press. Note also that, as late as 1965, Watson was compelled to have a section heading in *Molecular Biology of the Gene* which reads "Cells obey the laws of physics and chemistry". Interestingly, this continues to appear in later editions.

of experiments by Simmons and colleagues culminating in the demonstration that bats can discriminate reliably among echoes that differ by just $\sim$10–50 *nano*seconds in delay [31]. In these experiments, bats stand at the base of a Y with loudspeakers on the two arms. Their ultrasonic calls are monitored by microphones and returned through the loudspeakers with programmable delays. In a typical experiment, the "artificial echoes" produced by one side of the Y are at a fixed delay τ, while the other side alternately produces delays of $\tau \pm \delta\tau$. The bat is trained to take a step toward the side which alternates, and the question is how small we can make $\delta\tau$ and still have the bat make reliable decisions.

Early experiments from Simmons and coworkers suggested that delays differences of $\delta\tau \sim 1\,\mu$sec were detectable, and perhaps more surprisingly that delays of $\sim 35\,\mu$sec were less detectable. The latter result might make sense if the bat were trying to measure delays by matching the detailed waveforms of the call and echo, since these sounds have most of their power at frequencies near $f \sim 1/(35\,\mu\text{sec})$–the bat can be confused by delay differences which correspond to an integer number of periods in the acoustic waveform, and one can even see the $n = 2$ "confusion resonance" if one is careful.

The problem with these results on delay discrimination in the 1–50 μsec range is not that they are too precise but that they are not precise enough. One can measure the background acoustic noise level (or add noise so that the level is controlled) and given this noise level a detector which looks at the detailed acoustic waveform and integrates over the whole call should be able to estimate arrival times much more accurately than $\sim 1\,\mu$sec. Detailed calculations show that the smallest detectable delay differences should be tens of nanoseconds. I think this was viewed as so obviously absurd that it was grounds for throwing out the whole idea that the bat uses detailed waveform information[11]. In an absolutely stunning development, however, Simmons and company went back to their experiments, produced delays in the appropriate range–convincing yourself that you have control of acoustic and electronic delays with nanosecond precision is not so simple–and found that the bats could do what they should be able to do as ideal detectors. Further, they added noise in the background of the echoes and showed that performance of the bats tracked the ideal performance over a range of noise levels.

Problem 4: Head movements and delay accuracy. Just to be sure you understand the scale of things... When bats are asked to make "ordinary"

[11] The alternative is that the bat bases delay estimates on the envelope of the returning echo, so that one is dealing with structures on the millisecond time scale, seemingly much better matched to the intrinsic time scales of neurons.

discriminations in the Y apparatus, they move their head from arm to arm with each call. How accurately would they have to reposition their head to be sure that the second echo from one arm is not shifted by more than $\sim 1\,\mu$sec? By more than 10 nsec? Explain the behavioral strategy that bats could use to avoid this problem. Can you position (for example) your hand with anything like the precision that the bat needs for its head in these experiments?

Returning to vision, part of the problem with photon counting is that it almost seems inappropriate to dignify such a simple task as detecting a flash of light with the name "perception". Barlow and colleagues have studied a variety of problems that seem richer, in some cases reaching into the psychology literature for examples of gestalt phenomena–where our perception is of the whole rather than its parts [32]. One such example is the recognition of symmetry in otherwise random patterns. Suppose that we want to make a random texture pattern. One way to do this is to draw the contrast $C(\vec{\mathbf{x}})$ at each point $\vec{\mathbf{x}}$ in the image from some simple probability distribution that we can write down. An example is to make a Gaussian random texture, which corresponds to

$$P[C(\vec{\mathbf{x}})] \propto \exp\left[-\frac{1}{2}\int \mathrm{d}^2x \int \mathrm{d}^2x' C(\vec{\mathbf{x}})K(\vec{\mathbf{x}}-\vec{\mathbf{x}}')C(\vec{\mathbf{x}}')\right], \tag{3.1}$$

where $K(\vec{\mathbf{x}}-\vec{\mathbf{x}}')$ is the kernel or propagator that describe the texture. By writing K as a function of the difference between coordinates we guarantee that the texture is homogeneous; if we want the texture to be isotropic we take $K(\vec{\mathbf{x}}-\vec{\mathbf{x}}') = K(|\vec{\mathbf{x}}-\vec{\mathbf{x}}'|)$. Using this scheme, how do we make a texture with symmetry, say with respect to reflection across an axis?

The statement that texture has symmetry across an an axis is that for each point $\vec{\mathbf{x}}$ we can find the corresponding reflected point $\mathbf{R}\cdot\vec{\mathbf{x}}$, and that the contrasts at these two points are very similar; this should be true for every point. This can be accomplished by choosing

$$\begin{aligned} P_\gamma[C(\vec{\mathbf{x}})] \propto \exp\Bigg[&-\frac{1}{2}\int \mathrm{d}^2x \int \mathrm{d}^2x' C(\vec{\mathbf{x}})K(\vec{\mathbf{x}}-\vec{\mathbf{x}}')C(\vec{\mathbf{x}}') \\ &+\frac{\gamma}{2}\int \mathrm{d}^2x |C(\vec{\mathbf{x}}) - C(\mathbf{R}\cdot\vec{\mathbf{x}})|^2\Bigg], \end{aligned} \tag{3.2}$$

where γ measures the strength of the tendency toward symmetry. Clearly as $\gamma \rightarrow \infty$ we have an exactly symmetric pattern, quenching half of the degrees of freedom in the original random texture. On the other hand, as $\gamma \rightarrow 0$, the weakly symmetric textures drawn from P_γ become almost indistinguishable from a pure random texture ($\gamma = 0$). Given images of a certain size, and a known kernel K, there is a limit to the smallest value of γ that can be distinguished reliably from zero, and we can compare this statistical limit to the performance of human observers. This is more or less what Barlow did, although he used blurred random dots rather than the Gaussian textures considered here; the idea is the same (and must be formally the same in the limit of many dots). The result is that human observers come within a factor of two of the statistical limit for detecting γ or its analog in the random dot patterns.

One can use similar sorts of visual stimuli to think about motion, where rather than having to recognize a match between two halves of a possibly symmetric image we have to match successive frames of a movie. Here again human observers can approach the statistical limits [33], as long as we stay in the right regime: we seem not to make use of fine dot positioning (as would be generated if the kernel K only contained low order derivatives) nor can we integrate efficiently over many frames. These results are interesting because they show the potentialities and limitations of optimal visual computation, but also because the discrimination of motion in random movies is one of the places where people have tried to make close links between perception and neural activity in the (monkey) cortex [34]. In addition to symmetry and motion, other examples of optimal or near optimal performance include other visual texture discriminations and auditory identification of complex pitches in the auditory system; even bacteria can approach the limits set by physical noise sources as they detect and react to chemical gradients, and there is a species of French cave beetle that can sense milliKelvin temperature changes, almost at the limit set by thermodynamic temperature fluctuations in their sensors.

I would like to discuss one case in detail, because it shows how much we can learn by stepping back and looking for a simple example (in proper physics tradition). Indeed, I believe that one of the crucial things one must do in working at the interface of physics and biology is to take some particular biological system and dive into the details. However much we believe in general principles, we have to confront particular cases. In thinking about brains it would be nice to have some "simple system" that we can explore, although one must admit that the notion of a simple brain seems almost a non–sequitur. Humans tend to be interested in the brains of other humans, but as physicists we know that we are not at the center of the universe, and we might worry that excessive attention to our own brains reflects a sort of preCopernican prejudice. It behooves us, then, to look around the animal kingdom for accessible examples of what we hope are general phenomena. For a variety of reasons, our attention is drawn to invertebrates–animals without backbones–and to insects in particular.

First, most of the animals on earth are insects, or, more precisely, arthropods. Second, the nervous system of a typical invertebrate has far fewer neurons than found in a mammal or even a "lower vertebrate" such as a fish or frog. The fly's entire visual brain has roughly 5×10^5 cells, while just the primary visual cortex of a monkey has $\sim 10^9$. Third, many of the cells in the invertebrate nervous system are *identified*: cells of essentially the same structure occur in every individual, and that if one records the response of these cells to sensory stimuli (for example) these responses are reproducible from individual to individual[12]. Thus the cells can be named and numbered based on their structure or function in the neural circuit[13]. Finally, the overall physiology of invertebrates allows for very long, stable recordings of the electrical activity of their neurons. In short, experiments on invertebrate nervous systems look and feel like the physics experiments with which we all grew up–stable samples with quantitatively reproducible behavior.

Of course what I give here is meant to sound like a rational account of why one might choose the fly as a model system. In fact my own choice was driven by the good fortune of finding myself as a postdoc in Groningen some years ago, where in the next office Rob de Ruyter van Steveninck was working on his thesis. When Rob showed me the kinds of experiments he could do–recording from a single neuron for a week, or from photoreceptor cells all afternoon–we both realized that this was a golden opportunity to bring theory and experiment together in studying a variety of problems in neural coding and computation. Since this is a school, and hence the lectures have in part the flavor of advice to students, I should point out that (1) the whole process of theory/experiment collaboration was made easier by the fact that Rob himself is a physicist, and indeed the Dutch were then far ahead of the rest of the world in bringing biophysics into physics departments, but (2) despite all the positive factors, including the fact that as postdoc and student we had little else to do, it still took months for us to formulate a reasonable first step in our collaboration. I admit that it is only after having some success with Rob that I have have had the courage to venture into collaborations with real biologists.

What do flies actually do with their visual brains? If you watch a fly flying around in a room or outdoors, you will notice that flight paths tend to consist of rather straight segments interrupted by sharp turns. These observations can be quantified through the measurement of trajectories during

[12]This should not be taken to mean that the properties of neurons in the fly's brain are fixed by genetics, or that all individuals in the species are identical. Indeed, we will come to the question of individuality *vs.* universality in what follows. What is important here is that neural responses are sufficiently reproducible that one can speak meaningfully about the properties of corresponding cells in different individuals.

[13]If you want to know more about the structure of a fly's brain, there is a beautiful book by Strausfeld [35], but this is very hard to find. An alternative is the online flybrain project that Strausfeld and others have been building [36]. Another good general reference is the collection of articles edited by Stavenga & Hardie [37].

free [38, 39] or lightly tethered [40, 41] flight and in experiments where the fly is suspended from a torsion balance [42]. Given the aerodynamics for an object of the fly's dimensions, even flying straight is tricky. In the torsion balance one can demonstrate directly that motion across the visual field drives the generation of torque, and the sign is such as to stabilize flight against rigid body rotation of the fly. Indeed one can close the feedback loop by measuring the torque which the fly produces and using this torque to (counter)rotate the visual stimulus, creating an imperfect "flight simulator" for the fly in which the only cues to guide the flight are visual; under natural conditions the fly's mechanical sensors play a crucial role. Despite the imperfections of the flight simulator, the tethered fly will fixate small objects, thereby stabilizing the appearance of straight flight. Similarly, Land and Collett showed that aspects of flight behavior under free flight conditions can be understood if flies generate torques in response to motion across the visual field, and that this response is remarkably fast, with a latency of just $\sim$30 msec [38]. The combination of free flight and torsion balance experiments strongly suggests that flies can estimate their angular velocity from visual input alone, and then produce motor outputs based on this estimate [42].

When you look down on the head of a fly, you see–almost to the exclusion of anything else–the large compound eyes. Each little hexagon that you see on the fly's head is a separate lens, and in large flies there are $\sim$5000 lenses in each eye, with approximately 1 receptor cell behind each lens[14], and roughly 100 brain cells per lens devoted to the processing of visual information. The lens focuses light on the receptor, which is small enough to act as an optical waveguide. Each receptor sees only a small portion of the world, just as in our eyes; one difference between flies and us is that diffraction is much more significant for organisms with compound eyes–because the lenses are so small, flies have an angular resolution of about 1°, while we do about $100\times$ better. There is a beautiful literature on optimization principles for the design of the compound eye; the topic even makes an appearance in the Feynman lectures.

Voltage signals from the receptor cells are processed by several layers of the brain, each layer having cells organized on a lattice which parallels the

[14]This is the sort of sloppy physics speak which annoys biologists. The precise statement is different in different insects. For flies there are eight receptors behind each lens. Two provide sensitivity to polarization and some color vision, but these are not used for motion sensing. The other six receptors look out through the same lens in different directions, but as one moves to neighboring lenses one finds that there is one cell under each of six neighboring lenses which looks in the same direction. Thus these six cells are equivalent to one cell with six times larger photon capture cross section, and the signals from these cells are collected and summed in the first processing stage (the lamina). One can even see the expected six fold improvement in signal to noise ratio [43].

lattice of lenses visible from the outside of the fly. After passing through the lamina, the medulla, and the lobula, signals arrive at the lobula plate. Here there is a stack of about 50 cells which are are sensitive to motion [44, 45]. The cells all are identified in the sense defined above, and are specialized to detect different kinds of motion. If one kills individual cells in the lobula plate then the simple experiment of moving a stimulus and recording the flight torque no longer works [46], strongly suggesting that these cells are an obligatory link in the pathway from the retina to the flight motor. If one lets the fly watch a randomly moving pattern, then it is possible to "decode" the responses of the movement sensitive neurons and to reconstruct the time dependent angular velocity signal [47], as will be discussed below. Taken together, these observations support a picture in which the fly's brain uses photoreceptor signals to estimate angular velocity, and encodes this estimate in the activity of a few neurons[15]. Further, we can study the photoreceptor signals (and noise) as well as the responses of motion–sensitive neurons with a precision almost unmatched in any other set of neurons: thus we have a good idea of what the system is "trying" to do, and we have tremendous access to both inputs and outputs. I'll try to make several points:

- Sequences of a few action spikes from the H1 neuron allow for discrimination among different motions with a precision close to the limit set by noise in the photodetector array;
- With continuous motion, the spike train of H1 can be decoded to recover a running estimate of the motion signal, and the precision of this estimate is again within a factor of two of the theoretical limit;

[15]Let me emphasize that you should be skeptical of any specific claim about what the brain computes. The fact that flies can stabilize their flight using visual cues, for example, does *not* mean that they compute motion in any precise sense–they could use a form of "bang–bang" control that needs knowledge only of the algebraic sign of the velocity, although I think that the torsion balance experiments argue against such a model. It also is a bit mysterious why we find neurons with such understandable properties: one could imagine connecting photoreceptors to flight muscles via a network of neurons in which there is nothing that we could recognize as a motion–sensitive cell. Thus it is not obvious either that the fly must compute motion or that there must be motion–sensitive neurons (one might make the same argument about whether there needs to be a whole area of motion–sensitive neurons in primate cortex, as observed). As you will see, when the dust settles I will claim that flies in fact compute motion *optimally*. The direct evidence for this claim comes from careful examination of the responses of single neurons. We don't know why the fly goes to the trouble of doing this, and in particular it is hard to point to a behavior in which this precision has been demonstrated experimentally (or is plausibly necessary for survival). This is a first example of the laundry list problem: if the brain makes optimal estimates of x, y and z, then we have an opening to a principled theory of $x-$, $y-$, and $z-$perception and the corresponding neural responses, but we don't know why the system chooses to estimate x, y, z as opposed to x', y', and z'. Hang in there... we'll try to address this too!

- Analogies between the formal problem of optimal estimation and statistical physics help us to develop a theory of optimal estimation which predicts the structure of the computation that the fly *should* do to achieve the most reliable motion estimates;
- In several cases we can find independent evidence that the fly's computation has the form predicted by optimal estimation theory, including some features of the neural response that were found only after the theoretical predictions;

We start by getting an order–of–magnitude feel for the theoretical limits.

Suppose that we look at a pattern of typical contrast C and it moves by an angle $\delta\theta$. A single photodetector element will see a change in contrast of roughly $\delta C \sim C \cdot (\delta\theta/\phi_0)$, where ϕ_0 is the angular scale of blurring due to diffraction. If we can measure for a time τ, we will count an average number of photons $R\tau$, with R the counting rate per detector, and hence the noise can be expressed a fractional precision in intensity of $\sim 1/\sqrt{R\tau}$. But fractional intensity is what we mean by contrast, so $1/\sqrt{R\tau}$ is really the contrast noise in one photodetector. To get the signal to noise ratio we should compare the signal and noise in each of the $N_{\rm cells}$ detectors, then add the squares if we assume (as for photon shot noise) that noise is independent in each detector while the signal is coherent. The result is

$$SNR \sim N_{\rm cells} \cdot \left(\frac{\delta\theta}{\phi_0}\right)^2 C^2 R\tau. \tag{3.3}$$

This calculation is rough, and we can do a little better [4,48], but it contains the right ideas. Motion discrimination is hard for flies because they have small lenses and hence blurry images (ϕ_0 is large) and because they have to respond quickly (τ is small). Under reasonable laboratory conditions the optimal estimator would reach $SNR = 1$ at an angular displacement of $\delta\theta \sim 0.05°$.

We can test the precision of motion estimation in two very different ways. One is similar to the experiments described for photon counting or for bat echolocation: we create two alternatives and ask if we can discriminate reliably. For the motion sensitive neurons in the fly visual system Rob pursued this line by recording the responses of a single neuron (H1, which is sensitive, not surprisingly, to horizontal motions) to steps of motion that have either an amplitude θ_+ or an amplitude θ_- [49]. The cell responds with a brief volley of action potentials which we can label as occurring at times $t_1, t_2, \ldots$ We as observers of the neuron can look at these times and try to decide whether the motion had amplitude θ_+ or θ_-; the idea is exactly the same as in Problem 2, but here we have to measure the distributions $P_\pm(t_1, t_2, \ldots)$ rather than making assumptions about their form. Doing the integrals, one finds that looking at spikes generated in the first ~30 msec after the step (as in the fly's behavior) we can reach the reliability expected for $SNR = 1$ at a displacement $\delta\theta = |\theta_+ - \theta_-| \sim 0.12°$, within a factor of two of the theoretical limit set by noise in the photodetectors. These data are quite rich, and it is worth noting a few more points that emerged from the analysis:

- On the ~30 msec time scale of relevance to behavior, there are only a handful of spikes. This is partly what makes it possible to do the analysis so completely, but it also is a lesson for how we think about the neural representation of information in general;

- Dissecting the contributions of individual spikes, one finds that each successive spike makes a nearly independent contribution to the signal to noise ratio for discrimination, so there is essentially no redundancy;

- Even one or two spikes are enough to allow discrimination of motions much smaller than the lattice spacing on the retina or the nominal "diffraction limit" of angular resolution. Analogous phenomena have been known in human vision for more than a century and are called hyperacuity; see Section 4.2 in *Spikes* for a discussion [5].

The step discrimination experiment gives us a very clear view of reliability in the neural response, but as with the other discrimination experiments discussed above it's not a very natural task. An alternative is to ask what happens when the motion signal (angular velocity $\dot{\theta}(t)$) is a complex function of time. Then we can think of the signal to noise ratio in equation (3.3) as being equivalent to a spectral density of displacement noise $N_\theta^{\rm eff} \sim \phi_0^2/(N_{\rm cells}C^2R)$, or a generalization in which the photon counting rate is replaced by an effective (frequency dependent) rate related to the noise characteristics of the photoreceptors [48]. It seems likely, as discussed above, that the fly's visual system really does make a continuous or running estimate of the angular velocity, and that this estimate is encoded in the sequence of discrete spikes produced by neurons like H1. It is not clear that any piece of the brain ever "decodes" this signal in an explicit way, but if *we* could do such a decoding we could test directly whether the accuracy of our decoding reaches the limiting noise level set by noise in the photodetectors.

The idea of using spike trains to recover continuous time dependent signals started with this analysis of the fly visual system [47, 50, 51], and has since expanded to many different systems [5]. Generalizations of these ideas to decoding from populations of neurons [52] even have application to future prosthetic devices which might be able to decode the commands given in motor cortex to control robot arms [53]. Here our interest is not so much in the structure of the code, or in the usefulness of the decoding;

rather our goal is to use the decoding as a tool to characterize the precision of the underlying computations.

To understand how we can decode a continuous signal from discrete sequences of spike it is helpful to have an example, following Ref. [51]. Suppose that the signal of interest is $s(t)$ and the neuron we are looking at generates action potentials according to a Poisson process with a rate $r(s)$. Then the probability that we observe spikes at times $t_1, t_2, \ldots, t_N \equiv \{t_i\}$ given the signal $s(t)$ is (generalizing from Problem 1)

$$P[\{t_i\}|s(t)] = \frac{1}{N!} \exp\left[-\int \mathrm{d}t\, r(s(t))\right] \prod_{i=1}^{N} r(s(t_i)). \tag{3.4}$$

For simplicity let us imagine further that the signal s itself comes from a Gaussian distribution with zero mean, unit variance and a correlation time τ_c, so that

$$P[s(t)] \propto \exp\left[-\frac{1}{4\tau_\mathrm{c}} \int \mathrm{d}t\, \left(\tau_\mathrm{c}^2 \dot{s}^2 + s^2\right)\right] \cdot \tag{3.5}$$

Our problem is not to predict the spikes from the signal, but rather given the spikes to estimate the signal, which means that we are interested in conditional distribution $P[s(t)|\{t_i\}]$. From Bayes' rule,

$$P[s(t)|\{t_i\}] = \frac{P[\{t_i\}|s(t)]P[s(t)]}{P[\{t_i\}]} \tag{3.6}$$

$$\propto \exp\left[-\frac{\tau_\mathrm{c}}{4} \int \mathrm{d}t\, \dot{s}^2 - \int \mathrm{d}t V_\mathrm{eff}(s(t))\right] \prod_{i=1}^{N} r(s(t_i)), \tag{3.7}$$

where

$$V_\mathrm{eff}(s) = \frac{1}{4\tau_\mathrm{c}} s^2 + r(s). \tag{3.8}$$

I write the probability distribution in this form to remind you of the (Euclidean) path integral description of quantum mechanics, where the amplitude for a particle of mass m to move along a trajectory $x(t)$ in a potential $V(x)$ is given by

$$A[x(t)] \propto \exp\left[-\frac{m}{2} \int \mathrm{d}t\, \dot{x}^2 - \int \mathrm{d}t V(x(t))\right], \tag{3.9}$$

in units where $\hbar = 1$. If we want to estimate $s(t)$ from the probability distribution $P[s(t)|\{t_i\}]$, then we can compute the conditional mean, which will give us the best estimator in the sense that χ^2 between our estimate and the true signal will be minimized (see Problem 5 below). Thus the estimate at some particular time t_0 is given by

$$s_{\rm est}(t_0) = \int Ds(t)\, s(t_0) P[s(t)|\{t_i\}] \tag{3.10}$$

$$\propto \left\langle s(t_0) \prod_{i=1}^{N} r(s(t_i)) \right\rangle, \tag{3.11}$$

where $\langle \cdots \rangle$ stands for an expectation value over trajectories drawn from the distribution

$$P_{\rm eff}[s(t)] \propto \exp\left[-\frac{\tau_{\rm c}}{4} \int {\rm d}t\, \dot{s}^2 - \int {\rm d}t V_{\rm eff}(s(t)) \right] \cdot \tag{3.12}$$

Thus estimating the trajectory involves computing an $N+1$–point function in the quantum mechanics problem defined by the potential $V_{\rm eff}$.

Problem 5: Optimal estimators. Imagine that you observe y, which is related to another variable x that you actually would like to know. This relationship can be described by the joint probability distribution $P(x, y)$, which you know. Any estimation strategy can be described as computing some function $x_{\rm est} = F(y)$. For any estimate we can compute the expected value of χ^2,

$$\chi^2 = \int {\rm d}x{\rm d}y P(x, y) |x - F(y)|^2. \tag{3.13}$$

Show that the estimation strategy which minimizes χ^2 is the computation of the conditional mean,

$$F_{\rm opt}(y) = \int {\rm d}x\, x P(x|y). \tag{3.14}$$

If we took any particular model seriously we could in fact try to compute the relevant expectation values, but here (and in applications of these ideas to analysis of real neurons) I don't really want to trust these details; rather I want to focus on some general features. First one should note that trying to do the calculations in a naive perturbation theory will work only in some very limited domain. Simple forms of perturbation theory are equivalent to the statement that interesting values of the signal $s(t)$ do not sample strongly nonlinear regions of the input/output relation $r(s)$, and this is unlikely to be true in general. On the other hand, there is a chance to do something simple, and this is a cluster expansion:

$$\left\langle s(t_0) \prod_{i=1}^{N} r(s(t_i)) \right\rangle \approx \langle s(t_0) \rangle \prod_{i=1}^{N} \langle r(s(t_i)) \rangle$$

$$+ A \sum_{i=1}^{N} \langle \delta s(t_0) \delta r(s(t_i)) \rangle + \cdots , \tag{3.15}$$

where δs refers to fluctuations around around the mean $\langle s\rangle$, and similarly for r, while A is a constant. As with the usual cluster expansions in statistical physics, this does not require perturbations to be weak; in particular the relation $r(s)$ can have arbitrarily strong (even discontinuous) nonlinearities. What is needed for a cluster expansion to make sense is that the times which appear in the N–point functions be far apart when measured in units of the correlation times for the underlying trajectories. In the present case, this will happen (certainly) if the times between spikes are long compared with the correlation times of the signal. Interestingly, as the mean spike rate gets higher, it is the correlation times computed in the full $V_{\rm eff}(s)$ which are important, and these are smaller than the bare correlation time in many cases. At least in this class of models, then, there is a regime of low spike rate where we can use a cluster expansion, and this regime extends beyond the naive crossover determined by the number of spikes per correlation time of s. As explained in *Spikes,* there are good reasons to think that many neurons actually operate in this limit of spike trains which are "sparse in the time domain" [5].

What are the consequences of the cluster expansion? If we can get away with keeping the leading term, and if we don't worry about constant offsets in our estimates (which can't be relevant...), then we have

$$s_{\rm est}(t) = \sum_{i=1}^{N} f(t - t_i) + \cdots , \qquad (3.16)$$

where $f(t)$ again is something we could calculate if we trusted the details of the model. But this is very simple: we can estimate a continuous time dependent signal just by adding up contributions from individual spikes, or equivalently by filtering the sequence of spikes. If we don't want to trust a calculation of $f(t)$ we can just use experiments to find $f(t)$ by asking for that function which gives us the smallest value of χ^2 between our estimate and the real signal, and this optimization problem is also simple since χ^2 is quadratic in f [5, 47]. So the path is clear–do a long experiment on the response of a neuron to signals drawn from some distribution $P[s(t)]$, use the first part of the experiment to find the filter $f(t)$ such that the estimate in equation (3.16) minimizes χ^2, and then test the accuracy of our estimates on the remainder of the data. This is exactly what we did with H1 [47], and we found that over a broad range of frequencies the spectral density of errors in our estimates was within a factor of two of the limit set by noise in the photoreceptor cells. Further, we could change, for example, the image contrast and show that the resulting error spectrum scaled as expected from the theoretical limit [5].

To the extent that the fly's brain can estimate motion with a precision close to the theoretical limit, one thing we know is that the act of processing itself does not add too much noise. But being quiet is not enough: to make maximally reliable estimates of nontrivial stimulus features like motion one must be sure to do the correct computation. To understand how this works let's look at a simple example, and then I'll outline what happens when we use the same formalism to look at motion estimation. My discussion here follows joint work with Marc Potters [54].

Suppose that someone draws a random number x from a probability distribution $P(x)$. Rather than seeing x itself, you get to see only a noisy version, $y = x + \eta$, where η is drawn from a Gaussian distribution with variance σ^2. Having seen y, your job is to estimate x, and for simplicity let's say that the "best estimate" is best in the least squares sense, as above. Then from Problem 5 we know that the optimal estimator is the conditional mean,

$$x_{\rm est}(y) = \int \mathrm{d}x\, xP(x|y). \tag{3.17}$$

Now we use Bayes' rule and push things through:

$$\begin{aligned} x_{\rm est}(y) &= \int \mathrm{d}x\, xP(y|x)P(x)\frac{1}{P(y)} && (3.18)\\ &= \frac{1}{P(y)}\frac{1}{\sqrt{2\pi\sigma^2}}\int \mathrm{d}x\, xP(x)\exp\left[-\frac{1}{2\sigma^2}(y-x)^2\right] && (3.19)\\ &= \frac{1}{Z(y)}\int \mathrm{d}x\, x\exp\left[-\frac{V_{\rm eff}(x)}{k_{\rm B}T_{\rm eff}} + \frac{F_{\rm eff}x}{k_{\rm B}T_{\rm eff}}\right], && (3.20) \end{aligned}$$

where we can draw the analogy with statistical mechanics by noting the correspondences:

$$\begin{aligned} \frac{V_{\rm eff}(x)}{k_{\rm B}T_{\rm eff}} &= -\ln P(x) + \frac{x^2}{2\sigma^2} && (3.21)\\ k_{\rm B}T_{\rm eff} &= \sigma^2 && (3.22)\\ F_{\rm eff} &= y. && (3.23) \end{aligned}$$

Thus, making optimal estimates involves computing expectation values of position, the potential is (almost) the prior distribution, the noise level is the temperature, and the data act as an external force. The connection with statistical mechanics is more than a convenient analogy; it helps us in finding approximation schemes. Thus at large noise levels we are at high temperatures, and all other things being equal the force is effectively small so we can compute the response to this force in perturbation theory. On the other hand, at low noise levels the computation of expectation values must be dominated by ground state or saddle point configurations.

Statistical mechanics	**Optimal Estimation**
Temperature	Noise level
Potential	(log) Prior distribution
Average position	Estimate
External force	Input data

In the case of visual motion estimation, the data are the voltages produced by the photoreceptors $\{V_n(t)\}$ and the thing we are trying to estimate is the angular velocity $\dot{\theta}(t)$. These are functions rather than numbers, but this isn't such a big deal if we are used to doing functional integrals. The really new point is that $\dot{\theta}(t)$ does not directly determine the voltages. What happens instead is that even if the fly flies along a perfectly straight path, so that $\dot{\theta} = 0$, the world effectively projects a movie $C(\mathbf{x}, t)$ onto the retina and it is this movie which determines the voltages (up to noise). The motion $\theta(t)$ transforms this movie, and so enters in a strongly nonlinear way; if we take a one dimensional approximation we might write $C(x, t) \to C(x - \theta(t), t)$. Each photodetector responds to the contrast as seen through an aperture function $M(x - x_n)$ centered on a lattice point x_n in the retina, and for simplicity let's take the noise to be dominated by photon shot noise. Since we have independent knowledge of the $C(x, t)$, to describe the relationship between $\dot{\theta}(t)$ and $\{V_n(t)\}$ we have to integrate over all possible movies, weighted by their probability of occurrence $P[C]$.

Putting all of these things together with our general scheme for finding optimal estimates, we have

$$\dot{\theta}_{\rm est}(t_0) = \int D\theta\, \dot{\theta}(t_0) P[\theta(t)|\{V_n(t)\}] \tag{3.24}$$

$$P[\theta(t)|\{V_n(t)\}] = P[\{V_n(t)\}|\theta(t)] P[\theta(t)] \frac{1}{P[\{V_n(t)\}]} \tag{3.25}$$

$$P[\{V_n(t)\}|\theta(t)] \propto \int DC\, P[C] \exp\left[-\frac{R}{2} \sum_n \int {\rm d}t\, |V_n(t) - \bar{V}_n(t)|^2 \right] \tag{3.26}$$

$$\bar{V}_n(t) = \int {\rm d}x M(x - x_n) C(x - \theta(t), t). \tag{3.27}$$

Of course we can't do these integrals exactly, but we can look for approximations, and again we can do perturbation theory at low signal to noise levels and search for saddle points at high signal to noise ratios. When the dust settles we find [54]:

- At low signal to noise ratios the optimal estimator is quadratic in the receptor voltages,

$$\dot{\theta}_{\rm est}(t) \approx \sum_{nm} \int {\rm d}\tau \int {\rm d}\tau' V_n(t-\tau) K_{nm}(\tau,\tau') V_m(t-\tau'). \quad (3.28)$$

- At moderate signal to noise ratios, terms with higher powers of the voltage become relevant and "self energy" terms provide corrections to the kernel $K_{nm}(\tau,\tau')$.

- At high signal to noise ratios averaging over time becomes less important and the optimal estimator crosses over to

$$\dot{\theta}_{\rm est}(t) \approx \frac{\sum_n \dot{V}_n(t)[V_n(t) - V_{n-1}(t)]}{{\rm constant} + \sum_n [V_n(t) - V_{n-1}(t)]^2}, \quad (3.29)$$

  where constant depends on the typical contrast and dynamics in the movies chosen from $P[C(x,t)]$ and on the typical scale of angular velocities in $P[\theta(t)]$.

Before looking at the two limits in detail, note that the whole form of the motion computation depends on the statistical properties of the visual environment. Although the limits look very different, one can show that there is no phase transition and hence that increasing signal to noise ratio takes us smoothly from one limit to the other; although this is sort of a side point, it was a really nice piece of work by Marc. An obvious question is whether the fly is capable of doing these different computations under different conditions.

We can understand the low signal to noise ratio limit by realizing that when something moves there are correlations between what we see at the two space–time points (x,t) and $(x+v\tau, t+\tau)$. These correlations extend to very high orders, but as the background noise level increases the higher order correlations are corrupted first, until finally the only reliable thing left is the two–point function, and closer examination shows that near neighbor correlations are the most significant: we can be sure something is moving because signals in neighboring photodetectors are correlated with a slight delay. This form of "correlation based" motion computation was suggested long ago by Reichardt and Hassenstein based on behavioral experiments with beetles [55]; later work from Reichardt and coworkers explored the applicability of this model to fly behavior [42]. Once the motion sensitive neurons were discovered it was natural to check if their responses could be understood in these terms.

There are two clear signatures of the correlation model. First, since the receptor voltage is linear in response to image contrast, the correlation model confounds contrast with velocity: all things being equal, doubling the image contrast causes our estimate of the velocity to increase by a factor of four (!). This is an observed property of the flight torque that flies generate in response to visual motion, at least at low contrasts, and the same quadratic behavior can be seen in the rate at which motion sensitive neurons generate spikes and even in human perception (at very low contrast). Although this might seem strange, it's been known for decades. What is interesting here is that this seemingly disastrous confounding of signals occurs even in the optimal estimator: optimal estimation involves a tradeoff between systematic and random errors, and at low signal to noise ratio this tradeoff pushes us toward making larger systematic errors, apparently of a form made by real brains.

The second signature of correlation computation is that we can produce movies which have the right spatiotemporal correlations to generate a nonzero estimate $\dot{\theta}_{\rm est}$ but don't really have anything in them that we would describe as "moving" objects or features. Rob de Ruyter has a simple recipe for doing this [56], which is quite compelling (I recommend you try it yourself): make a spatiotemporal white noise movie $\psi(\vec{\mathbf{x}}, t)$,

$$\langle \psi(\vec{\mathbf{x}}, t)\psi(\vec{\mathbf{x}}', t')\rangle = \delta(\vec{\mathbf{x}} - \vec{\mathbf{x}}')\delta(t - t'), \tag{3.30}$$

and then add the movie to itself with a weight and an offset:

$$C(\vec{\mathbf{x}}, t) = \psi(\vec{\mathbf{x}}, t) + a\psi(\vec{\mathbf{x}} + \Delta\vec{\mathbf{x}}, t + \Delta t). \tag{3.31}$$

Composed of pure noise, there is nothing really moving here. If you watch the movie, however, there is no question that you think it's moving, and the fly's neurons respond too (just like yours, presumably). Even more impressive is that if you change the *sign* of the weight a... the direction of motion reverses, as predicted from the correlation model.

Because the correlation model has a long history, it is hard to view evidence for this model as a success of optimal estimation theory. The theory of optimal estimation also predicts, however, that the kernel $K_{nm}(\tau, \tau')$ should adapt to the statistics of the visual environment, and it does. Most dramatically one can just show random movies with different correlation times and then probe the transient response of H1 to step motions; the time course of transients (presumably reflecting the details of K) can vary over nearly two orders of magnitude from 30–300 msec in response to different correlation times [56]. All of this makes sense in light of optimal estimation theory but again perhaps is not a smoking gun. Closer to a real test is Rob's argument that the absolute values of the time constants seen in such adaptation experiments match the frequencies at which natural movies would fall below

$SNR = 1$ in each photoreceptor, so that filtering in the visual system is set (adaptively) to fit the statistics of signals and noise in the inputs [56, 57].

What about the high SNR limit? If we remember that voltages are linear in contrast, and let ourselves ignore the lattice in favor of a continuum limit, then the high SNR limit has a simple structure,

$$\dot{\theta}_{\rm est}(t) \approx \frac{\int {\rm d}x\,(\partial_t C)(\partial_x C)}{B + \int {\rm d}x\,(\partial_x C)^2} \rightarrow \frac{\partial_t C}{\partial_x C}, \tag{3.32}$$

where the last limit is at high contrasts. As promised by the lack of a phase transition, this starts as a quadratic function of contrast just like the correlator, but saturates as the ratio of temporal and spatial derivatives. Note that if $C(x,t) = C(x+vt)$, then this ratio recovers the velocity v exactly. This simple ratio computation is not optimal in general because real movies have dynamics other than rigid motion and real detectors have noise, but there is a limit in which it must be the right answer. Interestingly, the correlation model (with multiplicative nonlinearity) and the ratio of derivatives model (with a divisive nonlinearity) have been viewed as mutually exclusive models of how brains might compute motion. One of the really nice results of optimal estimation theory is to see these seemingly very different models emerge as different limits of a more general strategy. But, do flies know about this more general strategy?

The high SNR limit of optimal estimation predicts that the motion estimate (and hence, for example, the rate at which motion sensitive neurons generate spikes) should saturate as a function of contrast, but this contrast–saturated level should vary with velocity. Further, if one traces through the calculations in more detail, the the constant B and hence the contrast level required for (*e.g.*) half–saturation should depend on the typical contrast and light intensity in the environment. Finally, this dependence on the environment really is a response to the statistics of that environment, and hence the system must use some time and a little "memory" to keep track of these statistics–the contrast response function should reflect the statistics of movies in the recent past. All of these things are observed [56–58].

So, where are we? The fly's visual system makes nearly optimal estimates of motion under at least some conditions that we can probe in the lab. The theory of optimal estimation predicts that the structure of the motion computation ought to have some surprising properties, and many of these are observed–some were observed only when theory said to go look for them, which always is better. I would like to get a clearer demonstration that the system really takes a ratio, and I think we're close to doing that [59]. Meanwhile, one might worry that theoretical predictions depend too much on assumptions about the structure of the relevant probability distributions $P[C]$ and $P[\theta]$, so Rob is doing experiments where he walks through the woods with both a camera and gyroscope mounted on his head (!), sampling the joint distribution of movies and motion trajectories. Armed with these samples one can do all of the relevant functional integrals by Monte Carlo, which really is lovely since now we are in the real natural distribution of signals. I am optimistic that all of this will soon converge on a complete picture. I also believe that the success so far is sufficient to motivate a more general look at the problem of optimization as a design principle in neural computation.

4 Toward a general principle?

One attempt to formulate a general principle for neural computation goes back to the work of Attneave [60] and Barlow [61,62] in the 1950s. Focusing on the processing of sensory information, they suggested that an important goal of neural computation is to provide an efficient representation of the incoming data, where the intuitive notion of efficiency could be made precise using the ideas of information theory [63].

Imagine describing an image by giving the light intensity in each pixel. Alternatively, we could give a description in terms of objects and their locations. The latter description almost certainly is more efficient in a sense that can be formalized using information theory. The idea of Barlow and Attneave was to turn this around–perhaps by searching for maximally efficient representations of natural scenes we would be forced to discover and recognize the objects out of which our perceptions are constructed. Efficient representation would have the added advantage of allowing the communication of information from one brain region to another (or from eye to brain along the optic nerve) with a minimal number of nerve fibers or even a minimal number of action potentials. How could we test these ideas?

- If we make a model for the class of computations that neurons can do, then we could try to find within this class the one computation that optimizes some information theoretic measure of performance. This should lead to predictions for what real neurons are doing at various stages of sensory processing;

- We could try to make a direct measurement of the efficiency with which neurons represent sensory information;

- Because efficient representations depend on the statistical structure of the signals we are trying to represent, a truly efficient brain would adapt its strategies to track changes in these statistics, and we could search for this "statistical adaptation". Even better would be if we could show that the adaptation has the right form to optimize efficiency.

Before getting started on this program we need a little review of information theory itself[16]. Almost all statistical mechanics textbooks note that the entropy of a gas measures our lack of information about the microscopic state of the molecules, but often this connection is left a bit vague or qualitative. Shannon proved a theorem that makes the connection precise [63]: entropy is the unique measure of available information consistent with certain simple and plausible requirements. Further, entropy also answers the practical question of how much space we need to use in writing down a description of the signals or states that we observe.

Two friends, Max and Allan, are having a conversation. In the course of the conversation, Max asks Allan what he thinks of the headline story in this morning's newspaper. We have the clear intuitive notion that Max will "gain information" by hearing the answer to his question, and we would like to quantify this intuition. Following Shannon's reasoning, we begin by assuming that Max knows Allan very well. Allan speaks very proper English, being careful to follow the grammatical rules even in casual conversation. Since they have had many political discussions Max has a rather good idea about how Allan will react to the latest news. Thus Max can make a list of Allan's possible responses to his question, and he can assign probabilities to each of the answers. From this list of possibilities and probabilities we can compute an entropy, and this is done in exactly the same way as we compute the entropy of a gas in statistical mechanics or thermodynamics: If the probability of the nth possible response is p_n, then the entropy is

$$S = -\sum_n p_n \log_2 p_n \text{ bits.} \tag{4.1}$$

The entropy S measures Max's uncertainty about what Allan will say in response to his question. Once Allan gives his answer, all this uncertainty is removed–one of the responses occurred, corresponding to $p = 1$, and all the others did not, corresponding to $p = 0$–so the entropy is reduced to zero. It is appealing to equate this reduction in our uncertainty with the information we gain by hearing Allan's answer. Shannon proved that this is not just an interesting analogy; it is the *only* definition of information that conforms to some simple constraints.

[16]At Les Houches this review was accomplished largely by handing out notes based on courses given at MIT, Princeton and ICTP in 1998–99. In principle they will become part of a book to be published by Princeton University Press, tentatively titled *Entropy, Information and the Brain.* I include this here so that the presentation is self–contained, and apologize for the eventual self–plagiarism that may occur.

To start, Shannon assumes that the information gained on hearing the answer can be written as a function of the probabilities p_n[17]. Then if all N possible answers are equally likely the information gained should be a monotonically increasing function of N. The next constraint is that if our question consists of two parts, and if these two parts are entirely independent of one another, then we should be able to write the total information gained as the sum of the information gained in response to each of the two subquestions. Finally, more general multipart questions can be thought of as branching trees, where the answer to each successive part of the question provides some further refinement of the probabilities; in this case we should be able to write the total information gained as the weighted sum of the information gained at each branch point. Shannon proved that the only function of the $\{p_n\}$ consistent with these three postulates–monotonicity, independence, and branching–is the entropy S, up to a multiplicative constant.

If we phrase the problem of gaining information from hearing the answer to a question, then it is natural to think about a discrete set of possible answers. On the other hand, if we think about gaining information from the acoustic waveform that reaches our ears, then there is a continuum of possibilities. Naively, we are tempted to write

$$S_{\text{continuum}} = -\int \mathrm{d}x P(x) \log_2 P(x), \tag{4.2}$$

or some multidimensional generalization. The difficulty, of course, is that probability distributions for continuous variables [like $P(x)$ in this equation] have units–the distribution of x has units inverse to the units of x–and we should be worried about taking logs of objects that have dimensions. Notice that if we wanted to compute a difference in entropy between two distributions, this problem would go away. This is a hint that only entropy differences are going to be important[18].

[17] In particular, this "zeroth" assumption means that we must take seriously the notion of enumerating the possible answers. In this framework we cannot quantify the information that would be gained upon hearing a previously unimaginable answer to our question.

[18] The problem of defining the entropy for continuous variables is familiar in statistical mechanics. In the simple example of an ideal gas in a finite box, we know that the quantum version of the problem has a discrete set of states, so that we can compute the entropy of the gas as a sum over these states. In the limit that the box is large, sums can be approximated as integrals, and if the temperature is high we expect that quantum effects are negligible and one might naively suppose that Planck's constant should disappear from the results; we recall that this is not quite the case. Planck's constant has units of momentum times position, and so is an elementary area for each pair of conjugate position and momentum variables in the classical phase space; in the classical limit the entropy becomes (roughly) the logarithm of the occupied volume in

Returning to the conversation between Max and Allan, we assumed that Max would receive a complete answer to his question, and hence that all his uncertainty would be removed. This is an idealization, of course. The more natural description is that, for example, the world can take on many states W, and by observing data D we learn something but not everything about W. Before we make our observations, we know only that states of the world are chosen from some distribution $P(W)$, and this distribution has an entropy $S(W)$. Once we observe some particular datum D, our (hopefully improved) knowledge of W is described by the conditional distribution $P(W|D)$, and this has an entropy $S(W|D)$ that is smaller than $S(W)$ if we have reduced our uncertainty about the state of the world by virtue of our observations. We identify this reduction in entropy as the information that we have gained about W.

Problem 6: Maximally informative experiments. Imagine that we are trying to gain information about the correct theory T describing some set of phenomena. At some point, our relative confidence in one particular theory is very high; that is, $P(T = T_*) > F \cdot P(T \neq T_*)$ for some large F. On the other hand, there are many possible theories, so our absolute confidence in the theory T_* might nonetheless be quite low, $P(T = T_*) << 1$. Suppose we follow the "scientific method" and design an experiment that has a yes or no answer, and this answer is perfectly correlated with the correctness of theory T_*, but uncorrelated with the correctness of any other possible theory–our experiment is designed specifically to test or falsify the currently most likely theory. What can you say about how much information you expect to gain from such a measurement? Suppose instead that you are completely irrational and design an experiment that is irrelevant to testing T_* but has the potential to eliminate many (perhaps half) of the alternatives. Which experiment is expected to be more informative? Although this is a gross cartoon of the scientific process, it is not such a terrible model of a game like "twenty questions". It is interesting to ask whether people play such question games following strategies that might seem irrational but nonetheless serve to maximize information gain [64]. Related but distinct criteria for optimal experimental design have been developed in the statistical literature [65].

Perhaps this is the point to note that a single observation D is not, in fact, guaranteed to provide positive information, as emphasized by

phase space, but this volume is measured in units of Planck's constant. If we had tried to start with a classical formulation (as did Boltzmann and Gibbs, of course) then we would find ourselves with the problems of equation (4.2), namely that we are trying to take the logarithm of a quantity with dimensions. If we measure phase space volumes in units of Planck's constant, then all is well. The important point is that the problems with defining a purely classical entropy do *not* stop us from calculating entropy differences, which are observable directly as heat flows, and we shall find a similar situation for the information content of continuous ("classical") variables.

DeWeese & Meister [66]. Consider, for instance, data which tell us that all of our previous measurements have larger error bars than we thought: clearly such data, at an intuitive level, reduce our knowledge about the world and should be associated with a negative information. Another way to say this is that some data points D will increase our uncertainty about state W of the world, and hence for these particular data the conditional distribution $P(W|D)$ has a larger entropy than the prior distribution $P(D)$. If we identify information with the reduction in entropy, $I_D = S(W) - S(W|D)$, then such data points are associated unambiguously with negative information. On the other hand, we might hope that, on average, gathering data corresponds to gaining information: although single data points can increase our uncertainty, the average over all data points does not.

If we average over all possible data–weighted, of course, by their probability of occurrence $P(D)$, we obtain the average information that D provides about W,

$$I(D \to W) = S(W) - \sum_D P(D) S(W|D) \tag{4.3}$$

$$= \sum_W \sum_D P(W, D) \log_2 \left[\frac{P(W, D)}{P(W)P(D)} \right]. \tag{4.4}$$

Note that the information which D provides about W is symmetric in D and W. This means that we can also view the state of the world as providing information about the data we will observe, and this information is, on average, the same as the data will provide about the state of the world. This "information provided" is therefore often called the mutual information, and this symmetry will be very important in subsequent discussions; to remind ourselves of this symmetry we write $I(D; W)$ rather than $I(D \to W)$.

Problem 7: Positivity of information. Prove that the mutual information $I(D \to W)$, defined in equation (4.4), is positive.

One consequence of the symmetry or mutuality of information is that we can write

$$I(D; W) = S(W) - \sum_D P(D) S(W|D) \tag{4.5}$$

$$= S(D) - \sum_W P(W) S(D|W). \tag{4.6}$$

If we consider only discrete sets of possibilities then entropies are positive (or zero), so that these equations imply

$$I(D; W) \leq S(W) \tag{4.7}$$

$$I(D; W) \leq S(D). \tag{4.8}$$

The first equation tells us that by observing D we cannot learn more about the world then there is entropy in the world itself. This makes sense: entropy measures the number of possible states that the world can be in, and we cannot learn more than we would learn by reducing this set of possibilities down to one unique state. Although sensible (and, of course, true), this is not a terribly powerful statement: seldom are we in the position that our ability to gain knowledge is limited by the lack of possibilities in the world around us[19]. The second equation, however, is much more powerful. It says that, whatever may be happening in the world, we can never learn more than the entropy of the distribution that characterizes our data. Thus, if we ask how much we can learn about the world by taking readings from a wind detector on top of the roof, we can place a bound on the amount we learn just by taking a very long stream of data, using these data to estimate the distribution $P(D)$, and then computing the entropy of this distribution.

The entropy of our observations[20] thus limits how much we can learn no matter what question we were hoping to answer, and so we can think of the entropy as setting (in a slight abuse of terminology) the capacity of the data D to provide or to convey information. As an example, the entropy of neural responses sets a limit to how much information a neuron can provide about the world, and we can estimate this limit even if we don't yet understand what it is that the neuron is telling us (or the rest of the brain). Similarly, our bound on the information conveyed by the wind detector does not require us to understand how these data might be used to make predictions about tomorrow's weather.

Since the information we can gain is limited by the entropy, it is natural to ask if we can put limits on the entropy using some low order statistical properties of the data: the mean, the variance, perhaps higher moments or correlation functions... In particular, if we can say that the entropy has a maximum value consistent with the observed statistics, then we have placed a firm upper bound on the information that these data can convey.

The problem of finding the maximum entropy given some constraint again is familiar from statistical mechanics: the Boltzmann distribution

[19]This is not quite true. There is a tradition of studying the nervous system as it responds to highly simplified signals, and under these conditions the lack of possibilities in the world can be a significant limitation, substantially confounding the interpretation of experiments.

[20]In the same way that we speak about the entropy of a gas I will often speak about the entropy of a variable or the entropy of a response. In the gas, we understand from statistical mechanics that the entropy is defined not as a property of the gas but as a property of the distribution or ensemble from which the microscopic states of the gas are chosen; similarly we should really speak here about "the entropy of the distribution of observations", but this is a bit cumbersome. I hope that the slightly sloppy but more compact phrasing does not cause confusion.

is the distribution that has the largest possible entropy given the mean energy. More generally, let us imagine that we have knowledge not of the whole probability distribution $P(D)$ but only of some expectation values,

$$\langle f_i \rangle = \sum_D P(D) f_i(D), \tag{4.9}$$

where we allow that there may be several expectation values known ($i = 1, 2, ..., K$). Actually there is one more expectation value that we always know, and this is that the average value of one is one; the distribution is normalized:

$$\langle f_0 \rangle = \sum_D P(D) = 1. \tag{4.10}$$

Given the set of numbers $\{\langle f_0 \rangle, \langle f_1 \rangle, \ldots, \langle f_K \rangle\}$ as constraints on the probability distribution $P(D)$, we would like to know the largest possible value for the entropy, and we would like to find explicitly the distribution that provides this maximum.

The problem of maximizing a quantity subject to constraints is formulated using Lagrange multipliers. In this case, we want to maximize $S = -\sum P(D) \log_2 P(D)$, so we introduce a function $\tilde{S}$, with one Lagrange multiplier λ_i for each constraint:

$$\begin{aligned} \tilde{S}[P(D)] &= -\sum_D P(D) \log_2 P(D) - \sum_{i=0}^{K} \lambda_i \langle f_i \rangle & (4.11) \\ &= -\frac{1}{\ln 2} \sum_D P(D) \ln P(D) - \sum_{i=0}^{K} \lambda_i \sum_D P(D) f_i(D). & (4.12) \end{aligned}$$

Our problem is then to find the maximum of the function $\tilde{S}$, but this is easy because the probability for each value of D appears independently. The result is that

$$P(D) = \frac{1}{Z} \exp\left[-\sum_{i=1}^{K} \lambda_i f_i(D) \right], \tag{4.13}$$

where $Z = \exp(1 + \lambda_0)$ is a normalization constant.

Problem 8: Details. Derive equation (4.13). In particular, show that equation (4.13) provides a probability distribution which genuinely *maximizes* the entropy, rather than being just an extremum.

These ideas are enough to get started on "designing" some simple neural processes. Imagine, following Laughlin [67], that a neuron is responsible for representing a single number such as the light intensity $\mathcal{I}$ averaged over small patch of the retina (don't worry about time dependence). Assume that this signal will be represented by a continuous voltage V, which is true for the first stages of processing in vision. This voltage is encoded into discrete spikes only as a second or third step. The information that the voltage provides about the intensity is

$$I(V \to \mathcal{I}) = \int \mathrm{d}\mathcal{I} \int \mathrm{d}V\, P(V,\mathcal{I}) \log_2 \left[\frac{P(V,\mathcal{I})}{P(V)P(\mathcal{I})} \right] \tag{4.14}$$

$$= \int \mathrm{d}\mathcal{I} \int \mathrm{d}V\, P(V,\mathcal{I}) \log_2 \left[\frac{P(V|\mathcal{I})}{P(V)} \right]. \tag{4.15}$$

The conditional distribution $P(V|\mathcal{I})$ describes the process by which the neuron responds to its input, and so this is what we should try to "design".

Let us suppose that the voltage is on average a nonlinear function of the intensity, and that the dominant source of noise is additive (to the voltage), independent of light intensity, and small compared with the overall dynamic range of the cell:

$$V = g(\mathcal{I}) + \xi, \tag{4.16}$$

with some distribution $P_{\rm noise}(\xi)$ for the noise. Then the conditional distribution

$$P(V|\mathcal{I}) = P_{\rm noise}(V - g(\mathcal{I})), \tag{4.17}$$

and the entropy of this conditional distribution can be written as

$$S_{\rm cond} = -\int \mathrm{d}V\, P(V|\mathcal{I}) \log_2 P(V|\mathcal{I}) \tag{4.18}$$

$$= -\int \mathrm{d}\xi\, P_{\rm noise}(\xi) \log_2 P_{\rm noise}(\xi). \tag{4.19}$$

Note that this is a constant, independent both of the light intensity and of the nonlinear input/output relation $g(\mathcal{I})$. This is useful because we can write the information as a difference between the total entropy of the output variable V and this conditional or noise entropy, as in equation (4.6):

$$I(V \to \mathcal{I}) = -\int \mathrm{d}V P(V) \log_2 P(V) - S_{\rm cond}. \tag{4.20}$$

With $S_{\rm cond}$ constant independent of our "design", maximizing information is the same as maximizing the entropy of the distribution of output voltages. Assuming that there are maximum and minimum values for this voltage, but no other constraints, then the maximum entropy distribution is just the uniform distribution within the allowed dynamic range. But if the noise is small it doesn't contribute much to broadening $P(V)$ and we calculate this distribution as if there were no noise, so that

$$P(V)\mathrm{d}V = P(\mathcal{I})\mathrm{d}\mathcal{I}, \tag{4.21}$$

$$\frac{\mathrm{d}V}{\mathrm{d}\mathcal{I}} = \frac{1}{P(V)} \cdot P(\mathcal{I}). \tag{4.22}$$

Since we want to have $V = g(\mathcal{I})$ and $P(V) = 1/(V_{\max} - V_{\min})$, we find

$$\frac{\mathrm{d}g(\mathcal{I})}{\mathrm{d}\mathcal{I}} = (V_{\max} - V_{\min})P(\mathcal{I}), \tag{4.23}$$

$$g(\mathcal{I}) = (V_{\max} - V_{\min}) \int_{\mathcal{I}_{\min}}^{\mathcal{I}} \mathrm{d}\mathcal{I}' P(\mathcal{I}'). \tag{4.24}$$

Thus, the optimal input/output relation is proportional to the cumulative probability distribution of the input signals.

The predictions of equation (4.24) are quite interesting. First of all it makes clear that any theory of the nervous system which involves optimizing information transmission or efficiency of representation inevitably predicts that the computations done by the nervous system must be matched to the statistics of sensory inputs (and, presumably, to the statistics of motor outputs as well). Here the matching is simple: in the right units we could just read off the distribution of inputs by looking at the (differentiated) input/output relation of the neuron. Second, this simple model automatically carries some predictions about adaptation to overall light levels. If we live in a world with diffuse light sources that are not directly visible, then the intensity which reaches us at a point is the product of the effective brightness of the source and some local reflectances. As is it gets dark outside the reflectances don't change–these are material properties–and so we expect that the distribution $P(\mathcal{I})$ will look the same except for scaling. Equivalently, if we view the input as the log of the intensity, then to a good approximation $P(\log \mathcal{I})$ just shifts linearly along the $\log \mathcal{I}$ axis as mean light intensity goes up and down. But then the optimal input/output relation $g(\mathcal{I})$ would exhibit a similar invariant shape with shifts along the input axis when expressed as a function of $\log \mathcal{I}$, and this is in rough agreement with experiments on light/dark adaptation in a wide variety of visual neurons. Finally, although obviously a simplified version of the real problem facing even the first stages of visual processing, this calculation does make a quantitative prediction that would be tested if we measure both the input/output relations of early visual neurons and the distribution of light intensities that the animal encounters in nature.

Laughlin [67] made this comparison (20 years ago!) for the fly visual system. He built an electronic photodetector with aperture and spectral sensitivity matched to those of the fly retina and used his photodetector to scan natural scenes, measuring $P(\mathcal{I})$ as it would appear at the input to these neurons. In parallel he characterized the second order neurons of the fly visual system–the large monopolar cells which receive direct synaptic input from the photoreceptors–by measuring the peak voltage response to flashes of light. The agreement with equation (4.24) was remarkable, especially when we remember that there are no free parameters. While there are obvious open questions (what happened to time dependence?), this is a really beautiful result.

Laughlin's analysis focused on the nonlinear input/output properties of neurons but ignored dynamics. An alternative which has been pursued by several groups is to treat dynamics but to ignore nonlinearity [68, 69], and we tried to review some of these ideas in Section 5.3 of *Spikes* [5]. As far as I know there is not much work which really brings together dynamics and nonlinearity, although there are striking results about filtering and nonlinearity in the color domain [70]. While these model problems capture something about real neurons, it would be nice to confront the real systems more directly. In particular, most neurons in the brain represent signals through trains of action potentials. As noted in the introduction to this section, we'd like to make a direct measurement of the information carried by these spikes or of the efficiency with which the spikes represent the sensory world.

The first question we might ask is how much information we gain about the sensory inputs by observing the occurrence of just one spike at some time t_0 [71]. For simplicity let us imagine that the inputs are described just by one function of time $s(t)$, although this is not crucial; what will be crucial is that we can repeat exactly the same time dependence many times, which for the visual system means showing the same movie over and over again, so that we can characterize the variability and reproducibility of the neural response. In general, the information gained about $s(t)$ by observing a set of neural responses is

$$I = \sum_{\text{responses}} \int Ds(\tau) P[s(\tau)\,,\, \text{resp}] \log_2 \left(\frac{P[s(\tau)\,,\, \text{resp}]}{P[s(\tau)]P(\text{resp})} \right), \quad (4.25)$$

where information is measured in bits. In the present case, the response is a single spike, so summing over the full range of responses is equivalent to integrating over the possible spike arrival times t_0:

$$I_{1\,\mathrm{spike}} = \int \mathrm{d}t_0 \int Ds(\tau) P[s(\tau)\,,\,t_0] \log_2 \left(\frac{P[s(\tau)\,,\,t_0]}{P[s(\tau)]P[t_0]} \right) \tag{4.26}$$

$$= \int \mathrm{d}t_0 P[t_0] \int Ds(\tau) P[s(\tau)|t_0] \log_2 \left(\frac{P[s(\tau)|t_0]}{P[s(\tau)]} \right), \tag{4.27}$$

where by $P[s(\tau)|t_0]$ we mean the distribution of stimuli given that we have observed a spike at time t_0. In the absence of knowledge about the stimulus, all spike arrival times are equally likely, and hence $P[t_0]$ is uniform. Furthermore, we expect that the coding of stimuli is stationary in time, so that the conditional distribution $P[s(\tau)|t_0]$ is of the same shape for all t_0, provided that we measure the time dependence of the stimulus $s(\tau)$ relative to the spike time t_0: $P[s(\tau)|t_0] = P[s(\tau - \Delta t)|t_0 - \Delta t]$. With these simplifications we can write the information conveyed by observing a single spike at time t_0 as [50]

$$I_{1\,\mathrm{spike}} = \int Ds(\tau) P[s(\tau)|t_0] \log_2 \left(\frac{P[s(\tau)|t_0]}{P[s(\tau)]} \right). \tag{4.28}$$

In this formulation we think of the spike as "pointing to" certain regions in the space of possible stimuli, and of course the information conveyed is quantified by an integral that relates to the volume of these regions. The difficulty is that if we want to use equation (4.28) in the analysis of real experiments we will need a model of the distribution $P[s(\tau)|t_0]$, and this could be hard to come by: the stimuli are drawn from a space of very high dimensionality (a function space, in principle) and so we cannot sample this distribution thoroughly in any reasonable experiment. Thus computing information transmission by mapping spikes back into the space of stimuli involves some model of how the code works, and then this model is used to simplify the structure of the relevant distributions, in this case $P[s(\tau)|t_0]$. We would like an alternative approach that does not depend on such models[21].

From Bayes' rule we can relate the conditional probability of stimuli given spikes to the conditional probability of spikes given stimuli:

$$\frac{P[s(\tau)|t_0]}{P[s(\tau)]} = \frac{P[t_0|s(\tau)]}{P[t_0]}. \tag{4.29}$$

[21] I hope that it is clear where this could lead: if we can estimate information using a model of what spikes stand for, and also estimate information without such a model, then by comparing the two estimates we should be able to test our model of the code in the most fundamental sense–does our model of what the neural response represents capture all of the information that this response provides?

But we can measure the probability of a spike at t_0 given that we know the stimulus $s(\tau)$ by repeating the stimulus many times and looking in a small bin around t_0 to measure the fraction of trials on which a spike occurred. If we normalize by the size of the bin in which we look then we are measuring the probability per unit time that a spike will be generated, which is a standard way of averaging the response over many trials; the probability per unit time is also called the time dependent firing rate $r[t_0; s(\tau)]$, where the notation reminds us that the probability of a spike at one time depends on the whole history of inputs leading up to that time.

If we don't know the stimulus then the probability of a spike at t_0 can only be given by the average firing rate over the whole experiment, $\bar{r} = \langle r[t; s(\tau)] \rangle$, where the expectation value $\langle \cdots \rangle$ denotes an average over the distribution of stimuli $P[s(\tau)]$. Thus we can write

$$\frac{P[s(\tau)|t_0]}{P[s(\tau)]} = \frac{r[t_0; s(\tau)]}{\bar{r}}. \tag{4.30}$$

Furthermore, we can substitute this relation into equation (4.28) for the information carried by a single spike, and then we obtain

$$I_{1\,\text{spike}} = \left\langle \left(\frac{r[t_0; s(\tau)]}{\bar{r}} \right) \log_2 \left(\frac{r[t_0; s(\tau)]}{\bar{r}} \right) \right\rangle, \tag{4.31}$$

We can compute the average in equation (4.31) by integrating over time, provided that the stimulus we use runs for a sufficiently long time that it provides a fair (ergodic) sampling of the true distribution $P[s(\tau)]$ from which stimuli are drawn.

Explicitly, then, if we sample the ensemble of possible stimuli by choosing a single time dependent stimulus $s(t)$ that runs for a long duration T, and then we repeat this stimulus many times to accumulate the time dependent firing rate $r[t; s(\tau)]$, the information conveyed by a single spike is given exactly by an average over this firing rate:

$$I_{1\,\text{spike}} = \frac{1}{T} \int_0^T \mathrm{d}t \left(\frac{r[t; s(\tau)]}{\bar{r}} \right) \log_2 \left(\frac{r[t; s(\tau)]}{\bar{r}} \right). \tag{4.32}$$

This is an exact formula,independent of any model for the structure of the neural code. It makes sense that the information carried by one spike should be related to the firing rate, since the the rate *vs.* time gives a complete description of the "one body" or one spike statistics of the spike train, in the same way that the single particle density describes the one body statistics of a gas or liquid.

Problem 9: Poisson model and lower bounds. Prove that equation (4.32) provides a lower bound to the information per spike transmitted if the entire spike train is a modulated Poisson process [48].

Another view of the result in equation (4.32) is in terms of the distribution of times at which a single spike might occur. First we note that the information a single spike provides about the stimulus must be the same as the information that knowledge of the stimulus provides about the occurrence time of a single spike–information is mutual. Before we know the precise trajectory of the stimulus $s(t)$, all we can say is that if we are looking for one spike, it can occur anywhere in our experimental window of size T, so that the probability is uniform, $p_0(t) = 1/T$ and the entropy of this distribution is just $\log_2 T$. Once we know the stimulus, we can expect that spikes will occur preferentially at times where the firing rate is large, so the probability distribution should be proportional to $r[t; s(\tau)]$; with proper normalization we have $p_1(t) = r[t; s(\tau)]/(T\bar{r})$. Then the conditional entropy is

$$S_1 = -\int_0^T \mathrm{d}t p_1(t) \log_2 p_1(t) \tag{4.33}$$

$$= -\frac{1}{T}\int_0^T \mathrm{d}t \frac{r[t; s(\tau)]}{\bar{r}} \log_2 \left(\frac{r[t; s(\tau)]}{\bar{r}T}\right). \tag{4.34}$$

The reduction in entropy is the gain in information, so

$$I_{1\,\mathrm{spike}} = S_0 - S_1 \tag{4.35}$$

$$= \frac{1}{T}\int_0^T \mathrm{d}t \left(\frac{r[t; s(\tau)]}{\bar{r}}\right) \log_2 \left(\frac{r[t; s(\tau)]}{\bar{r}}\right), \tag{4.36}$$

as before.

A crucial point about equation (4.32) is that when we derive it we do not make use of the fact that t_0 is the time of a single spike: it could be any event that occurs at a well defined time. There is considerable interest in the question of whether "synchronous" spikes from two neurons provide special information in the neural code. If we define synchronous spikes as two spikes that occur within some fixed (small) window of time then this compound event can also be given an arrival time (*e.g.*, the time of the later spike), and marking these arrival times across repeated presentations of the same stimulus we can build up the rate $r_E[t; s(\tau)]$ for these events of class E in exactly the same way that we build up an estimate of the spike rate. But if we have compound events constructed from two spikes, it makes sense to compare the information carried by a single event I_E with the information that would be carried independently by two spikes, $2I_{1\,\mathrm{spike}}$. If the compound event conveys more information than the sum of its parts, then this compound event indeed is a special symbol in the code. The same arguments apply to compound events constructed from temporal patterns of spikes in one neuron.

If a compound event provides an amount of information exactly equal to what we expect by adding up contributions from the components, then we say that the components or elementary events convey information independently. If there is less than independent information we say that the elementary events are redundant, and if the compound event provides more than the sum of its parts we say that there is *synergy* among the elementary events.

When we use these ideas to analyze Rob's experiments on the fly's H1 neuron [71], we find that the occurrence of a single spike can provide from 1 to 2 bits of information, depending on the details of the stimulus ensemble. More robustly we find that pairs of spikes separated by less than 10 msec can provide more–and sometimes vastly more–information than expected just by adding up the contributions of the individual spikes. There is a small amount of redundancy among spikes with larger separations, and if stimuli have a short correlation time then spikes carry independent information once they are separated by more than 30 msec or so. It is interesting that this time scale for independent information is close to the time scales of behavioral decisions, as if the fly waited long enough to see all the spikes that have a chance of conveying information synergistically.

We'd like to understand what happens as all the spikes add up to give us a fuller representation of the sensory signal: rather than thinking about the information carried by particular events, we want to estimate the information carried by long stretches of the neural response. Again the idea is straightforward [72, 73]: use Shannon's definitions to write the mutual information between stimuli and spikes in terms of difference between two entropies, and then use a long experiment to sample the relevant distributions and thus estimate these entropies. The difficulty is that when we talk not about single events but about "long stretches of the neural response", the number of possible responses is (exponentially) larger, and sampling is more difficult. Much of the effort in the original papers thus is in convincing ourselves that we have control over these sampling problems.

Let us look at segments of the spike train with length T, and within this time we record the spike train with time resolution $\Delta\tau$; these parameters are somewhat arbitrary, and we will need to vary them to be be sure we understand what is going on. In this view, however, the response is a "word" with $K = T/\Delta\tau$ letters; for small $\Delta\tau$ there can be only one or zero spikes in a bin and so the words are binary words, while for poorer time resolution we have a larger alphabet. If we let the fly watch a long movie, many different words will be produced and with a little luck we can get a good estimate of the probability distribution of these words, $P(W)$. This distribution has an entropy

$$S_{\text{total}}(T, \Delta\tau) = -\sum_W P(W) \log_2 P(W) \tag{4.37}$$

which measures the size of the neuron's vocabulary and hence the capacity of the code given our parameters T and $\Delta\tau$ [*cf.* Eq. (4.8) and the subsequent discussion]. While a large vocabulary is a good thing, to convey information I have to associate words with particular things in a reproducible way. Here we can show the same movie many times, and if we look across the many trials at a moment t relative to the start of the movie we again will see different words (since there is some noise in the response), and these provide samples of the distribution $P(W|t)$. This distribution in turn has an entropy which we call the noise entropy since any variation in response to the same inputs constitutes noise in the representation of those inputs[22]:

$$S_{\text{noise}} = \left\langle -\sum_W P(W|t) \log_2 P(W|t) \right\rangle_t , \tag{4.38}$$

where $\langle \cdots \rangle$ denotes an average over t and hence (by ergodicity) over the ensemble of sensory inputs $P[s]$. Finally, the information that the neural response provides about the sensory input is the difference between the total entropy and the noise entropy,

$$I(T, \Delta\tau) = S_{\text{total}}(T, \Delta\tau) - S_{\text{noise}}(T, \Delta\tau), \tag{4.39}$$

as in equation (4.6). A few points worth emphasizing:

- By using time averages in place of ensemble averages we can measure the information that the response provides about the sensory input without any explicit coordinate system on the space of inputs and hence without making any assumptions about which features of the input are most relevant;

- If we can sample the relevant distributions for sufficiently large times windows T, we expect that entropy and information will become extensive quantities, so that it makes sense to define entropy rates and information rates;

[22]This is not to say that such variations might not provide information about something else, but in the absence of some other signal against which we can correlate this ambiguity cannot be resolved. It is good to keep in mind, however, that what we call noise could be signal, while what we call signal really does constitute information about the input, independent of any further hypotheses.

- As we improve our time resolution, making $\Delta\tau$ smaller, the capacity of the code S_{total} must increase, but it is an experimental question whether the brain has the timing accuracy to make efficient use of this capacity.

The fly's H1 neuron provides an ideal place to do all of this because of the extreme stability of the preparation. In an effort to kill off any concerns about sampling and statistics, Rob did a huge experiment with order one thousand replays of the same long movie [72, 73]. With this large data set we were able to see the onset of extensivity, so we extracted information and entropy rates (although this really isn't essential) and we were able to explore a wide range of time resolutions, $800 > \Delta\tau > 2$ ms. Note that $\Delta\tau = 800$ ms corresponds to counting spikes in bins that contain typically thirty spikes, while $\Delta\tau = 2$ ms corresponds to timing each spike to within 5% of the typical interspike interval. Over this range, the entropy of the spike train varies over a factor of roughly 40, illustrating the increasing capacity of the system to convey information by making use of spike timing. The information that the spike train conveys about the visual stimulus increases in approximate proportion to the entropy, corresponding to $\sim$50% efficiency, although we start to see some saturation of information at the very highest time resolutions. Interestingly, this level of efficiency (and its approximate constancy as a function of time resolution) confirms an earlier measurement of efficiency in mechanosensor neurons from frogs and crickets that used the ideas of decoding discussed in the previous section [74].

What have we learned from this? First of all, the earliest experiments on neural coding showed that the rate of spiking encodes stimulus amplitude for static stimuli, but this left open the question of whether the precise timing of spikes carries additional information. The first application of information theory to neurons (as far as I know) was MacKay and McCulloch's calculation of the capacity of neurons to carry information given different assumptions about the nature of the code [75], and of course they drew attention to the fact that the capacity increases as we allow fine temporal details of the spike sequence to become distinguishable symbols in the code. Even MacKay and McCulloch were skeptical about whether real neurons could use a significant fraction of their capacity, however, and other investigators were more than skeptical. The debate about whether spike timing is important raged on, and I think that one of the important contributions of an information theoretic approach has been to make precise what we might mean by "timing is important".

There are two senses in which the timing of action potentials could be important to the neural code. First there is the simple question of whether marking spike arrival times to higher resolution really allows us to extract more information about the sensory inputs. We know (following MacKay and McCulloch) that if we use higher time resolution the entropy of the spike train increases, and hence the capacity to transmit information also increases. The question is whether this capacity is used, and the answer (for one neuron, under one set of conditions...) is in Ref. [73]: yes, unambiguously.

A second notion of spike timing being important is that temporal patterns of spikes may carry more information than would be expected by adding the contributions of single spikes. Again the usual setting for this sort of code is in a population of neurons, but the question is equally well posed for patterns across time in a single cell. Another way of asking the question is whether the high information rates observed for the spike train as a whole are "just" a reflection of rapid, large amplitude modulations in the spike rate[23]. Equation (4.32) makes clear that "information carried by rate modulations" is really the information carried by single spikes. The results of Ref. [71] show that pairs of spikes can carry more than twice the single spike information, and the analysis of longer windows of responses shows that this synergy is maintained, so that the spike train as a whole is carrying 30–50% more information than expected by summing the contributions of single spikes. Thus the answer to the question of whether temporal patterns are important, or whether there is "more than just rate" is again: yes, unambiguously.

These results on coding efficiency and synergy are surprisingly robust [76]: if we analyze the responses of H1 neurons from many different flies, all watching the same movie, it is easy to see that the flies are very different–average spike rates can vary by a factor of three among individuals, and by looking at the details of how each fly associates stimuli and responses we can distinguish a typical pair of flies from just 30 msec of data. On the other hand, if we look at the coding efficiency–the information divided by the total entropy–this is constant to within 10% across the population of flies in our experiments, and this high efficiency always has a significant contribution from synergy beyond single spikes.

The direct demonstration that the neural code is efficient in this information theoretic sense clearly depends on using complex, dynamic sensory inputs rather than the traditional quasistatic signals. We turned to these dynamic inputs not because they are challenging to analyze but rather because we thought that they would provide a better model for the problems encountered by the brain under natural conditions. This has become part of a larger effort in the community to analyze the way in which the nervous

[23]Note that in many cases this formulation obscures the fact that the a single "large amplitude modulation" is a bump in the firing rate with area of order one, so what might be called a large rate change is really one spike.

system deals with natural signals, and it probably is fair to point out that this effort has not been without its share of controversies [77]. One way to settle the issue is to strive for ever more natural experimental conditions. I think that Rob's recent experiments hold the record for "naturalness": rather than showing movies in the lab, he has taken his whole experiment outside into the woods where the flies are caught and recorded from H1 while rotating the fly through angular trajectories like those observed for freely flying flies [78]. This is an experimental tour de force (I can say this without embarrassment since I'm a theorist) because you have to maintain stable electrical recordings of neural activity while the sample is spinning at thousands of degrees per second and and accelerating to reverse direction within 10 msec. The reward is that spike timing is even more reproducible in the "wild" than in the lab, coding efficiencies and information rates are higher and are maintained to even smaller values $\Delta\tau$.

All of the results above point to the idea that spike trains really do provide an efficient representation of the sensory world, at least in one precise information theoretic sense. Many experimental groups are exploring whether similar results can be obtained in other systems, in effect asking if these theoretically appealing features of the code are universal. Here I want to look at a different question, namely whether this efficient code is fixed in the nervous system or whether it adapts and develops in response to the surroundings. The ideas and results that we have on these issues are, I think, some of the clearest evidence available for optimization in the neural code. Once again all the experiments are done in Rob de Ruyter's lab using H1 as the test case, and most of what we currently know is from work by Brenner & Fairhall [79, 80].

There are two reasons to suspect that the efficiency we have measured is achieved through adaptation rather than through hard wiring. First, one might guess that a fixed coding scheme could be so efficient and informative only if we choose the right ensemble of inputs, and you have to trust me that we didn't search around among many ensembles to find the results that I quoted. Second, under natural conditions the signals we want to encode are intermittent–the fly may fly straight, so that typical angular velocities are $\sim 50°/\text{s}$, or may launch into acrobatics where typical velocities are $\sim 2000°/\text{s}$, and there are possibilities in between. At a much simpler level, if we look across a natural scene, we find regions where the variance of light intensity or contrast is small, and nearby regions in which it is large [81]. Observations on contrast variance in natural images led to the suggestion that neurons in the retina might adapt in real time to this variance, and this was confirmed [82]. Here we would like to look at the parallel issue for velocity signals in the fly's motions sensitive neurons.

In a way what we are looking for is really contained in Laughlin's model problem discussed above. Suppose that we could measure the strategy that the fly actually uses for converting continuous signals into spikes; we could characterize this by giving the probability that a spike will be generated at t given the signal $s(\tau < t)$, which is what we have called the firing rate $r[t; s(\tau)]$. We are hoping that the neuron sets this coding strategy using some sort of optimization principle, although it is perhaps not so clear what constraints are relevant once we move from the model problem to the real neurons. On the other hand, we *can* say something about the nature of such optimization problems if we think about scaling: when we plot $r[s]$, what sets the scale along the s axis?

We know from the previous section that there is a limit to the smallest motion signals that can be reliably estimated, and of course there is a limit to the highest velocities that the system can deal with (if you move sufficiently fast everything blurs and vision is impossible). Happily, most of the signals we (and the fly) deal with are well away from these limits, which are themselves rather far apart. But this means that there is nothing intrinsic to the system which sets the scale for measuring angular velocity and encoding it in spikes, so if $r[s]$ is to emerge as the solution to an optimization problem then the scale along the s axis must be determined by outside world, that is by the distribution $P[s]$ from which the signals are drawn. Further, if we scale the distribution of inputs, $P[s] \to \lambda P[\lambda s]$ then the optimal coding strategy also should scale, $r[s] \to r[\lambda s]$[24]. The prediction, then, is that if the system can optimize its coding strategy in relation to the statistical structure of the sensory world, then we should be able to draw signals from a family of scaled distributions and see the input/output relations of the neurons scale in proportion to the input dynamic range. To make a long story short, this is exactly what we saw in H1 [79][25].

---

[24]I'm being a little sloppy here: s is really a function of time, not just a single number, but I hope the idea is clear.

[25]There is one technical issue in the analysis of Ref. [79] that I think is of broader theoretical interest. In trying to characterize the input/output relation $r[t; s(\tau)]$ we face the problem that the inputs $s(\tau)$ really live in a function space, or in more down to earth terms a space of very large dimensionality. Clearly we can't just plot the function $r[t; s(\tau)]$ in this space. Further, our intuition is that neurons are not equally sensitive to all of the dimensions or features of their inputs. To make progress (and in fact to get the scaling results) we have to make this intuition precise and find the relevant dimensions in stimulus space; along the way it would be nice to provide some direct evidence that the number of dimensions is actually small (!). If we are willing to consider Gaussian $P[s(\tau)]$, then we can show that by computing the right correlation functions between $s(\tau)$ and the stream of spikes $\rho(t) = \sum_i \delta(t - t_i)$ we can both count the number of relevant dimensions and provide an explicit coordinate system on the relevant subspace [59, 79]. These techniques are now being used to analyze other systems as well, and we are trying to understand if we can make explicit use of information theoretic tools to move beyond

The observation of scaling behavior is such a complex system certainly warms the hearts of physicists who grew up in a certain era. But Naama realized that one could do more. The observation of scaling tells us (experimentally!) that the system has a choice among a one parameter family of input/output relations, and so we can ask why the system chooses the one that it does. The answer is striking: the exact scale factor chosen by the system is the one that maximizes information transmission.

If the neural code is adapting in response to changes of the input distribution, and further if this adaptation serves to maximize information transmission, then we should be able to make sudden a change between two very different input distributions and "catch" the system using the wrong code and hence transmitting less than the maximum possible information. As the system collects enough data to be sure that the distribution has changed, the code should adapt and information transmission should recover. As with the simpler measurements of information transmission in steady state, the idea here is simple enough but finding an experimental design that actually gives enough data to avoid all statistical problems is a real challenge, and this is what Adrienne did in Ref. [80]. The result is clear: when we switch from one $P[s]$ to another we can detect the drop in efficiency of information transmission associated with the use of the old code in the new distribution, and we can measure the time course of information recovery as the code adapts. What surprised us (although it shouldn't have) was the speed of this recovery, which can be complete in much less than 100 msec. In fact, for the conditions of these experiments, we can actually calculate how rapidly an optimal processor could make a reliable decision about the change in distribution, and when the dust settles the answer is that the dynamics of the adaptation that we see in the fly are running within a factor of two of the maximum speed set by these limits to statistical inference[26].

I have the feeling that my presentation of these ideas mirrors their development. It took us a long time to build up the tools that bring information theory and optimization principles into contact with real experiments on

the analysis of Gaussian inputs and low order correlation functions. I find the geometrical picture of neurons as selective for a small number of dimensions rather attractive as well being useful, but it is a bit off the point of this discussion.

[26] Again there is more to these experiments than what I have emphasized here. The process of adaptation in fact has multiple time scales, ranging from tens of milliseconds out to many minutes. These rich dynamics offer possibilities for longer term statistical properties of the spike train to resolve the ambiguities (how does the fly know the absolute scale of velocity if it is scaled away?) that arise in any adaptive coding scheme. The result is that while information about the scaled stimulus recovers quickly during the process of adaptation, some information about the scale itself is preserved and can be read out by simple algorithms.

the neural coding of complex, dynamic signals. Once we have the tools, however, the results are clear, at least for this one system where we (or, more precisely, Rob) can do almost any experiment we want:

- Spike trains convey information about natural input signals with $\sim$50% efficiency down to millisecond time resolution;
- This efficiency is enhanced significantly by synergistic coding in which temporal patterns of spikes stand for more than the sum of their parts;
- Although the detailed structure of the neural code is highly individualized, these basic features are strikingly constant across individuals;
- Coding efficiency and information rates are higher under more natural conditions;
- The observed information transmission rates are the result of an adaptive coding scheme which takes a simple scaling form in response to changes in the dynamic range of the inputs;
- The precise choice of scale by the real code is the one which maximizes information transmission;
- The dynamics of this adaptation process are almost as fast as possible given the need to collect statistical evidence for changes in the input distribution.

I think it is fair to say that this body of work provides very strong evidence in support of information theoretic optimization as a "design principle" within which we can understand the phenomenology of the neural code.

5 Learning and complexity

The world around us, thankfully, is a rather structured place. Whether we are doing a careful experiment in the laboratory or are gathering sense data on a walk through the woods, the signals that arrive at our brains are far from random noise; there appear to be some underlying regularities or rules. Surely one task our brain must face is the learning or extraction of these rules and regularities. Perhaps the simplest example of learning a rule is fitting a function to data–we believe in advance that the rule belongs to a class of possible rules that can be parameterized, and as we collect data we learn the values of the parameters. This simple example introduces us to many deep issues:

- If there is noise in the data then really we are trying to learn a probability distribution, not just a functional relation;

- We would like to compare models (classes of possible rules) that have different numbers of parameters, and incorporate the intuition that "simpler" models are better;
- We might like to step outside the restrictions of finite parameterization and consider the possibility that the data are described by functions that are merely "smooth" to some degree;
- We would like to quantify how much we are learning (or how much *can* be learned) about the underlying rules.

In the last decade or so, a rich literature has emerged, tackling these problems with a sophistication far beyond the curve fitting exercises that we all performed in our student physics laboratories. I will try to take a path through these developments, emphasizing the connections of these learning problems to problems in statistical mechanics and the implications of this statistical approach for an information theoretic characterization of how much we learn. Most of what I have to say on this subject is drawn from collaborations with Nemenman & Tishby [83–85]; in particular the first of these papers is long and has lots of references to more standard things which I will outline here without attribution.

Let's just plunge in with the classic example: We observe two streams of data x and y, or equivalently a stream of pairs (x_1, y_1), (x_2, y_2), $\dots$, (x_N, y_N). Assume that we know in advance that the x's are drawn independently and at random from a distribution $P(x)$, while the y's are noisy versions of some function acting on x,

$$y_n = f(x_n; \boldsymbol{\alpha}) + \eta_{\rm n}, \tag{5.1}$$

where $f(x; \boldsymbol{\alpha})$ is one function from a class of functions parameterized by $\boldsymbol{\alpha} \equiv \{\alpha_1, \alpha_2, \cdots, \alpha_K\}$ and $\eta_{\rm n}$ is noise, which for simplicity we will assume is Gaussian with known standard deviation σ. We can even start with a *very* simple case, where the function class is just a linear combination of basis functions, so that

$$f(x; \boldsymbol{\alpha}) = \sum_{\mu=1}^{K} \alpha_\mu \phi_\mu(x). \tag{5.2}$$

The usual problem is to estimate, from N pairs $\{x_i, y_i\}$, the values of the parameters $\boldsymbol{\alpha}$; in favorable cases such as this we might even be able to find an effective regression formula. Probably you were taught that the way to do this is to compute χ^2,

$$\chi^2 = \sum_n [y_n - f(x_n; \boldsymbol{\alpha})]^2, \tag{5.3}$$

and then minimize to find the correct parameters $\boldsymbol{\alpha}$. You may or may not have been taught *why* this is the right thing to do.

With the model described above, the probability that we will observe the pairs $\{x_i, y_i\}$ can be written as

$$P(\{x_i, y_i\}|\boldsymbol{\alpha}) = \exp\left[-\frac{N}{2}\ln(2\pi\sigma^2) - \frac{\chi^2}{2\sigma^2}\right]\prod_n P(x_n), \tag{5.4}$$

assuming that we know the parameters. Thus finding parameters which minimize χ^2 also serves to maximize the probability that our model could have given rise to the data. But why is this a good idea?

We recall that the entropy is the expectation value of $-\log P$, and that it is possible to encode signals so that the amount of "space" required to specify each signal uniquely is on average equal to the entropy[27]. With a little more work one can show that each possible signal s drawn from $P(s)$ can be encoded in a space of $-\log_2 P(s)$ bits. Now any model probability distribution implicitly defines a scheme for coding signals that are drawn from that distribution, so if we make sure that our data have high probability in the distribution (small values of $-\log P$) then we also are making sure that our code or representation of these data is compact. What this means is that good old fashioned curve fitting really is all about finding efficient representations of data, precisely the principle enunciated by Barlow for the operation of the nervous system (!).

If we follow this notion of efficient representation a little further we can do better than just maximizing χ^2. The claim that a model provides a code for the data is not complete, because at some point I have to represent my knowledge of the model itself. One idea is to do this explicitly–estimate how accurately you know each of the parameters, and then count how many bits you'll need to write down the parameters to that accuracy and add this to the length of your code; this is the point of view taken by Risannen and others in a set of ideas called "minimum description length" or MDL. Another idea is more implicit–the truth is that I don't really know the parameters, all I do is estimate them from the data, so it's not so obvious that I should separate coding the data from coding the parameters (although this might emerge as an approximation). In this view what we should do is to integrate over all possible values of the parameters, weighted by some

[27]This is obvious for uniform probability distributions with 2^n alternatives, since then the binary number representing each alternative is this code we want. For nonuniform distributions we need to think about writing things down many times and taking an average of the space we use each time, and the fact that the answer comes out the same (as the entropy) hinges on the "typicality" of such data streams, which is the information theorist's way of talking about the equivalence of canonical and microcanonical ensembles.

prior knowledge (maybe just that the parameters are bounded), and thus compute the probability that our data could have arisen from the *class* of models we are considering.

To carry out this program of computing the total probability of the data given the model class we need to do the integral

$$P(\{x_i, y_i\}|\text{class}) = \int \mathrm{d}^K\alpha\, P(\boldsymbol{\alpha}) P[\{x_i, y_i\}|\boldsymbol{\alpha}] \tag{5.5}$$

$$= \left[\prod_n P(x_n)\right] \times \int \mathrm{d}^K\alpha\, P(\boldsymbol{\alpha}) \exp\left[-\frac{N}{2}\ln(2\pi\sigma^2) - \frac{\chi^2}{2\sigma^2}\right]. \tag{5.6}$$

But remember that χ^2 as we have defined it is a sum over data points, which means we expect it (typically) will be proportional to N. This means that at large N we are doing an integral in which the exponential has terms proportional to N–and so we should use a saddle point approximation. The saddle point of course is close to the place where χ^2 is minimized, and then we do the usual Gaussian (one loop) integral around this point; actually if we stick with the simplest case of equation (5.2) then this Gaussian approximation becomes exact. When the dust settles we find

$$-\ln P(\{x_i, y_i\}|\text{class}) = -\sum_n \ln P(x_n) + \frac{\chi^2_{\text{min}}}{2\sigma^2} + \frac{K}{2}\ln N + \cdots, \tag{5.7}$$

and we recall that this measures the length of the shortest code for $\{x_i, y_i\}$ that can be generated given the class of models. The first term averages to N times the entropy of the distribution $P(x)$, which makes sense since by hypothesis the x's are being chosen at random. The second term is as before, essentially the length of the code required to describe the deviations of the data from the predictions of the best fit model; this also grows in proportion to N. The third term must be related to coding our knowledge of the model itself, since it is proportional to the number of parameters. We can understand the $(1/2)\ln N$ because each parameter is determined to an accuracy of $\sim 1/\sqrt{N}$, so if we start with a parameter space of size ~ 1 there is a reduction in volume by a factor of $\sqrt{N}$ and hence a decrease in entropy (gain in information) by $(1/2)\ln N$. Finally, the terms$\cdots$ don't grow with N.

What is crucial about the term $(K/2)\ln N$ is that it depends explicitly on the number of parameters. In general we expect that by considering models with more parameters we can get a better fit to the data, which means that χ^2 can be reduced by considering more complex model classes. But we know intuitively that this has to stop–we don't want to use arbitrarily complex models, even if they do provide a good fit to what we have seen. It is attractive, then, that if we look for the shortest code which can be generated by a class of models, there is an implicit penalty or coding cost for increased complexity. It is interesting from a physicist's point of view that this term emerges essentially from consideration of phase space or volumes in model space. It thus is an entropy–like quantity in its own right, and the selection of the best model class could be thought of as a tradeoff between this entropy and the "energy" measured by χ^2. If we keep going down this line of thought we can imagine a thermodynamic limit with large numbers of parameters and data points, and there can be "aha!" types of phase transitions from poor fits to good fits as we increase the ratio N/K [86].

The reason we need to control the complexity of our models is because the real problem of learning is neither the estimation of parameters nor the compact representation of the data we have already seen. The real problem of learning is *generalization*: we want to extract the rules underlying what we have seen because we believe that these rules will continue to be true and hence will describe the relationships among data that we will observe in the future. Our experience is that overly complex models might provide a more accurate description of what we have seen so far but do a bad job at predicting what we will see next. This suggests that there are connections between predictability and complexity.

There is in fact a completely different motivation for quantifying complexity, and this is to make precise an impression that some systems, such as life on earth or a turbulent fluid flow, evolve toward a state of higher complexity; one might even like to classify these states. These problems traditionally are in the realm of dynamical systems theory and statistical physics. A central difficulty in this effort is to distinguish complexity from randomness–trajectories of dynamical systems can be regular, which we take to mean "simple" in the intuitive sense, or chaotic, but what we mean by complex is somewhere in between. The field of complexology (as Magnasco likes to call it) is filled with multiple definitions of complexity and confusing remarks about what they all might mean. In this noisy environment, there is a wonderful old paper by Grassberger [87] which gives a clear signal: systems with regular or chaotic/random dynamics share the property that the entropy of sample trajectories is almost exactly extensive in the length of the trajectory, while for systems that we identify intuitively as being complex there are large corrections to extensivity which can even diverge as we take longer and longer samples. In the end Grassberger suggested that these subextensive terms in the entropy really do quantify our intuitive notions of complexity, although he made this argument by example rather than axiomatically.

We can connect the measures of complexity that arise in learning problems with those that arise in dynamical systems by noticing that the subextensive components of entropy identified by Grassberger in fact determine the information available for making predictions[28]. This also suggests a connection to the importance or value of information, especially in a biological or economic context: information is valuable if it can be used to guide our actions, but actions take time and hence observed data can be useful only to the extent that those data inform us about the state of the world at later times. It would be attractive if what we identify as "complex" in a time series were also the "useful" or "meaningful" components.

While prediction may come in various forms, depending on context, information theory allows us to treat all of them on the same footing. For this we only need to recognize that all predictions are probabilistic, and that, even before we look at the data, we know that certain futures are more likely than others. This knowledge can be summarized by a prior probability distribution for the futures. Our observations on the past lead us to a new, more tightly concentrated distribution, the distribution of futures conditional on the past data. Different kinds of predictions are different slices through or averages over this conditional distribution, but information theory quantifies the "concentration" of the distribution without making any commitment as to which averages will be most interesting.

Imagine that we observe a stream of data $x(t)$ over a time interval $-T < t < 0$; let all of these past data be denoted by the shorthand $x_{\rm past}$. We are interested in saying something about the future, so we want to know about the data $x(t)$ that will be observed in the time interval $0 < t < T'$; let these future data be called $x_{\rm future}$. In the absence of any other knowledge, futures are drawn from the probability distribution $P(x_{\rm future})$, while observations of particular past data $x_{\rm past}$ tell us that futures will be drawn from the conditional distribution $P(x_{\rm future}|x_{\rm past})$. The greater concentration of the conditional distribution can be quantified by the fact that it has smaller entropy than the prior distribution, and this reduction in entropy is the information that the past provides about the future. We can write the average of this *predictive information* as

$$\mathcal{I}_{\rm pred}(T,T') = \left\langle \log_2 \left[\frac{P(x_{\rm future}|x_{\rm past})}{P(x_{\rm future})}\right]\right\rangle \tag{5.8}$$

$$= -\langle \log_2 P(x_{\rm future})\rangle - \langle \log_2 P(x_{\rm past})\rangle - \left[-\langle \log_2 P(x_{\rm future}, x_{\rm past})\rangle\right], \tag{5.9}$$

[28]The text of the discussion here follows Ref. [84] rather closely, and I thank my colleagues for permission to include it here.

where $\langle \cdots \rangle$ denotes an average over the joint distribution of the past and the future, $P(x_{\text{future}}, x_{\text{past}})$.

Each of the terms in equation (5.9) is an entropy. Since we are interested in predictability or generalization, which are associated with some features of the signal persisting forever, we may assume stationarity or invariance under time translations. Then the entropy of the past data depends only on the duration of our observations, so we can write $-\langle \log_2 P(x_{\text{past}}) \rangle = S(T)$, and by the same argument $-\langle \log_2 P(x_{\text{future}}) \rangle = S(T')$. Finally, the entropy of the past and the future taken together is the entropy of observations on a window of duration $T + T'$, so that $-\langle \log_2 P(x_{\text{future}}, x_{\text{past}}) \rangle = S(T + T')$. Putting these equations together, we obtain

$$\mathcal{I}_{\text{pred}}(T, T') = S(T) + S(T') - S(T + T'). \tag{5.10}$$

In the same way that the entropy of a gas at fixed density is proportional to the volume, the entropy of a time series (asymptotically) is proportional to its duration, so that $\lim_{T\to\infty} S(T)/T = \mathcal{S}_0$; entropy is an extensive quantity. But from equation (5.10) any extensive component of the entropy cancels in the computation of the predictive information: *predictability is a deviation from extensivity.* If we write

$$S(T) = \mathcal{S}_0 T + S_1(T), \tag{5.11}$$

then equation (5.10) tells us that the predictive information is related *only* to the nonextensive term $S_1(T)$.

We know two general facts about the behavior of $S_1(T)$. First, the corrections to extensive behavior are positive, $S_1(T) \geq 0$. Second, the statement that entropy is extensive is the statement that the limit

$$\lim_{T\to\infty} S(T)/T = \mathcal{S}_0 \tag{5.12}$$

exists, and for this to be true we must also have $\lim_{T\to\infty} S_1(T)/T = 0$. Thus the nonextensive terms in the entropy must be *sub*extensive, that is they must grow with T less rapidly than a linear function. Taken together, these facts guarantee that the predictive information is positive and subextensive. Further, if we let the future extend forward for a very long time, $T' \to \infty$, then we can measure the information that our sample provides about the entire future,

$$I_{\text{pred}}(T) = \lim_{T'\to\infty} \mathcal{I}_{\text{pred}}(T, T') = S_1(T), \tag{5.13}$$

and this is precisely equal to the subextensive entropy.

If we have been observing a time series for a (long) time T, then the total amount of data we have collected in is measured by the entropy $S(T)$, and at large T this is given approximately by $\mathcal{S}_0 T$. But the predictive information that we have gathered cannot grow linearly with time, even if we are making predictions about a future which stretches out to infinity. As a result, of the total information we have taken in by observing x_{past}, only a vanishing fraction is of relevance to the prediction:

$$\lim_{T\to\infty} \frac{\text{Predictive Information}}{\text{Total Information}} = \frac{I_{\text{pred}}(T)}{S(T)} \to 0. \tag{5.14}$$

In this precise sense, most of what we observe is irrelevant to the problem of predicting the future.

Consider the case where time is measured in discrete steps, so that we have seen N time points $x_1, x_2, \ldots, x_N$. How much is there to learn about the underlying pattern in these data? In the limit of large number of observations, $N \to \infty$ or $T \to \infty$, the answer to this question is surprisingly universal: predictive information may either stay finite, or grow to infinity together with T; in the latter case the rate of growth may be slow (logarithmic) or fast (sublinear power).

The first possibility, $\lim_{T\to\infty} I_{\text{pred}}(T) = \text{constant}$, means that no matter how long we observe we gain only a finite amount of information about the future. This situation prevails, in both extreme cases mentioned above. For example, when the dynamics are very regular, as for a purely periodic system, complete prediction is possible once we know the phase, and if we sample the data at discrete times this is a finite amount of information; longer period orbits intuitively are more complex and also have larger I_{pred}, but this doesn't change the limiting behavior $\lim_{T\to\infty} I_{\text{pred}}(T) = \text{constant}$.

Similarly, the predictive information can be small when the dynamics are irregular but the best predictions are controlled only by the immediate past, so that the correlation times of the observable data are finite. This happens, for example, in many physical systems far away from phase transitions. Imagine, for example, that we observe $x(t)$ at a series of discrete times $\{t_n\}$, and that at each time point we find the value x_n. Then we always can write the joint distribution of the N data points as a product,

$$P(x_1, x_2, \ldots, x_N) = P(x_1)P(x_2|x_1)P(x_3|x_2, x_1)\ldots. \tag{5.15}$$

For Markov processes, what we observe at t_n depends only on events at the previous time step t_{n-1}, so that

$$P(x_n|\{x_{1\le i\le n-1}\}) = P(x_n|x_{n-1}), \tag{5.16}$$

and hence the predictive information reduces to

$$I_{\text{pred}} = \left\langle \log_2 \left[\frac{P(x_n|x_{n-1})}{P(x_n)} \right] \right\rangle. \tag{5.17}$$

The maximum possible predictive information in this case is the entropy of the distribution of states at one time step, which in turn is bounded by the logarithm of the number of accessible states. To approach this bound the system must maintain memory for a long time, since the predictive information is reduced by the entropy of the transition probabilities. Thus systems with more states and longer memories have larger values of $I_{\rm pred}$.

Problem 10: Brownian motion of a spring. Consider the Brownian motion of an overdamped particle bound a spring. The Langevin equation describing the particle position $x(t)$ is

$$\gamma \frac{\mathrm{d}x(t)}{\mathrm{d}t} + \kappa x(t) = F_{\rm ext}(t) + \delta F(t), \tag{5.18}$$

where γ is the damping constant, κ is the stiffness of the spring, $F_{\rm ext}(t)$ is an external force that might be applied to the particle, and $\delta F(t)$ is the Langevin force. The Langevin force is random, with Gaussian statistics and white noise correlation properties,

$$\langle \delta F(t) \delta F(t') \rangle = 2\gamma k_{\rm B} T \delta(t - t'). \tag{5.19}$$

Show that the correlation function has a simple exponential form,

$$\langle x(t) x(t') \rangle = \langle x^2 \rangle \exp(-|t - t'|/\tau_{\rm c}), \tag{5.20}$$

and evaluate the correlation time. Now take the original Langevin equation and form a discrete version, introducing a small time step Δt; be sure that your discretization preserves exactly the observable variance $\langle x^2 \rangle$. You should be able to find a natural discretization in which the evolution of x is Markovian, and then you can compute the predictive information for the time series $x(t)$. How does this result depend on temperature? Why? Express the dependence on Δt in units of the correlation time $\tau_{\rm c}$. Is there a well defined limit as $\Delta t \to 0$? Again, why (or why not)?

More interesting are those cases in which $I_{\rm pred}(T)$ diverges at large T. In physical systems we know that there are critical points where correlation times become infinite, so that optimal predictions will be influenced by events in the arbitrarily distant past. Under these conditions the predictive information can grow without bound as T becomes large; for many systems the divergence is logarithmic, $I_{\rm pred}(T \to \infty) \propto \log T$.

Long range correlation also are important in a time series where we can learn some underlying rules, as in the discussion of curve fitting that started this section. Since we saw that curve fitting with noisy data really involves a probabilistic model, let us talk explicitly about the more general problem of learning distributions. Suppose a series of random vector variables $\{\vec{x}_i\}$ are drawn independently from the same probability distribution $Q(\vec{x}|\boldsymbol{\alpha})$, and this distribution depends on a (potentially infinite dimensional) vector of parameters $\boldsymbol{\alpha}$. The parameters are unknown, and before the series starts they are chosen randomly from a distribution $\mathcal{P}(\boldsymbol{\alpha})$. In this setting, at least implicitly, our observations of $\{\vec{x}_i\}$ provide data from which we can learn the parameters $\boldsymbol{\alpha}$. Here we put aside (for the moment) the usual problem of learning–which might involve constructing some estimation or regression scheme that determines a "best fit" $\boldsymbol{\alpha}$ from the data $\{\vec{x}_i\}$–and treat the ensemble of data streams $P[\{\vec{x}_i\}]$ as we would any other set of configurations in statistical mechanics or dynamical systems theory. In particular, we can compute the entropy of the distribution $P[\{\vec{x}_i\}]$ even if we can't provide explicit algorithms for solving the learning problem.

As shown in [83], the crucial quantity in such analysis is the density of models in the vicinity of the target $\bar{\boldsymbol{\alpha}}$–the parameters that actually generated the sequence. For two distributions, a natural distance measure is the Kullback–Leibler divergence,

$$D_{\mathrm{KL}}(\bar{\boldsymbol{\alpha}}||\boldsymbol{\alpha}) = \int \mathrm{d}\vec{x} Q(\vec{x}|\bar{\boldsymbol{\alpha}}) \log \left[\frac{Q(\vec{x}|\bar{\boldsymbol{\alpha}})}{Q(\vec{x}|\boldsymbol{\alpha})}\right], \tag{5.21}$$

and the density is

$$\rho(D;\bar{\boldsymbol{\alpha}}) = \int \mathrm{d}^K\alpha \mathcal{P}(\boldsymbol{\alpha})\delta[D - D_{\mathrm{KL}}(\bar{\boldsymbol{\alpha}}||\boldsymbol{\alpha})]. \tag{5.22}$$

If ρ is large as $D \to 0$, then one easily can get close to the target for many different data; thus they are not very informative. On the other hand, small density means that only very particular data lead to $\bar{\boldsymbol{\alpha}}$, so they carry a lot of predictive information. Therefore, it is clear that the density, but not the number of parameters or any other simplistic measure, characterizes predictability and the complexity of prediction. If, as often is the case for $\dim \boldsymbol{\alpha} < \infty$, the density behaves in the way common to finite dimensional systems of the usual statistical mechanics,

$$\rho(D \to 0, \bar{\boldsymbol{\alpha}}) \approx A D^{(K-2)/2}, \tag{5.23}$$

then the predictive information to the leading order is

$$I_{\mathrm{pred}}(N) \approx \frac{K}{2} \log N. \tag{5.24}$$

The modern theory of learning is concerned in large part with quantifying the complexity of a model class, and in particular with replacing a simple count of parameters with a more rigorous notion of dimensionality for the space of models; for a general review of these ideas see Ref. [88], and for a discussion close in spirit to this one see Ref. [89]. The important point here is that the dimensionality of the model class, and hence the complexity of the class in the sense of learning theory, emerges as the coefficient of the logarithmic divergence in $I_{\rm pred}$. Thus a measure of complexity in learning problems can be derived from a more general dynamical systems or statistical mechanics point of view, treating the data in the learning problem as a time series or one dimensional lattice. The logarithmic complexity class that we identify as being associated with finite dimensional models also arises, for example, at the Feigenbaum accumulation point in the period doubling route to chaos [87].

As noted by Grassberger in his original discussion, there are time series for which the divergence of $I_{\rm pred}$ is stronger than a logarithm. We can construct an example by looking at the density function ρ in our learning problem above: finite dimensional models are associated with algebraic decay of the density as $D \to 0$, and we can imagine that there are model classes in which this decay is more rapid, for example

$$\rho(D \to 0) \approx A \exp\left[-B/D^{\mu}\right], \quad \mu > 0. \tag{5.25}$$

In this case it can be shown that the predictive information diverges very rapidly, as a sublinear power law,

$$I_{\rm pred}(N) \sim N^{\mu/(\mu+1)}. \tag{5.26}$$

One way that this scenario can arise is if the distribution $Q(\vec{x})$ that we are trying to learn does not belong to any finite parameter family, but is itself drawn from a distribution that enforces a degree of smoothness [90]. Understandably, stronger smoothness constraints have smaller powers (less to predict) than the weaker ones (more to predict). For example, a rather simple case of predicting a one dimensional variable that comes from a continuous distribution produces $I_{\rm pred}(N) \sim \sqrt{N}$.

As with the logarithmic class, we expect that power–law divergences in $I_{\rm pred}$ are not restricted to the learning problems that we have studied in detail. The general point is that such behavior will be seen in problems where predictability over long scales, rather then being controlled by a fixed set of ever more precisely known parameters, is governed by a progressively more detailed description–effectively increasing the number of parameters–as we collect more data. This seems a plausible description of what happens in language, where rules of spelling allow us to predict forthcoming letters of long words, grammar binds the words together, and compositional unity of the entire text allows predictions about the subject of the last page of the book after reading only the first few. Indeed, Shannon's classic experiment on the predictability of English text (by human readers!) shows this behavior [91, 92], and more recently several groups have extracted power–law subextensive components from the numerical analysis of large corpora of text.

Interestingly, even without an explicit example, a simple argument ensures existence of exponential densities and, therefore, power law predictive information models. If the number of parameters in a learning problem is not finite then in principle it is impossible to predict anything unless there is some appropriate regularization. If we let the number of parameters stay finite but become large, then there is *more* to be learned and correspondingly the predictive information grows in proportion to this number. On the other hand, if the number of parameters becomes infinite without regularization, then the predictive information should go to zero since nothing can be learned. We should be able to see this happen in a regularized problem as the regularization weakens: eventually the regularization would be insufficient and the predictive information would vanish. The only way this can happen is if the predictive information grows more and more rapidly with N as we weaken the regularization, until finally it becomes extensive (equivalently, drops to zero) at the point where prediction becomes impossible. To realize this scenario we have to go beyond $I_{\rm pred} \propto \log T$ with $I_{\rm pred} \propto N^{\mu/(\mu+1)}$; the transition from increasing predictive information to zero occurs as $\mu \to 1$.

This discussion makes it clear that the predictive information (the subextensive entropy) distinguishes between problems of intuitively different complexity and thus, in accord to Grassberger's definitions [87], is probably a good choice for a universal complexity measure. Can this intuition be made more precise?

First we need to decide whether we want to attach measures of complexity to a particular signal $x(t)$ or whether we are interested in measures that are defined by an average over the ensemble $P[x(t)]$. One problem in assigning complexity to single realizations is that there can be atypical data streams. Second, Grassberger [87] in particular has argued that our visual intuition about the complexity of spatial patterns is an ensemble concept, even if the ensemble is only implicit. The fact that we admit probabilistic models is crucial: even at a colloquial level, if we allow for probabilistic models then there is a simple description for a sequence of truly random bits, but if we insist on a deterministic model then it may be very complicated to generate precisely the observed string of bits. Furthermore, in the context of probabilistic models it hardly makes sense to ask for a dynamics that generates a particular data stream; we must ask for dynamics that generate the data with reasonable probability, which is more or less equivalent to asking that the given string be a typical member of the ensemble generated by the model. All of these paths lead us to thinking not about single strings but about ensembles in the tradition of statistical mechanics, and so we shall search for measures of complexity that are averages over the distribution $P[x(t)]$.

Once we focus on average quantities, we can provide an axiomatic proof (much in the spirit of Shannon's [63] arguments establishing entropy as a unique information measure) that links $I_{\rm pred}$ to complexity. We can start by adopting Shannon's postulates as constraints on a measure of complexity: if there are N equally likely signals, then the measure should be monotonic in N; if the signal is decomposable into statistically independent parts then the measure should be additive with respect to this decomposition; and if the signal can be described as a leaf on a tree of statistically independent decisions then the measure should be a weighted sum of the measures at each branching point. We believe that these constraints are as plausible for complexity measures as for information measures, and it is well known from Shannon's original work that this set of constraints leaves the entropy as the only possibility. Since we are discussing a time dependent signal, this entropy depends on the duration of our sample, $S(T)$. We know of course that this cannot be the end of the discussion, because we need to distinguish between randomness (entropy) and complexity. The path to this distinction is to introduce other constraints on our measure.

First we notice that if the signal x is continuous, then the entropy is not invariant under transformations of x, even if these reparamterizations do not mix points at different times. It seems reasonable to ask that complexity be a function of the process we are observing and not of the coordinate system in which we choose to record our observations. However, it is not the whole function $S(T)$ which depends on the coordinate system for x; it is only the extensive component of the entropy that has this noninvariance. This can be seen more generally by noting that subextensive terms in the entropy contribute to the mutual information among different segments of the data stream (including the predictive information defined here), while the extensive entropy cannot; mutual information is coordinate invariant, so all of the noninvariance must reside in the extensive term. Thus, any measure complexity that is coordinate invariant must discard the extensive component of the entropy.

If we continue along these lines, we can think about the asymptotic expansion of the entropy at large T. The extensive term is the first term in this series, and we have seen that it must be discarded. What about the other terms? In the context of predicting in a parameterized model, most of the terms in this series depend in detail on our prior distribution in parameter space, which might seem odd for a measure of complexity. More generally, if we consider transformations of the data stream $x(t)$ that mix points within a temporal window of size τ, then for $T >> \tau$ the entropy $S(T)$ may have subextensive terms which are constant, and these are not invariant under this class of transformations. On the other hand, if there are divergent subextensive terms, these *are* invariant under such temporally local transformations. So if we insist that measures of complexity be invariant not only under instantaneous coordinate transformations, but also under temporally local transformations, then we can discard both the extensive and the finite subextensive terms in the entropy, leaving only the divergent subextensive terms as a possible measure of complexity.

To illustrate the purpose of these two extra conditions, we may think of measuring the velocity of a turbulent fluid flow at a given point. The condition of invariance under reparameterizations means that the complexity is independent of the scale used by the speedometer. On the other hand, the second condition ensures that the temporal filtering due to the finite inertia of the speedometer's needle does not change the estimated complexity of the flow.

I believe that these arguments (or their slight variation also presented in [83]) settle the question of the unique definition of complexity. Not only is the divergent subextensive component of the entropy the unique complexity measure, but it is also a universal one since it is connected in a straightforward way to many other measures that have arisen in statistics and in dynamical systems theory. In my mind the really big open question is whether we can connect *any* of these theoretical developments to experiments on learning by real animals (including humans).

I have emphasized the problem of learning probability distributions or probabilistic models rather than learning deterministic functions, associations or rules. In the previous section we have discussed examples where the nervous system adapts to the statistics of its inputs; these experiments can be thought of as a simple example of the system learning a parameterized distribution. When making saccadic eye movements, human subjects alter their distribution of reaction times in relation to the relative probabilities of different targets, as if they had learned an estimate of the relevant likelihood ratios [93]. Humans also can learn to discriminate almost optimally between random sequences (fair coin tosses) and sequences that are correlated or anticorrelated according to a Markov process; this learning can be accomplished from examples alone, with no other feedback [94]. Acquisition of language may require learning the joint distribution of successive phonemes, syllables, or words, and there is direct evidence for learning of conditional probabilities from artificial sound sequences, both by infants and by adults [95, 96].

Classical examples of learning in animals–such as eye blink conditioning in rabbits–also may harbor evidence of learning probability distributions. The usual experiment is to play a brief sound followed by a puff of air to the eyes, and then the rabbit learns to blink its eye at the time when the air puff is expected. But if the time between a sound and a puff of air to the eyes is chosen from a probability distribution, then rabbits will perform graded movements of the eyelid that seem to more or less trace the shape of the distribution, as if trying to have the exposure of the eye matched to the (inverse) likelihood of the noxious stimulus [97]. These examples, which are not exhaustive, indicate that the nervous system can learn an appropriate probabilistic model, and this offers the opportunity to analyze the dynamics of this learning using information theoretic methods: what is the entropy of N successive reaction times following a switch to a new set of relative probabilities in the saccade experiment? How much information does a single reaction time provide about the relevant probabilities?

Using information theory to characterize learning is appealing because the predictive information in the data itself (that is, in the data from which the subject is being asked to learn) sets a limit on the generalization power that the subject has at his or her disposal. In this sense I_{pred} provides an absolute standard against which to measure learning performance in the same way that spike train entropy provides a standard against which to measure the performance of the neural code. I'm not really sure how to do this yet, but I can imagine that an information theoretic analysis of learning would thus lead to a measurement of learning efficiency [1] that parallels the measurement of coding efficiency or even detection efficiency in photon counting. Given our classification of learning tasks by their complexity, it would be natural to ask if the efficiency of learning were a critical function of task complexity: perhaps we can even identify a limit beyond which efficient learning fails, indicating a limit to the complexity of the internal model used by the brain during a class of learning tasks.

6 A little bit about molecules

It would be irresponsible to spend this many hours (or pages) on the brain without saying something that touches the explosion in our knowledge of what happens at the molecular level. Electrical signals in neurons are carried by ions, such as potassium or sodium, flowing through water or through specialized conducting pores. These pores, or channels, are large molecules (proteins) embedded in the cell membrane, and can thus respond to the electric field or voltage across the membrane. The coupled dynamics of channels and voltages makes each neuron into a nonlinear circuit, and this seems to be the molecular basis for neural computation. Many cells have the property that these nonlinear dynamics select stereotyped pulses that can propagate from one cell to another; these action potentials are the dominant form of long distance cell to cell communication in the brain, and our understanding of how these pulses occur is one the triumphs of the (now) classical "biophysics". Signals also can be carried by small molecules, which trigger various chemical reactions when they arrive at their targets. In particular, signal transmission across the synapse, or connection between two neurons, involves such small molecule messengers called neurotransmitters. Calcium ions can play both roles, moving in response to voltage gradients and regulating a number of important biochemical reactions in living cells, thereby coupling electrical and chemical events. Chemical events can reach into the cell nucleus to regulate which protein molecules–which ion channels and transmitter receptors–the cell produces. We will try to get a feeling for this range of phenomena, starting on the back of an envelope and building our way up to the facts.

Ions and small molecules diffuse freely through water, but cells are surrounded by a membrane that functions as a barrier to diffusion. In particular, these membranes are composed of lipids, which are nonpolar, and therefore cannot screen the charge of an ion that tries to pass through the membrane. The water, of course, is polar and does screen the charge, so pulling an ion out of the water and pushing it through the membrane would require surmounting a large electrostatic energy barrier. This barrier means that the membrane provides an enormous resistance to current flow between the inside and the outside of the cell. If this were the whole story there would be no electrical signalling in biology. In fact, cells construct specific pores or channels through which ions can pass, and by regulating the state of these channels the cell can control the flow of electric current across the membrane.

Ion channels are themselves molecules, but very large ones–they are proteins composed of several thousand atoms in very complex arrangements. Let's try, however, to ask a simple question: if we open a pore in the cell membrane, how quickly can ions pass through? More precisely, since the ions carry current and will move in response to a voltage difference across the membrane, how large is the current in response to a given voltage? We recall that the ratio of current to voltage is called conductance, so we are really asking for the conductance of an open channel. Again we only want an order of magnitude estimate, not a detailed theory.

Imagine that one ion channel serves, in effect, as a hole in the membrane. Let us pretend that ion flow through this hole is essentially the same as through water. The electrical current that flows through the channel is

$$J = q_{\rm ion} \cdot [\text{ionic flux}] \cdot [\text{channel area}], \tag{6.1}$$

where $q_{\rm ion}$ is the charge of one ion, and we recall that "flux" measures the current across a unit area, so that

$$\begin{aligned} \text{ionic flux} &= \frac{\text{ions}}{\text{cm}^2\text{s}} = \frac{\text{ions}}{\text{cm}^3} \cdot \frac{\text{cm}}{\text{s}} && (6.2)\\ &= [\text{ionic concentration}] \cdot [\text{velocity of one ion}] && (6.3)\\ &= cv. && (6.4)\end{aligned}$$

Major current carriers like sodium and potassium are at $c \sim 100$ milliMolar, or $c \sim 6\times 10^{19}$ ions/cm^3. The average velocity is related to the applied force through the mobility μ, the force on an ion is in turn equal to the electric field times the ionic charge, and the electric field is (roughly) the voltage difference V across the membrane divided by the thickness ℓ of the membrane:

$$v = \mu q_{\rm ion} E \sim \mu q_{\rm ion}\frac{V}{\ell} \sim \frac{D}{k_{\rm B}T} q_{\rm ion}\frac{V}{\ell}, \tag{6.5}$$

where in the last step we recall the Einstein relation between mobility and diffusion constant. Putting the various factors together we find the current

$$\begin{aligned} J &= q_{\rm ion} \cdot [\text{ionic flux}] \cdot [\text{channel area}] \\ &= q_{\rm ion} \cdot [cv] \cdot [\pi d^2/4] && (6.6)\\ &\sim \frac{\pi}{4} q_{\rm ion} \cdot \frac{cd^2 D}{\ell} \cdot \frac{q_{\rm ion} V}{k_{\rm B}T}, && (6.7)\end{aligned}$$

where the channel has a diameter d. If we assume that the ion carries one electronic charge, as does sodium, potassium, or chloride, then $q_{\rm ion} = 1.6 \times 10^{-19}$ C and

$$\frac{q_{\rm ion} V}{k_{\rm B}T} = \frac{V}{25\ \text{mV}}. \tag{6.8}$$

Typical values for the channel diameter should be comparable to the diameter of a single ion, $d \sim 0.3$ nm, and the thickness of the membrane is $\ell \sim 5$ nm. Diffusion constants for ions in water are $D \sim 2 \times 10^{-9}$ m^2/s, or $\sim 2\,(\mu\text{m})^2$/s, which is a more natural unit. Plugging in the numbers,

$$\begin{aligned} J &= gV && (6.9)\\ g &\sim 2 \times 10^{-11}\ \text{Amperes/Volt} = 20\ \text{picoSiemens}. && (6.10)\end{aligned}$$

So our order of magnitude argument leads us to predict that the conductance of an open channel is roughly 20 pS, which is about right experimentally.

Empirically, cell membranes have resistances of $R_{\rm m} \sim 10^3\,\text{ohm/cm}^2$, or conductances of $G_{\rm m} \sim 10^{-3}\,\text{S/cm}^2$. If each open channel contributes roughly 10 pS, then this membrane conductance corresponds to an average density of $\sim 10^8$ open channels per cm^2, or roughly one channel per square micron. This is correct but misleading. First, they are many channels which are, at one time, not open; indeed the dynamics with which channels open and close is very important, as we shall see momentarily. Second, channels are not distributed uniformly over the cell surface. At the synapse, or connection between two neurons, the postsynaptic cell has channels that open in response to the binding of the transmitter molecules released by the presynaptic cell. These "receptor channels" form a nearly close packed crystalline array in the small patch of cell membrane that forms the closest contact with the presynaptic cell, and there are other examples of great concentrations of channels in other regions of the cell.

Problem 11: Membrane capacitance. From the facts given above, estimate the capacitance of the cell membrane. You should get $C \sim 1\,\mu\text{F}/\text{cm}^2$.

Channels are protein molecules: heteropolymers of amino acids. As discussed by other lecturers here, there are twenty types of amino acid and a protein can be anywhere from 50 to 1000 units in length. Channels tend to be rather large, composed of several hundred amino acids; often there are several subunits, each of this size. For physicists, the "protein folding problem" is to understand what it is about real proteins that allows them to collapse into a unique structure. This is, to some approximation, a question about the equilibrium state of the molecule, since for many proteins we can "unfold" the molecule either by heating or by chemical treatment and then recover the original structure by returning to the original condition[29]. At present, this problem is attracting considerable attention in the statistical mechanics community. For a biologist, the protein folding problem is slightly different: granting that proteins fold into unique structures, one would like to understand the mapping from the linear sequence of amino acids in a particular protein into the three dimensional structure of the folded state. Again, this is a very active–but clearly distinct–field of research.

We actually need a little more than a unique folded state for proteins. Most proteins have a few rather similar structures which are stable, and the energy differences between these structures are several (up to ~ 10) $k_{\rm B}T$, which means that the molecule can be pushed from one state to another by interesting perturbations, such as the binding of a small molecule. For channels, there is a more remarkable fact, namely that (for most channels) out of several accessible states, only one is "open" and conducting. The other states are closed or (and this is different!) inactivated. If we think about arranging the different states in a kinetic scheme, we might write

$$C_1 \leftrightarrow C_2 \leftrightarrow O, \tag{6.11}$$

[29]Of course there are interesting exceptions to this rule.

which corresponds to two closed states and one open state, with the constraint that the molecule must pass through the second closed state in order to open. If the open state also equilibrates with an "inactive" state I that is connected to C_1,

$$C_1 \leftrightarrow C_2 \leftrightarrow O \leftrightarrow I \leftrightarrow C_1, \tag{6.12}$$

then depending on the rate constants for the different transitions the channel can be forced to pass through the inactive state and then through all of the closed states before opening again. This is interesting because the physical processes of "closing" and "inactivating" are often different, and this means that the transition rates can differ by orders of magnitude: there are channels that can flicker open and closed in a millisecond, but require minutes to recover from inactivation. If we imagine that channels open in response to certain inputs to the cell, this process of inactivation endows the cell with a memory of how many of these inputs have occurred over the past minute–the states of individual molecules are keeping count, and the cell can read this count because the molecular states influence the dynamics of current flow across the membrane.

Individual amino acids have dipole moments, and this means that when the protein makes a slight change in structure (say $C_2 \to O$) there will be a change in the dipole moment of the protein unless there is an incredible coincidence. But this has the important consequence that the energy differences among the different states of the channel will be modulated by the electric field and hence by the voltage across the cell membrane. If the difference in dipole moment were equivalent to moving one elementary charge across the membrane, then we could shift the equilibrium between the two states by changing the voltage over $\sim k_{\rm B}T/q_{\rm e} = 25\,{\rm mV}$, while if there are order ten charges transferred the channel will switch from one state to another over just a few mV. While molecular rearrangements within the channel protein do not correspond to charge transfer across the whole thickness of the membrane, the order of magnitude change in dipole moment is in this range.

It is important to understand that one can measure the current flowing through single channels in a small patch of membrane, and hence one can observe the statistics of opening and closing transitions in a single molecule. From such experiments one can build up kinetic models like that in equation (6.12), and these provide an essentially exact description of the dynamics at the single molecule level. The arrows in such kinetic schemes are to be interpreted not as macroscopic chemical reaction rates but rather as probabilities per unit time for transitions among the states, and from long records of single channel dynamics one can extract these probabilities and their voltage dependences. Again, this is not easy, in part because one can distinguish only the open state–different closed or inactivated states all have zero conductance and hence are indistinguishable when measuring current–so that multiple closed states have to be inferred from the distribution of times between openings of the channel. This is a very pretty subject, driven by the ability to do extremely quantitative experiments; it is even possible to detect the shot noise as ions flow through the open channel, as well as a small amount of excess noise due to the "breathing" of the channel molecule while it is in the open state. The first single channel experiments were by Neher & Sakmann [98], and a modern summary of what we have learned is given in textbooks [99, 100].

Problem 12: Closed time distributions. For a channel with just two states, show that the distribution of times between one closing of the channel and the next opening is exponential in form. How is this distribution changed if there are two closed states? Can you distinguish a second closed state ("before" opening) from an inactive state ("after" opening)?

We would like to pass from a description of single channels to a description of a macroscopic piece of membrane, perhaps even the whole cell. If we can assume that the membrane is homogeneous and isopotential then there is one voltage V across the whole membrane, and each channel has the same stochastic dynamics. If the region we are talking about has enough channels, we can write approximately deterministic equations for the number of channels in each state. These equations have coefficients (the transition probabilities) that are voltage dependent, and of course the voltage across the membrane has a dynamics driven by the currents that pass through the open channels. Let's illustrate this with the simplest case.

Consider a neuron that has one type of ion channel that is sensitive to voltage, and a "leak" conductance (some channels that we haven't studied in detail, and which don't seem to open and close in the interesting range of voltages). Let the channel have just two states, open and closed, and a conductance g when it is open. Assume that the number of open channels $n(t)$ relaxes to its equilibrium value $n_{\rm eq}(V)$ with a time constant $\tau(V)$. In addition assume that the gated channel is (perfectly) selective for ions that have a chemical potential difference of $V_{\rm ion}$ across the membrane, while the leak conductance $G_{\rm leak}$ pulls the membrane potential back to its resting level $V_{\rm rest}$. Finally, assume that the cell has a capacitance C, and allow for the possibility of injecting a current $I_{\rm ext}$ across the membrane. Then the equations of motion for the coupled dynamics of channels and voltage are

$$C\frac{{\rm d}V}{{\rm d}t} = -gn(V - V_{\rm ion}) - G_{\rm leak}(V - V_{\rm rest}) + I_{\rm ext}, \qquad (6.13)$$

$$\frac{{\rm d}n}{{\rm d}t} = -\frac{1}{\tau(V)}[n - n_{\rm eq}(V)]. \qquad (6.14)$$

These equations already have a lot in them:

- If we linearize around a steady state we find that the effect of the channels can be thought of as adding some new elements to the effective circuit describing the membrane. In particular these elements can include an (effective) inductance and a negative resistance;
- Inductances of course make for resonances, which actually can be tuned by cells to build arrays of channel–based electrical filters [101]. If the negative resistance is large enough, however, the filter goes unstable and one gets oscillations;
- One can also arrange the activation curve $n_{\rm eq}(V)$ relative to $V_{\rm ion}$ so that the system is bistable, and the switch from one state to the other can be triggered by a pulse of current. In an extended structure like the axon of a neuron this switching would propagate as a front at some fixed velocity;
- In realistic models there is more than one kind of channel, and the nonlinear dynamics which selects a propagating front instead selects a propagating pulse, which is the action potential or spike generated by that neuron.

It is worth recalling the history of these ideas, at least briefly. In a series of papers, Hodgkin & Huxley [102–105] wrote down equations similar to equations (6.13, 6.14) as a phenomenological description of ionic current flow across the cell membrane. They studied the squid giant axon, which is a single nerve cell that is a small gift from nature, so large that one can insert a wire along its length! This axon, like that in all neurons, exhibits propagating action potentials, and the task which Hodgkin and Huxley set themselves was to understand the mechanism of these spikes. It is important to remember that action potentials provide the only mechanism for long distance communication among specific neurons, and so the question of how action potentials arise is really the question of how information gets from one place to another in the brain. The first step taken by Hodgkin and Huxley was to separate space and time: suspecting that current flow along the length of the axon involved only passive conduction through the fluid, they "shorted" this process by inserting a wire and thus forcing the entire axon to become isopotential. By measuring the dynamics of current flow between the wire and an electrode placed outside of the cell they were then measuring the average properties of current flow across a patch of membrane.

It was already known that the conductance of the cell membrane changes during an action potential, and Hodgkin and Huxley studied this systematically by holding the voltage across the membrane at one value and then stepping to another. With dynamics of the form in equation (6.14), the fraction of open channels will relax exponentially... and after some effort one should be able to pull out the equilibrium fraction of open channels and the relaxation rates, each as functions of of voltage; again it is important to have the physical picture that the channels in the membrane are changing state in response to voltage (or, more naturally, electric field) and hence the dynamics are simple if the voltage is (piecewise) constant.

There are two glitches in this simple picture. First, the relaxation of conductance or current is not exponential. Hodgkin and Huxley interpreted this (again, phenomenologically!) by saying that the equations for elementary "gates" were as in equation (6.14) but that conductance of ions trough a pore might require that several independent gates are open. So instead of writing

$$C\frac{\mathrm{d}V}{\mathrm{d}t} = -gn(V - V_{\text{ion}}) - G_{\text{leak}}(V - V_{\text{rest}}) + I_{\text{ext}},$$

they wrote, for example,

$$C\frac{\mathrm{d}V}{\mathrm{d}t} = -gn^4(V - V_{\text{ion}}) - G_{\text{leak}}(V - V_{\text{rest}}) + I_{\text{ext}}, \qquad (6.15)$$

which is saying that four gates need to open in order for the channel to conduct (their model for the potassium channel). To model the inactivation of sodium channels they used equations in which the number of open channels was proportional to m^3h, where m and h each obey equations like equation (6.14), but the voltage dependences $m_{\text{eq}}(V)$ and $h_{\text{eq}}(V)$ have opposite behaviors–thus a step change in voltage can lead to an increase in conductance as m relaxes toward its increased equilibrium value, then a decrease as h starts to relax to its decreased equilibrium value. In modern language we would say that the channel molecule has more than two states, but the phenomenological picture of multiple gates works quite well; it is interesting that Hodgkin and Huxley themselves were careful not to take too seriously any particular molecular interpretation of their equations. The second problem in the analysis is that there are several types of channels, although this is easier in the squid axon because "several" turns out to be just two–one selective for sodium ions and one selective for potassium ions.

The great triumph of Hodgkin and Huxley was to show that, having described the dynamics of current flow across a single patch of membrane, they could predict the existence, structure, and speed of propagating action potentials. This was a milestone, not least because it represents one of the few cases where a fundamental advance in our understanding of biological systems was marked by a successful quantitative prediction. Let me remind you that, in 1952, the idea that nonlinear partial differential equations like the HH equations would generate propagating stereotyped pulses was by no means obvious; the numerical methods used by Hodgkin and Huxley were not so different from what we might use today, while rigorous proofs came only much later.

Of course I have inverted the historical order in this presentation, describing the properties of ion channels (albeit crudely) and then arguing that these can be put together to construct the macroscopic dynamics of ionic currents. In fact the path from Hodgkin and Huxley to the first observation of single channels took nearly twenty five years. There were several important steps. First, the HH model makes definite predictions about the magnitude of ionic currents flowing during an action potential, and in particular the relative contributions of sodium and potassium; these predictions were confirmed by measuring the flux of radioactive ions[30]. Second, as mentioned already, the transitions among different channels states are voltage dependent only because these different states have different dipole moments. This means that changes in channel state should be accompanied by capacitive currents, called "gating currents", which persist even if conduction of ions through the channel is blocked, and this is observed. The next crucial step is that if we have a patch of membrane with a finite number of channels, then it should be possible to observe fluctuations in current flow due to the fluctuations in the number of open channels–the opening and closing of each channel is an independent, thermally activated process. Kinetic models make unambiguous predictions about the spectrum of this noise, and again these predictions were confirmed both qualitatively and quantitatively; noise measurements also led to the first experimental estimates of the conductance through a single open channel. Finally, observing the currents through single channels required yet better amplifiers and improved contact between the electrode and the membrane to insure that the channel currents are not swamped by Johnson noise in stray conductance paths.

Problem 13: Independent opening and closing. The remark that channels open and close independently is a bit glib. We know that different states have different dipole moments, and you might expect that these dipoles would interact. Consider an area A of membrane with N channels that each have two states. Let the two states differ by an effective displacement of charge q_{gate} across the membrane, and this charge interacts with the voltage V across the membrane in the usual way. In addition, there is an energy associated with the voltage itself,

[30]It also turns out that the different types of channels can be blocked, more or less independently, by various molecules. Some of the most potent channel blockers are neurotoxins, such as tetrodotoxin from puffer fish, which is a sodium channel blocker. These different toxins allow a pharmacological "dissection" of the molecular contributions to ionic current flow.

since the membrane has a capacitance. If we represent the two states of each channel by an Ising spin σ_n, convince yourself that the energy of the system can be written as

$$E = \frac{1}{2}CV^2 + \frac{1}{2}\sum_{n=1}^{N}(\epsilon + q_{\rm gate}V)\sigma_n. \tag{6.16}$$

Set up the equilibrium statistical mechanics of this system, and average over the voltage fluctuations. Show that the resulting model is a mean field interaction among the channels, and state the condition that this interaction be weak, so that the channels will gate independently. Recall that both the capacitance and the number of channels are proportional to the area. Is this condition met in real neurons? In what way does this condition limit the "design" of a cell? Specifically, remember that increasing $q_{\rm gate}$ makes the channels more sensitive to voltage changes, since they make their transitions over a voltage range $\delta V \sim k_{\rm B}T/q_{\rm gate}$; if you want to narrow this range, what do you have to trade in order to make sure that the channels gate independently? And why, by the way, is it desirable to have independent gating?

So, in terms of Hodgkin–Huxley style models we would describe a neuron by equations of the form

$$C\frac{{\rm d}V}{{\rm d}t} = -\sum_i G_i m_i^{\mu_i} h_i^{\nu_i}(V - V_i) - G_{\rm leak}(V - V_{\rm rest}) + I_{\rm ext}, \tag{6.17}$$

$$\frac{{\rm d}m_i}{{\rm d}t} = -\frac{1}{\tau^i_{\rm act}}[m_i - m^i_{\rm eq}(V)], \tag{6.18}$$

$$\frac{{\rm d}h_i}{{\rm d}t} = -\frac{1}{\tau^i_{\rm inact}}[h_i - h^i_{\rm eq}(V)], \tag{6.19}$$

where i indexes a class of channels specific for ions with an equilibrium potential V_i and we have separate kinetics for activation and inactivation. Of course there have been many studies of such systems of equations. What is crucial is that by doing, for example, careful single channel experiments on patches of membrane from the cell we want to study, we measure essentially every parameter of these equations except for the total number of each kind of channel. This is a level of detail that is not available in any other biological system as far as I know.

If we agree that the activation and inactivation variables run from zero to unity, representing probabilities, then the number of channels is in the parameters G_i which are the conductances we would observe if all channels of class i were to open. With good single channel measurements, these are the only parameters we don't know.

For many years it was a standard exercise to identify the types of channels in a cell, then try to use these Hodgkin–Huxley style dynamics to explain what happens when you inject currents etc. It probably is fair to say that this program was successful beyond the wildest dreams of Hodgkin and Huxley themselves–myriad different types of channel from many different types of neuron have been described effectively by the same general sorts of equations. On the other hand (although nobody ever said this) you have to hunt around to find the right G_is to make everything work in any reasonably complex cell. It was Larry Abbott, a physicist, who realized that if this is a problem for his graduate student then it must also be a problem for the cell (which doesn't have graduate students to whom the task can be assigned). So, Abbott and his colleagues realized that there must be regulatory mechanisms that control the channel numbers in ways that stabilized desirable functions of the cell in the whole neural circuit [106]. This has stimulated a beautiful series of experiments by Turrigiano and collaborators, first in "simple" invertebrate neurons [107] and then in cortical neurons [108], showing that indeed these different cells can change the number of each different kind of channel in response to changes in the environment, stabilizing particular patterns of activity or response. Mechanisms are not yet clear. I believe that this is an early example of the robustness problem [109, 110] that was emphasized by Barkai and Leibler for biochemical networks [111]; they took adaptation in bacterial chemotaxis as an example (*cf.* the lectures here by Duke) but the question clearly is more general. For more on these models of self–organization of channel densities see [112–114].

The problem posed by Abbott and coworkers was, to some approximation, about homeostasis: how does a cell hold on to its function in the network, keeping everything stable. In the models, the "correct" function is defined implicitly. The fact that we have seen adaptation processes which serve to optimize information transmission or coding efficiency makes it natural to ask if we can make models for the dynamics which might carry out these optimization tasks. There is relatively little work in this area [115], and I think that any effort along these lines will have to come to grips with some tough problems about how cells "know" they are doing the right thing (by any measure, information theoretic or not).

Doing the right thing, as we have emphasized repeatedly, involves both the right deterministic transformations and proper control of noise. We know a grat deal about noise in ion channels, as discussed above, but I think the conventional view has been that most neurons have lots of channels and so this source of noise isn't really crucial for neural function. In recent work Schneidman and collaborators have shown that this dismissal of channel noise may have been a bit too quick [116]: neurons operate in a regime where the number of channels that participate in the "decision" to generate an action potential is vastly smaller than the total number of channels, so that fluctuation effects are much more important that expected naively. In particular, realistic amounts of channel noise may serve to jitter the timing of spikes on time scales which are comparable to the degree of reproducibility observed in the representation of sensory signals (as discussed in Sect. 4). In this way the problems of molecular level noise and the optimization of information transmission may be intertwined [109].

It should be emphasized that the molecular components we have been discussing are strikingly universal. Thus we can recognize homologous potassium channels in primate cortex (the stuff we think with) and in the nerves of an earthworm. There are vast numbers of channels coded in the genome, and these can be organized into families of proteins that probably have common ancestors [99]. With such a complete molecular description of electrical signalling in single cells, one would imagine that we could answer a deceptively simple question: what do individual neurons compute? In neural network models, for example, neurons are cartooned as summing their inputs and taking a threshold. We could make this picture a bit more dynamical by using an "integrate and fire" model in which input currents are filtered by the RC time constant of the cell membrane and all the effects of channels are summarized by saying that when the resulting voltage reaches threshold there is a spike and a resetting to a lower voltage. We would like to start with a more realistic model and show how one can identify systematically some computational functions, but really we don't know how to do this. One attempt is discussed in Refs. [117, 118], where we use the ideas of dimensionsality reduction [50, 59, 71] to pass from the Hogdkin–Huxley model to a description of the neuron as projecting dynamic input currents onto a low dimensional space and then performing some nonlinear operations to determine the probablity of generating a spike. If a simple summation and threshold picture (or a generalized "filter and fire" model) were correct, this approach would find it, but it seems that even with two types of channels neurons can do something richer than this. Obviously this is just a start, and understanding will require us to face the deeper question of how we can indentify the computational function of a general dynamical system.

In this discussion I have focused on the dynamics of ion channels within one neuron. To build a brain we need to make connections or synapses between cells, and of course these have their own dynamics and molecular mechanisms. There are also problems of noise, not least because synapses are very small structures, so that crucial biochemical events are happening in cubic micron volumes or less. The emergence of optical techniques that allow us to look into these small volumes, deep in the living brain, will quite literally bring into focus a number of questions about noise in biochemical networks that are of interest both because they relate to how we learn and remember things and because they are examples of problems that all cells must face as they carry out essential functions.

7 Speculative thoughts about the hard problems

It is perhaps not so surprising that thinking like a physicist helps us to understand how rod cells count single photons, or helps to elucidate the molecular events that underlie the electrical activity of single cells. A little more surprising, perhaps, is that physical principles are still relevant when we go deeper into a fly's brain and ask about how that brain extracts interesting features such as motion from a complex array of data in the retina, or how these dynamic signals are encoded in streams of action potentials. As we come to the problems in learning, we have built an interesting theoretical structure with clear roots in statistical physics, but we don't yet know how to connect these ideas with experiment. Behind this uncertainty is a deeper and more disturbing question: maybe as we progress from sensory inputs toward the personal experiences of that world created by our brains we will encounter a real boundary where physics stops and biology or psychology begins. My hope, as you might guess, is that this is not the case, and that we eventually will understand perception, cognition and learning from the same principled mathematical point of view that we now understand the inanimate parts of the physical world. This optimism was shared, of course, by Helmholtz and others more than a century ago. In this last lecture I want to collect some of my reasons for keeping faith despite obvious problems.

In Shannon's original work on information theory, he separated the problem of transmitting information from the problem of ascribing meaning to this information [63]:

> Frequently the messages have *meaning;* that is they refer to or are correlated according to some system with certain physical or conceptual entities. These semantic aspects of communication are irrelevant to the engineering problem.

This quote is from the second paragraph of a very long paper; italics appeared in the original. Arguably this is *the* major stumbling block in the use of information theory or any other "physical" approach to analyze cognitive phenomena: our brains presumably are interested only in information that has meaning or relevance, and if we are in a framework that excludes such notions then we can't even get started.

Information theory is a statistical approach, and there is a widespread belief that there must be "more than just statistics" to our understanding of the world. The clearest formulation of this claim was by Chomsky [119], in a rather direct critique of Shannon and his statistical approach to the description of English. Shannon had used N^{th} order Markov approximations to the distribution of letters or words, and other people used this N-gram method in a variety of ways, including the amusing "creative writing" exercises of Pierce and others. Chomsky claims that all of this is hopeless, for several reasons:

1. The significance of words or phrases is unrelated to their frequency of occurrence;

2. Utterances can be arbitrarily long, with arbitrarily long range dependences among words, so that no finite $N^{\rm th}$ order approximation is adequate;

3. A rank ordering of sentences by their probability in such models will have grammatical and ungrammatical utterances mixed, with little if any tendency for the grammatical sentences to be more probable.

There are several issues here[31], and while I am far from being an expert on language I think if we try to dissect these issues we'll get a feeling for the general problems of thinking about the brain more broadly in information theoretic terms.

First we have the distinction between the true probability distribution of sentences (for example) and any finite $N^{\rm th}$ order approximation. There are plenty of cases in physics where analogous approximations fail, so this shouldn't bother us, nor is it a special feature of language. Nonetheless, it *is* important to ask how we can go beyond these limited models. There is a theoretical question of how to characterize statistics beyond N–grams, and there is an experimental issue of how to measure these long range dependencies in real languages or, more subtly, in people's knowledge of languages. I think that we know a big part of the answer to the first question, as explained above: the crucial measure of long range correlation is a divergence in the *predictive information* $I_{\rm pred}(N)$, that is the information that a sequence of N characters or words provides about the remainder of the text. We can distinguish logarithmic divergence, which means roughly that the sequence of words allows us to learn a model with a finite number of parameters (the coefficient of the log then counts the dimensionality of the parameter space), from a power law divergence, which is what happens when longer and longer sequences allow us to learn a more and more detailed description of the underlying model. There are hints that language is in the more complex power law class.

---

[31]Some time after writing an early draft of these ideas I learned that Abney [120] had expressed similar thoughts about the nature of the Chomsky/Shannon debate; he is concerned primarily with the first of the issues below. I enjoyed especially his introduction to the problem: "in one's introductory linguistics course, one learns that Chomsky disabused the field once and for all of the notion that there was anything of interest to statistical models of language. But one usually comes away a little fuzzy on the question of what, precisely, he proved".

A second question concerns the learnability of the relevant distributions. It might be that the true distribution of words and phrases contains everything we want to know about the language, but that we cannot learn this distribution from examples. Here it is important that "we" could be a child learning a language, or a group of scientists trying to analyze a body of data on language as it is used. Although learnability is a crucial issue, I think that there is some confusion in the literature. Thus, even in very recent work, we find comments that confuse the frequency of occurrence of examples that we have seen with the estimate that an observer might make of the underlying distribution[32]. The easiest way to see this is to think about distributions of continuous variables, where obviously we have to interpolate or smooth so that our best estimate of the probability is not zero at unsampled points nor is it singular at the sampled points. There are many ways of doing this, and I think that developments of the ideas in Ref. [90] are leading us toward a view of this problem which at least seems principled and natural from a physicist's point of view [123–127]. On the other hand, the question of how one does such smoothing or regularization in the case of discrete distributions (as for words and phrases) is much less clear (see, for example, [128]).

Even if we can access the full probability distribution of utterances (leaving aside the issue of learning this distribution from examples), there is a question of whether this distribution captures the full structure of the language. At one level this is trivial: if we really have the full distribution we can generate samples, and there will be no statistical test that will distinguish these samples from real texts. Note again that probability distributions are "generative" in the sense that Chomsky described grammar, and hence that no reasonable description of the probability distribution is limited to generating sequences which were observed in some previous finite sampling or learning period. Thus, if we had an accurate model of the probability distribution for texts, we could pass some sort of Turing test. The harder question is whether this description of the language would contain any notions of syntax or semantics. Ultimately we want to ask about

[32]In particular, a widely discussed paper by Marcus *et al.* [121] makes the clear statement that all unseen combinations of words must be assigned probability zero in a statistical learning scheme, and this simply is wrong. The commentary on this paper by Pinker [122] has some related confusions about what happens when one learns a distribution from examples. He notes that we can be told "that a whale is not a fish... overriding our statistical experience..." In the same way that reasonable learning algorithms have to deal with unobserved combinations, they also have to deal with outliers in distributions; the existence of outliers, or the evident difficulty in dealing with them, has nothing to do with the question of whether our categories of fish and mammals are built using a probabilistic approach. The specific example of whales may be a red herring: does being told that a whale is not a fish mean that "all the fish in the sea" cannot refer to whales?

meaning: is it possible for a probabilistic model to encode the meanings of words and sentences, or must these be supplied from the outside? Again the same question arises in other domains: in what sense does knowing the distribution of all possible natural movies correspond to "understanding" what we are seeing?

Recently, Tishby, Pereira and I have worked on the problem of defining and extracting *relevant information*, trying to fill the gap left by Shannon [85]. Briefly, the idea is that we observe some signal $x \in X$ but are interested in another signal $y \in Y$. Typically a full description of x requires many more bits than are relevant to the determination of y, and we would like to separate the relevant bits from the irrelevant ones. Formally we can do this by asking for a compression of x into some new space $\tilde{X}$ such that we keep as much information as possible about Y while throwing away as much as possible about X. That is, we want to find a mapping $x \rightarrow \tilde{x}$ that maximizes the information about Y while holding the information about X fixed at some small value. This problem turns out to be equivalent to a set of self–consistent equations for the mapping $x \rightarrow \tilde{x}$, and is very much like a problem of clustering. It is important that, unlike most clustering procedures, there is no need to specify a notion of similarity or distance among points in the X or Y spaces–all notions of similarity emerge directly from the joint statistics of X and Y.

To see a little of how this works, let's start with a somewhat fanciful question: what is the information content of the morning newspaper? Since entropy provides the only measure of information that is consistent with certain simple and plausible constraints (as emphasized above), it is tempting to suggest that the information content of a news article is related to the entropy of the distribution from which newspaper texts are drawn. This is troublesome–more random texts have higher entropy and hence would be more informative–and also incorrect. Unlike entropy, information always is *about* something. We can ask how much an article tells us about, for example, current events in France, or about the political bias of the editors, and in a foreign country we might even use the newspaper to measure our own comprehension of the language. In each case, our question of interest has a distribution of possible answers, and (on average) this distribution shifts to one with a lower entropy once we read the news; this decrease in entropy is the information that we gain by reading. This relevant information typically is much smaller than the entropy of the original signal: information about the identity of a person is much smaller than the entropy of images that include faces, information about words is much smaller than the entropy of the corresponding speech sounds, and so on. Our intuitive notion of "understanding" these high entropy input signals corresponds to isolating the relevant information in some reasonably compact form: summarizing the news article, replacing the image with a name, enumerating the words that have been spoken.

When we sit down to read, we have in mind some question(s) that we expect to have answered by the newspaper text. We can enumerate all of the possible answers to these questions, and label the answers by $y \in Y$: this is the relevant variable, the information of value to us. On the other hand, we can also imagine the ensemble of all possible newspaper texts, and label each possible text by $x \in X$: these are the raw data that we have to work with. Again, there are many more bits in the text x than are relevant to the answers y, and understanding the text means that we must separate these relevant bits from the majority of irrelevant bits. In practice this means that we can "summarize" the text, and in the same way that we enumerate all possible texts we can also enumerate all possible summaries, and labelling them by $\tilde{x} \in \tilde{X}$. If we can generate good or efficient summaries then we can construct a mapping of the raw data x into the summaries $\tilde{x}$ such that we discard most of the information about the text but preserve as much information as possible about the relevant variable y.

The statement that we want our summaries to be compact or efficient means that we want to discard as much information as possible about the original signal. Thus, we want to "squeeze" or minimize the information that the summary provides about the raw data, $I(\tilde{x}; x)$. On the other hand, if the summary is going to capture our understanding, then it must preserve information about y, so we want to maximize $I(\tilde{x}; y)$. More precisely, there is going to be some tradeoff between the level of detail $[I(\tilde{x}; x)]$ that we are willing to tolerate and the amount of relevant information $[I\tilde{x}; y)]$ that we can preserve. The optimal procedure would be to find rules for generating summaries which provide the maximum amount of relevant information given their level of detail. The way to do this is to maximize the weighted difference between the two information measures,

$$-\mathcal{F} = I(\tilde{x}; y) - TI(\tilde{x}; x), \tag{7.1}$$

where T is a dimensionless parameter that measures the amount of extra detail we are willing to accept for a given increase in relevant information. We will refer to this parameter as the temperature, for reasons that become clear below. So, to find optimal summaries we want to search all possible rules for mapping $x \to \tilde{x}$ until we find a maximum of $-\mathcal{F}$, or equivalently a minimum of the "free energy" $\mathcal{F}$. Note that the structure of the optimal procedure generating summaries will evolve as we change the temperature T; there is no "correct" value of the temperature, since different values correspond to different ways of striking a balance between detail and effectiveness in the summaries.

There are several different interpretations of the principle that we should minimize $\mathcal{F}$. One view is that we are optimizing the weighted difference of the two informations, counting one as a benefit and one as a cost. Alternatively, we can see minimizing $\mathcal{F}$ as maximizing the relevant information while holding fixed the level of detail in the summary, and in this case we interpret T as a Lagrange multiplier that implements the constraint holding $I(\tilde{x};x)$ fixed. Similarly, we can divide through by T and interpret our problem as one of squeezing the summary as much as possible–minimizing $I(\tilde{x};x)$–while holding fixed the amount of relevant information that the summaries convey; in this case $1/T$ serves as the Lagrange multiplier.

It turns out that the problem of minimizing the free energy $\mathcal{F}$ can solved, at least formally. To begin we need to say what we mean by searching all possible rules for mapping $x \to \tilde{x}$. We consider here only the case where the summaries forma discrete set, and for simplicity we (usually) assume that the data x and the relevant variable y also are drawn from discrete sets of possibilities. The general mapping from x to $\tilde{x}$ is probabilistic, and the set of mapping rules is given completely if we specify the set of probabilities $P(\tilde{x}|x)$ that any raw data point x will be assigned to the summary $\tilde{x}$. These probabilities of course must be normalized, so we must enforce

$$\sum_{\tilde{x}\in\tilde{X}} P(\tilde{x}|x) = 1 \tag{7.2}$$

for each $x \in X$. We can do this by introducing a Lagrange multiplier $\Lambda(x)$ for each x and then solving the constrained optimization problem

$$\min_{P(\tilde{x}|x)} \left[\mathcal{F} - \sum_{x\in X} \Lambda(x) \sum_{\tilde{x}\in\tilde{X}} P(\tilde{x}|x) \right], \tag{7.3}$$

and at the end we have choose the values of $\Lambda(X)$ to satisfy the normalization condition in equation (7.2).

As shown in Ref. [85], the Euler–Lagrange equations for this variational problem are equivalent to a set of self–consistent equations for the probability distribution $P(\tilde{x}|x)$:

$$P(\tilde{x}|x) = \frac{P(\tilde{x})}{Z(x,T)} \exp\left\{ -\frac{1}{T} \sum_{y\in Y} P(y|x) \ln\left[\frac{P(y|x)}{P(y|\tilde{x})}\right] \right\} \tag{7.4}$$

$$\begin{aligned} P(y|\tilde{x}) &= \sum_{x\in X} P(y|x)P(x|\tilde{x}) \\ &= \frac{1}{P(\tilde{x})} \sum_{x\in X} P(y|x)P(\tilde{x}|x)P(x). \end{aligned} \tag{7.5}$$

Up to this point, the set of summaries $\tilde{X}$ is completely abstract. If we choose a fixed number of possible summaries then the evolution with temperature is continuous, and as we lower the temperature the summaries become progressively more detailed [$I(\tilde{x};x)$ is increasing] and more informative [$I(\tilde{x};y)$ is increasing]; the local coefficient that relates the extra relevant information per increment of detail is the temperature itself.

If X is discrete, then the detail in the summary can never exceed the entropy $S(X)$, and of course the relevant information provided by the summaries can never exceed the relevant information in the original signal. This means that there is a natural set of normalized coordinates in the information plane $I(\tilde{x};y)$ *vs.* $I(\tilde{x};x)$, and different signals are characterized by different trajectories in these normalized coordinates. If signals are "understandable" in the colloquial sense, it must be that most of the available relevant information can be captured by summaries that are very compact, so that $I(\tilde{x};y)/I(x;y)$ is near unity even when $I(\tilde{x};x)/S(X)$ is very small. At the opposite extreme are signals that have been encrypted (or texts which are so convoluted) so that no small piece of the original data contains any significant fraction of the relevant information.

Throughout most of the information plane the optimal solution has a probabilistic structure–the assignment rules $P(\tilde{x}|x)$ are not deterministic. This means that our problem of providing informative but compact summaries is very different from the usual problems in classification or recognition, where if we ask for assignment rules that minimize errors we will always find that the optimal solution is deterministic (recall Problem 2). Thus the information theoretic approach encompasses automatically a measure of confidence in which the optimal strategy involves (generically) a bit of true random guessing when faced with uncertainty. Returning to our example of the newspaper, this has an important consequence. If asked to provide a summary of the front page news, the optimal summaries have probabilistic assignments to the text–if asked several times, even an "optimal reader" will have a finite probability of giving different answers each time she is asked. The fact that assignment rules are probabilistic means also that these rules can be perturbed quantitatively by finite amounts of additional data, so that small amounts of additional information about, for example, the *a priori* likelihood of different interesting events in the world can influence the optimal rules for summarizing the text. It is attractive that a purely objective procedure, which provides an optimal extraction of relevant information, generates these elements of randomness and subjectivity.

Extracting relevant information has bene called the "information bottleneck" because we squeeze the signal X through a narrow channel $\tilde{X}$ while trying to preserve information about Y. This approach has been used, at least in preliminary form, in several cases of possible interest for the analysis of language. First, we can take X to be one word and Y to be the next word. If we insist that there be very few categories for $\tilde{X}$–we squeeze through a very narrow bottleneck–then to a good approximation the mapping from X into $\tilde{X}$ constitutes a clustering of words into sensible syntactic categories (parts of speech, with a clear separation of proper nouns). It is interesting that in the cognitive science literature there is also a discussion of how one might acquire syntactic categories from such "distributional information" (see, for example, [129]), although this discussion seems to use somewhat arbitrary metrics on the space of conditional distributions.

If we allow for $\tilde{X}$ to capture more bits about X in the next word problem, then we start to see the general part of speech clusters break into groups that have some semantic relatedness, as noted also by Redington *et al.* [129]. A more direct demonstration was given by Pereira *et al.* [130], who took X and Y to be the noun and verb of each sentence[33]. Now one has the clear impression that the clusters of words (either nouns or verbs) have similar meanings, although this certainly is only a subjective remark at this point. Notice, however, that in this formulation the absolute frequency of occurrence of individual words or even word pairs is not essential (connecting to Chomsky's point #1 above); instead the clustering of words with apparently similar meanings arises from the total structure of the set of conditional distributions $P(y|x)$.

Yet another possible approach to "meaning" involves taking X as the identity of a document Y as a word in the document. Slonim & Tishby [131] did this for documents posted to twenty different news groups on the web, of course hiding any information that directly identifies the news group in the document. The result is that choosing roughly twenty different values for $\tilde{X}$ captures most of the mutual information between X and Y, and these twenty clusters have a very strong overlap with the actual newsgroups. This procedure–which is "unsupervised" since the clustering algorithm does not have access to labels on the documents–yields a categorization that is competitive with state of the art methods for supervised learning of the newsgroup identities. While one may object that some of these tasks are too easy, these results at least go in the direction of suggesting that analysis of word statistics alone can identify the "topic" of a document as it was assigned by the author.

[33]This was done before the development of the bottleneck ideas so we need to be a bit careful. Tishby *et al.* proposed clustering X according the conditional distributions $P(y|x)$ and suggested the use of the Kullback–Leibler divergence ($D_{\rm KL}$) as a natural measure of distance. In the bottleneck approach there is no need to postulate a distance measure, but what emerges from the analysis is essentially the soft clustering based on $D_{\rm KL}$ as suggested by Tishby *et al.*

I think that some reasonably concrete questions emerge from all of this:

- How clear is the evidence that language–or other (colloquially) complex natural signals–fall into the power-law class defined through the analysis of predictive information?
- On a purely theoretical matter, can we regularize the problem of learning distributions over discrete variables in a (principled) way which puts such learning problems in the power–law class?
- Can we use the information bottleneck ideas to find the features of words and phrases that efficiently represent the large amounts of predictive information that we find in texts?
- Can we test, in psycholinguistic experiments, the hypothesis that this clustering of words and phrases through the bottleneck collects together items with similar meaning?
- If we believe that meanings are related to statistical structure, can we shift our perceptions of meaning by exposure to texts with different statistics?

This last experiment would, I think, be quite compelling (or at least provocative). When we set out to test the idea that neural codes are "designed" to optimize information transmission, the clear qualitative prediction was that the code would have to change in response to the statistics of sensory inputs, and this would have to work in ways beyond the standard forms of adaptation that had been characterized previously. This prediction was confirmed, and it would be quite dramatic if we could design a parallel experiment on our perception of meaning. Surely this is the place to stop.

The text and reference list make clear the enormous debt I owe to my colleagues and collaborators. I especially want to thank Rob de Ruyter, since it is literally because of his experiments that I had the courage to "think about the brain" in the sense which I tried to capture in these lectures. The adventure which Rob and I have had in thinking about fly brains in particular has been one we enjoyed sharing with many great students, postdocs and collaborators over the years: A. Zee, F. Rieke, D.K. Warland, M. Potters, G.D. Lewen, S.P. Strong, R. Koberle, N. Brenner, E. Schneidman, and A.L. Fairhall (in roughly historical order). While all of this work blended theory and experiment to the point that there are few papers which are "purely" one or the other, I've also been interested in some questions which, as noted, have yet to make contact with data. These ideas also have developed in a very collaborative way with C.G. Callan, I. Nemenman, F. Pereira, and N. Tishby (alphabetically). Particularly for the more theoretical topics, discussions with my current collaborators J. Miller and S. Still have been crucial. I am grateful to the students at Les Houches for making the task of lecturing so pleasurable, and to the organizers both for creating this opportunity and for being so patient with me as I struggled to complete these notes. My thanks also to S. Still for reading through a (very) late draft and providing many helpful suggestions.

References

[1] H.B. Barlow, *Nature* **304** (1983) 207–209.

[2] H.B. Barlow, *Proc. R. Soc. Lond. Ser. B* **212** (1981) 1–34.

[3] W. Bialek, *Ann. Rev. Biophys. Biophys. Chem.* **16** (1987) 455–478.

[4] W. Bialek, Optimal signal processing in the nervous system, in *Princeton Lectures on Biophysics,* edited by W. Bialek (World Scientific, Singapore, 1992) pp. 321–401.

[5] F. Rieke, D. Warland, R. de Ruyter van Steveninck and W. Bialek, *Spikes: Exploring the Neural Code* (MIT Press, Cambridge, 1997).

[6] M. Bouman, in *Sensory Communication,* edited by W. Rosenblith (MIT Press, Cambridge, 1960).

[7] F. Rieke and D.A. Baylor, *Revs. Mod. Phys.* **70** (1998) 1027–1036.

[8] S. Hecht, S. Shlaer and M.H. Pirenne, *J. Gen. Physiol.* **25** (1942) 819–840.

[9] H.A. van der Velden, *Physica* **11** (1944) 179–189.

[10] D.A. Baylor, T.D. Lamb and K.-W. Yau, *J. Physiol. (Lond.)* **288** (1979) 613–634.

[11] D.A. Baylor, B.J. Nunn and J.F. Schnapf, *Macaca fascicularis, J. Physiol. (Lond.)* **357** (1984) 575–607.

[12] H.B. Barlow, *J. Opt. Soc. Am.* **46** (1956) 634–639.

[13] D.M. Green and J.A. Swets, *Signal Detection Theory and Psychophysics* (Wiley, New York, 1966).

[14] M.C. Teich, P.R. Prucnal, G. Vannucci, M.E. Breton and W.J. McGill, *Biol. Cybern.* **44** (1982) 157–165.

[15] D.A. Baylor, G. Matthews and K.–W. Yau, *J. Physiol. (Lond.)* **309** (1980) 591–621.

[16] A.–C. Aho, K. Donner, C. Hydén, L.O. Larsen and T. Reuter, *Nature* **334** (1988) 348–350.

[17] A.G. Doukas, M.R. Junnarkar, R.R. Alfano, R.H. Callender, T. Kakitani and B. Honig, *Proc. Nat. Acad. Sci. (USA)* **81** (1984) 4790–4794.

[18] R.W. Schoenlein, L.A. Peteanu, R.A. Mathies and C.V. Shank, *Science* **254** (1991) 412–415.

[19] Q. Wang, R.W. Schoenlein, L.A. Peteanu, R.A. Mathies and C.V. Shank, *Science* **266** (1994) 422–424.

[20] W. Bialek, R.F. Goldstein and A. Kivelson, Simple models for the dynamics of biomolecules: How far can we go?, *in Structure, Dynamics and Function of Biomolecules: The First EBSA Workshop,* edited by A. Ehrenberg, R. Rigler, A. Graslund and L.J. Nilsson (Springer–Verlag, Berlin, 1987), pp. 65–69.

[21] F.L.J. Vos, D.P. Aalberts and W. van Saarloos, *Phys. Rev. B* **53** (1996) 14922–14928.

[22] D.P. Aalberts, M.S.L. du Croo de Jongh, B.F. Gerke and W. van Saarloos, *Phys. Rev. A* **61** (2000) 040701.

[23] M.H. Vos, J.–C. Lambry, S.J. Robles, D.C. Youvan, J. Breton and J.–L. Martin, *Proc. Nat. Acad. Sci. (USA)* **88** (1991) 8885–8889.

[24] M.H. Vos, M.R. Jones, C.N. Hunter, J. Breton and J.–L. Martin, *Proc. Nat. Acad. Sci. (USA)* **91** (1994) 12701–12705.

[25] A.–C. Aho, K. Donner, S. Helenius, L.O. Larsen and T. Reuter, *J. Comp. Physiol. A* **172** (1993) 671–682.

[26] F. Rieke and D.A. Baylor, *Biophys. J.* **75** (1998) 1836–1857.

[27] P.B. Detwiler, S. Ramanathan, A. Sengupta and B.I. Shraiman, *Biophys. J.* **79** (2000) 2801–2817.

[28] F. Rieke and D.A. Baylor, *Biophys. J.* **71** (1996) 2553–2572.

[29] H.B. Barlow, W. Levick and M. Yoon, *Vision Res. Suppl.* **3** (1971) 87–101.

[30] A.B. Valbo, Single afferent neurons and somatic sensation in humans, in *The Cognitive Neurosciences,* edited by M. Gazzaniga (MIT Press, Cambridge, 1995), pp. 237–252.

[31] J.A. Simmons, M. Ferragamo, C.F. Moss, S.B. Stevenson and R.A. Altes, *J. Comp. Physiol. A* **167** (1990) 589–616.

[32] H.B. Barlow, *Philos. Trans. R. Soc. Lond. Ser. B* **290** (1980) 71–82.

[33] H. Barlow and S.P. Tripathy, *J. Neurosci.* **17** (1997) 7954–7966.

[34] W.T. Newsome, M.N. Shadlen, E. Zohary, K.H. Britten and J.A. Movshon, Visual motion: Linking neuronal activity to psychophysical performance, in *The Cognitive Neurosciences,* edited by M. Gazzaniga (MIT Press, Cambridge, 1995), pp. 401–414.

[35] N. Strausfeld, *Atlas of an Insect Brain* (Springer–Verlag, Berlin, 1976).

[36] `http://flybrain.neurobio.arizona.edu/Flybrain/html/`

[37] D.G. Stavenga and R.C. Hardie, *Facets of Vision* (Springer–Verlag, Berlin, 1989).

[38] M.F. Land and T.S. Collett, *J. Comp. Physiol.* **89** (1974) 331–357.

[39] H. Wagner, *Phil. Trans. R. Soc.Ser. B* **312** (1986) 527–595.

[40] C. Schilstra and J.H. van Hateren, *J. Exp. Biol.* **202** (1999) 1481–1490.

[41] J.H. van Hateren and C. Schilstra, *J. Exp. Biol.* **202** (1999) 1491–1500.

[42] W. Reichardt and T. Poggio, *Q. Rev. Biophys.* **9** (1976) 311-375.

[43] R.R. de Ruyter van Steveninck and S.B. Laughlin, *Nature* **379** (1996) 642–645.

[44] K. Hausen, The lobular complex of the fly: structure, function, and significance in behavior, in *Photoreception and vision in invertebrates,* edited by M. Ali (Plenum, New York, 1984), pp. 523–559.

[45] K. Hausen and M. Egelhaaf, Neural mechanisms of visual course control in insects, in [37], (1989) pp. 391–424.

[46] K. Hausen and C. Wehrhahn, *Proc. R. Soc. Lond. B* **21** (1983) 211–216.

[47] W. Bialek, F. Rieke, R.R. de Ruyter van Steveninck and D. Warland, *Science* **252** (1991) 1854–1857.

[48] W. Bialek, Theoretical physics meets experimental neurobiology, in *1989 Lectures in Complex Systems, SFI Studies in the Sciences of Complexity, Lect. Vol. II,* edited by E. Jen (Addison–Wesley, Menlo Park CA, 1990), pp. 513–595.

[49] R.R. de Ruyter van Steveninck and W. Bialek, *Phil. Trans. R. Soc. Lond.* **348** (1995) 321–340.

[50] R. de Ruyter van Steveninck and W. Bialek, *Proc. R. Soc. London Ser. B* **234** (1988) 379–414.

[51] W. Bialek and A. Zee, *J. Stat. Phys.* **59** (1990) 103–115.

[52] D.K. Warland, P. Reinagel and M. Meister, *J. Neurophysiol.* **78** (1997) 2336–2350.

[53] J. Wessberg, C.R. Stambaugh, J.D. Kralik, P.D. Beck, M. Laubach, J.K. Chapin, J. Kim, S.J. Biggs, M.A. Srinivasan and M.A.L. Nicolelis, *Nature* **408** (2000) 361–365.

[54] M. Potters and W. Bialek, *J. Phys. I France* **4** (1994) 1755–1775.

[55] W. Reichardt, Autocorrelation, a principle for the evaluation of sensory information by the central nervous system, in *Principles of sensory communication,* edited by W.A. Rosenblith (Wiley, New York, 1961), pp. 303–317.

[56] R.R. de Ruyter van Steveninck, W. Bialek, M. Potters, R.H. Carlson and G.D. Lewen, Adaptive movement computation by the blowfly visual system, in *Natural and Artificial Parallel Computation: Proceedings of the Fifth NEC Research Symposium,* edited by D.L. Waltz (SIAM, Philadelphia, 1996), pp. 21–41.

[57] R.R. de Ruyter van Steveninck and W. Bialek, Optimality and adaptation in motion estimation by the blowfly visual system, *Proceedings of the IEEE 22nd Annual Northeast Bioengineering Conference* (1996) pp. 40–41.

[58] R.R. de Ruyter van Steveninck, W. Bialek, M. Potters and R.H. Carlson, Statistical adaptation and optimal estimation in movement computation by the blowfly visual system, in *Proc. I.E.E.E. Conf. Sys. Man Cybern.* (1994) pp. 302–307.

[59] W. Bialek and R.R. de Ruyter van Steveninck (in preparation). Features and dimensions: Motion estimation in fly vision.

[60] F. Attneave, *Psych. Rev.* **61** (1954) 183–193.

[61] H.B. Barlow, Sensory mechanisms, the reduction of redundancy and intelligence, in *Proceedings of the Symposium on the Mechanization of Thought Processes, Vol. 2,* edited by D.V. Blake and A.M. Uttley (H.M. Stationery Office, London, 1959), pp. 537–574.

[62] H.B. Barlow, Possible principles underlying the transformation of sensory messages, in *Sensory Communication*, edited by W. Rosenblith (MIT Press, Cambridge, 1961), pp. 217–234.

[63] C.E. Shannon (1948), A mathematical theory of communication, *Bell Sys. Tech. J.* **27**, 379–423 & 623–656. Reprinted in C.E. Shannon and W. Weaver, *The Mathematical Theory of Communication* (University of Illinois Press, Urbana, 1949).

[64] I. Ginzburg and T.J. Sejnowski (1996). Dynamics of Rule Induction by Making Queries: Transition Between Strategies, in *18th Annual Conference of the Cognitive Science Society,* pp. 121–125 (Lawrence Erlbaum, Mahwah NJ), see also `http://www.cnl.salk.edu/CNL/annual--reps/annual--rep95.html`

[65] V.V. Fedorov, V. V. *Theory of Optimal Experimental Design,* translated and edited by W.J. Studden and E.M. Klimko (Academic Press, New York, 1972).

[66] M.R. DeWeese and M. Meister, *Network* **10** (1999) 325–340.

[67] S.B. Laughlin, *Z. Naturforsch.* **36c** (1981) 910–912.

[68] J.J. Atick, Could information theory provide an ecological theory of sensory processing?, in *Princeton Lectures on Biophysics*, edited by W. Bialek (World Scientific, Singapore, 1992), pp. 223–289.

[69] J.H. van Hateren, *J. Comp. Physiol. A* **171** (1992) 157–170.

[70] D.L. Ruderman, T.W. Cronin and C.C. Chiao, *J. Opt. Soc. Am. A* **15** (1998) 2036–2045.

[71] N. Brenner, S.P. Strong, R. Koberle, R. de Ruyter van Steveninck and W. Bialek, *Neural Comp.* **12** (2000) 1531–1552.

[72] R.R. de Ruyter van Steveninck, G.D. Lewen, S.P. Strong, R. Koberle and W. Bialek, *Science* **275** (1997) 1805–1808.

[73] S.P. Strong, R. Koberle, R. de Ruyter van Steveninck and W. Bialek, *Phys. Rev. Lett.* **80** (1998) 197–200.

[74] F. Rieke, D. Warland and W. Bialek, *Europhys. Lett.* **22** (1993) 151–156.

[75] D. MacKay and W.S. McCulloch, *Bull. Math. Biophys.* **14** (1952) 127–135.

[76] E. Schneidman, N. Brenner, N. Tishby, R.R. de Ruyter van Steveninck and W. Bialek, Universality and individuality in a neural code, to appear in *Advances in Neural Information Processing 13,* edited by T.K. Leen, T.G. Dietterich and V. Tresp (MIT Press, Cambridge, 2001), pp. 159–165, see also `physics/0005043`.

[77] R. de Ruyter van Steveninck, A. Borst and W. Bialek, Real time encoding of motion: Answerable questions and questionable answers from the fly's visual system, in *Processing Visual Motion in the Real World: A Survey of Computational, Neural and Ecological Constraints,* edited by J.M. Zanker and J. Zeil (Springer–Verlag, Berlin, 2001), pp. 279–306, see also `physics/0004060`.

[78] G.D. Lewen, W. Bialek and R.R. de Ruyter van Steveninck, *Network* **12** (2001) 317–329, see also `physics/0103088`.

[79] N. Brenner, W. Bialek and R. de Ruyter van Steveninck, *Neuron* **26** (2000) 695–702.

[80] A.L. Fairhall, G.D. Lewen, W. Bialek and R.R. de Ruyter van Steveninck, *Nature* **412** (2001) 787–792.

[81] D.L. Ruderman and W. Bialek, *Phys. Rev. Lett.* **73** (1994) 814–817.

[82] S. Smirnakis, M.J. Berry II, D.K. Warland, W. Bialek and M. Meister, *Nature* **386** (1997) 69–73.

[83] W. Bialek, I. Nemenman and N. Tishby, *Neural Comp.* **13** (2001) 2409–2463, see also `physics/0007070`.

[84] W. Bialek, I. Nemenman and N. Tishby, *Physica A* **302** (2001) pp. 89–99, see also `physics/0103076`.

[85] N. Tishby, F. Pereira and W. Bialek, The information bottleneck method, in *Proceedings of the 37th Annual Allerton Conference on Communication, Control and Computing*, edited by B. Hajek and R.S. Sreenivas (University of Illinois, 1999), pp. 368–377, see also `physics/0004057`.

[86] H.S. Seung, H. Sompolinsky and N. Tishby, *Phys. Rev. A* **45** (1992) 6056–6091.

[87] P. Grassberger, *Int. J. Theor. Phys.* **25** (1986) 907–938.

[88] V. Vapnik, *Statistical Learning Theory* (John Wiley & Sons, New York, 1998).

[89] V.Balasubramanian, *Neural Comp.* **9** (1997) 349–368, see also `cond-mat/9601030`.

[90] W. Bialek, C.G. Callan and S.P. Strong, *Phys. Rev. Lett.* **77** (1996) 4693–4697, see also `cond-mat/9607180`.

[91] C.E. Shannon, Prediction and entropy of printed English, *Bell Sys. Tech. J.* **30**, 50–64. Reprinted in edited by N.J.A. Sloane and A.D. Wyner, *Claude Elwood Shannon: Collected papers* (IEEE Press, New York, 1993).

[92] W. Hilberg, *Frequenz* **44** (1990) 243–248.

[93] R.H.S. Carpenter and M.L.L. Williams, *Nature* **377** (1995) 59–62.

[94] L.L. Lopes and G.C. Oden, *J. Exp. Psych.: Learning, Memory, and Cognition* **13** (1987) 392–400.

[95] J.R. Saffran, R.N. Aslin and E.L. Newport, *Science* **274** (1996) 1926–1928.

[96] J.R. Saffran, E.K. Johnson, R.H. Aslin and E.L. Newport, *Cognition* **70** (1999) 27–52.

[97] M.D. Mauk and B.P. Ruize, *Behav. Neurosci.* **106** (1992) 666–681.

[98] E. Neher and B. Sakmann, *Nature* **260** (1976) 799–802.

[99] B. Hille, *Ionic Channels of Excitable Membranes,* 2d ed. (Sinauer Associates, Sunderland MA, 1992).

[100] D. Johnston and S.–M. Wu, *Foundations of Cellular Neurophysiology* (MIT Press, Cambridge, 1995).

[101] E.M.C. Jones, M. Gray–Keller and R. Fettiplace, *J. Physiol. (Lond.)* **518** (1999) 653–665.

[102] A.L. Hodgkin and A.F. Huxley, *J. Physiol.* **116** (1952a) 449–472.

[103] A.L. Hodgkin and A.F. Huxley, *J. Physiol.* **116** (1952b) 473–496.

[104] A.L. Hodgkin and A.F. Huxley, *J. Physiol.* **116** (1952c) 497–506.

[105] A.L. Hodgkin and A.F. Huxley, *J. Physiol.* **117** (1952d) 500–544.

[106] G. LeMasson, E. Marder and L.F. Abbott, *Science* **259** (1993) 1915–1917.

[107] G. Turrigiano, L.F. Abbott and E. Marder, *Science* **264** (1994) 974–977.

[108] N.S. Desai, L.C. Rutherford and G.G. Turrigiano, *Nature Neurosci.* **2** (1999) 489–491.

[109] E. Schneidman, I. Segev and N. Tishby, Information capacity and robustness of stochastic neuron models, in *Advances in Neural Information Processing 12,* edited by S.A. Solla, T.K. Leen and K.–R. Müller (MIT Press, Cambridge, 2000), pp. 178–184.

[110] M.S. Goldman, J. Golowasch, E. Marder and L.F. Abbott, *J. Neurosci.* **21** (2001) 5229–5238.

[111] N. Barkai and S. Leibler, *Nature* **387** (1997) 913–917.

[112] Z. Liu, J. Golowasch, E. Marder and L.F. Abbott, *J. Neurosci.* **18** (1998) 2309–2320.

[113] J. Golowasch, M. Casey, L.F. Abbott and E. Marder, *Neural Comp.* **11** (1999) 1079–1096.

[114] C. Soto–Trovino, K.A. Thoroughman, E. Marder and L.F. Abbott, *Nature Neurosci.* **4** (2001) 297–303.

[115] M. Stemmler and C. Koch, *Nature Neurosci.* **2** (1999) 521–527.

[116] E. Schneidman, B. Freedman and I. Segev, *Neural Comp.* **10** (1998) 1679–1703.

[117] B. Agüera y Arcas, A. Fairhall and W. Bialek, What can a single neuron compute?, in *Advances in Neural Information Processing 13,* edited by T.K. Leen, T.G. Dietterich and V. Tresp (MIT Press, Cambridge, 2001), pp. 75–81.

[118] B. Agüera y Arcas, A. Fairhall and W. Bialek (in preparation), Computation in a single neuron: The Hodkgin–Huxley model.

[119] N. Chomsky, *I. R. E. Trans. Inf. Theory* **IT–2** (1956) 113–124.

[120] S. Abney, Statistical methods and linguistics, in *The Balancing Act: Combining Statistical and Symbolic Approaches to Language,* edited by J.L. Klavans and P. Resnik (MIT Press, Cambridge, 1996), pp. 1–26.

[121] G.F. Marcus, S. Vijayan, S. Bandi Rao and P.M. Vishton, *Science* **283** (1999) 77–80.

[122] S. Pinker, *Science* **283** (1999) 40–41.

[123] V. Periwal, *Phys. Rev. Lett.* **78** (1997) 4671–4674, see also `hep-th/9703135`.

[124] T.E. Holy, *Phys. Rev. Lett.* **79** (1997) 3545–3548, see also `physics/9706015`.

[125] V. Periwal (1998), Geometrical statistical inference, preprint. Available at `adap-org/9801001`.

[126] T. Aida, *Phys. Rev. Lett.* **83** (1999) 3554–3557, see also `cond-mat/9911474`.

[127] I. Nemenman and W. Bialek, *Phys. Rev. E* **65** (2002) 026137, see also `cond-mat/0009165`.

[128] I. Nemenman, F. Shafee and W. Bialek, Entropy and inference, revisited, to appear in *Advances in Neural Information Processing 14,* edited by T.G. Dietterich, S. Becker and Z. Ghahramani (MIT Press, Cambridge, 2002), see also `physics/0108025`.

[129] M. Redington, N. Chater and S. Finch, *Cog. Sci.* **22** (1998) 425–469.

[130] F. Pereira, N. Tishby and L. Lee, Distributional clustering of english words, *30th Annual Meeting of the Association for Computational Linguistics,* (1993) pp. 183–190.

[131] N. Slonim and N. Tishby, Document clustering via word clusters using the information bottleneck method, in *Proc. of the 23rd Annual ACM SIGIR Conference on Research and Development in Information Retrieval* (2000) pp. 208–215.

Short talks by participants

N. Goddard: *Sequence dependent rigidity of ssDNA*
M. Mueller: *Nature of the glassy phase and sequence disorder in RNA folding*
N. Dekker: *Single molecule studies of topoisomerase I*
Z. Gueroui: *Stretching DNA in order to image transcription*
K. Kruse: *Actively interacting polar filaments*
I. Derenyi: *Formation of membrane tubes and protein segregation along them*
A. Zilman: *Thermodynamics and structure of self-assembling networks*
Z. Frenkel: *Molecular dynamics study of protein folding*
R. Horwath: *Biosensors based on planar optical waveguides*
P. Galajda: *Light-driven micro-machines and their possible use in biology*
N. Gadegaard: *Cell response to nano- and microstructures of biological and non-biological origin*
J. Tegenfeld: *A barcode reader for DNA based on near field optical detection*
M. Dahan: *Visualizing the motion of individual membrane receptors*
Y. Orlov: *Computer system "Gene Discovery" for retrieving organization patterns in eukaryotic regulatory systems*
T. Liverpool: *Viscoelasticity of solutions of motile polymers*
D. Biron: *Cooperative reproduction at single cell level – the midwife*

Poster and e-poster sessions

V. Baulin: *Neutral water soluble polymers and their interaction with colloids*
D. Dan: *Noise-induced transport in rocked inhomogeneous ratchets*
H. Delanoe: *Nucleation and growth of contact between biological cells*
C. Gabel: *Response of E. Coli flagellar motor to changes in its energy source*
F. Gabel: *Global dynamics in Thylakoid membranes*
I. Kosinska: *Stochastic resonance in discrete maps*
G. Koster: *Spacial organization of lipid bilayers by motor proteins and microtubules*
C. Prinz: *Cell lysis and DNA extraction of a microfabricated device*
C. Tanase: *A Brownian ratchet model for microtubule force production*
I. Tessmer: *Optical tweezers studies of the effect of DNA-binding drugs on single molecules of DNA*
I.-M. Tolic: *Measuring the traction forces exerted by cells*
S. Tolic-Nørrelykke: *Single molecule study of RNA polymerase*
A. Vifan: *Self-tuned Hopf bifurcation in auditory hair cells*

M. Weiss: *A in silico approach to protein sorting in the Golgi apparatus*
Z. Kora: *The maximal force generated by a molecular motor*
C. Gorba: *Brownian dynamics simulation of multi-protein systems*
Y. Orlov: *Transcription regulatory region database (TRRD)*

Achevé d'imprimer par Corlet, Imprimeur, S.A. - 14110 Condé-sur-Noireau (France)
N° d'Imprimeur : 60745 - Dépôt légal : septembre 2002 - Imprimé en U.E.